Springer Proceedings in Physics

Volume 240

Indexed by Scopus

The series Springer Proceedings in Physics, founded in 1984, is devoted to timely reports of state-of-the-art developments in physics and related sciences. Typically based on material presented at conferences, workshops and similar scientific meetings, volumes published in this series will constitute a comprehensive up-to-date source of reference on a field or subfield of relevance in contemporary physics. Proposals must include the following:

- name, place and date of the scientific meeting
- a link to the committees (local organization, international advisors etc.)
- scientific description of the meeting
- list of invited/plenary speakers
- an estimate of the planned proceedings book parameters (number of pages/ articles, requested number of bulk copies, submission deadline).

More information about this series at http://www.springer.com/series/361

Alexander D. Pogrebnjak · Oleksandr Bondar
Editors

Microstructure and Properties of Micro- and Nanoscale Materials, Films, and Coatings (NAP 2019)

Selected Articles from the International Conference on Nanomaterials: Applications and Properties, (NAP 2019)

Springer

Editors
Alexander D. Pogrebnjak
Sumy State University
Sumy, Ukraine

Oleksandr Bondar
Sumy State University
Sumy, Ukraine

ISSN 0930-8989 ISSN 1867-4941 (electronic)
Springer Proceedings in Physics
ISBN 978-981-15-1741-9 ISBN 978-981-15-1742-6 (eBook)
https://doi.org/10.1007/978-981-15-1742-6

This Springer imprint is published by the registered company Springer Nature Singapore Pte Ltd.
The registered company address is: 152 Beach Road, #21-01/04 Gateway East, Singapore 189721, Singapore

Preface

Nanocrystalline coatings on the basis of nitrides, carbides, and semiconducting materials are under great interest in modern physics and materials science nowadays. A variety of properties are actively studied, including but not limited to protective, electrical, magnetical, photoluminescent, etc. Scientists from all over the world took part in the International Conference on Nanomaterials: Applications and Properties (NAP 2019) with oral and plenary talks, and the appropriate extended reports are presented in this book. In general, the book is devoted to studies of microstructure and properties of multilayer nitride coatings, high-entropy alloys, polymeric and nanocomposite materials, semiconducting materials based on ZnS, ZnTe, SnO_2, etc. Peculiarities of electrospark doping of alloys and steels are discussed, as well as deposition of powder coatings by high-speed plasma fluxes is studied. Laser irradiation and electron beams are used in some reports for surface modification and fabrication of nanocrystal structures. Special attention is paid to several theoretical works, devoted to first-principles molecular dynamics calculations and self-organization processes. The book will be interesting for master and Ph.D. students, researchers, and specialists in physics, materials science, and semiconductor physics.

Sumy, Ukraine

Alexander D. Pogrebnjak
Oleksandr Bondar

Contents

Contributors

Tatiana S. Antonenko Semenenko Institute of Geochemistry, Mineralogy and Ore Formation, National Academy of Science of Ukraine, Kyiv, Ukraine

B. Antoszewski Kielce University of Technology, Kielce, Poland

E. M. Baba Department of Solar Energy, Institute for Energy Technology, Kjeller, Norway

A. A. Bagdasaryan Sumy State University, Sumy, Ukraine

V. A. Barbash National Technical University of Ukraine “Igor Sikorsky Kyiv Polytechnic Institute”, Kiev, Ukraine

A. Belous Sumy National Agrarian University, Sumy, Ukraine

A. V. Bendak Uzhhorod National University, Uzhhorod, Ukraine

Olena Berdnikova E. O. Paton Electric Welding Institute of the National Academy of Sciences of Ukraine, Kiev, Ukraine

Vyacheslav Beresnev V. N. Karazin Kharkiv National University, Kharkiv, Ukraine

Z. Bloshenko V. N. Karazin Kharkiv National University, Kharkiv, Ukraine

Sergey Bogatyrenko V. N. Karazin Kharkiv National University, Kharkiv, Ukraine

Oleksandr Bondar Sumy State University, Sumy, Ukraine

O. G. Bordunova Sumy National Agrarian University, Sumy, Ukraine

Vadym Borysiuk Sumy State University, Sumy, Ukraine

Aleksandr B. Brik Semenenko Institute of Geochemistry, Mineralogy and Ore Formation, National Academy of Science of Ukraine, Kyiv, Ukraine

Volodymir Buturlim Department of Condensed Matter Physics, Faculty of Mathematics and Physics, Charles University, Prague, Czech Republic

A. Čerškus Department of Electronics State Research Center, Center for Physical Sciences and Technologies, Vilnius, Lithuania

O. M. Chernenko Dnipro State Agrarian and Economic University, Dnipro, Ukraine

Vadym Chernyavsky Igor Sikorsky Kyiv Polytechnic Institute, Kiev, Ukraine

I. V. Cheshko Sumy State University, Sumy, Ukraine

V. D. Chivanov Institute of Applied Physics (IAP), NAS of Ukraine, Sumy, Ukraine

A. M. Chornous Department of Physics, Sumy State University, Sumy, Ukraine

Maciej Chrobak Faculty of Physics and Applied Computer Science, Academic Centre for Materials and Nanotechnology, AGH University of Science and Technology, Cracow, Poland

M. Chuprina National Technical University of Ukraine "Igor Sikorsky KPI", Kyiv, Ukraine

I. Churilov V. N. Karazin Kharkiv National University, Kharkiv, Ukraine

Alexander Dahov V. N. Karazin Kharkiv National University, Kharkiv, Ukraine

L. V. Dekhtyaruk Kharkiv National University of Construction and Architecture, Kharkiv, Ukraine

V. Demchenko Institute of Macromolecular Chemistry of the NAS of Ukraine, Kiev, Ukraine;
E. O. Paton Electric Welding Institute, of the NAS of Ukraine, Kiev, Ukraine

V. E. Diyuk Taras Shevchenko National University of Kyiv, Kyiv, Ukraine

R. V. Dolbanosova Sumy National Agrarian University, Sumy, Ukraine

S. R. Dolgov-Gordiichuk Department of Physics, Sumy State University, Sumy, Ukraine

M. Dovzhyk Sumy National Agrarian University, Sumy, Ukraine

O. O. Drozdenko Sumy State University, Sumy, Ukraine

Nataliia O. Dudchenko Semenenko Institute of Geochemistry, Mineralogy and Ore Formation, National Academy of Science of Ukraine, Kyiv, Ukraine

Stanislav Dudin V. N. Karazin Kharkiv National University, Kharkiv, Ukraine

S. V. Dukarov School of Physics, V. N. Karazin Kharkiv National University, Kharkiv, Ukraine

M. Dumanchuk Sumy National Agrarian University, Sumy, Ukraine

K. Dyadyura Sumy State University, Sumy, Ukraine

M. V. Fedorischeva Institute of Strength Physics and Materials Science, Siberian Branch of the Russian Academy of Sciences, Tomsk, Russia

O. Gapon G.K. Parts Group LTD, Kyiv, Ukraine

O. Gaponova Sumy State University, Sumy, Ukraine

V. Gerasimenko Sumy National Agrarian University, Sumy, Ukraine

Georgy Grigorenko E. O. Paton Electric Welding Institute of the National Academy of Sciences of Ukraine, Kiev, Ukraine

L. M. Grishchenko Taras Shevchenko National University of Kyiv, Kyiv, Ukraine

B. P. Gritsenko Institute of Strength Physics and Materials Science, Siberian Branch of the Russian Academy of Sciences, Tomsk, Russia;
Tomsk Polytechnic University, Tomsk, Russia

Vasyl Gritsyna Department of Physics and Technology, V. N. Karazin Kharkiv National University, Kharkiv, Ukraine

Ladislav Havela Department of Condensed Matter Physics, Faculty of Mathematics and Physics, Charles University, Prague, Czech Republic

D. S. Horodetska Taras Shevchenko National University of Kyiv, Kyiv, Ukraine

M. Iurzhenko Institute of Macromolecular Chemistry of the NAS of Ukraine, Kiev, Ukraine;
E. O. Paton Electric Welding Institute, of the NAS of Ukraine, Kiev, Ukraine

L. A. Ivashchenko Institute for Problems of Materials Sciences, NAS of Ukraine, Kiev, Ukraine

V. I. Ivashchenko Institute for Problems of Materials Sciences, NAS of Ukraine, Kiev, Ukraine

V. Yu. Izai Uzhhorod National University, Uzhhorod, Ukraine

Olexander O. Kalenyk Department of Physics, Taras Shevchenko National University of Kyiv, Kiev, Ukraine

A. I. Kalinichenko National Science Center "Kharkov Institute of Physics and Technology", Kharkiv, Ukraine

Victor Karachevtsev Department of Molecular Biophysics, B. Verkin ILTPE of NASU, Kharkiv, Ukraine

S. Karazhanov Department of Solar Energy, Institute for Energy Technology, Kjeller, Norway

Yurij Kazarinov Institute of High-energy Physics and Nuclear Physics, NSC Kharkiv Institute of Physics and Technology, Kharkiv, Ukraine

A. L. Khrypunova Materials Engineering for Solar Cells Department (MESC), National Technical University "Kharkiv Polytechnic Institute" (NTU "KhPI"), Kharkiv, Ukraine

I. V. Khrypunova Materials Engineering for Solar Cells Department (MESC), National Technical University "Kharkiv Polytechnic Institute" (NTU "KhPI"), Kharkiv, Ukraine

S. M. Khursenko Sumy National Agrarian University, Sumy, Ukraine

N. -T. H. Kim-Ngan Institute of Physics, Pedagogical University of Cracow, Cracow, Poland

M. V. Kirichenko Materials Engineering for Solar Cells Department (MESC), National Technical University "Kharkiv Polytechnic Institute" (NTU "KhPI"), Kharkiv, Ukraine

K. S. Klepikova Materials Engineering for Solar Cells Department (MESC), National Technical University "Kharkiv Polytechnic Institute" (NTU "KhPI"), Kharkiv, Ukraine

N. P. Klochko Materials Engineering for Solar Cells Department (MESC), National Technical University "Kharkiv Polytechnic Institute" (NTU "KhPI"), Kharkiv, Ukraine

O. V. Klymov Department of Applied Fisica, University of Valencia, Valencia, Spain

M. N. Kolchevskaya Belarussian State University, Minsk, Belarus

M. M. Kolesnyk Sumy State University, Sumy, Ukraine

O. Kolisnichenko E. O. Paton Electric Welding Institute of the NAS of Ukraine, Kiev, Ukraine

F. F. Komarov A. N. Sevchenko Institute of Applied Physics Problems, Belarusian State University, Minsk, Belarus;
National University of Science and Technology MISIS, Moscow, Russia

Ie. Konoplianchenko Sumy National Agrarian University, Sumy, Ukraine

S. V. Konstantinov A. N. Sevchenko Institute of Applied Physics Problems, Belarusian State University, Minsk, Belarus

V. R. Kopach Materials Engineering for Solar Cells Department (MESC), National Technical University “Kharkiv Polytechnic Institute” (NTU “KhPI”), Kharkiv, Ukraine

I. Korotash Institute for Metal Physics NASU, Kyiv, Ukraine

Valery Kostin E. O. Paton Electric Welding Institute of the National Academy of Sciences of Ukraine, Kiev, Ukraine

Y. R. Kostyuchenko Materials Engineering for Solar Cells Department (MESC), National Technical University “Kharkiv Polytechnic Institute” (NTU “KhPI”), Kharkiv, Ukraine

N. Kovalchuk Joint Stock Company “Integral”, Minsk, Belarus

A. O. Kozak Institute for Problems of Materials Sciences, NAS of Ukraine, Kiev, Ukraine

D. P. Koziarskyi Department of Electronics and Power Engineering, Yuriy Fedkovych Chernivtsi National University, Chernivtsi, Ukraine

I. P. Koziarskyi Department of Physics of Semiconductors and Nanostructures, Yuriy Fedkovych Chernivtsi National University, Chernivtsi, Ukraine

Yaroslav Kravchenko Sumy State University, Sumy, Ukraine

A. S. Krivets Department of Nanoelectronics, Sumy State University, Sumy, Ukraine

Aleksandr Kryshtal AGH University of Science and Technology, Kraków, Poland

Cz. Kundera Kielce University of Technology, Kielce, Poland;
Sumy State University, Sumy, Ukraine

D. I. Kurbatov Sumy State University, Sumy, Ukraine

Nikita Kurnosov Department of Molecular Biophysics, B. Verkin ILTPE of NASU, Kharkiv, Ukraine

Alla M. Kuryliuk Department of Physics, Taras Shevchenko National University of Kyiv, Kiev, Ukraine

Vasyl V. Kuryliuk Department of Physics, Taras Shevchenko National University of Kyiv, Kiev, Ukraine

P. Kúš Department of Experimental Physics, Comenius University, Bratislava, Slovakia

O. Kushnaryova E. O. Paton Electric Welding Institute of the NAS of Ukraine, Kiev, Ukraine

Olena M. Lavrynenko Frantsevich Institute for Problems of Material Science, National Academy of Science of Ukraine, Kyiv, Ukraine

V. V. Lisnyak Taras Shevchenko National University of Kyiv, Kyiv, Ukraine

V. B. Loboda Sumy National Agrarian University, Sumy, Ukraine

A. M. Lohvynov Sumy State University, Sumy, Ukraine

A. A. Luchaninov National Science Center "Kharkov Institute of Physics and Technology", Kharkiv, Ukraine

O. R. Ludchik Belarussian State University, Minsk, Belarus

A. Lyalka O. L. Pshenychky Vysochanska Secondary School, Kharkiv, Ukraine

V. M. Lyubov Materials Engineering for Solar Cells Department (MESC), National Technical University "Kharkiv Polytechnic Institute" (NTU "KhPI"), Kharkiv, Ukraine

E. V. Maistruk Department of Electronics and Power Engineering, Yuriy Fedkovych Chernivtsi National University, Chernivtsi, Ukraine

Yurii N. Makogon Department of Physics of Metals, National Technical University of Ukraine "Igor Sikorsky Kyiv Polytechnic Institute", Kiev, Ukraine

Olga Maksakova Sumy State University, Sumy, Ukraine

Pavlo V. Makushko Department of Physics of Metals, National Technical University of Ukraine "Igor Sikorsky Kyiv Polytechnic Institute", Kiev, Ukraine

V. S. Manzhara Institute of Physics, NAS of Ukraine, Kiev, Ukraine

V. Martsynkovskyy Sumy National Agrarian University, Sumy, Ukraine

Alexandr Matveev Igor Sikorsky Kyiv Polytechnic Institute, Kiev, Ukraine

A. V. Melezhik Tambov State Technical University, Tambov, Russia

N. R. Memetov Tambov State Technical University, Tambov, Russia

K. Merkotan Odesa National Polytechnic University, Odesa, Ukraine

P. Mikheenko Department of Physics, University of Oslo, Oslo, Norway

M. Mikulina Sumy National Agrarian University, Sumy, Ukraine

O. V. Milchanin A. N. Sevchenko Institute of Applied Physics Problems, Minsk, Belarus

Alexey Minenkov V. N. Karazin Kharkiv National University, Kharkiv, Ukraine

R. B. Miranovich Belarussian State University, Minsk, Belarus

A. V. Mischanchuk Chuiko Institute of Surface Chemistry of the National Academy of Science of Ukraine, Kyiv, Ukraine

V. N. Mizernik Scientific Physical-Technologic Center of MES and NAS of Ukraine, Kharkiv, Ukraine

A. I. Mostovyi Department of Electronics and Power Engineering, Yuriy Fedkovych Chernivtsi National University, Chernivtsi, Ukraine

A. Mudryi Scientific and Practical Materials Research Center, National Academy of Sciences of Belarus, Minsk, Belarus

Bohdan Natalich Sumy State University, Sumy, Ukraine

G. Zh. Nogaibekova Institute of Strength Physics and Materials Science, Siberian Branch of the Russian Academy of Sciences, Tomsk, Russia;
Tomsk Polytechnic University, Tomsk, Russia

E. N. Odarenko School of Electronics and Biomedical Engineering, Kharkiv National University of Radio Electronics, Kharkiv, Ukraine

Mykola V. Okonchuk Department of Physics, Taras Shevchenko National University of Kyiv, Kiev, Ukraine

A. S. Opanasyuk Sumy State University, Sumy, Ukraine

I. D. Parfimovich A. N. Sevchenko Institute of Applied Physics Problems, Minsk, Belarus

I. Parkhomenko Belarussian State University, Minsk, Belarus

Olesia Yu. Pavlenko Frantsevich Institute for Problems of Material Science, National Academy of Science of Ukraine, Kyiv, Ukraine

I. M. Pazukha Department of Physics, Sumy State University, Sumy, Ukraine

S. I. Petrushenko School of Physics, V. N. Karazin Kharkiv National University, Kharkiv, Ukraine

Alexander D. Pogrebnjak Sumy State University, Sumy, Ukraine

D. Polotskiy Institute for Metal Physics NASU, Kyiv, Ukraine

Ye. Polovetskiy E. O. Paton Electric Welding Institute of the NAS of Ukraine, Kiev, Ukraine

A. Yu. Popov Omsk State Technical University, Omsk, Russia

O. K. Porada Institute for Problems of Materials Sciences, NAS of Ukraine, Kiev, Ukraine

H. V. Potsiluiko-Hryhoriak Automatisation, Publishing and Polygraphy, Electrical Engineering Division, Chernivtsi Industrial College, Chernivtsi, Ukraine

S. I. Protsenko Department of Physics, Sumy State University, Sumy, Ukraine

D. S. Rechenko Omsk State Technical University, Omsk, Russia

E. N. Reshetnyak National Science Center “Kharkov Institute of Physics and Technology”, Kharkiv, Ukraine

S. Riabov Institute of Macromolecular Chemistry of the NAS of Ukraine, Kiev, Ukraine

E. A. Rogachev Omsk State Technical University, Omsk, Russia

I. Romanov Belarussian State University, Minsk, Belarus

E. Rudenko Institute for Metal Physics NASU, Kyiv, Ukraine

N. Rybalchenko D. K. Zabolotny Institute of Microbiology and Virology of the NAS of Ukraine, Kiev, Ukraine

Galina V. Saenko Department of Physics, Taras Shevchenko National University of Kyiv, Kiev, Ukraine

Y. A. Samokhina Sumy National Agrarian University, Sumy, Ukraine

B. Sarzhanov Sumy National Agrarian University, Sumy, Ukraine

O. Semernya Sumy National Agrarian University, Sumy, Ukraine

V. P. Sergeev Institute of Strength Physics and Materials Science, Siberian Branch of the Russian Academy of Sciences, Tomsk, Russia;
Tomsk Polytechnic University, Tomsk, Russia

Yu. M. Shabelnyk Sumy State University, Sumy, Ukraine

Mark N. Shamis Department of Physics of Metals, National Technical University of Ukraine “Igor Sikorsky Kyiv Polytechnic Institute”, Kiev, Ukraine

Artem Shchibrya V. N. Karazin Kharkiv National University, Kharkiv, Ukraine

V. V. Shchotkin Department of Physics, Sumy State University, Sumy, Ukraine

Yu. O. Shkurdoda Department of Physics, Sumy State University, Sumy, Ukraine

A. A. Shmat’ko School of Radiophysics, Biomedical Electronics and Computer Systems, V. N. Karazin Kharkiv National University, Kharkiv, Ukraine

Yuriy S. Shukin Frantsevich Institute for Problems of Material Science, National Academy of Science of Ukraine, Kyiv, Ukraine

Sergiy I. Sidorenko Department of Physics of Metals, National Technical University of Ukraine “Igor Sikorsky Kyiv Polytechnic Institute”, Kiev, Ukraine

F. Sizov V. E. Lashkaryov Institute of Semiconductor Physics NASU, Kyiv, Ukraine

Yu. Skyriya Sumy State University, Sumy, Ukraine

G. Smolyarov Sumy National Agrarian University, Sumy, Ukraine

K. V. Smyrnova Sumy State University, Sumy, Ukraine

A. M. Solomon Institute of Electron Physics, National Academy of Sciences, Uzhhorod, Ukraine

Sylwia Sowa Institute of Physics, Pedagogical University of Cracow, Cracow, Poland;
Łukasiewicz Research Network—Institute for Sustainable Technologies, Radom, Poland

A. E. Stetsko Faculty of Computer Engineering Printing, Ukrainian Academy of Printing, UAP, Lviv, Ukraine

Y. T. Stetsko Faculty of Computer Engineering Printing, Ukrainian Academy of Printing, UAP, Lviv, Ukraine

Vyacheslav Stolbovoy National Science Center Kharkiv Institute of Physics and Technology, Kharkiv, Ukraine

R. A. Stolyarov Tambov State Technical University, Tambov, Russia

V. E. Strel'nitskij National Science Center "Kharkov Institute of Physics and Technology", Kharkiv, Ukraine

I. P. Studenyak Uzhhorod National University, Uzhhorod, Ukraine

V. I. Studenyak Uzhhorod National University, Uzhhorod, Ukraine

V. Sukhov V. N. Karazin Kharkiv National University, Kharkiv, Ukraine

K. Svezhentsova V. E. Lashkaryov Institute of Semiconductor Physics NASU, Kyiv, Ukraine

O. M. Sydor Department of Layered Crystals, Frantsevich Institute for Problems of Materials Science, (Chernivtsi Department), Chernivtsi, Ukraine

Maxim Sysoyev Igor Sikorsky Kyiv Polytechnic Institute, Kiev, Ukraine

O. O. Sytikov Institute for Problems of Materials Sciences, NAS of Ukraine, Kiev, Ukraine

N. Tarelnyk Sumy National Agrarian University, Sumy, Ukraine

V. Tarelnyk Sumy National Agrarian University, Sumy, Ukraine

Zbigniew Tarnawski Faculty of Physics and Applied Computer Science, AGH University of Science and Technology, Cracow, Poland

Ye. Titkov E. O. Paton Electric Welding Institute of the NAS of Ukraine, Kiev, Ukraine

O. P. Tkach Sumy State University, Sumy, Ukraine

A. G. Tkachev Tambov State Technical University, Tambov, Russia

A. Tolbatov Sumy National Agrarian University, Sumy, Ukraine

V. Tolbatov Sumy National Agrarian University, Sumy, Ukraine

T. V. Tomila Institute for Problems of Materials Sciences, NAS of Ukraine, Kiev, Ukraine

Tatiana L. Tsaregradskaya Department of Physics, Taras Shevchenko National University of Kyiv, Kiev, Ukraine

Z. Tsybrii V. E. Lashkaryov Institute of Semiconductor Physics NASU, Kyiv, Ukraine

Yu. Tyurin E. O. Paton Electric Welding Institute of the NAS of Ukraine, Kiev, Ukraine

A. V. Vakaliuk Taras Shevchenko National University of Kyiv, Kyiv, Ukraine

O. Vasilenko Sumy National Agrarian University, Sumy, Ukraine

V. V. Vasyliev National Science Center "Kharkov Institute of Physics and Technology", Kharkiv, Ukraine

O. A. Vasylieva National Technical University of Ukraine "Igor Sikorsky Kyiv Polytechnic Institute", Kiev, Ukraine

Tetiana I. Verbytska Department of Physics of Metals, National Technical University of Ukraine "Igor Sikorsky Kyiv Polytechnic Institute", Kiev, Ukraine

L. Vlasukova Belarussian State University, Minsk, Belarus

A. A. Voznyi Sumy State University, Sumy, Ukraine

A. P. Voznyi Sumy State University, Sumy, Ukraine

M. Vuichyk V. E. Lashkaryov Institute of Semiconductor Physics NASU, Kyiv, Ukraine

Stanislav Yakovin V. N. Karazin Kharkiv National University, Kharkiv, Ukraine

O. V. Yashchenko National Technical University of Ukraine "Igor Sikorsky Kyiv Polytechnic Institute", Kiev, Ukraine

A. V. Yatsymyrskyi Taras Shevchenko National University of Kyiv, Kyiv, Ukraine

Nina Yefymenko V. N. Karazin Kharkiv National University, Kharkiv, Ukraine

Yu. S. Yeromenko Sumy State University, Sumy, Ukraine

Alexandra Yurkova Igor Sikorsky Kyiv Polytechnic Institute, Kiev, Ukraine

O. V. Yushchenko Department of Nanoelectronics, Sumy State University, Sumy, Ukraine

A. N. Zaderko Taras Shevchenko National University of Kyiv, Kyiv, Ukraine

D. O. Zhadan Materials Engineering for Solar Cells Department (MESC), National Technical University "Kharkiv Polytechnic Institute" (NTU "KhPI"), Kharkiv, Ukraine

V. Zhivulko Scientific and Practical Materials Research Center, National Academy of Sciences of Belarus, Minsk, Belarus

Victor Zukov E. O. Paton Electric Welding Institute of the National Academy of Sciences of Ukraine, Kiev, Ukraine

Pawel Zukowski Lublin University of Technology, Lublin, Poland

Alexander Zykov V. N. Karazin Kharkiv National University, Kharkiv, Ukraine

Chapter 1
Electrical and Magnetic Behavior of GdOH Thin Films: A Search for Hydrogen Anion Superconductivity

P. Mikheenko, E. M. Baba and S. Karazhanov

Abstract Anomalous resistive and magnetic behavior of GdOH thin films, which belong to the novel class of materials known as oxyhydrides, is reported. The oxyhydrides contain hydrogen in its rare negatively-charged anion state in combination with oxygen, which is also in the anion state. A range of GdOH films prepared on glass and single-crystalline substrates demonstrate a resistive transition from an insulating to conducting state with decrease of resistance starting as high as at about 200 K. At room temperature, the resistance per square area for the best GdOH films of the thickness of 200 nm is about 100 Ω, which is close to the resistance of the films of high-temperature superconductors of similar thickness. Apparent zero resistance is observed at about 40 K. Magneto-optical imaging registers effect of trapping magnetic flux typical for the superconducting state. The possible anion superconductivity is discussed in connection with recently published papers on near-room-temperature superconductivity in hydrides at high pressure and hydrogen-based superconductivity in biological systems.

1.1 Introduction

Superconductivity is a macroscopic quantum phenomenon of enormous importance. In theory, it represents quantum mechanics in most pure form while its applications range from magnets for Large Hadron Collider and International Thermonuclear Experimental Reactor to recently developed quantum computers. The material in the superconducting state can be compared with giant molecule, in which all electrons are paired and condensed in one state. Any attempt to change energy of one electron requires change in energy of all electrons. This leads to dissipation-free electrical current, ideal diamagnetism, quantization of magnetic flux, ability of current to flow

P. Mikheenko (✉)
Department of Physics, University of Oslo, Postboks 1048, Blindern, 0316 Oslo, Norway
e-mail: pavlo.mikheenko@fys.uio.no

E. M. Baba · S. Karazhanov
Department of Solar Energy, Institute for Energy Technology, Postboks 40, 2027 Kjeller, Norway

A. D. Pogrebnjak and O. Bondar (eds.), *Microstructure and Properties of Micro- and Nanoscale Materials, Films, and Coatings (NAP 2019)*,
Springer Proceedings in Physics 240, https://doi.org/10.1007/978-981-15-1742-6_1

without dissipation through a barrier and generation of coherent electromagnetic radiation in the presence of voltage on the barrier.

Such an ideal state of matter, however, is delicate, relatively rare and takes place mainly at low temperatures. There are several mechanisms able to combine electrons into pairs. Most known is electron-phonon interaction. It is, however, not able to provide high critical temperature (T_c), the temperature, below which material becomes superconducting. There are multiple attempts to extend T_c to room temperature involving exotic materials and different pairing mechanisms. A detailed account of these attempts is given in [1]. Most recently, these attempts resulted in the extension of T_c to nearly room temperature in the hydrides H_3S [2] and LaH_x [3, 4] at very high pressure. In these materials, electrons are paired through interaction with high-energy optical phonons.

It is interesting that superconductivity with T_c even higher than room temperature was predicted in hydrogen-based quasi-one-dimensional chains of organic molecules [5], and first experimental data start to appear confirming this prediction [6]. According to these data, superconductivity could be common in biological systems and most likely resides in microtubules filled with water, or rather tightly-confined hydrogen-oxygen chains [6]. If, like in other hydrides, hydrogen in these chains could be in anion form, it may allow, due to the presence of oxygen, its fluctuation between cation and anion states resulting in oscillations of charge and electron-electron interaction necessary for very high T_c, as in the model of Little [5]. From this point of view, investigation of materials with both oxygen and hydrogen in anion state is very important. Such materials are named oxyhydrides, and they are now a subject of very intensive research [7, 8]. They are, however, very delicate and difficult to study, as anion hydrogen is weakly bonded to the crystal lattice and easily lost being replaced by oxygen. GdOH is one of oxyhydrides. A range of GdOH thin films was prepared in this work, and their properties were investigated by electrical transport and magneto-optical measurements searching for the superconductivity in the material.

1.2 Experiment Details

1.2.1 Samples Preparation

GdHO films were prepared by a two-step process. In the first step, GdH_2 film was deposited by reactive sputtering in a chamber containing a mixture of argon and hydrogen using Leybold Optics A550V7 sputter unit. The deposition was done from a commercial Gd target of 99.9% purity. In the second step, GdH_2 film was taken from the chamber allowing for its natural oxidation for a certain period of time, after which it was sealed in nitrogen. Three types of substrates have been used for the deposition: borosilicate glass (Menzel–Gläser microscope slides) and two types of single-crystalline substrates—Si and $SrTiO_3$.

1.2.2 Measurements Techniques

The main measurement technique was a technique for electrical transport measurements allowing to record temperature dependence of the resistance of the samples. These measurements were performed in a protective nitrogen or helium atmosphere at a pressure close to ambient. The sample with a closely-spaced temperature sensor was mounted on an electrically-insulated rod inserted into liquid nitrogen or liquid helium dewar. The wires of the electrical circuit were attached to the sample with pressed indium contacts. The temperature was changed by slowly varying the position of the sample in the dewar. The measurement circuit consisted of power supply, temperature controller and two voltmeters to measure separately voltage and current through the sample. All devices were automatically controlled using a Matlab script with continuous record of a large number of data points in a designated time interval.

Before and between the measurements, the samples were outside of the protective atmosphere for a very short time only, which was necessary to attach contacts and wires. Typically, two current leads were permanently attached to the sample, while potential leads were frequently changed to explore different parts of the films.

Another technique used in this work was magneto-optical imaging [9]. In this technique, the sample is mounted on a cold finger of a vacuum optical cryostat and covered by a transparent indicator film with a mirror between the film and the sample. The indicator film is illuminated with a beam of linear-polarized light, and it changes the polarization of the light in the presence of magnetic field. An analyzer outside the cryostat detects change in the polarization and maps distribution of magnetic field in the indicator, i.e. in the sample, which is close to it. It is important that the sample itself is not illuminated by the light, as it is behind the mirror. Therefore, it is not seen in magneto-optics if it is non-magnetic. This technique is based on the Faraday effect and it is widely used to investigate superconducting and magnetic materials [9].

1.3 Results and Discussion

The GdOH films deposited on borosilicate glass were the first, which were investigated in this paper. After the oxidation, they visually changed appearance indicating the development of an inhomogeneous polycrystalline structure with small grain size. The films showed an insulating behavior with a strong increase of resistance with a decrease in temperature, as it is shown in Fig. 1.1. However, in some areas, the decrease of resistance at temperature below 60 K has been observed. One could assume that high resistance is caused by either grain boundaries, or the regions of over-oxidation, in which anion hydrogen was removed from the structure. The decrease in resistance may indicate a phase transition, including development of local superconductivity with a large spread in T_c. To better isolate the phase responsible for

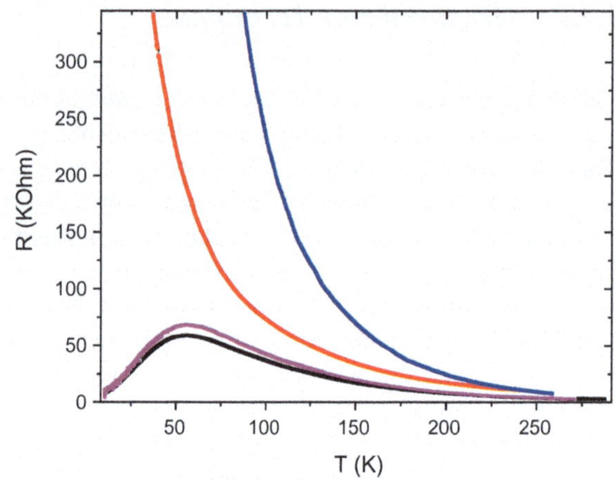

Fig. 1.1 Temperature dependence of resistance in different areas of GdOH films (shown in different color) deposited on borosilicate glass. A decrease in resistance at temperature below 60 K has been observed in some areas

the decrease in resistance, GdOH films were further prepared on single-crystalline $SrTiO_3$ and Si substrates.

An important result for GdOH films deposited on $SrTiO_3$ substrates was a very large drop in resistance compared with the films deposited on borosilicate glass. For the films of the thickness of 200 nm, resistance was about 100 Ω, which is close to the resistance of the films of high-temperature superconductors of similar thickness. Moreover, in spite of still present background increase of resistance at the reduction of temperature, its decrease is registered starting at temperatures as high as about 180 K (see Fig. 1.2). Such temperature is close to the superconducting transition temperature for hydrides at high pressure [2–4]. In this case, however, a decrease of resistance takes place at ambient pressure, but the transition is very wide, which is not characteristic for a homogeneous superconductor. In one of the curves in Fig. 1.2 plotted in black, resistance at low temperatures is close to zero.

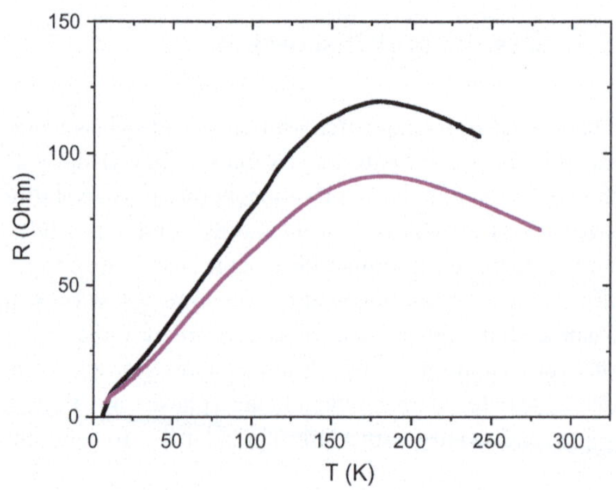

Fig. 1.2 Temperature dependence of resistance for different areas of GdOH films deposited on $SrTiO_3$ single-crystalline substrate shown in different color. A decrease in resistance has been observed at temperature below 180 K

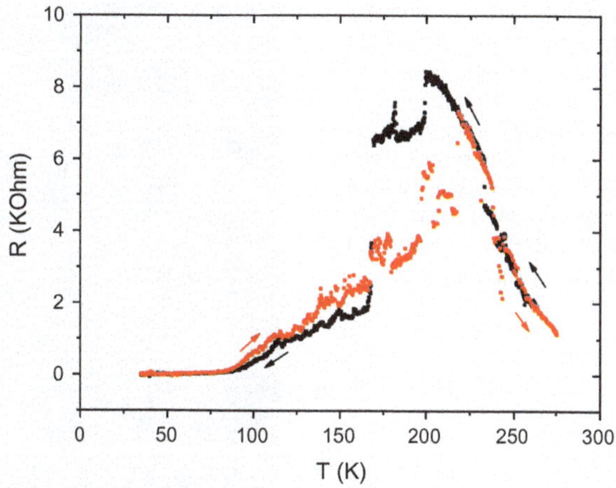

Fig. 1.3 Temperature dependence of resistance for GdOH film deposited on Si single-crystalline substrate. In spite of hysteresis and irreproducible jumps coming from the contacts, an apparent zero resistance is observed at temperatures below 40 K

The GdOH films deposited on single-crystalline Si substrates showed more peculiar behavior. Although their resistance was very high, similar to that observed in films deposited on borosilicate glass, decrease in resistance started at higher temperature than in other films, at about 200 K, and there was an apparent zero resistance at temperatures below 40 K. Such a temperature is close to T_c in other oxyhydrides reported recently [10, 11]. It was difficult to attach contacts to the films deposited on Si. Therefore, there were hysteresis and irreproducible jumps in the measurements, as it is seen in Fig. 1.3, in which points, recorded a decrease of temperature are shown in black and at increase—in red, with the arrows of the corresponding color.

The observations above suggest local superconductivity in GdOH films. However, an additional study is necessary to confirm this result. A robust technique for detecting superconductivity is magneto-optical imaging (MOI) described in the experimental part.

MOI allows observing the screening of the magnetic field by superconductor or trapped in its magnetic flux. It only works when superconducting current is of considerable magnitude. Here this technique is used on one of the GdOH films deposited on $SrTiO_3$. In the experiment, magnetic field of 18.7 mT was applied perpendicular to the film at temperature of 5.9 K. After that, it was reduced to zero allowing trap of magnetic field and the development of persisting supercurrent in the case the film is superconducting. This was followed by the increase of temperature and record of MOI images at selected temperatures. Since sensitivity of MOI indicator films is limited, differential imaging was used subtracting MOI images recorded at different temperatures.

In Fig. 1.4, two differential images are shown. The top image shows MOI map recorded at 5.9 K, from which the MOI map recorded at 20 K was subtracted, and the bottom image shows MOI map recorded at 16 K, from which the same 20 K MOI map was subtracted too. The outline of the square sample behind the indicator film is clearly seen on the top image (pointed by white arrow), while it is not seen on the bottom image. The form of the outline is typical for a superconducting film with a

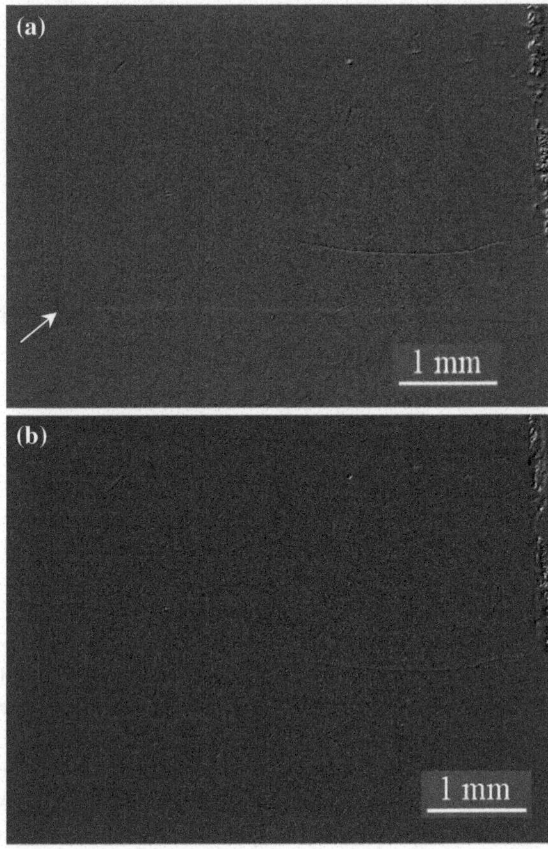

Fig. 1.4 Differential images for a GdOH film deposited on $SrTiO_3$ single-crystalline substrate. **a** MOI map recorded at 5.9 K, from which the map recorded at 20 K was subtracted. **b** MOI map recorded at 16 K, from which MOI map recorded at 20 K was subtracted too

trapped magnetic field. Two images are shown to ensure that there is no instrumental artifact, in which case the outline of the sample would be present on the bottom image too.

The recorded MOI signal in the top image is not strong. It is much weaker than, for example, in epitaxial films of high-temperature superconductors. In spite of that, this is an important result as GdOH might be the first oxyhydride with reasonably high T_c, superconductivity in which is confirmed by MOI. The weakness of the signal, however, indicates that conditions of GdOH film preparation and oxidation still need to be improved.

It is currently not clear what critical temperature of the optimally prepared GdOH would be. A double transition typical to superconducting granular systems is observed in Figs. 1.1, 1.2 and 1.3, but the top transition at a temperature of about 200 K could not be related to superconductivity. In the MOI, the trapped magnetic field effect disappears at temperatures about 16 K, but this technique is not very sensitive. This suggests a necessity for a more extensive experimental study of GdOH films with better control over preparation and handling, and perhaps with the deposition of protective layer, to not allow over-oxidation and loss of hydrogen.

In conclusion, electrical transport and magneto-optical measurements on hydride films of GdOH reveal interesting features that could be related to anion superconductivity in a novel class of superconducting materials from the family of oxyhydrides with suggested dominant electron-electron interaction providing the formation of Cooper pairs.

Acknowledgements E. M. B and S. Z. K has received funding from M-ERA.net project "TESTIMONIES" from the Research Council of Norway.

References

1. V.Z. Kresin, Paths to room-temperature superconductivity, J. Supercond. Novel Magn. **31**, 611 (2017)
2. A.P. Drozdov, M.I. Eremets, I.A. Troyan, V. Ksenofontov, S.I. Shylin, Conventional superconductivity at 203 kelvin at high pressures in the sulfur hydride system, Nature **525**, 73 (2015)
3. A.P. Drozdov, V.S. Minkov, S.P. Besedin, P.P. Kong, M.A. Kuzovnikov, D.A. Knyazev, M.I. Eremets, Condensed Matter. arXiv (2018). https://arxiv.org/ftp/arxiv/papers/1808/1808.07039.pdf
4. M. Somayazulu, M. Ahart, A.K. Mishra, Z.M. Geballe, M. Baldini, Y. Meng, V.V. Struzhkin, R.J. Hemley, Evidence for Superconductivity above 260 K in Lanthanum Superhydride at Megabar Pressures, Phys. Rev. Lett. **122**, 027001 (2019)
5. W.A. Little, Possibility of synthesizing an organic superconductor, Phys. Rev. **134**, 1416 (1964)
6. P. Mikheenko, Possible superconductivity in the brain, J. Supercond. Novel Magn. **32**, 1121 (2019)
7. H. Kageyama, K. Hayashi, K. Maeda, J.P. Attfield, Z. Hiroi, J.M. Rondinelli, K.R. Poeppelmeier, New chemistry of transition metal oxyhydrides, Nat. Commun. **9**, 772 (2018)
8. Y. Kobayashi, O. Hernandez, C. Tassel, H. Kageyama, Expanding frontiers in materials chemistry and physics with multiple anions, Sci. Technol. Adv. Mater. **18**, 905 (2017)
9. T.H. Johansen, D.V. Shantsev (eds.), *Magneto-Optical Imaging* (Kluwer Academic Publishers, Dordrecht, 2004)
10. H. Hosono, S. Matsuishi, Superconductivity induced by hydrogen anion substitution in 1111-type iron arsenides, Curr. Opin. Solid St. M. **17**, 49 (2013)
11. J. Matsumoto, K. Hanzawa, M. Sasase, S. Haindl, T. Katase, H. Hiramatsu, H. Hosono, Condensed Matter. arXiv (2019). https://arxiv.org/ftp/arxiv/papers/1903/1903.11819.pdf

Chapter 2
Electrical Properties of Heterojunction *n*-MoO_x/*p*-$Cd_3In_2Te_6$

I. P. Koziarskyi, E. V. Maistruk, D. P. Koziarskyi, A. I. Mostovyi, O. M. Sydor and H. V. Potsiluiko-Hryhoriak

Abstract Heterostructures *n*-MoO_x/*p*-$Cd_3In_2Te_6$ were manufactured by sputtering a thin film MoO_x (*n*-type) onto $Cd_3In_2Te_6$ substrates (*p*-type) by the method of reactive magnetron sputtering. I–V-characteristics of the structure were investigated in the temperature range 288–373 K. It is established that they have a straightening character. The height of the potential barrier φ_0 varies from 0.34 to 0.09 eV, and the rectifying ratio RR is from 900 to 16 with increasing temperature. Investigation of current transfer mechanisms under direct inclusions showed that the generation and recombination process of current transfer dominates, and in the reverse inclusion, the tunneling mechanism of transport of charge carriers through a potential barrier is dominant.

2.1 Introduction

At present, it is interesting to create heterostructures based on metal nitrides [1–5] and oxides [6–10], which can be promising for the creation of functional elements of photovoltaic and solar cells [11–16]. Molybdenum and molybdenum oxide are promising materials [17–21], given that they have high values of work function

I. P. Koziarskyi (✉)
Department of Physics of Semiconductors and Nanostructures, Yuriy Fedkovych Chernivtsi National University, Chernivtsi 58012, Ukraine
e-mail: i.koziarskyi@chnu.edu.ua

E. V. Maistruk · D. P. Koziarskyi · A. I. Mostovyi
Department of Electronics and Power Engineering, Yuriy Fedkovych Chernivtsi National University, Chernivtsi 58012, Ukraine

O. M. Sydor
Department of Layered Crystals, Frantsevich Institute for Problems of Materials Science, (Chernivtsi Department), Chernivtsi 58000, Ukraine

H. V. Potsiluiko-Hryhoriak
Automatisation, Publishing and Polygraphy, Electrical Engineering Division, Chernivtsi Industrial College, Chernivtsi 58000, Ukraine

A. D. Pogrebnjak and O. Bondar (eds.), *Microstructure and Properties of Micro- and Nanoscale Materials, Films, and Coatings (NAP 2019)*, Springer Proceedings in Physics 240, https://doi.org/10.1007/978-981-15-1742-6_2

and as a consequence of electron affinity. This greatly simplifies the creation of heterojunctions [22–27]. Crystals p-$Cd_3In_2Te_6$ refer to radiation-resistant materials with stoichiometric vacancies [28–32]. Such a combination of base material and conductive transparent oxide greatly enhances the possibilities of the investigated heterostructures [33–38].

Heterostructures *n*-MoO_x/*p*-$Cd_3In_2Te_6$ were made by sputtering molybdenum oxide onto substrates of defective *p*-$Cd_3In_2Te_6$ crystals by direct-current reactive magnetron sputtering [16, 39, 40]. The width of the band gap of these semiconductor crystals 1.3 eV. That meets the requirements for solar cells and explains the use of these semiconductors as the base material for heterostructures as promising materials for solar cells [41–49].

2.2 Experiment Details

MoO_x thin films were sprayed onto pre-cleaned substrates of $Cd_3In_2Te_6$ crystals [50] using DC reactive magnetron sputtering of pure molybdenum targets in an atmosphere of a mixture of argon and oxygen in a Leybold-Heraeus L560 universal vacuum unit.

The substrates were placed above the magnetron with the subsequent rotation of the table to ensure the uniformity of the films in thickness. The vacuum chamber was pumped to a residual pressure of 5×10^{-3} Pa before the spraying process started. Short-term ion etching of the target surface and the substrate with argon was used to remove uncontrolled contamination.

Conditions have been created for the formation of a gas mixture in an arbitrary proportion during the spraying process (from two independent sources so that both gases are fed simultaneously) in order to facilitate the selection of technological parameters and film application modes.

The partial vents in the vacuum chamber were 0.24 Pa for argon and 0.034 Pa for oxygen for MoO film spraying. Magnetron power was 120 W. The spray process lasted for 10 min at a substrate temperature of ~570 K.

2.3 Results and Discussion

There is the I–V curve of heterojunction *n*-MoO_x/*p*-$Cd_3In_2Te_6$ at the different temperatures in the half-logarithmic scale represented on the Fig. 2.1.

On the basis of the obtained I–V curves, the extrapolation of the straight-line sections of the straight branches to the intersection with the voltage axis determined the value of the height of the potential barrier and its dependence on the temperature was constructed (Fig. 2.2a) from which $\varphi_0(0\ K) = 1.18$ eV and coefficient $\beta_\varphi = -2.93 \times 10^{-3}$ eV $\times$ K^{-1}.

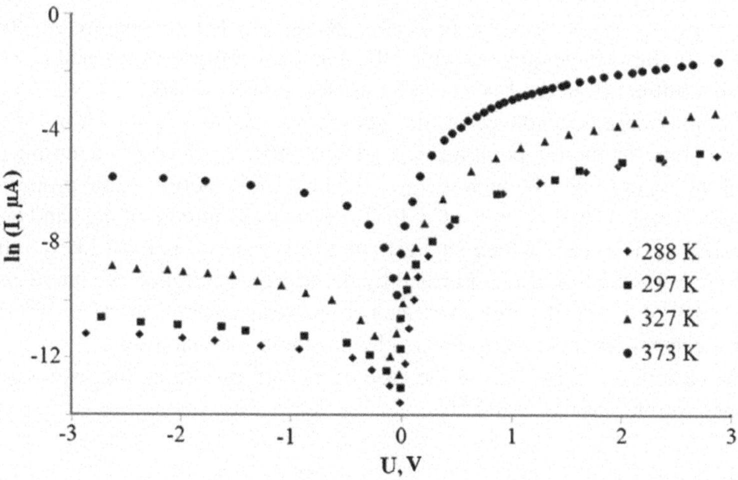

Fig. 2.1 Dark I–V curve n-MoO_x/p-$Cd_3In_2Te_6$ heterostructure at different temperatures

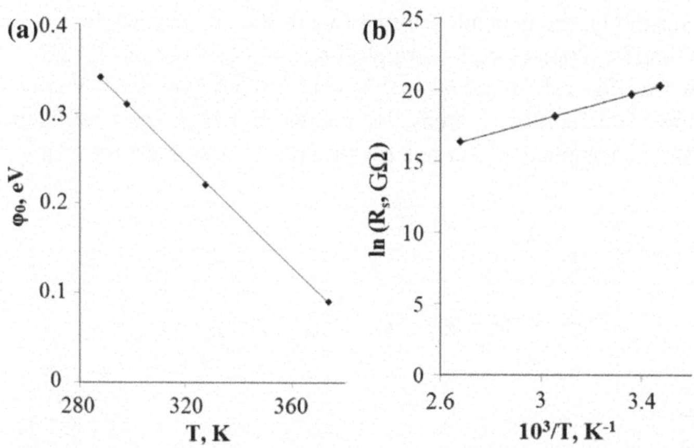

Fig. 2.2 The dependence of the height of the potential barrier φ_k on temperature (**a**) and the dependence of the successive resistance on the temperature in half-logarithmic coordinates (**b**) for the heterostructure n-MoO_x/p-$Cd_3In_2Te_6$

The explored I–V curves of the heterostructure have some expressed diode characteristics (Fig. 2.1), rectification ratio RR ~10^3 at r.t. and U = 5 V and decreases with the temperature increasing. This, in turn, is associated with an increase of the main charge carriers in the base material (holes) p-$Cd_3In_2Te_6$ with the temperature increase, which also confirms the decrease of the consistent resistance of the structure with the growth of T (Fig. 2.2b).

The value of the successive resistance R_s of the heterostructure *n*-MoO_x/*p*-$Cd_3In_2Te_6$ in the voltage range at which the barrier is completely opened U > 0.5 V, can be determined from the I–V curve by the formula $R_s = U/I$.

The temperature dependence of the successive resistance is shown in Fig. 2.2b. From the slope of the temperature dependence $\ln(R_s) = f(10^3/T)$ determined the depth of the acceptor level which is $E_A = 0.54$ eV, and determines the properties of the base material. This level in the $Cd_3In_2Te_6$ is near the middle of the band gap (E_g ($Cd_3In_2Te_6$) = 1.16 eV), which suggests that this material is most likely intrinsic semiconductor, which is also confirmed by the low concentration of charge carriers ($N = 2 \times 10^{11}$ cm^{-3}) and a high specificelectrical resistance ($\rho = 0.41 \times 10^6$ Ω cm). Acceptor centers are most likely formed by vacancies of cadmium.

In the case of a reverse bias in the case of a sharp transition, the expression for the tunnel current has the form [51]:

$$I_{rev}^{t} \approx a_0 \exp\left(\frac{b_0}{\sqrt{\varphi_0(T) - eV}}\right), \qquad (1)$$

where a_0, b_0—parameters that are not voltage dependent.

According to (1), the dependence of I (V) in the coordinates $\ln I_{rev}^{t} = f(\varphi_0 - qV)^{-1/2}$ should be represented by straight lines, as observed in Fig. 2.3.

As shown in Fig. 2.1 at voltages (0 V < U < 0.3 V) on the I–V-characteristics there are observed rectilinear areas, by means of which the coefficient of non-ideality n was determined which has a value of 2.4 and does not change with the

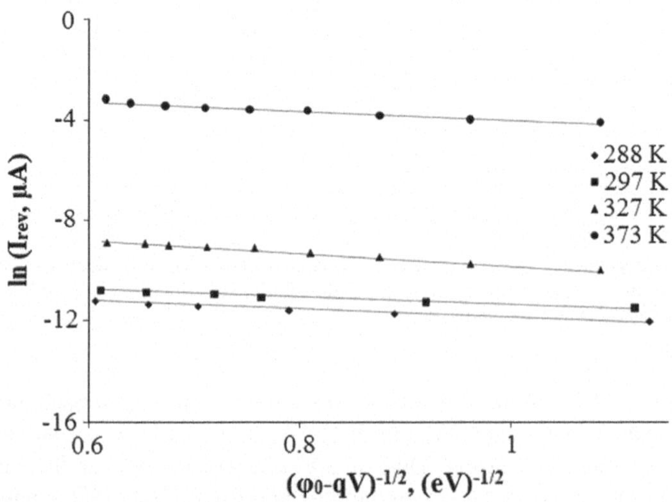

Fig. 2.3 Dependencies characterizing the tunneling mechanism of charge carrier movement through a potential barrier with reverse displacement

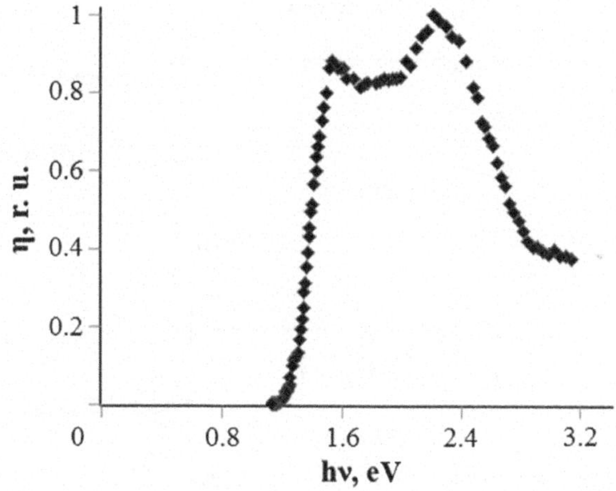

Fig. 2.4 Spectral distribution of the quantum efficiency of the heterojunction n-MoO_x/p-$Cd_3In_2Te_6$

temperature, indicating that in the heterojunction under investigation the generation-recombination current transfer mechanism with the participation of surface states on the metallurgical limit of division at small direct displacements.

The spectral distribution of quantum efficiency η (λ) was determined as the ratio of the number of charge carriers, which form the short-circuit current to the number of falling photons of fixed energy in the heterojunction of n-MoO_x/p-$Cd_3In_2Te_6$ from MoO_x. The measured spectral distribution η (λ) is characterized by the width of the band $\delta_{1/2} = 1.36$ eV at its half-height (Fig. 2.4).

The decrease in quantum efficiency in the short-wave region is due to the small thickness of the MoO_x film. The determining parameter for quantum efficiency in the long-wavelength region is the diffusion length of minority carriers in the base material.

So, reducing the quantum efficiency with increasing wavelength can be explained by a decrease in diffusion length due to recombination of electrons in the neutral region because energy levels are in the forbidden zone.

Energy diagram of anisotypic heterostructures n-MoO_x/p-$Cd_3In_2Te_6$in accordance with the Anderson model, excluding surface electric conditions and the dielectric layer, was built on the values of the parameters of semiconductors power (Fig. 2.5). The value of energy needed to build an energy heterojunction diagram for n-MoO_x were taken from Wang et al. [38], and for p-$Cd_3In_2Te_6$, they were determined from the studies of kinetic and optical properties (the output for $Cd_3In_2Te_6$ was taken as for CdTe since the molar content of the latter in the three-component compound was 75%).

As was said above, the band gap width of $Cd_3In_2Te_6$ is 1.16 eV (from optical studies) and the concentration of charge carriers is 10^{11} cm^{-3}, under such conditions it can be assumed that this semiconductor is intrinsic, and therefore the Fermi level is in the middle of the band gap. Given the lack of accurate data on the Fermi level position and the work of the electron output in the $Cd_3In_2Te_6$, the results are estimated character, but despite this, set value from the energy diagram $\varphi_0 = 0.34$ eV, coincides

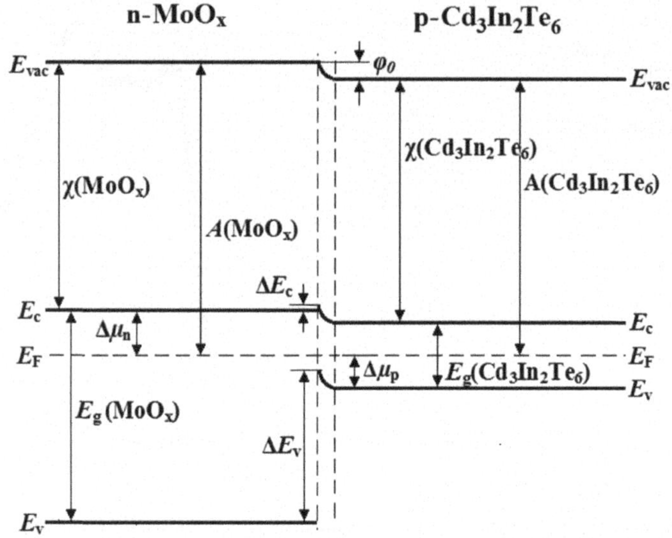

Fig. 2.5 Energy diagram of the heterostructure *n*-MoO_x/*p*-$Cd_3In_2Te_6$: $E_g^{MoOx} = 3.7$ eV; $\chi_{MoOx} = 4.4$ eV; $A_{MoOx} = 5.2$ eV; $E_g^{Cd3In2Te6} = 1.16$ eV; $\chi_{Cd3In2Te6} = 4.28$ eV; $A_{Cd3In2Te6} = 4.86$ eV; $\Delta\mu_n = -.8$ eV; $\Delta\mu_p = 0.58$ eV; $\Delta E_c = 0.12$ eV; $\Delta E_v = 2.66$ eV; $\varphi_0 = 0.34$ eV

with the experimentally found in the study of I–V-characteristics values φ_0, which is also 0.34 eV at room temperature.

2.4 Conclusions

For the first time, heterostructures of *n*-MoO_x/*p*-$Cd_3In_2Te_6$ were obtained by applying a thin MoO_x film (*n*-type) on a *p*-type $Cd_3In_2Te_6$ substrate by DC magnetron sputtering. I–V-characteristics of the structure were investigated in the temperature range 288–373 K. It is established that they have a straightening character. The height of the potential barrier φ_0 varies from 0.34 to 0.09 eV, and the rectifying ratio RR is from 900 to 16 with increasing temperature. Investigation of current transfer mechanisms under direct inclusions showed that the generation and recombination process of current transfer dominates, and in the reverse inclusion, the tunneling mechanism of transport of charge carriers through a potential barrier is dominant.

References

1. X. Liu, Y. Chen, D. Li, S.-W. Wang, C.-C. Ting, L. Chen, K.-W. Ang, C.-W. Qiu, Y.-L. Chueh, X. Sun, H.-C. Kuo, Nearly lattice-matched molybdenum disulfide/gallium nitride heterostructure

enabling high-performance phototransistors. Photonics Res. **7**, 311 (2019). https://doi.org/10.1364/PRJ.7.000311

2. M. Moun, M. Kumar, M. Garg, R. Pathak, R. Singh, Understanding of MoS_2/GaN heterojunction diode and its photodetection properties. Sci. Rep. **8**, 11799 (2018). https://doi.org/10.1038/s41598-018-30237-8
3. F. Qian, Y. Li, S. Gradečak, D. Wang, C.J. Barrelet, C.M. Lieber, Gallium nitride-based nanowire radial heterostructures for nanophotonics. Nano Lett. **4**, 1975–1979 (2004). https://pubs.acs.org/doi/full/10.1021/nl0487774
4. M. Stutzmann, G. Steinhoff, M. Eickhoff, O. Ambacher, C.E. Nebel, J. Schalwig, R. Neuberger, G. Müller, GaN-based heterostructures for sensor applications. Diam. Relat. Mater. **11**, 886–891 (2002). https://doi.org/10.1016/S0925-9635(02)00026-2
5. C. Zhu, P. Yang, D. Chao, X. Wang, X. Zhang, S. Chen, B.K. Tay, H. Huang, H. Zhang, W. Mai, H.J. Fan, All metal nitrides solid-state asymmetric supercapacitors. Adv. Mater. **27**, 4566–4571 (2015). https://doi.org/10.1002/adma.201501838
6. D. Li, Z. Xiao, H.R. Golgir, L. Jiang, V.R. Singh, K. Keramatnejad, K.E. Smith, X. Hong, L. Jiang, J.-F. Silvain, Y. Lu, Large-area 2D/3D MoS_2-MoO_2 heterostructures with thermally stable exciton and intriguing electrical transport behaviors. Adv. Electron. Mater. **3**, 1600335 (2017). https://doi.org/10.1002/aelm.201600335
7. H. Liu, Recent progress in atomic layer deposition of multifunctional oxides and two-dimensional transition metal dichalcogenides. J. Mol. Eng. Mater. **4**, 1640010 (2016). https://doi.org/10.1142/S2251237316400104
8. C. Liu, E.F. Chor, L.S. Tan, Investigations of HfO_2/AlGaN/GaN metal-oxide-semiconductor high electron mobility transistors. Appl. Phys. Lett. **88**, 173504 (2006). https://doi.org/10.1063/1.2198507
9. H. Qin, W. Li, Y. Xia, T. He, Photocatalytic activity of heterostructures based on ZnO and N-doped ZnO. ACS Appl. Mater. Interfaces. **3**, 3152–3156 (2011). https://doi.org/10.1021/am200655h
10. J. Shang, G. Liu, H. Yang, X. Zhu, X. Chen, H. Tan, B. Hu, L. Pan, W. Xue, R.-W. Li, Thermally stable transparent resistive random access memory based on all-oxide heterostructures. Adv. Funct. Mater. **24**, 2171–2179 (2013). https://doi.org/10.1002/adfm.201303274
11. T.T. Kovaliuk, M.M. Solovan, O.A. Parfenyuk, V.V. Brus, I.P. Koziarskyi, P.D. Maryanchuk, in *Proceedings of SPIE. Thirteenth International Conference on Correlation Optics*, vol 10612, ed. by O.V. Angelsky (SPIE, Bellingham, 2018). http://dx.doi.org/10.1117/12.2304772
12. I.P. Koziarskyi, E.V. Maistruk, D.P. Koziarskyi, P.D. Maryanchuk, in *Proceedings of SPIE. Thirteenth International Conference on Correlation Optics*, vol 10612, ed. by O.V. Angelsky (SPIE, Bellingham, 2018). https://doi.org/10.1117/12.2304879
13. I.P. Koziarskyi, E.V. Maistruk, D.P. Koziarskyi, P.D. Maryanchuk, Electric properties of thin films $Cu_2ZnSnSe_4$ and $Cu_2ZnSnSe_2Te_2$ (S_2) obtained by thermal vacuum deposition. J. Nano Electron. Phys. **10**, 01028 (2018).https://doi.org/10.21272/jnep.10(1).01028
14. I.P. Koziarskyi, E.V. Maistruk, D.P. Koziarskyi, in *Proceedings of 2019 IEEE 2nd Ukraine Conference on Electrical and Computer Engineering (UKRCON)*, vol. CFP19K03-ART, ed. by M.S. Antyufeyeva (IEEE, Danvers, 2019)
15. E.V. Maistruk, I.G. Orletsky, M.I. Ilashchuk, I.P. Koziarskyi, D.P. Koziarskyi, P.D. Marianchuk, O.A. Parfenyuk, Influence of heat treatment of the base material on the electrical properties of anisotyped heterojunctions *n*-ZnO:Al/*p*-CdZnTe. Semicond. Sci. Tech. **34**, 045016 (2019). https://doi.org/10.1088/1361-6641/ab0a1c
16. E.V. Maistruk, I.P. Koziarskyi, D.P. Koziarskyi, P.D. Marianchuk, V.V. Brus, in *Proceedings of SPIE. Thirteenth International Conference on Correlation Optics*, Chernivtsi, Ukraine, 11–15 Sept 2018. https://doi.org/10.1117/12.2304801
17. J. Baltrusaitis, B. Mendoza-Sanchez, V. Fernandez, R. Veenstra, N. Dukstiene, A. Roberts, N. Fairley, Generalized molybdenum oxide surface chemical state XPS determination via informed amorphous sample model. Appl. Surf. Sci. **326**, 151–161 (2015). https://doi.org/10.1016/j.apsusc.2014.11.077

18. T. He, J. Yao, Photochromism of molybdenum oxide. J. Photochem. Photobiol **4**, 125–143 (2003). https://doi.org/10.1016/S1389-5567(03)00025-X
19. X. Ma, J. Gong, S. Wang, N. Gao, D. Wang, X. Yang, F. He, Reactivity and surface properties of silica supported molybdenum oxide catalysts for the transesterification of dimethyl oxalate with phenol. Catal. Commun. **5**, 101–106 (2004). https://doi.org/10.1016/j.catcom.2003.12.001
20. V.S. Saji, C.-W. Lee, Molybdenum, molybdenum oxides, and their electrochemistry. Chemsuschem **5**, 1146–1161 (2012). https://doi.org/10.1002/cssc.201100660
21. J. Zhou, N.-S. Xu, S.-Z. Deng, J. Chen, J.-C. She, Z.-L. Wang, Large-area nanowire arrays of molybdenum and molybdenum oxides: synthesis and field emission properties. Adv. Mater. **15**, 1835–1840 (2003). https://doi.org/10.1002/adma.200305528
22. S.Y. Davydov, On the electron affinity of silicon carbide polytypes. Semiconductors **41**, 696–698 (2007). https://doi.org/10.1134/S1063782607060152
23. F.C. Jamieson, E.B. Domingo, T. McCarthy-Ward, M. Heeney, N. Stingelin, J.R. Durrant, Fullerene crystallisation as a key driver of charge separation in polymer/fullerene bulk heterojunction solar cells. Chem. Sci. **3**, 485–492 (2012). https://doi.org/10.1039/C1SC00674F
24. J. Nelson, Polymer: fullerene bulk heterojunction solar cells. Mater. Today **14**, 462–470 (2011). https://doi.org/10.1016/S1369-7021(11)70210-3
25. M. Scharber, D. Mühlbacher, M. Koppe, P. Denk, C. Waldauf, A. Heeger, C. Brabec, Design rules for donors in bulk-heterojunction solar cells—towards 10% energy-conversion efficiency. Adv. Mater. **18**, 789–794 (2006). https://doi.org/10.1002/adma.200501717
26. M.C. Scharber, N.S. Sariciftci, Efficiency of bulk-heterojunction organic solar cells. Prog. Polym. Sci. **38**, 1929–1940 (2013). https://doi.org/10.1016/j.progpolymsci.2013.05.001
27. L. Zhao, C.L. Zhou, H.L. Li, H.W. Diao, W.J. Wang, Role of the work function of transparent conductive oxide on the performance of amorphous/crystalline silicon heterojunction solar cells studied by computer simulation. Phys. Stat. Sol. A **205**, 1215–1221 (2008). https://doi.org/10.1002/pssa.200723276
28. I.P. Koziarskyi, E.V. Maistruk, D.P. Koziarskyi, in *Proceedings of SPIE. Eleventh International Conference on Correlation Optics*, vol 9066, ed. by O.V. Angelsky (SPIE, Bellingham, 2013). https://doi.org/10.1117/12.2053538
29. I.P. Koziarskyi, E.V. Maistruk, D.P. Koziarskyi, P.D. Maryanchuk, Charge transport and mechanisms of electron scattering in $(HgSe)_3(In_2Se_3)$ crystals doped with 3d transition metals. Inorg. Mater. **50**, 447–451 (2014). https://doi.org/10.1134/S0020168514050070
30. I.P. Koziarskyi, V.T. Maslyuk, P.D. Maryanchuk, E.V. Maistruk, D.P. Koziarskyi, I.G. Megela, G.V. Lashkarev, Radiation Resistance of $(HgSe)_3(In_2Se_3)$ <Mn>. Russ. Phys. J. **61**, 1189–1193 (2018).https://link.springer.com/article/10.1007/s11182-018-1516-1
31. P.D. Maryanchuk, D.P. Kozyarskii, Mechanisms of electron scattering in $(3HgSe)_{1-x}(Al_2Se_3)_x$ crystals doped with manganese. Russ. Phys. J. **52**, 1355–1357 (2009). https://doi.org/10.1007/s11182-010-9377-2
32. P.D. Maryanchuk, D.P. Kozyarskii, Electrical and optical properties of manganese-doped $(3HgSe)_{1--x}(Al_2Se_3)_x$ crystals. Inorg. Mater. **46**, 460–463 (2010). https://doi.org/10.1134/S0020168510050043
33. J.A. Aranovich, D. Golmayo, A.L. Fahrenbruch, R.H. Bube, Photovoltaic properties of ZnO/CdTe heterojunctions prepared by spray pyrolysis. J. Appl. Phys. **51**, 4260–4268 (1980). https://doi.org/10.1063/1.328243
34. P.P. Edwards, A. Porch, M.O. Jones, D.V. Morgan, R.M. Perks, Basic materials physics of transparent conducting oxides. Dalton Trans. **19**, 2995–3002 (2004). https://doi.org/10.1039/B408864F
35. J.D. Major, R. Tena-Zaera, E. Azaceta, L. Bowen, K. Durose, Development of ZnO nanowire based CdTe thin film solar cells. Sol. Energ. Mat. Sol. C **160**, 107–115 (2017). https://doi.org/10.1016/j.solmat.2016.10.024
36. U.N. Roy, R.M. Mundle, G.S. Camarda, Y. Cui, R. Gul, A. Hossain, G. Yang, A.K. Pradhan, R.B. James, Novel ZnO: Al contacts to CdZnTe for X- and gamma-ray detectors. Sci. Rep. **6**, 26384 (2016). https://doi.org/10.1038/srep26384

37. F. Ruske, in *Physics and Technology of Amorphous-Crystalline Heterostructure Silicon Solar Cells. Engineering Materials*, ed. by W.G.J.H.M. van Sark, L. Korte, F. Roca (Springer, Berlin, Heidelberg, 2012), pp. 301–330. https://doi.org/10.1007/978-3-642-22275-7_9
38. F. Wang, Z. Tan, Y. Li, Solution-processable metal oxides/chelates as electrode buffer layers for efficient and stable polymer solar cells. Energ. Environ. Sci. **8**, 1059–1091 (2015). https://doi.org/10.1039/C4EE03802A
39. E.V. Maistruk, I.P. Koziarskyi, D.P. Koziarskyi, G.O. Andrushchak, in *Proceedings of the 2017 IEEE 7th International Conference on Nanomaterials: Applications & Properties (NAP-2017)*, Zatoka, Ukraine, 10–15 Sept 2017. https://doi.org/10.1109/NAP.2017.8190163
40. E.V. Maistruk, I.P. Koziarskyi, D.P. Koziarskyi, G.O. Andrushchak, in *Proceedings of SPIE. Thirteenth International Conference on Correlation Optics*, Chernivtsi, Ukraine, 11–15 Sept 2018. https://doi.org/10.1117/12.2304328
41. K. Alfaramawi, A. Sweyllam, S. Abboudy, N.G. Imam, H.A. Motaweh, Interface states-induced-change in the energy band diagram and capacitance-voltage characteristics of isotype ZnTe/CdTe heterojunctions. Int. J. Mod. Phys. B **24**, 4717–4725 (2010). https://doi.org/10.1142/S0217979210056165
42. V. Consonni, S. Renet, J. Garnier, P. Gergaud, L. Artús, J. Michallon, L. Rapenne, E. Appert, A. Kaminski-Cachopo, Improvement of the physical properties of ZnO/CdTe core-shell nanowire arrays by CdCl2 heat treatment for solar cells. Nanoscale Res. Lett. **9**, 222 (2014). https://doi.org/10.1186/1556-276X-9-222
43. A. Crossay, S. Buecheler, L. Kranz, J. Perrenoud, C.M. Fella, Y.E. Romanyuk, A.N. Tiwari, Spray-deposited Al-doped ZnO transparent contacts for CdTe solar cells. Sol. Energ. Mat. Sol. C **101**, 283–288 (2012). https://doi.org/10.1016/j.solmat.2012.02.008
44. S.M. Sze, K.N. Kwok, *Physics of Semiconductor Devices* (Wiley, 2006)
45. L. Vasanelli, A. Valentini, A. Quirini, A.M. Mancini, A. Losacco, in *Proceedings of the 7th E. C. Photovoltaic Solar Energy Conference*, Sevilla, Spain, 27–31 Oct 1986. https://doi.org/10.1007/978-94-009-3817-5
46. G. Wary, A. Rahman, Indian J. Pure Ap. Phy. **41**, 474 (2003)
47. J. Zhu, Y. Yang, Y. Gao, D. Qin, H. Wu, L. Hou, W. Huang, Enhancement of open-circuit voltage and the fill factor in CdTe nanocrystal solar cells by using interface materials. Nanotechnology **25**, 365203 (2014). https://doi.org/10.1088/0957-4484/25/36/365203
48. I.P. Koziarskyi, E.V. Maistruk, D.P. Koziarskyi, in *Advances in Thin Films, Nanostructured Materials, and Coatings. Lecture Notes in Mechanical Engineering*, ed. by A. Pogrebnjak, V. Novosad (Springer, Singapore, 2019), pp. 267–275. https://doi.org/10.1007/978-981-13-6133-3_26
49. J.-J. Wang, T. Ling, S.-Z. Qiao, X.-W. Du, Double open-circuit voltage of three-dimensional ZnO/CdTe solar cells by a balancing depletion layer. ACS Appl. Mater. Interfaces. **6**, 14718–14723 (2014). https://doi.org/10.1021/am5041219
50. I.P. Koziarskyi, E.V. Maistruk, D.P. Koziarskyi, P.D. Maryanchuk, Electrical properties of the $Cu_2O/Cd_{1-x}Zn_xTe$ heterostructure. J. Nano Electron. Phys. **11**, 02007 (2019). https://doi.org/10.21272/jnep.11(2).02007
51. B.L. Sharma, R.K. Purohit, *Semiconductor Heterojunctions* (Pergamon Press, 1974)

Chapter 3
Transport Properties of Cubic Cuprous Iodide Films Deposited by Successive Ionic Layer Adsorption and Reaction

N. P. Klochko, K. S. Klepikova, D. O. Zhadan, V. R. Kopach, Y. R. Kostyuchenko, I. V. Khrypunova, V. M. Lyubov, M. V. Kirichenko, A. L. Khrypunova, S. I. Petrushenko and S. V. Dukarov

Abstract In this work we investigate crystal structure, morphology, and composition of CuI films produced in different modes of SILAR on glass plates and on flexible poly(ethylene terephthalate) (PET) substrates in connection with their transport properties, and with electrical and thermoelectric properties as a whole. Temperature dependences of resistivity for most CuI films have the crossover from semiconducting to metallic behavior with increasing temperature. The semiconductor carrier transport occurs in CuI films through nearest neighboring hopping. Metallic transport in CuI films carried out in accordance with the ionized impurity scattering and the carrier–carrier scattering model.

3.1 Introduction

Since an idea of invisible circuits conceived, transparent semiconductor devices have gained remarkable interest over the last years [1–8]. Another functionality added if taking into account transparent thermoelectric devices for power generation working at near room temperatures. Cubic cuprous iodide (γ-CuI, hereinafter CuI) is a promising intrinsically *p*-type conducting material both for transparent optoelectronics due to its high band gap of $\approx$3 eV at room temperature [1–7], and for thermoelectrics [1, 3, 6, 7]. To this day CuI is by far the best transparent *p*-type thermoelectric material with a room-temperature thermoelectric figure of merit of $ZT \approx 0.21$ [1]. As described in [1–7], CuI thin films are heavily doped and disordered degenerate semiconductors possessing properties of *p*-type transparent conductors. For example, a

N. P. Klochko (✉) · K. S. Klepikova · D. O. Zhadan · V. R. Kopach · Y. R. Kostyuchenko · I. V. Khrypunova · V. M. Lyubov · M. V. Kirichenko · A. L. Khrypunova
Materials Engineering for Solar Cells Department (MESC), National Technical University "Kharkiv Polytechnic Institute" (NTU "KhPI"), Kharkiv, Ukraine
e-mail: klochko.np16@gmail.com

S. I. Petrushenko · S. V. Dukarov
School of Physics, V. N. Karazin Kharkiv National University, Kharkiv, Ukraine

A. D. Pogrebnjak and O. Bondar (eds.), *Microstructure and Properties of Micro- and Nanoscale Materials, Films, and Coatings (NAP 2019)*, Springer Proceedings in Physics 240, https://doi.org/10.1007/978-981-15-1742-6_3

record low room-temperature resistivity $\rho \approx 6.4 \times 10^{-5}$ Ω m has been shown in [6] for as-deposited and $\rho \approx 3.5 \times 10^{-5}$ Ω m for I-doped polycrystalline CuI thin films obtained in vacuum by reactive sputtering.

Moreover, thin films CuI are of particular interest for flexible wearable electronic and thermoelectric applications due to its unique properties and since they can be made on flexible substrates at near-room temperature. It is remarkable that CuI films suitable for barrier heterostructures [9] and for thermoelectric generators [10] can be synthesized from abundant resources using an inexpensive and suitable for large-scale production wet chemical Successive Ionic Layer Adsorption and Reaction (SILAR) method.

However, although CuI is the first discovered transparent conductive material, until now there are only a few publications regarding the analysis and understanding of its transport properties [1, 6, 7]. As can be seen from the temperature (T) dependence of the resistivity (ρ) reported in [1, 6, 7, 11], transport properties of the p-type CuI thin films are quite unusual. A crossover from semiconducting to metallic behavior with increasing temperature and a typical ρ minimum in the temperature-dependent resistivity graph has been shown in [1, 6, 11] for nanocrystalline CuI thin films obtained through different techniques, including vacuum evaporation [6] and SILAR [11]. Authors [6] note, that the observed crossover from semiconducting to metallic behavior cannot be explained by thermal activation of the carriers, rather, the granular nanocrystalline CuI structure should be the reason for this behavior. In [1] investigation on the transport mechanism for polycrystalline CuI thin films by studying temperature-dependent resistivity for over ten samples with different carrier concentrations allowed to construct a conclusive barrier model, which is applicable to describe the conductivity in CuI thin films by the presence of a tunneling process, which model was proposed by Sheng [12]. In [12] a theoretical expression for the tunneling conductivity was derived, which displays thermally activated characteristics at high temperatures, but becomes identical to the temperature-independent simple elastic tunneling at low temperatures. Between the two limiting behaviors, the shape of the tunneling barrier controls the temperature dependence of the conductivity.

In accordance with [1, 6], the modeling of the transport properties, when CuI film demonstrates a metallic behavior (for the branch of the resistivity vs. temperature graph where $d\rho/dT > 0$) leads to an insight into the properties of the interior of the CuI grains. A simple power law was used in [1, 6] for the metallic CuI film, which resistivity ρ_m reasonable explanation is the scattering on ionized impurities explicitly derived for degenerate semiconductors, and also carrier–carrier scattering. According to [1], both processes are justified due to the high carrier densities in CuI and produce a similar quadratic temperature dependence of the resistivity:

$$\rho_m(T) = \rho_0 + \rho_{02} T^{\gamma}. \tag{1}$$

Here ρ_0 is a residual offset resistivity at $T = 0$ K, ρ_{02} is a temperature-independent prefactor, and γ is a numerical exponent (according to [1], fitting with $\gamma = 2$ reproduces the data fairly well). Similarly, it states in [6] that within the metallic region, the conductivity follows the power law, $\sigma_m(T) \sim T^{-2}$, according to the ionized impurity scattering model for degenerate semiconductors.

The semiconducting temperature dependence of the CuI resistivity ρ_s (in the branch of the resistivity vs. temperature graph where $d\rho/dT < 0$) is explained, according to [1, 6], by the granular structure of the CuI films, and a model capable of reproducing the semiconducting behavior of CuI thin films is a fluctuation-induced tunneling conductivity (FITC) model developed by Sheng [12]. According to [12], consider a region of close approach between two large conducting segments in a disordered material. Since the electron tunneling probability depends exponentially on the insulating barrier thickness, it is expected that practically all tunneling occurs within the small surface areas. We can approximate such a tunnel junction by a parallel-plate capacitor. Owing to the random thermal motion of electrons in the conducting region, there can be transient excess or deficit of charges on the tunnel junction surfaces, resulting in voltage fluctuations across the junction. Since the internal tunnel junctions in disordered materials are usually small in area, the resulting large voltage fluctuations across the junctions are expected to play an important role in modifying the electron tunneling probability. As shown in [1, 6, 7], FITC model is applicable for degenerate CuI thin films, when most of the conduction electrons are delocalized and free to move up to several atomic distances inside the crystal grains. Along with this, FITC model assumes energetic barriers between conducting regions of CuI grains, which are overcome by a tunneling process. This process takes place at the point of closest approach between the conducting regions that is, across the grain boundaries. The tunneling current applied for CuI films has an additional temperature dependence due to thermal voltage fluctuations at the barriers, which results in a typical temperature dependence of the conductivity for FITC model described in [1, 6] as follows:

$$\sigma_S(T) = \sigma_1 \exp[-T_0/(T + T_{0S})], \tag{2}$$

where σ_1 is a prefactor that is only weakly temperature-dependent, two temperatures T_0 and T_{0S} have a physical interpretation: T_0 corresponds to the energy needed to overcome the barrier, while well below T_{0S} the thermal voltage fluctuations become insignificant and the resistivity stays almost temperature independent.

As a special case of FITC [1, 6, 12], it is proved in [7] that transport of charge carriers in polycrystalline CuI films evaporated in a vacuum takes place via the localized states, and this transition through localized states is named as hopping. The conduction mechanism at near-room temperatures is nearest neighboring hopping (NNH). In NNH conduction, hole hops the nearest neighboring empty sites according to [7] and the temperature dependence of the resistivity is as follows:

$$\rho_S(T) = \exp(T_0/T), \tag{3}$$

where T_0 is a characteristic temperature (i.e. T_0 is the hopping energy in Kelvin degree). As stated in [7], the condition for NNH is, NNH distance (r) $\gg$Bohr exciton radius, which should be satisfied in polycrystalline CuI films (CuI Bohr exciton radius is 1.5 nm [13]). Thus, for the branch of the resistivity versus temperature graph where $d\rho/dT < 0$, the slope of the linear fitting of $\ln(\rho_s)$ vs $1/T$ gives hopping energy T_0, which is lesser than the activation energy required for the thermally activated transition of charge carriers to the valence band. The nearest neighboring hopping model fits the resistivity of the thermally evaporated cubic cuprous iodide films at near-room temperatures [7] with $T_0 \approx 38.6$ K, that equals to $T_0 \approx 3.3$ meV.

As it was found from the literature survey, no work has been done on the transport properties of CuI thin films deposited by SILAR, despite the fact that the development and improvement of these materials require thorough knowledge about their electrical and thermoelectric properties. Hence, issues remain to require further experiments to gain a complete picture on hole conductivity in these CuI films and to govern the electronic and transport properties of the CuI films deposited both on rigid and flexible substrates via different SILAR modes. In present work, we investigate an interconnection of crystal structure, morphology, and composition of CuI films produced in different modes of SILAR on glass plates and on flexible poly(ethylene terephthalate) (PET) substrates with their transport properties, and with electrical and thermoelectric properties as a whole.

3.2 Experimental Procedure

In this study, CuI thin films were synthesized via SILAR method both on soda lime glass substrates and on the flexible and transparent 20 μm thick PET films. The deposition of copper iodide films was carried out using an aqueous solution containing 0.1 M $CuSO_4$ and 0.1 M $Na_2S_2O_3$ as a cationic precursor, in which a copper (I) thiosulfate complex $Na[Cu(S_2O_3)]$ was formed, from which Cu^+ ions were released into solution. The glass plate and PET film inserted in a frame "back-to-back" immersed into the cationic precursor for 20 s. Then, the substrates washed in distilled water for 10 s. For the reaction of the firmly adsorbed Cu^+ ions on the glass and PET surfaces with I^- ions to obtain some CuI monolayers, the substrates were then immersed for 20 s into aqueous NaI solution (anionic precursor), which concentration was 0.05, 0.075 or 0.1 M. After that, the glass and PET substrates washed in distilled water for 10 s. The listed procedure was one SILAR cycle of CuI film deposition. Such SILAR cycles were repeated 25–40 times. Thereafter, CuI films deposited on the one side of each substrate, namely CuI/glass and CuI/PET samples, taken out of the frame, washed with distilled water and dried by air stream. As seen in photo of CuI/PET sample in Fig. 3.1a, b, the resulting CuI films are semi-transparent, yellowish in color, and well bonded to the substrates. The thickness of the CuI films in the 100–820 nm range determined gravimetrically, taking for a calculation the bulk CuI density 5.67 g/cm^3.

The morphology of the CuI films observed by scanning electron microscopy (SEM) in secondary electron mode. The SEM instrument (Tescan Vega 3 LMH) operated at an accelerating voltage of 30 kV without the use of additional conductive coatings. Elemental analysis of the CuI films was carried out by X-ray fluorescence

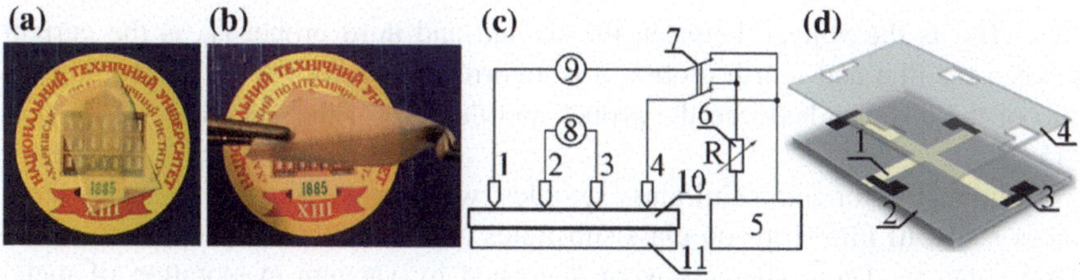

Fig. 3.1 **a**, **b** Two images of 200 nm thick CuI film deposited by SILAR (anionic precursor 0.075 M NaI, 25 SILAR cycles) on PET on the background of National Technical University "Kharkiv Polytechnic Institute" logo. **c** Schematic of apparatus for a studying of CuI film temperature-dependent resistivity using the four-probe method: 1–4—the system of probes; 5—current source; 6—variable resistor for regulating the operating mode of the power supply of the measuring circuit; 7—current direction switch; 8—voltmeter for U_{23} measuring; 9—current I_{14} measuring electrometer; 10—CuI/glass or CuI/PET sample; 11—resistive heater. **d** Schematic of CuI/glass sample for measurement of resistance R of CuI film strip (1) on glass substrate (2) connected with ohmic metal contacts (3), which were deposited by vacuum evaporation of metal through a shadow mask (4) on the surface of CuI

(XRF) microanalysis using an energy dispersive spectrometry (EDS) system "Bruker XFlash 5010". Energy dispersion spectra were taken from the 50 × 50 μm CuI film areas. Quantification of the spectra carried out in the self-calibrating detector mode.

To analyze phase composition and structural parameters of the CuI films, we recorded X-ray diffraction patterns by a "DRON-4" diffractometer as in [10] (XRD method). The microstrains ε in the polycrystalline CuI films $\varepsilon = \Delta d/d$ (where d is the crystal interplanar spacing according to the reference database JCPDS, and Δd is the difference between the corresponding experimental and reference interplanar spacings) were estimated by an analyzing the broadening of the X-ray diffraction peaks using the Scherrer's method [7]. The average crystallite size (D) was calculated using Scherrer's semi-empirical formula [7]:

$$D = (0.9 \cdot \lambda)/(\beta \cdot \cos\theta), \tag{4}$$

where λ is X-ray wavelength; $\beta = (B - b)$, when B is the observed Full Width at Half Maximum (FWHM), and b is the broadening in the peak due to the instrument in radians, θ denotes the Bragg's angle of the X-ray diffraction peak.

The conductivity type of the obtained CuI films determined using a standard hot-probe method.

The resistivity ρ of CuI films in all CuI/glass and CuI/PET samples was measured at temperatures T from 290 to 390 K in accordance with [14] by means four-point collinear probe resistivity method using a home-made apparatus for a studying the temperature-dependent resistivity presented in Fig. 3.1c. The resistivity was calculated according to [14] as follows:

$$\rho = (\pi\, t\, \delta\, U_{23})/(I_{14} \ln(2)), \tag{5}$$

where U_{23} is the voltage between the second and third probe; I_{14} is the current between the first and fourth probes; δ is a correction factor for the accounting the ratio of the distance between the probes and the size of the substrate; $\pi\delta/\ln(2) \approx 4.45$.

Additionally, for some CuI/glass samples, we measured resistivity using resistances R of CuI film strips on glass substrates connected with two opposite ohmic metal contacts. These contacts were deposited by vacuum evaporation of metal through a shadow mask on the surface of CuI as shown in Fig. 3.1d. To study the temperature-dependent resistivity, the CuI/glass sample was heated in the T range 290–345 K using a hot plate of a Peltier element "TEC1-12730", on which it was located. Resistivity value was obtained by the expression:

$$\rho = twRl^{-1}, \tag{6}$$

where t is CuI film thickness, w is *the* width of CuI film strip; R is resistance between two metal contacts, l is the length of CuI film strip.

To assess the thermoelectric performances of the obtained CuI films, the in-plane Seebeck coefficients S at the 295–395 K temperature range were obtained according to [14] as thermoelectric voltages induced in response to the temperature gradients ΔT along the CuI/glass and CuI/PET samples. As shown earlier [14], thermoelectric voltages were measured when a distance between hot and cold probes in the form of gold rings in the home-made setup was 2.3 cm.

3.3 Results and Discussion

Figures 3.2a and 3.3a show the morphology of CuI films prepared on glass and on PET substrates, respectively, by SILAR depending on the anionic precursor concentration. A tiny network like fibrous CuI structure for the SILAR deposited film is shown, especially on the glass substrate, when using as the anionic precursor of the dilute solution 0.05 NaI. The elevated anion precursor concentrations provide an increase in the thickness of copper iodide films and a change in the morphology of their surface. The smoothest dense fine-grained CuI layers were obtained using an aqueous solution of 0.075 mM NaI as the anion precursor, the further increase of the NaI concentration up to 0.1 M resulted in a coarse-grained CuI, especially on the glass substrate.

Figures 3.2b and 3.3b demonstrate experimental XRD pattern for these samples of copper iodide films deposited by SILAR on glass and on PET substrates, respectively. As can be seen, all CuI films are single-phase, polycrystalline and have a cubic Marshite copper iodide structure (zincblende, γ-CuI, JCPDS#06-0246). Calculations of the CuI crystallite sizes (coherent-scattering domains) yielded the 15–50 nm range, so, the films are nanocrystalline. Analysis of structural parameters of the obtained copper iodide films has revealed tensile microstrains $\varepsilon = (2.7\text{–}12.0) \times 10^{-3}$ arb. un.

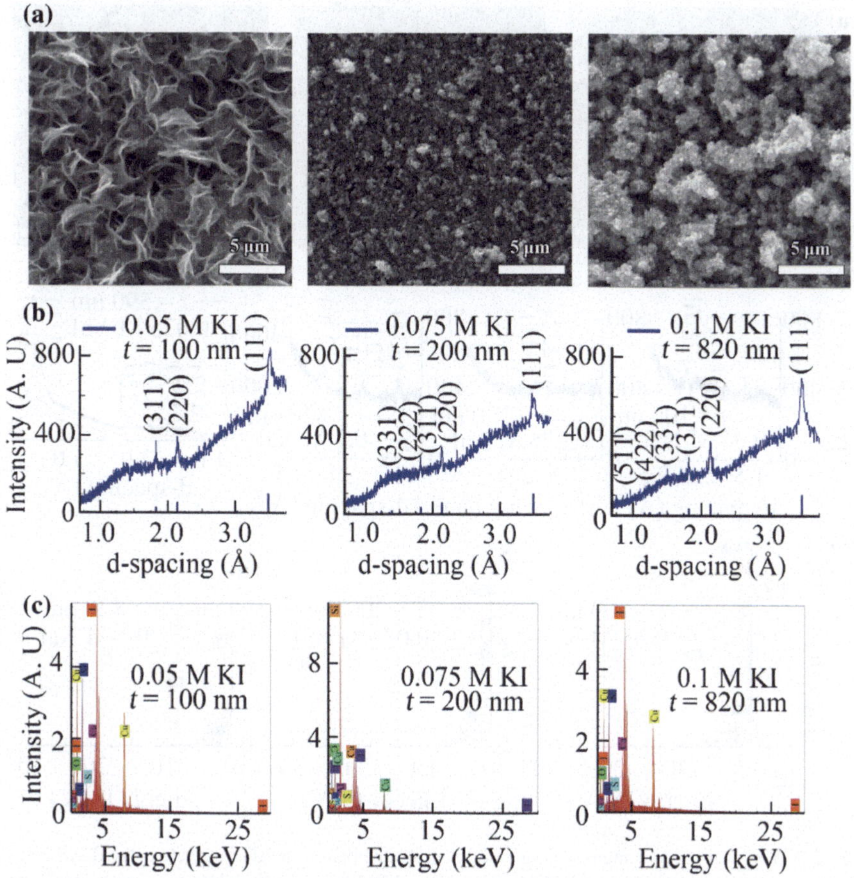

Fig. 3.2 Top-view SEM images of copper iodide films deposited on glass substrates via SILAR using anionic precursors with different NaI concentrations. XRD patterns (**b**) and the corresponding X-ray fluorescence spectra (**c**) of CuI films deposited on glass substrates via SILAR method using anionic precursors with different NaI concentrations

No significant effect of the substrate on the CuI crystal structure was found. CuI/glass and CuI/PET samples deposited by SILAR using 0.075 M NaI have shown the best CuI crystal structure.

The elemental X-ray fluorescence microanalysis of these CuI films (Figs. 3.2c, 3.3c) has revealed, that the elemental ratio (at. %) of the Cu/I is more than one (Cu/I ratio in the 1.2–1.4 range). Some other elements have also been found in the XRF spectra. Among them, it is established that Si, Na, Mg, Ca, O in Fig. 3.2c belong to the glass substrate. C and O belong to the PET Fig. 3.3c or might be attributed for both samples to H_2O, O_2, and CO_2 adsorbed on the surface of the CuI crystals in the air. A small peak of aluminum observed in the energy dispersion spectra is apparently generated by aluminum attachments of the vacuum chamber, in which the XRF

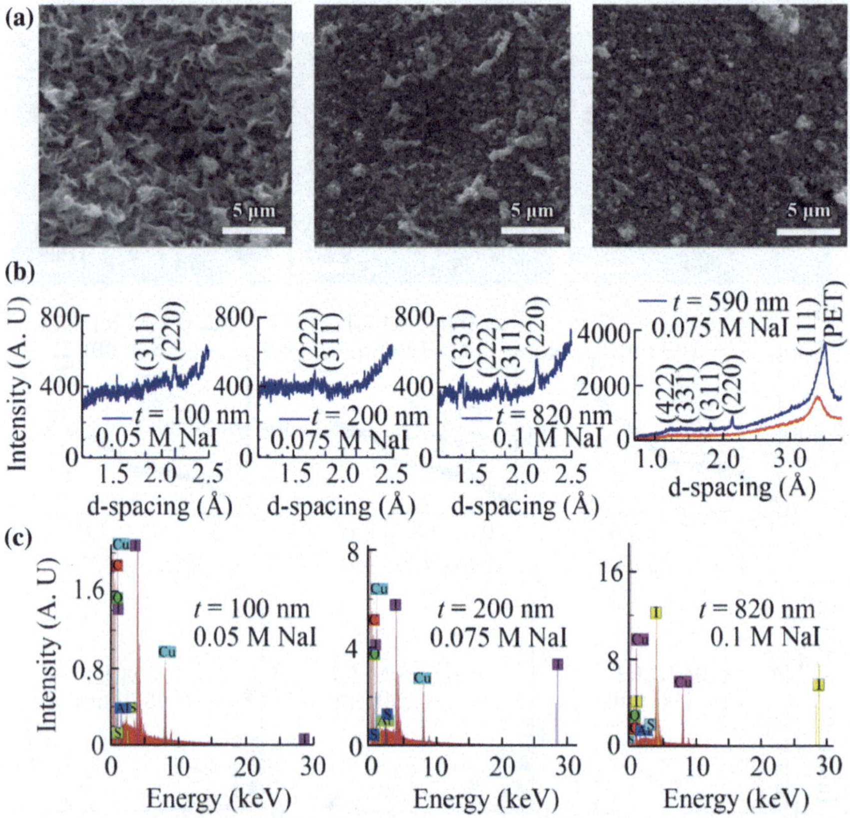

Fig. 3.3 Top-view SEM images of copper iodide films deposited on PETsubstrates via SILAR using anionic precursors with different NaI concentrations. XRD patterns (**b**) and the corresponding X-ray fluorescence spectra (**c**) of CuI films deposited on glass substrates via SILAR method using anionic precursors with different NaI concentrations, and XRD pattern of PET substrate

microanalysis was carried out, primarily, by the table and the holder for the samples. A distinctive feature of all obtained by us via the SILAR technique CuI films is the presence of 0.3–2.0 at.% of sulfur, the source of which, probably, is a chemically unstable compound sodium thiosulfate $Na_2S_2O_3$ from the cationic precursor solution. Despite the fact, that we could not find such information in the literature, to our opinion, sulfur, which is an element of the sixth group, can form acceptor levels in the copper iodide. Thus, sulfur can increase the number of free charge carriers (holes) in the *p*-CuI films making them degenerate semiconductors without formation of a distinguishable via X-ray diffractometry analysis copper sulfide phases.

The conventional hot probe method has identified that all obtained by means SILAR CuI films are *p*-type semiconductors. Comparison of Figs. 3.4, 3.5 and 3.6 shows that, in general, the resistivity of CuI films is low, but ρ for CuI/glass samples is greater than for CuI/PET ones. For our CuI/glass samples, ρ is in the range of

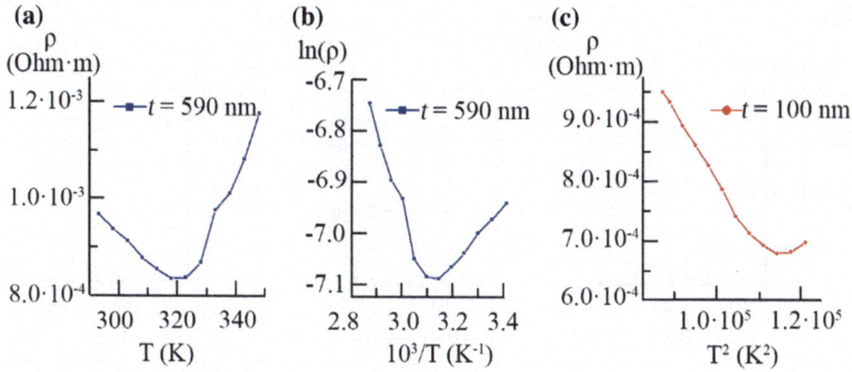

Fig. 3.4 Temperature dependences of resistivity ρ for CuI/glass samples in the form of 100 nm and 590 nm thick CuI strips connected with metal contacts on glass substrates: **a** resistivity versus temperature graphs; **b** ln ρ versus 1000/T graphs; **c** ρ versus T^2 graphs

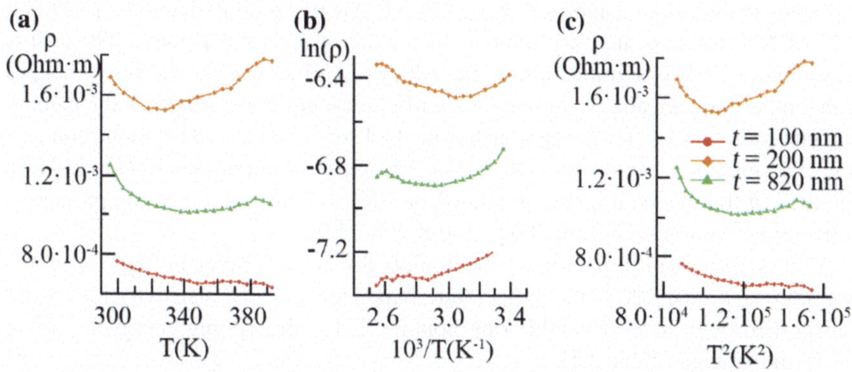

Fig. 3.5 Temperature dependences of resistivity ρ measured by means four-point probe method for CuI/glass samples with 100 nm, 200 nm and 890 nm thick CuI films: **a** resistivity versus temperature graphs; **b** ln ρ versus 1000/T graphs; **c** ρ versus T^2 graphs

6.8×10^{-4}–1.8×10^{-3} Ω m at 290–390 K. The CuI/PET samples have ρ in the 3.0×10^{-4}–1.0×10^{-3} Ω m range in the *T* interval 290–380 K.

The results of the transport properties analysis of the cubic cuprous iodide films are presented in Figs. 3.4, 3.5 and 3.6. It is seen that the substrate material affects the character of CuI resistivity changes with temperature to a certain extent, but only for the thinnest CuI films. With increasing of the CuI film thickness in both CuI/glass and CuI/PET samples, the effect of the substrate on the transport of carriers decreases, as can be seen in Figs. 3.4, 3.5 and 3.6. This is explained by the layer-by-layer growth of the films in the SILAR method. For 200–820 nm thick CuI films the crossover is observed from ρ decreasing with *T* increasing to ρ growth with *T*. The first behavior (semiconducting) is more characteristic for CuI/glass samples (Figs. 3.4, 3.5), and the second trend (metallic) dominates for CuI/PET samples (Fig. 3.6). Thin 100 nm

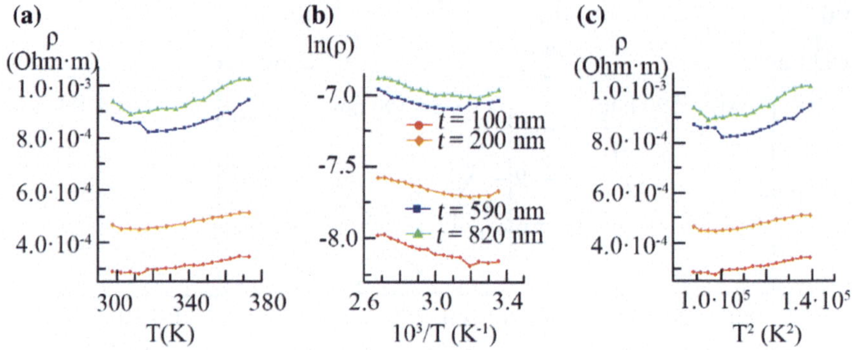

Fig. 3.6 Temperature dependences of resistivity ρ measured by means four-point probe method for CuI/PET samples with 100, 200, 590 and 890 nm thick CuI films: **a** resistivity versus temperature graphs; **b** ln ρ versus 1000/T graphs; **c** ρ versus T^2 graphs

CuI films in CuI/glass samples prepared by SILAR using diluted anionic precursor 0.05 M NaI demonstrate only semiconductor behavior at the *T* range 290–340 K (Figs. 3.4, 3.5). It is significant that the values of CuI resistivity and their changes with temperature are almost the same for CuI/glass samples regardless of the method of ρ measurement, i.e. for the approach using CuI strips connected by ohmic contacts (Fig. 3.4), and for a four-probe method (Fig. 3.5). On the contrary, the 100 nm CuI film obtained in the identical mode of SILAR on PET exhibits only metallic properties in the temperature range from 300 to 380 K (Fig. 3.6).

As it is seen from ln ρ versus 1000/*T* plots for most samples in Figs. 3.4b, 3.5, and 3.6b, at *T* near 290 K the temperature dependence of the resistivity fits carrier transport through nearest neighboring hopping [7] with hopping energy T_0 in the 10–76 meV range (Table 3.1).

Simultaneously, at *T* near 380–390 K most CuI films demonstrate metallic transport (Figs. 3.4, 3.5 and 3.6). It is seen from ρ versus T^2 plots in Figs. 3.4c, 3.5 and 3.6c, for all CuI/PET samples, and also for CuI/glass samples with 200–820 nm thick CuI films. Their resistivity follows the power law, $\rho_m(T) \sim T^2$ according to the ionized impurity scattering and carrier–carrier scattering model for degenerate semiconductors [1, 6].

The larger ρ_{02} values have revealed the more intense scattering on ionized impurities and carrier–carrier scattering for CuI/glass samples as compared with CuI/PET samples (Table 3.1) at *T* near 380–390 K. At the same time, the hole transport at *T* near 290 K is limited by barriers at the grain boundaries in the CuI/glass samples to a greater extent, since they have more height barriers (T_0 in the 27–76 meV range) compared to CuI/PET samples (T_0 in the 10–42 meV range).

As seen in Figs. 3.2 and 3.3, the observed transport properties of CuI films correspond to their nanocrystalline structure and nanostructured surfaces. However, it is impossible to explain the difference in transport properties of CuI/glass and CuI/PET samples, based only on our data on the morphology of the surfaces, the crystal structure and the composition of the CuI films. At the same time, the data on thermal

Table 3.1 Transport properties and Seebeck coefficients of CuI films produced by the different modes of SILAR method on glass and PET substrates

SILAR mode	CuI film carrier transport parameters and Seebeck coefficient							
Cationic precursor	Anionic precursor	Thickness t, nm	PET substrate	Glass substrate				
			ρ_{02}, ΩmK^{-2}	T_0, meV	S, μV/K	ρ_{02}, ΩmK^{-2}	T_0, meV	S, μV/K
0.1 M $CuSO_4$, 0.1 M $Na_2S_2O_3$	0.05 M NaI	100	2×10^{-9}	15	115	3×10^{-9} C	71 C	207
						–	27	
	0.075 M NaI	200	3×10^{-9}	10	85	6×10^{-9}	33	188
		590	2×10^{-9}	20	123	2×10^{-8} C	49 C	251
	0.1 M NaI	820	4×10^{-9}	42	113	2×10^{-9}	76	179

C—CuI strips connected with metal contacts on glass substrates

electromotive forces (Seebeck coefficients) given in Table 3.1 demonstrate the lower *S* values for all CuI/PET samples compared to CuI/glass samples, which is well correlated with their transport properties, as the Seebeck coefficient is a property, which depends on the electronic structure near the Fermi-level of degenerate semiconductors. According to [15], the *S* decreasing in CuI/PET samples is explained due to the fact, that these degenerate semiconducting CuI films behave more like metals than semiconductors.

3.4 Conclusions

The paper demonstrates the transport properties of nanocrystalline cubic cuprous iodide films deposited by SILAR on glass and PET substrates. Temperature dependences of resistivity for most CuI films in CuI/glass and CuI/PET samples have the crossover from semiconducting to metallic behavior with *T* increasing, which is typical for degenerate semiconductors. It has been confirmed that the semiconductor carrier transport occurs at *T* near 290 K through the nearest neighboring hopping. Metallic transport in CuI films carried out at *T* near 380–390 K in accordance with the ionized impurity scattering and the carrier–carrier scattering model for degenerate semiconductors. Comparison of thermoelectric properties is yet another confirmation of the fact that the degree of degeneration of CuI semiconductors in CuI/PET samples is higher than in CuI/glass samples deposited by SILAR method.

References

1. M. Kneiß, C. Yang, J. Barzola-Quiquia, G. Benndorf, H. von Wenckstern, P. Esquinazi, M. Lorenz, M. Grundmann, Suppression of grain boundary scattering in multifunctional p-type transparent γ-CuI thin films due to interface tunneling currents. Adv. Mater. Interfaces **5**(6), 1701411-1 (2018)
2. M. Grundmann, F.-L. Schein, M. Lorenz, T. Böntgen, J. Lenzner, H. Wenckstern, Cuprous iodide—a p-type transparent semiconductor: history and novel applications. Phys. Status Solidi A **210**, 1671 (2013)
3. N. Yamada, R. Ino, Y. Ninomiya, Truly transparent p-type γ-CuI thin films with high hole mobility. Chem. Mater. **28**(14), 4971 (2016)
4. F.-L. Schein, H. von Wenckstern, M. Grundmann, Transparent p-CuI/n-ZnO heterojunction diodes. Appl. Phys. Lett. **102**, 092109-1 (2013)
5. S. Inudo, M. Miyake, T. Hirato, Electrical properties of CuI films prepared by spin coating. Phys. Status Solidi A **210**(11), 2395 (2013)
6. C. Yang, M. Kneiβ, M. Lorenz, M. Grundmann, in *Proceedings of the National Academy of Sciences of the United States of America*, ed. by A. Rockett (USA, 2016)
7. D.K. Kaushik, M. Selvaraj, S. Ramu, A. Subrahmanyam, Thermal evaporated copper iodide (CuI) thin films: a note on the disorder evaluated through the temperature dependent electrical properties. Sol. Energ. Mat. Sol. C **165**, 52 (2017)
8. A. Pishtshev, S.Z. Karazhanov, Structure-property relationships in cubic cuprous iodide: a novel view on stability, chemical bonding, and electronic properties. J. Chem. Phys. **146**(6), 064706-1 (2017)
9. N.P. Klochko, V.R. Kopach, G.S. Khrypunov, V.E. Korsun, N.D. Volkova, V.N. Lyubov, M.V. Kirichenko, A.V. Kopach, D.O. Zhadan, A.N. Otchenashko, n-ZnO/p-CuI barrier heterostructure based on zinc-oxide nanoarrays formed by pulsed electrodeposition and SILAR copper-iodide films. Semiconductors **51**, 789 (2017)
10. N.P. Klochko, K.S. Klepikova, V.R. Kopach, I.I. Tyukhov, D.O. Zhadan, G.S. Khrypunov, S.I. Petrushenko, S.V. Dukarov, V.M. Lyubov, M.V. Kirichenko, A.L. Khrypunova, Semitransparent p-CuI and n-ZnO thin films prepared by low temperature solution growth for thermoelectric conversion of near-infrared solar light. Sol. Energy **171**, 704 (2018)
11. N.P. Klochko, V. Kopach, G. Khrypunov, V. Korsun, V. Lyubov, O. Otchenashko, D. Zhadan, M. Kirichenko, V. Nikitin, M. Maslak, A. Khrypunova, in *Proceedings of the 2017 IEEE 7th International Conference on Nanomaterials: Applications and Properties (NAP)*, Odessa, Ukraine, 2017, Part 2 (IEEE, 2017), p. 02NTF01
12. P. Sheng, Fluctuation-induced tunneling conduction in disordered materials. Phys. Rev. B **21**(6), 2180 (1980)
13. Y. Yang, S. Liu, K. Kimura, Synthesis of well-dispersed CuI nanoparticles from an available solution precursor. Chem. Lett. **34**(8), 1158 (2005)
14. N.P. Klochko, V.R. Kopach, I.I. Tyukhov, G.S. Khrypunov, V.E. Korsun, V.O. Nikitin, V.M. Lyubov, M.V. Kirichenko, O.N. Otchenashko, D.O. Zhadan, M.O. Maslak, A.L. Khrypunova, Wet chemical synthesis of nanostructured semiconductor layers for thin-film solar thermoelectric generator. Sol. Energy **157**, 657 (2017)
15. A. Shakouri, Recent developments in semiconductor thermoelectric physics and materials. Annu. Rev. Mater. Res. **41**(1), 399 (2011)

Chapter 4
Optical Absorption and Refractive Index of X-ray Irradiated Cu_6PSe_5I-Based Thin Film

I. P. Studenyak, A. V. Bendak, V. Yu. Izai, V. I. Studenyak, A. M. Solomon and P. Kúš

Abstract Cu_6PSe_5I-based thin film was deposited by magnetron sputtering. As-deposited thin films were irradiated by X-ray at different exposition times. Optical transmission spectra of X-ray irradiated $Cu_{5.5}P_{1.2}Se_{5.0}I_{1.3}$ thin film were measured depending on irradiation time. Urbach absorption edge and dispersion of refractive index for X-ray irradiated $Cu_{5.5}P_{1.2}Se_{5.0}I_{1.3}$ thin film were studied. Dependences of optical parameters of Urbach absorption edge and refractive index on irradiation time were analyzed. The nonlinear increase of energy pseudogap and Urbach energy, as well as the nonlinear decrease of refractive index with an increase of X-ray irradiation time, were revealed. The comparison of optical parameters of X-ray irradiated $Cu_{5.6}P_{1.7}S_{4.9}I_{0.8}$ thin film and $Cu_{5.5}P_{1.2}Se_{5.0}I_{1.3}$ thin film is performed.

4.1 Introduction

Cu_6PSe_5I crystals belong to the compounds with argyrodite structure [1, 2]. Such materials are known as superionic conductors or solid electrolytes. They have attracted great interest not only in the view of the possibility of their application in high-energy-density batteries and sensors but also due to their remarkable properties.

It should be noted that the physical properties of Cu_6PSe_5I single crystals are well known [3], while the investigation of physical properties of thin films on their basis only begins [4]. The most extensively studied among argyrodite-type thin films are phosphorus-containing argyrodites [5]. Electrical and optical studies of Cu_6PS_5I-based thin films with different copper atoms content were performed in [6].

I. P. Studenyak (✉) · A. V. Bendak · V. Yu. Izai · V. I. Studenyak
Uzhhorod National University, Uzhhorod, Ukraine
e-mail: studenyak@dr.com

A. M. Solomon
Institute of Electron Physics, National Academy of Sciences, Uzhhorod, Ukraine

P. Kúš
Department of Experimental Physics, Comenius University, Bratislava, Slovakia

A. D. Pogrebnjak and O. Bondar (eds.), *Microstructure and Properties of Micro- and Nanoscale Materials, Films, and Coatings (NAP 2019)*,
Springer Proceedings in Physics 240, https://doi.org/10.1007/978-981-15-1742-6_4

$Cu_{5.5}P_{1.2}Se_{5.0}I_{1.3}$ thin film was deposited onto a silicate glass substrate by non-reactive radio frequency magnetron sputtering [6]. The x-ray diffraction pattern shows the films to be amorphous. It is revealed that in two-dimensional $Cu_{5.5}P_{1.2}Se_{5.0}I_{1.3}$ thin film the Urbach absorption edge is formed by strong electron-phonon interaction. It should be noted that the transition from Cu_6PSe_5I single crystal to $Cu_{5.5}P_{1.2}Se_{5.0}I_{1.3}$ thin film caused the substantial increase of the Urbach energy, enhance the electron-phonon interaction and increase of the effective phonon energy as well as the increase of the relative contribution of static structural disordering into Urbach energy [6].

The influence of X-ray irradiation on optical absorption edge parameters of $Cu_{5.6}P_{1.7}S_{4.9}I_{0.8}$ thin film was investigated in [4]. This paper is devoted to the optical studies of X-ray irradiated $Cu_{5.5}P_{1.2}Se_{5.0}I_{1.3}$ thin films. Here we report on the influence of X-ray irradiation on the optical transmission spectra, Urbach absorption edge parameters and refractive indexes in $Cu_{5.5}P_{1.2}Se_{5.0}I_{1.3}$ thin films. Results of similar investigations for the different types of thin films were presented in [7, 8].

4.2 Experiment Details

The thin film was deposited from a 2-inch polycrystalline Cu_6PSe_5I target onto silicate glass substrate by non-reactive radio frequency magnetron sputtering with 90 W of RF power applied to the target. Thus, the film growth rate was 3 nm/min. The deposition was carried out at room temperature in Ar atmosphere. The structure of the deposited film was analyzed by X-ray diffraction. Diffraction pattern, obtained using DRON-3 diffractometer (conventional $\theta - 2\theta$ scanning technique, Bragg angle $2\theta \cong 10$–$60°$, Cu K_α, Ni filtered radiation), shows the film to be amorphous. Structural studies were performed for the thin film (Fig. 4.1) using a scanning electron microscopy technique (Hitachi S-4300). The thin film chemical composition

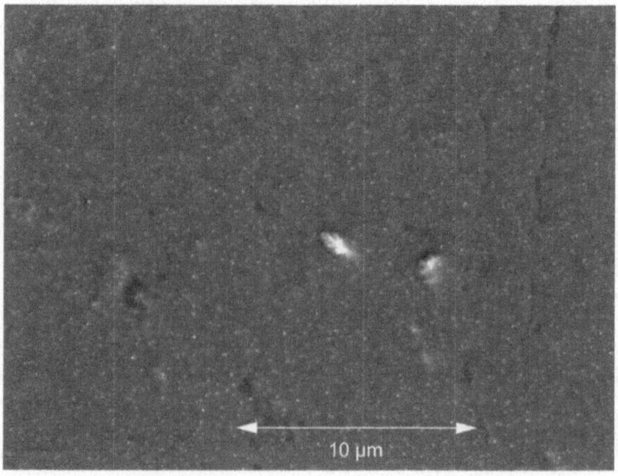

Fig. 4.1 SEM image of $Cu_{5.5}P_{1.2}Se_{5.0}I_{1.3}$ thin film

$Cu_{5.5}P_{1.2}Se_{5.0}I_{1.3}$ was determined by energy-dispersive X-ray spectroscopy studies which enabled us to check the chemical composition in different points of film surface. The deposited thin film is observed to be depleted by copper and enriched by phosphorous and iodine.

X-ray irradiation was performed for the different exposition times (30 and 60 min) using wideband radiation of Cu-anode X-ray tube with approx 400 W of power applied (33 kV, 13 mA).

The optical transmission spectra of $Cu_{5.5}P_{1.2}Se_{5.0}I_{1.3}$ thin film was studied at room temperature by an MDR-3 grating monochromator. From interference transmission spectra the spectral dependences of the absorption coefficient, as well as dispersion dependences of the refractive index, were derived [9].

4.3 Results and Discussions

With irradiation time increase, the slight blue shift of short-wavelength part of transmission spectra and short-wavelength shift of interference maxima are observed. Both non-irradiated and X-ray irradiated $Cu_{5.5}P_{1.2}Se_{5.0}I_{1.3}$ thin film exhibit the exponential behavior of optical absorption edge. In [6] it is shown that the optical absorption edge for as-deposited $Cu_{5.5}P_{1.2}Se_{5.0}I_{1.3}$ thin film in the region of its exponential behavior is described by Urbach rule [10, 11]

$$\alpha(h\nu, T) = \alpha_0 \cdot \exp\left[\frac{h\nu - E_0}{E_U(T)}\right], \quad (4.1)$$

where $E_U(T)$ is the Urbach energy, α_0 and E_0 are the coordinates of the convergence point of the Urbach bundle, $h\nu$ and T are the photon energy and temperature, respectively. In the X-ray irradiated $Cu_{5.5}P_{1.2}Se_{5.0}I_{1.3}$ thin film we also observed the Urbach behavior of the optical absorption edge (Fig. 4.2). It should be noted that

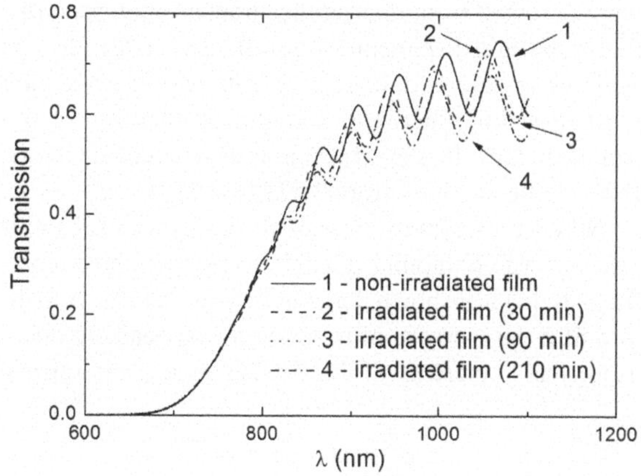

Fig. 4.2 Optical transmission spectra of as-deposited (1) and X-ray irradiated $Cu_{5.5}P_{1.2}Se_{5.0}I_{1.3}$ thin films at various irradiation time: (2) 30, (3) 90 and (4) 210 min

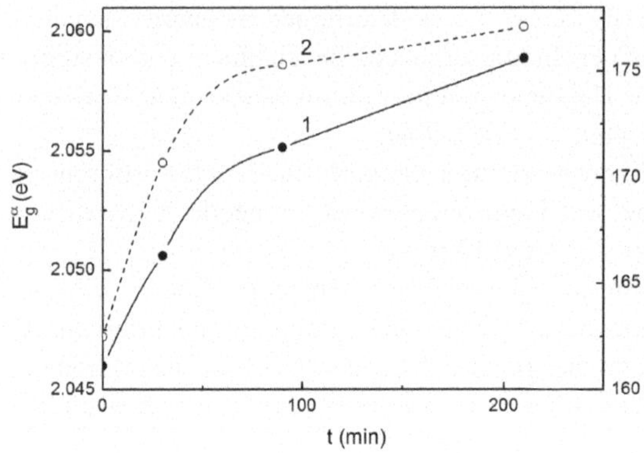

Fig. 4.3 Dependences of the energy pseudogap E_g^α ($\alpha = 10^4$ cm^{-1}) and Urbach energy E_U on X-ray irradiation time for $Cu_{5.5}P_{1.2}Se_{5.0}I_{1.3}$ thin film

the optical absorption edge for X-ray irradiated $Cu_{5.5}P_{1.2}Se_{5.0}I_{1.3}$ thin films is slight shifted to the high-energy region with irradiation time increase.

For the characterization of absorption edge energy position, such parameter as energy pseudogap E_g^α (E_g^α is the energy position of the exponential absorption edge) at a fixed absorption coefficient value α was determined. We used the E_g^α values taken at $\alpha = 10^4 \times$ cm^{-1} for thin films. The observed variation of the optical absorption edge leads to E_g^α and E_U values increase (Fig. 4.3) with irradiation time increase (from 2.046 to 2.0534 eV and from 162.5 to 175.3 meV, respectively). The Urbach energy E_U increase is the evidence of structural disordering due to X-ray irradiation.

It is well-known that Urbach energy characterizes the disordering degree in the investigated system and is described by the equation [12, 13]

$$E_U = (E_U)_T + (E_U)_X, \tag{4.2}$$

where $(E_U)_T$ and $(E_U)_X$ are the contributions of temperature and structural disordering to E_U, respectively. The Urbach energy increase is the evidence of the structural disordering increase caused by X-ray irradiation.

Dispersion dependences of the refractive index for the $Cu_{5.5}P_{1.2}Se_{5.0}I_{1.3}$ thin film were obtained from interference transmission spectra (Fig. 4.4). In the transparency region, a slight dispersion of the refractive index for X-ray irradiated $Cu_{5.5}P_{1.2}Se_{5.0}I_{1.3}$ thin film is observed, increasing with approaching the optical absorption edge. With irradiation time increase, a nonlinear decrease of refractive index value in X-ray irradiated $Cu_{5.5}P_{1.2}Se_{5.0}I_{1.3}$ thin film is revealed (X-ray irradiation leads to refractive index value decrease from 2.572 to 2.519).

Now we attempt to compare the influence of X-ray irradiation on the optical properties of Se-containing ($Cu_{5.5}P_{1.2}Se_{5.0}I_{1.3}$) and S-containing ($Cu_{5.6}P_{1.7}S_{4.9}I_{0.8}$) thin films. In [4] it is shown that contrary to $Cu_{5.5}P_{1.2}Se_{5.0}I_{1.3}$ thin film in $Cu_{5.6}P_{1.7}S_{4.9}I_{0.8}$ thin film the increase of irradiation time leads to decrease of the energy pseudogap as well as to increase of the Urbach energy and refractive index.

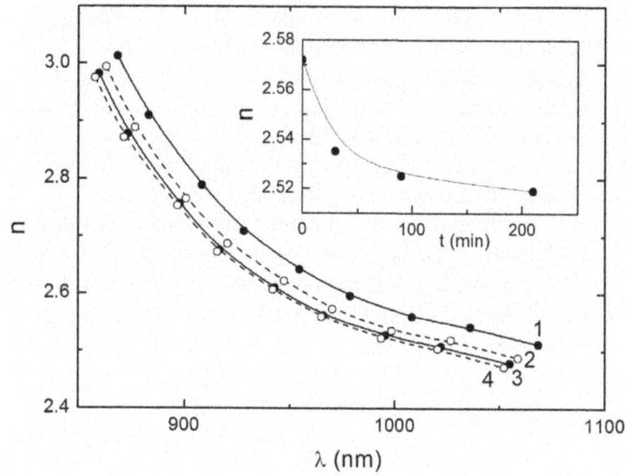

Fig. 4.4 Refractive index dispersions of as-deposited (1) and X-ray irradiated $Cu_{5.5}P_{1.2}Se_{5.0}I_{1.3}$ thin films at various irradiation time: (2) 30, (3) 90 and (4) 210 min. The inset shows the dependence of refractive index on X-ray irradiation time at $\lambda = 1\ \mu m$

4.4 Conclusions

$Cu_{5.5}P_{1.2}Se_{5.0}I_{1.3}$ thin films are deposited onto silicate glass substrates by non-reactive radio frequency magnetron sputtering. The influence of X-ray irradiation on the optical properties of $Cu_{5.5}P_{1.2}Se_{5.0}I_{1.3}$ thin films was investigated. With irradiation time increase the nonlinear increase of the energy pseudogap and Urbach energy, as well as the nonlinear decrease of refractive index, were observed. Besides, the increase of Urbach energy is the evidence of an increase in structural disordering contribution, which is caused by X-ray irradiation.

References

1. W.F. Kuhs, R. Nitsche, K. Scheunemann, The argyrodites – a new family of the tetrahedrally close-packed structures. Mater. Res. Bull. **14**, 241 (1979)
2. T. Nilges, A. Pfitzner, A structural differentiation of quaternary copper argyrodites: Structure - property relations of high temperature ion conductors. Z. Kristallogr. **220**, 281 (2005)
3. I.P. Studenyak, M. Kranjčec, M.V. Kurik, Urbach rule and disordering processes in Cu6P(S1-xSex)5Br1-yIy superionic conductors. J. Phys. Chem. Solids **67**, 807 (2006)
4. I.P. Studenyak, M.M. Kutsyk, A.V. Bendak, V.Yu. Izai, P. Kúš, M. Mikula, Influence of X-ray irradiation on optical absorption edge and refractive index dispersion in Cu6PS5I-based thin film deposited by magnetron sputtering. Semicond. Physics, Quant. Electron. Optoelectron. **20**, 246 (2017)
5. I.P. Studenyak, VYu. Izai, A.V. Bendak, P.P. Guranich, YuM Azhniuk, P. Kúš, D.R.T. Zahn, Optical and electrical properties of Cu6PS5I-based thin films versus copper content variation. Ukr. J. Phys. Opt. **18**, 232 (2017)
6. I.P. Studenyak, M.M. Kutsyk, A.V. Bendak, V.Yu. Izai, V.V. Bilanchuk, P. Kúš, M. Mikula, Structural and optical studies of Cu6PSe5I-based thin film deposited by magnetron sputtering. Semicond. Physics, Quant. Electron. Optoelectron. **20**, 64 (2017)
7. M.M. El-Nahass, A.F. El-Deeb, H.S. Metwally, H.E.A. El-Sayed, A.M. Hassanien, Influence of X-ray irradiation on the optical properties of iron (III) chloride tetraphenyporphyrin thin films. Solid State Sci. **12**, 552 (2010)

8. R. Imai, A. Fujimoto, M. Okada, S. Matsui, T. Yokogawa, E. Miura, T. Yamasaki, T. Suzuki, K. Kanda, Soft X-ray irradiation effect on the surface and material properties of highly hydrogenated diamond-like carbon thin films. Diam. Relat. Mater. **44**, 8 (2014)
9. R. Swanepoel, Determination of the thickness and optical constants of amorphous silicon. J. Phys. E: Sci. Instrum. **16**, 1214 (1983)
10. F. Urbach, The long-wavelength edge of photographic sensitivity and of the electronic absorption of solids. Phys. Rev. **92**, 1324 (1953)
11. M.V. Kurik, Urbach rule (Review). Phys. Status Solidi A **8**, 9 (1971)
12. G.D. Cody, T. Tiedje, B. Abeles, B. Brooks, Y. Goldstein, Disorder and the optical-absorption edge of hydrogenated amorphous silicon. Phys. Rev. Lett. **47**, 1480 (1981)
13. G.D. Cody, Urbach edge of crystalline and amorphous silicon: a personal review. J. Non-Cryst. Solids **141**, 3 (1992)

Chapter 5
Study of the Correlations Between the Dynamics of Thermal Destruction and the Morphological Parameters of Biogenic Calcites by the Method of Thermoprogrammed Desorption Mass Spectrometry (TPD-MS)

O. G. Bordunova, V. B. Loboda, Y. A. Samokhina, O. M. Chernenko, R. V. Dolbanosova and V. D. Chivanov

Abstract By the method of thermoprogrammed mass spectrometry (TPD-MS) were studied the biogenic calcites thermal desorption spectra (natural limestone chalk, bird eggshells: chicken (*Gallus gallus domesticus*), domestic turkey (*Meleagris gallopavo*), domestic geese (*Anser anser domesticus*), domestic duck (*Cairina moschata*), mollusk shells (*Anadara inaequivalvis*), and cephalopod fossils Belemnite (Pachyteuthis Bayle), as well as calcite nanoparticles. It was shown that the structure of the spectrum correlates with morphological parameters and is a function of samples dispersity degree of biogenic calcites. The increase in the content of nano-, ultra- and microdispersed components in calcite-based biocomposites leads to a significant change in the form of the thermal desorption spectrum, manifested in the appearance of additional temperature desorption areas (peaks) in thermograms and their displacement to lower temperatures area.

O. G. Bordunova (✉) · V. B. Loboda · Y. A. Samokhina · R. V. Dolbanosova
Sumy National Agrarian University, 160, Herasym Kondratiev Str., Sumy 40021, Ukraine
e-mail: bordunova_olga@Rambler.ru

O. M. Chernenko
Dnipro State Agrarian and Economic University, 25, S. Efremov St., Dnipro 49600, Ukraine

V. D. Chivanov
Institute of Applied Physics (IAP), NAS of Ukraine, 58, Petropavlovskaya Str., Sumy 40001, Ukraine

A. D. Pogrebnjak and O. Bondar (eds.), *Microstructure and Properties of Micro- and Nanoscale Materials, Films, and Coatings (NAP 2019)*,
Springer Proceedings in Physics 240, https://doi.org/10.1007/978-981-15-1742-6_5

5.1 Introduction

The problem of creation on the biomimetic principle of inexpensive and practical nanobiomaterial from common secondary raw materials is a serious challenge for basic and applied researchers [1–3]. The extreme degree of heterogeneity of such materials, which does not make it possible to carry out reliable prognostic calculations of the physicochemical and technological parameters of these materials, still remains only a partially solved problem [2].

A typical example is the still incomplete study of structural features and, as a consequence, the lack of calculated methods for reliable prediction of physicochemical, and especially mechanical, characteristics of biogenic calcite ($CaCO_3$) [4, 5].

Calcite (chalk, limestone, shells of fossil and living mollusks, the shell of bird eggs and eggs of lizards and snakes) is widely represented in the world around and is also widely used in industry in huge volumes as a building material, absorber-adsorbent of harmful substances in the technologies of protection environment, an effective catalyst in the field of catalytic chemistry, plastic filler, component of electric batteries, etc. [2, 6, 7].

However, there are still no reliable methods for predicting, in particular, the strength characteristics of natural biogenic calcites and their biomimetic analogs [4], which determine the use of various methods of physicochemical analysis to these objects [8].

One of them is a fairly common method of gas analysis—the method of thermoprogrammed desorption mass spectrometry (TPD-MS), which is based on the semi-quantitative determination of gases released from organic/inorganic samples when heated in a programmed mode in vacuum or in the flow of carrier gas [9–13].

The time dependence of the amount of the test gas on temperature (thermogram) is characterized by a complex non-linear character and, as some researchers believe, indirectly reflects the features of the micro- and macrostructures of heterogeneous solid-phase samples [14–17].

Based on the above, the objective of this study was to find possible correlative dependencies between the thermograms obtained in the study of individual biogenic calcites (chalk, bird eggshell, shells of common mollusks, calcite artifacts of fossil cephalopods (belemnites) and micro-macrostructural characteristics of the latter.

Of particular practical importance of research is the need to develop new "green" methods to prevent infectious diseases of agricultural birds during the hatching period [18–21] and to improve the technology for the production of edible eggs, which are characterized by increased strength during transportation and storage [18] as well as the eggshell waste processing of edible eggs (according to rough estimates, all large food concerns daily process more than 1 million eggs [22].

In the last 25–30 years, the scientific and technical direction, which originally appeared in the physics of thin films, has been rapidly developing, associated with the use of various film nanostructures as functional coatings (film protective coatings) [1–5] and functional elements of micro- and nanoelectronics (film structures with specific electrical and magnetic properties) [6–9]. In the last 25–30 years, the scientific and technical direction, which originally appeared in the physics of thin films, has been rapidly developing, associated with the use of various film nanostructures as both functional coatings (film protective coatings) [22, 23] and functional elements of micro- and nanoelectronics (film structures with specific electrical and magnetic properties).

5.2 Experiment Details

Samples of biogenic calcites were used in the work: (chalk as a typical natural variety of limestone taken from a deposit in the Sumy region; shell of fresh unincubated bird eggs: chicken (*Gallus gallus domesticus*), domestic turkey (*Meleagris gallopavo*), domestic geese (*Anser anser domesticus*), domestic duck (*Cairina moschata*), mollusk shells (*Anadara inaequivalvis*) and rostra of cephalopod fossils—Belemnite (*Pachyteuthis Bayle, Belemnitella Orbigny*).

Temperature-programmed desorption mass spectrometry experiments were carried out with ionization MKh-7304A monopole mass spectrometer (Ukraine) that was adapted for measurements [11, 24]. About 1–5 mg of samples were used for each run. At the beginning of the measurements all samples were degassed to ca. 5×10^{-3} Pa at temperature 20 °C after which they were heated to a temperature of 900 °C. The heating rate was 0.25 °C s^{-1}.

Microscopic research was performed using REMMA-102 and REM-106i scanning electron microscopes (Ukraine) and an optical research microscope Carl Zeiss MicroImaging GmbH. For digital processing was used the program Femtoscan (trial version), and for the thermal desorption spectra—software package Origin 8.1.

X-ray diffraction research was performed on X-ray diffractometer "DRON-4M" (Russia).

5.3 Results and Discussion

The source experiments were conducted by us with the use of chalk—carbonate sedimentary rock of white color, fine-grained, weakly cemented, soft and crumbly, insoluble in the water, of organic (biogenic) origin. The basis of the chemical composition of chalk is calcium carbonate (91–98.5%) with a small amount of magnesium carbonate, although a non-carbonate part is also present, mainly metal oxides. For natural chalk, the absence of recrystallization and layering is typical. In the chalk column there is a development of large sustained cracks—layered and vertical, filled with chalky flour. On the surface of chalk pieces the network of cracks is strongly condensed. When chalk samples are impregnated with oil, there are hidden wiry structures reveal in the form of interlacing tiny cracks. In all the chalk deposits in different areas (horizons), the chalk differs both in chemical composition and in physical-mechanical properties. In Fig. 5.1 the spectrum of thermal desorption (thermogram) of a chalk sample is shown.

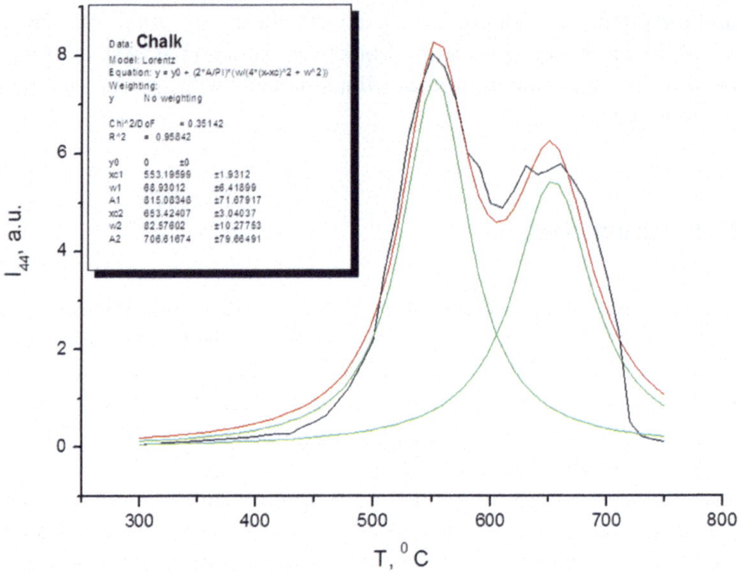

Fig. 5.1 Thermogram of CO_2 release/release (m/z = 44) by thermal decomposition of sample of chalk; in this Fig. and below the initial thermogram is marked in black (black line), in red-results of curve fitting,) in green—result of the total thermogram decomposition into Gaussian curves (Fit Multi-Peaks ⇒ Gaussian

As can be seen, the release of carbon dioxide CO_2 as a result of the reaction $CaCO_3$ (s) → CaO (s) + CO_2 (g), 178 kJ/ mol, begins at a temperature of 440–450 °C and ends at 720–750 °C. We note the nonlinear nature of the dependence of the partial pressure of CO_2 in a quartz cell on temperature with two clearly pronounced peaks of 550–560 and 640–660 °C, while the peak in the low-temperature region being significantly higher.

Approximately in the same temperature range of 460–720 °C, takes place the thermal destruction of the sample of the poultry shell (*Gallus gallus domesticus*); however, two pronounced peaks shift to high temperatures 590–610 and 670–690 °C (Fig. 5.2). The thermogram for the sample of the turkey shell is a lot more difficult. Although the range of intense CO_2 emissions is almost identical with the above samples (480–730 °C), the Gaussian decomposition of the total thermal destruction curve showed at least four peaks: 560, 580, 618 and 690 °C (Fig. 5.3). In Fig. 5.4 the thermogram of a domestic geese (*Anser anser domesticus*) domestic goose shell sample is shown, which is characterized by an even more complex dependency between the peaks of intense CO_2 emission and temperature—with interval of an indicated gas release (480–730 °C), using the mathematical processing of the total thermal destruction curve, we can identify six peaks: 520, 560, 600, 660 and 690 °C (Fig. 5.4 shows only two reliable peaks that remain in place with samples changing). A possible explanation for this phenomenon is the increased looseness of the shell of this bird species in combination with an increased amount of the organic component (over

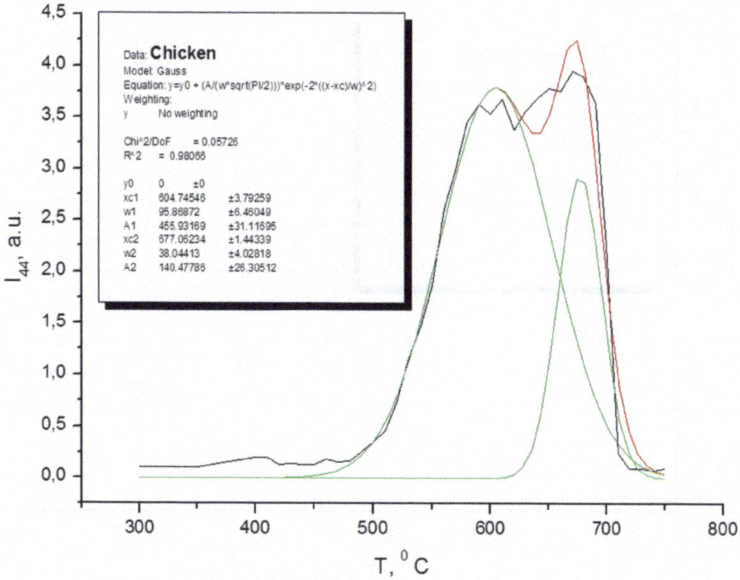

Fig. 5.2 Thermogram of CO_2 release/elimination (m/z = 44) by thermal decomposition of sample of egg shell chicken (*Gallus gallus domesticus*); unincubated eggs are used in the experiments

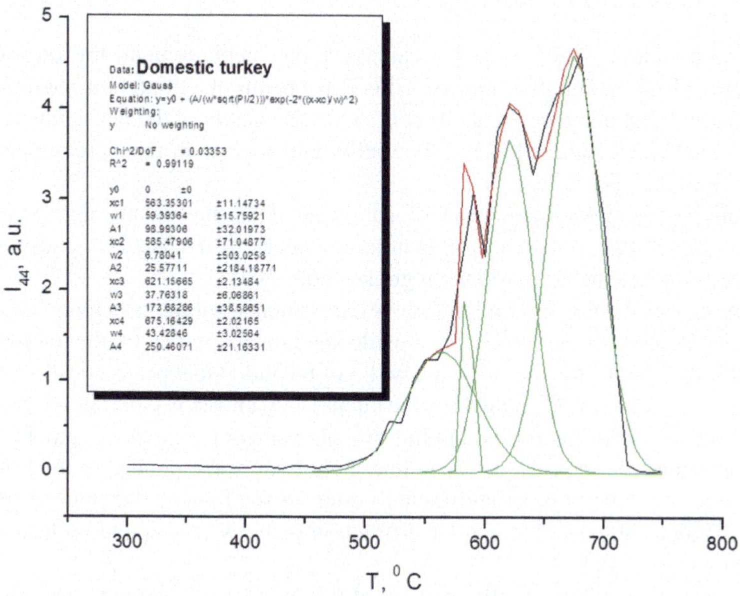

Fig. 5.3 Thermogram of CO_2 release/elimination (m/z = 44) by thermal decomposition of sample of egg shell domestic turkey (*Meleagris gallopavo*)

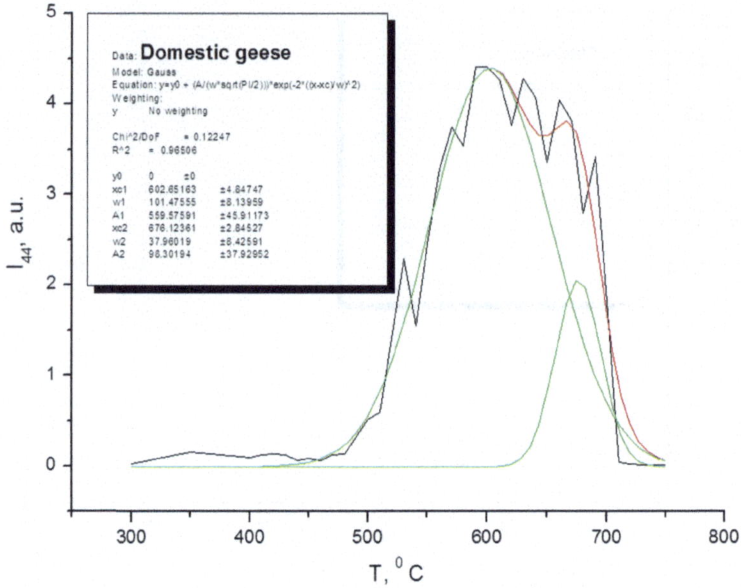

Fig. 5.4 Thermogram of CO_2 release/elimination (m/z = 44) by thermal decomposition of sample of egg shell domestic geese (*Anser anser domesticus*)

shell and membrane shell), as well as carcass "rebar" peptides in the thickness of the shell layer [19, 20]. Direct evidence of this is the result of a digital micrograph processing sample of home goose shells (Fig. 5.5). The nature of the three-dimensional image of the biocrystalline layer of the shell seems to support the increased level of calcite layer disorder.

Finally, in Fig. 5.6 a thermogram of a domestic duck shell sample domestic duck (*Cairina moschata*) is shown. In this case, the peaks of intense CO_2 emission are strongly shifted to the right—to the region of 660, 740 °C.

Summing up the results of all the above experiments, we can conclude that for all dissimilar samples of biogenic calcite while keeping an interval on the temperature scale of 440–750 °C, the intensity and width of the individual peaks constituting the summary thermogram and induced by the intense CO_2 emission is extremely variable. The working hypothesis for explaining this phenomenon was the assumption that the coordinates of the peaks on the temperature scale corresponded to the levels of dispersion of calcite crystals and their location in the biomaterial. Indeed, almost all the studied samples give similar diffraction patterns corresponding to calcium carbonate (Fig. 5.7).

We note that similar, slightly different diffractograms can be obtained from the polarly morphologically different samples of biogenic calcite from the eggshell of domestic chickens (Figs. 5.8, 5.9). If the suggestion about the critical effect of the morphology of a sample of biogenic calcite with the unconditional preservation of the basic crystalline phase (Fig. 5.7) is correct, then experiments with dense samples of

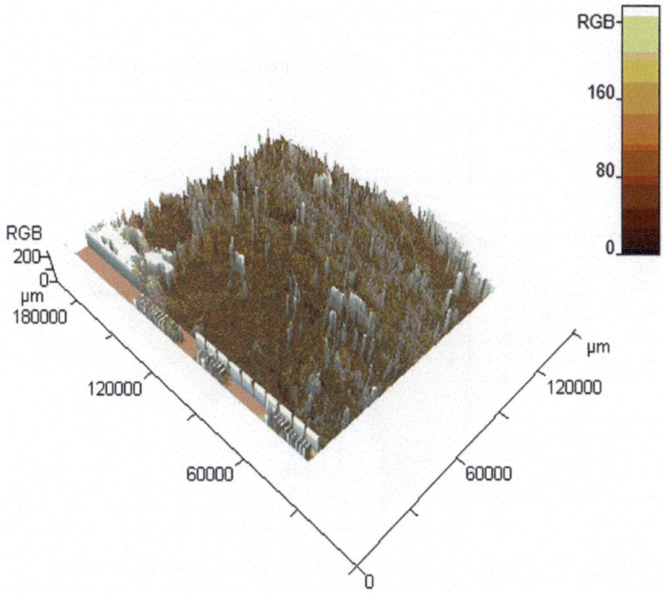

Fig. 5.5 Digital image of egg shell domestic geese (*Anser anser domesticus*); Femtoscan

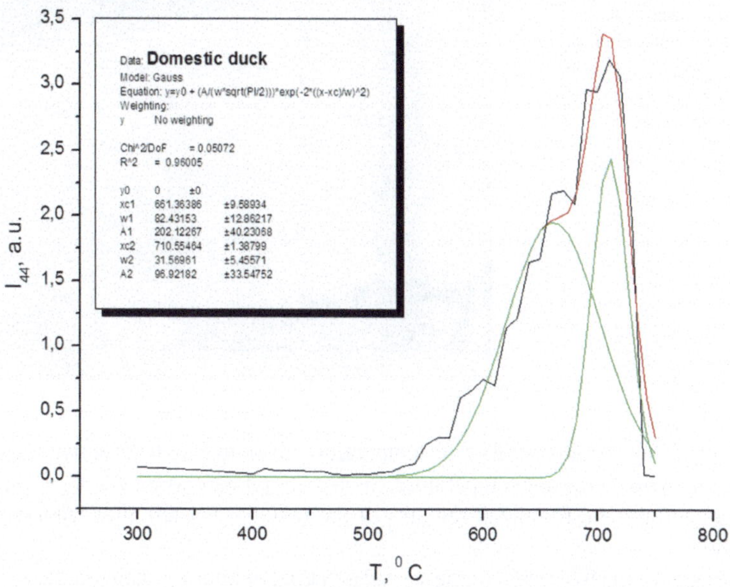

Fig. 5.6 Thermogram of CO_2 release/elimination (m/z = 44) by thermal decomposition of a sample of egg shell domestic duck (*Cairina moschata*)

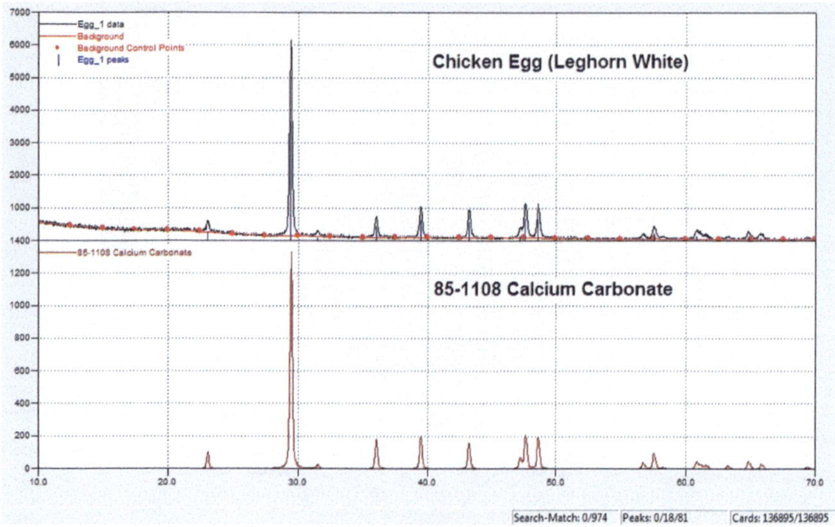

Fig. 5.7 Diffractogram of sample of egg shell chicken (*Gallus gallus domesticus*)

Fig. 5.8 Electron microscopic image of high-quality dense shell of healthy laying hens sample of egg shell chicken (*Gallus gallus domesticus*)

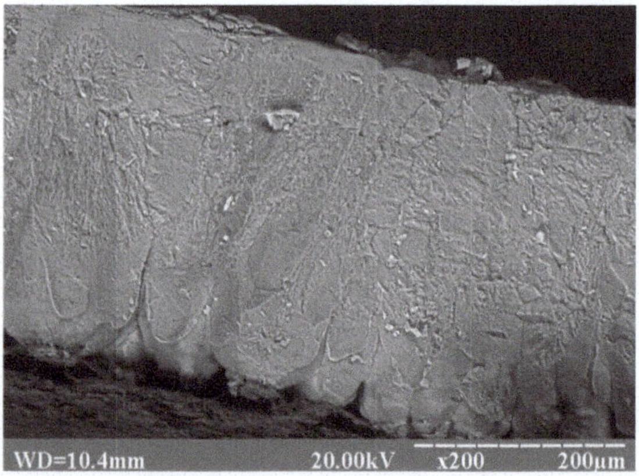

calcite must show a narrowing of the temperature range in which the active release of CO_2 occurs with increasing temperature. Such material we chose a coarse-crystalline calcite material—the fossilized remains of the rosters of belemnites (*Belemnitella Orbigny*).

The roster of the living belemnite served as a kind of internal skeleton. It consisted of radially dispersing needles of calcite. As can be seen from Figs. 5.10 and 5.11, whole and crushed samples of chicken eggshell and belemnite rosters are visually radically different—bird eggs calcite is a very loose conglomerate of $CaCO_3$ microcrystals, while growth substance is very close to natural crystalline calcite—highly ordered transparent substance. No wonder the thermogram obtained by heating an

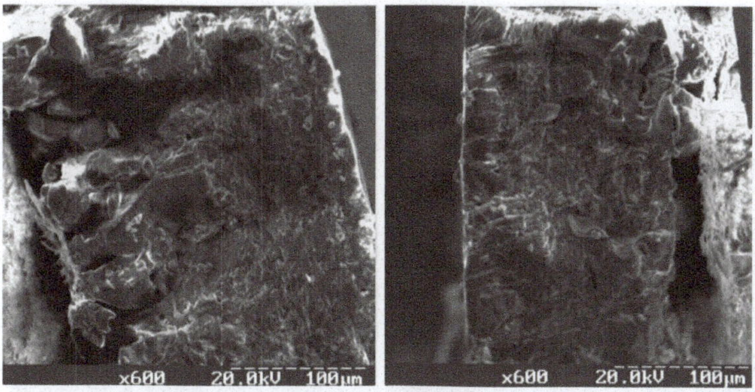

Fig. 5.9 Electron microscopic image of low-quality loose shells of laying hens exposed to stress and infected with infectious diseases sample of egg shell chicken (*Gallus gallus domesticus*)

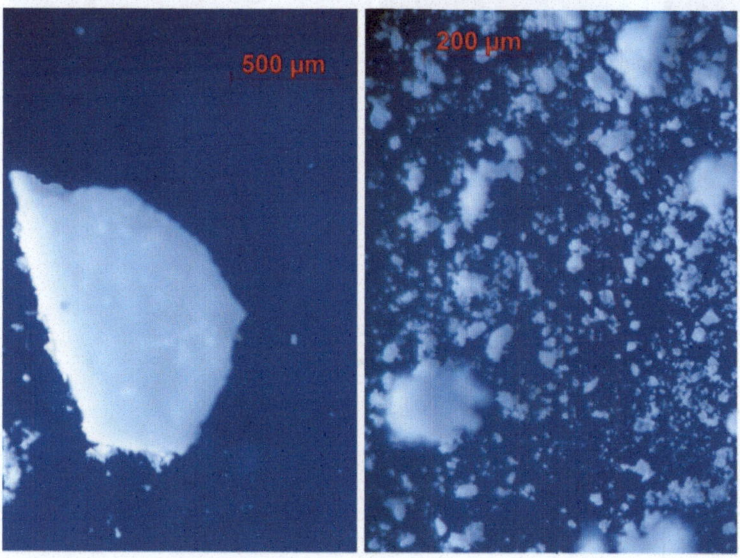

Fig. 5.10 Micrograph of a sample of the shell of a domestic chicken egg: native shell (left), crushed in a mortar (right) of a sample of egg shell chicken (*Gallus gallus domesticus*)

undivided portion of the crystal sample of the rostra belemnite showed a sharp difference from the similar thermogram for a typical native sample of the eggshells of domestic chicken (Fig. 5.2). On the thermogram of the belemnite sample, a symmetric intense peak is distinguished (range 480–750 °C, peak of 630 °C).

The following suggestion was that for the widening of the temperature range of destruction and an increase in the number of peaks of intense CO_2 emission from chalk and bird egg samples are responsible the heterogeneity of the micro- and

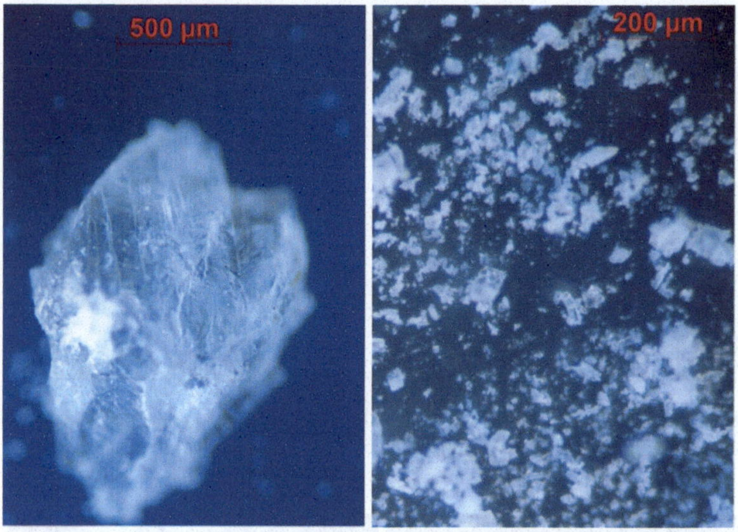

Fig. 5.11 Micrograph of a belemnite: whole roster section (left), rostrum ground in a mortar (right) of sample of belemnite (*Belemnitella Orbigny*)

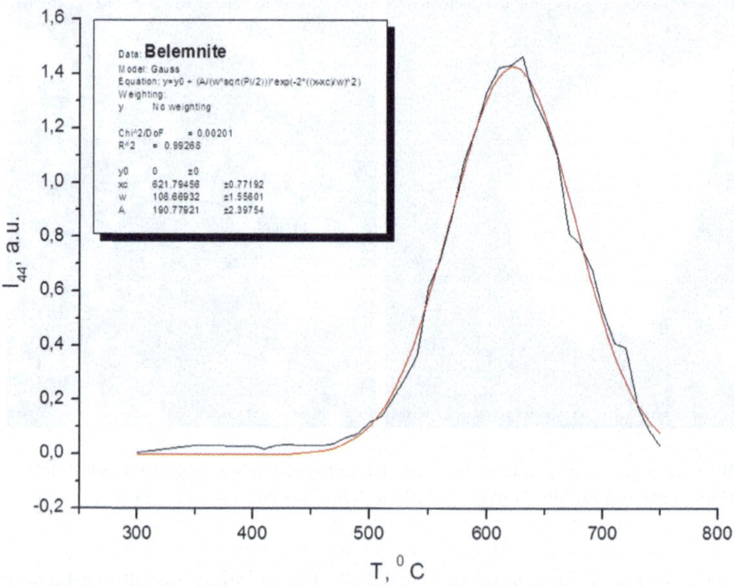

Fig. 5.12 Thermogram of CO_2 release/elimination (m/z = 44) by thermal decomposition of sample of belemnite (*Belemnitella Orbigny*)

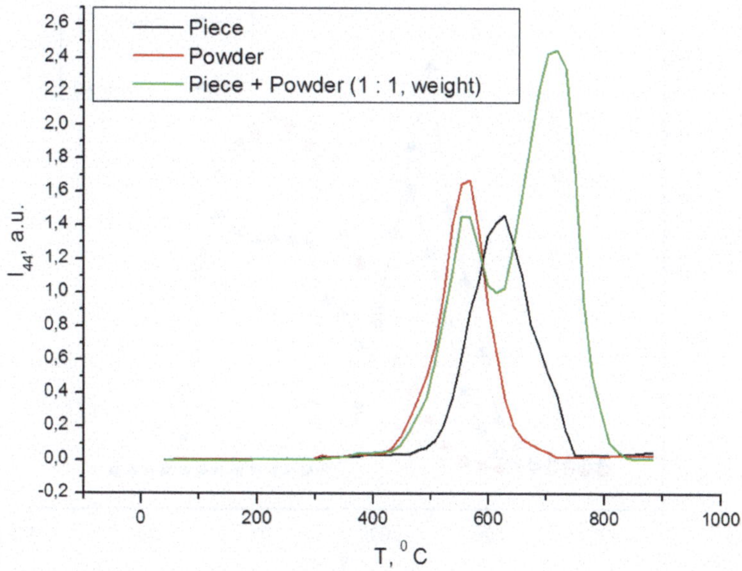

Fig. 5.13 Thermograms of CO_2 release/elimination (m/z = 44) by thermal decomposition of different sample of belemnite (*Belemnitella Orbigny*) **a** piece; **b** powder, **c** piece + powder

macrostructures of calcite-based biocomposites. In this case, the preliminary grinding of the belemnite sample must lead to a change in the type of thermogram, namely, to narrow the destruction interval for fine-grained (>5–10 μm) and to corresponding widening of the specified interval in the case of combining large- and fine-grained calcite fractions in one sample. As can be seen from Fig. 5.13, the fraction of the crushed sample forms on the thermogram a peak of intense CO_2 emission at 550 °C (control (native whole belemnite)—630 °C), while combining the crushed and native fraction of the sample in one sample leads to a strong widening of the thermogram and the appearance of clearly separated peaks at 550 and 730 °C, which give a good reason for the assumption that the broadening of the temperature ranges of the intense release of CO_2 in the thermograms of biogenic calcites is responsible for their heterogeneity. As can be seen from Fig. 5.14, almost similar results were obtained in the case of the study of the eggshell of domestic chicken (*Gallus gallus domesticus*) (Fig. 5.12).

Finally, the results of the experiment with $CaCO_3$ nanoparticles obtained using the electrolytic method confirmed our assumptions: a clear high-intensity peak at 520–530 °C was noted on the thermogram (Fig. 5.15).

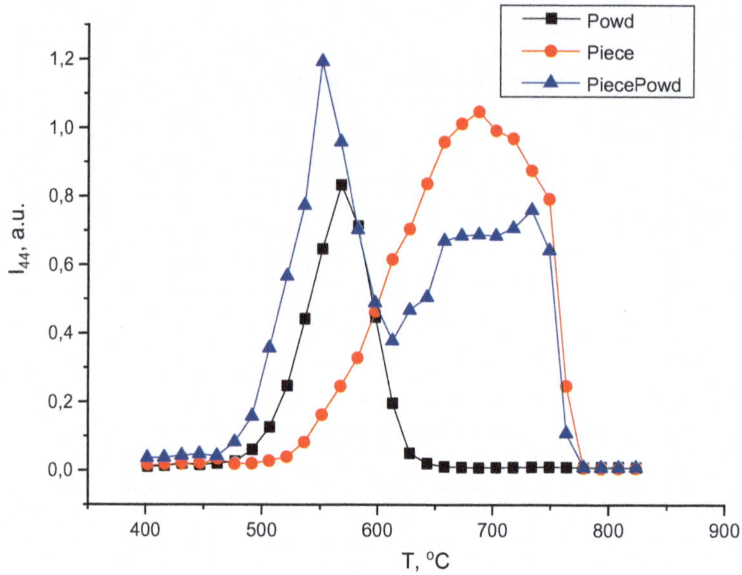

Fig. 5.14 Thermograms of CO_2 release/elimination (m/z = 44) by thermal decomposition of sample of egg shell chicken (*Gallus gallus domesticus*); 1—fine powder; 2—small piece; 3—combination of the small piece and fine powder

5.4 Conclusions

By the method of thermoprogrammed desorption mass spectrometry (TPD-MS) were studied the thermal desorption spectra of biogenic calcites (natural limestone chalk, bird egg shells: domestic chicken (*Gallus gallus domesticus*), domestic turkey (*Meleagris gallopavo*), domestic geese (*Anser anser domesticus*), domestic duck (*Cairina moschata*), mollusk shells (*Anadara inaequivalvis*) and cephalopod fossil belemnite (*Pachyteuthis Bayle*), as well as calcite nanoparticles. As shown that the spectrum structure correlates with morphological parameters and is a function of the degree of dispersion and samples of biogenic calcites.

An increase in the content of nano-, ultra- and microdispersed components in calcite-based biocomposites leads to a significant change in the form of the thermal desorption graph, occurs in the appearance of additional temperature desorption regions (peaks) and their total displacement to low temperatures area.

Thus, by thermal desorption spectra, obtained by TPD-MS, it is possible to undertake a prior assessment of the morphological parameters (biocrystalline layers degree ordering of biocomposites, the relative content of their components in terms of dispersion degree) of biogenic calcites samples of different origin.

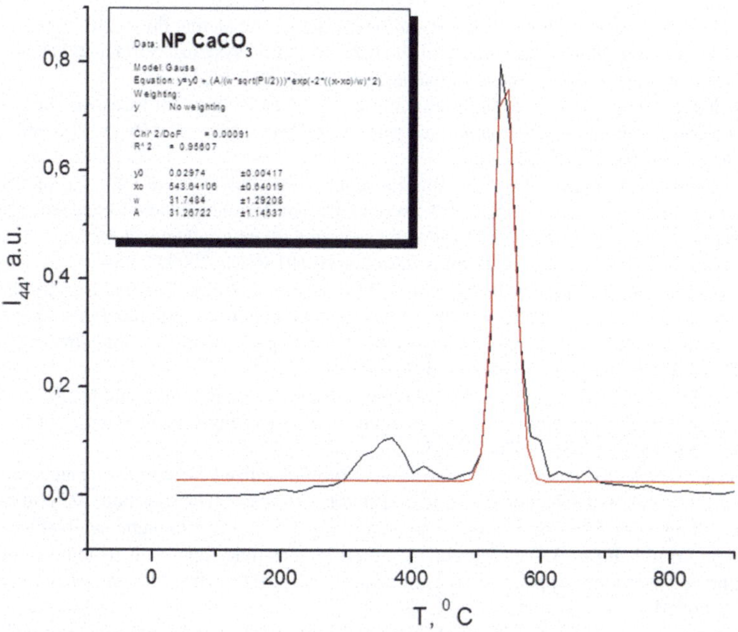

Fig. 5.15 Thermograms of CO_2 release/elimination (m/z = 44) by thermal decomposition of a sample of nanoparticles $CaCO_3$ (NP $CaCO_3$) obtained by electrolytic technique from egg shell chicken (*Gallus gallus domesticus*) (electrolysis media: acetic acid, 20%)

Acknowledgements The work has been performed under the financial support of the Ministry of Education and Science of Ukraine (Introduction of nanocomposite materials in innovative technologies for the incubation of poultry eggs; state registration number 0119U100551).

References

1. K. Endo, T. Kogure, H. Nagasawa, *Biomineralization: From Molecular and Nano-structural Analyses to Environmental Science* (Springer, Singapore, 2018), p. 413
2. E. Di Masi, L.B. Gower, *Biomineralization Sourcebook: Characterization of Biominerals and Biomimetic Materials* (CRC Press, 2014), p. 420
3. A. Shakeel, B. Krishna, I. Saiqa, K. *Suvardhan, Biocomposites: Biomedical and Environmental Applications* (Pan Stanford Publishing, CRC Press, 2018), p. 496
4. J.A.H. Oates, *Lime and Limestone: Chemistry and Technology, Production and Uses* (Wiley-VCH Verlag GmbH, 1998), p. 460
5. R. Zhang, L. Xie, Z. Yan, *Biomineralization Mechanism of the Pearl Oyster* (Springer, Singapore, 2019), p. 737
6. A. Laca, A. Laca, M. Diaz, Eggshell waste as catalyst: a review. J. Environ. Management **197**, 351–359 (2017). https://doi.org/10.1016/j.jenvman.2017.03.088
7. P.S. Guru, S. Dash, Sorption on eggshell waste–a review on ultrastructure, biomineralization and other applications. Adv. Colloid Interface Sci. **209**, 49–67 (2014). https://doi.org/10.1016/j.cis.2013.12.013

8. J.J. De Yoreo (ed.), in *Methods in Enzymology*, vol 532 (Academic Press, 2013), p. 614
9. V.A. Pokrovskij, Mass spectrometry of the nanostructured systems, Surface **2**(17), 63 (2010) (Ukrainian) https://dspace.nbuv.gov.ua/handle/123456789/39323
10. T.V. Kulik, Use of TPD–MS and linear free energy relationships for assessing the reactivity of aliphatic carboxylic acids on a silica surface. J. Phys. Chem. C **116**(1), 570–580 (2012). https://doi.org/10.1021/jp204266
11. T.V. Kulik, N.A. Lipkovska, V.N. Barvinchenko, B.B. Palyanytsya, O.A. Kazakova, O.A. Dovbiy, V.K. Pogorelyi, Interactions between bioactive ferulic acid and fumed silica by UV–vis spectroscopy, FT–IR, TPD MS investigation and quantum chemical methods. J. Colloid Interface Sci. **339**, 60–68 (2009). https://doi.org/10.1016/j.jcis.2009.07.055
12. N. Nastasiienko, B. Palianytsia, M. Kartel, M. Larsson, T, Kulik, Thermal transformation of caffeic acid on the nanoceria surface studied by temperature programmed desorption mass-spectrometry, thermogravimetric analysis and FT–IR spectroscopy. Colloids Interfaces **3**, 34 (2019). https://doi.org/10.3390/colloids3010034
13. T. Hatakeyama, H. Hatakeyama, Thermal properties of green polymers and biocomposites, in *Hot Topics in Thermal Analysis and Calorimetry,* vol 4 (Springer, Netherlands, 2005), p. 336. https://doi.org/10.1007/1-4020-2354-5
14. Y. Tsuboi, N. Koga, Thermal decomposition of biomineralized calcium carbonate: correlation between the thermal behavior and structural characteristics of avian eggshell. ACS Sustainable Chem. Eng. **6**(4), 5283–5295 (2018). https://doi.org/10.1021/acssuschemeng.7b04943
15. C.A.de Araujo Filho, D.Yu. Murzin, A structure sensitivity approach to temperature programmed desorption. Appl. Catal. A-Gen. **550**, 48–56 (2018). https://doi.org/10.1016/j.apcata.2017.11.001
16. M. Mohamed, S. Yusup, S. Maitra, Decomposition study of calcium carbonate in cockle shell, Journal of Engineering Science and Technology (JESTEC) School of Engineering, Taylor's University **7**(1), 1–10 (2012)
17. V.N. Kuznetsov, A.A. Yanovska, S.V. Novikov, V.V. Starikov, T.G. Kalinichenko, A.V. Kochenko, A.G. Ryabyshev, Ya.V. Khyzhnya, S.N. Danilchenko, J. Nano-Electron. Phys. **7**(3), 03034 (2015)
18. P.Y. Hester (ed.), *Egg Innovations and Strategies for Improvements* (Elsevier Inc., San Diego, 2017), p. 625
19. M.M. Bain, K. Mcdade, R. Burchmore, A. Law, P.W. Wilson, M. Schmutz, R. Preisinger, I.C. Dunn, Enhancing the egg's natural defence against bacterial penetration by increasing cuticle deposition. Anim. Genet. **44**(6), 661–668 (2013). https://doi.org/10.1111/age.12071
20. P.W. Wilson, C.S. Suther, M.M. Bain, W. Icken, A. Jones, F. Quinlan-Pluck, V. Olori, J. Gautron, I.C. Dunn, Understanding avian egg cuticle formation in the oviduct: a study of its origin and deposition. Biol. Reprod. **97**(1), 39–49 (2017). https://doi.org/10.1093/biolre/iox070
21. O. Bordunova, Dissertation, Mykolaiv Agrarian University, 2016
22. M.N. Freire, J.N.F. Holanda, Characterization of avian eggshell waste aiming its use in a ceramic wall tile paste. Cerâmica **52**(324) (2006). https://doi.org/10.1590/s0366-69132006000400004
23. A.D. Pogrebnjak, O.V. Sobol, V.M. Beresnev, in *Ceramic Engineering and Science Proceedings*, ed. by S. Mathur, S. Sinha Ray, T. Ohji. Nanostructured Materials and Nanotechnology IV, vol 31, (The American Ceramic Society, 2010), p. 127
24. S.N. Danilchenko, V.D. Chivanov, A.G. Ryabishev, S.V. Novikov, A.A. Stepanenko, V.N. Kuznetsov, E.V. Mironets, A.V. Mariychuk, A.A. Yanovska, O.G. Bordunova, A.N. Bugay, The study of thermal decomposition of natural calcium carbonate by the temperature-programmed mass spectrometry technique. J. Nano-Electron. Phys. **8**(4), 04031 (2016). https://doi.org/10.21272/jnep.8(4(1)).04031

Chapter 6
Floquet-Bloch Theory for Semiconductor Bragg Structure

A. A. Shmat'ko, V. N. Mizernik, E. N. Odarenko, A. S. Krivets and O. V. Yushchenko

Abstract The problem of Floquet-Bloch waves propagation in a semiconductor magnetophotonic crystal with a transverse magnetic field was solved. The fundamental solutions of the Hill equation in layers that based on the Floquet theory were obtained in an analytical form. The dispersion equation and its roots are found explicitly. The analysis of the dispersion properties of the structures depending on the material parameters of the layers was carried out. The parameters of gyrotropic layers for the full transmission and reflection of a plane wave for different frequencies through a limited magnetophotonic crystal in modes of surface and bulk waves are determined.

6.1 Introduction

In recent years, a lot of theoretical and experimental works [1–12] has been devoted to the problem of the terahertz range electromagnetic waves propagation through the

A. A. Shmat'ko (✉)
School of Radiophysics, Biomedical Electronics and Computer Systems, V. N. Karazin Kharkiv National University, Kharkiv, Ukraine
e-mail: alexandr.a.shmatko@univer.kharkov.ua

V. N. Mizernik
Scientific Physical-Technologic Center of MES and NAS of Ukraine, Kharkiv, Ukraine
e-mail: viknikm@ukr.net

E. N. Odarenko
School of Electronics and Biomedical Engineering, Kharkiv National University of Radio Electronics, Kharkiv, Ukraine
e-mail: yevhen.odarenko@nure.ua

A. S. Krivets · O. V. Yushchenko
Department of Nanoelectronics, Sumy State University, Sumy, Ukraine
e-mail: a.krivets@phe.sumdu.edu.ua

O. V. Yushchenko
e-mail: o.yushchenko@phe.sumdu.edu.ua

A. D. Pogrebnjak and O. Bondar (eds.), *Microstructure and Properties of Micro- and Nanoscale Materials, Films, and Coatings (NAP 2019)*, Springer Proceedings in Physics 240, https://doi.org/10.1007/978-981-15-1742-6_6

periodic thin films (one-dimensional photonic crystals). At the same time, photonic crystals have been already widely used in different applications of modern science and technology of the terahertz, microwave and optical ranges. One of the promising photonic crystals applications is the new terahertz radiation sources developing with frequency tuning possibility [13–17]. Majority of PhC theoretical investigations are based on the characteristic (dispersion) equation solution and on the periodical structure transfer matrix. Due to this approach, the properties of isotropic photonic crystals are well studied both for TE and TM modes [1, 2, 18].

An alternative treatment to the research of PhCs is the Floquet-Bloch theory [19–22], that based on the fundamental solutions of Hill's equations and allows to find in the analytic form not only the dispersion characteristics of PhCs but also expressions for the fields in each PhC layer [18, 23–28]. However, such researches have been carried out only for isotropic photonic crystals based on two-layers periodic dielectric structures. Recently, magnetophotonic crystals (MPhCs) which based on gyrotropic elements with a controlled transverse magnetic field have attracted the particular attention of researchers. In the case of gyrotropy presence in the medium, its material parameters are tensors. The presence of gyrotropic layers in the such structure provides opportunity to change the material parameters values due to the applied magnetic field magnitude and, ultimately, to realize electric-type control for the dispersion properties of the MPhCs and for the wave propagation characteristics. In this case, depending on the direction of the applied magnetic field to the gyromagnetic media, various effects can be observed: the Faraday effect, the magnetic two-beam refraction, the rotation of the polarization plane, the nonreciprocal phenomena for forward and backward waves, the presence of surface gyrotropic waves.

Investigations of MPhCs with gyrotropic elements in general case was carried out mainly by the matrix approach [3, 7, 9, 11]. The Floquet theory approach for gyrotropic MPCs have not been considered. In this paper the problem based on the Floquet theory and new fundamental solutions of the Hill's equation for own waves in MPhC with gyrotropic semiconductor layers was represented.

6.2 Floquet—Bloch Waves Theory

Let us consider the propagation of electromagnetic waves in a stratified two-layers periodic structure in general form with gyrotropic discrete layers (one-dimensional MPhC) (Fig. 6.1). Each of the two layers on the structure period $L = a + b$ is a gyrotropic semiconductor medium, the dielectric constant of which is characterized by a tensor of the standard form with the material parameters $\overleftrightarrow{\varepsilon}_j$, $(j = 1, 2)$. The width of one layer is a, and the other one is b.

The dielectric constant tensor of the hyrotropic semiconductor medium of each layer $(j = 1, 2)$ has a standard form [29]:

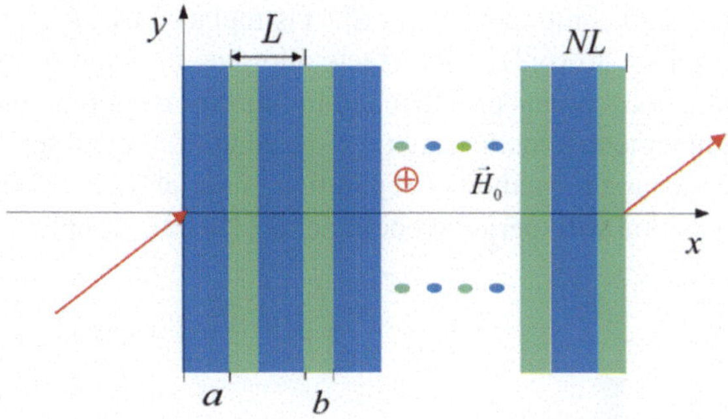

Fig. 6.1 The model of a gyrotropic MphC

$$\overset{\leftrightarrow}{\varepsilon}_j = \begin{Vmatrix} \varepsilon_j & -i\varepsilon_{aj} & 0 \\ i\varepsilon_{aj} & \varepsilon_j & 0 \\ 0 & 0 & \varepsilon_{\parallel j} \end{Vmatrix},$$

Traditionally, the analysis of wave propagation in MPhC is carried out on the basis of Maxwell's equations. In the two- a dimensional case $\left(\frac{\partial}{\partial z} = 0\right)$ from the Maxwell equations two independent Helmholtz equations can be obtained, each of which describes TE (transverse electric) or TM (transverse magnetic) waves. For TE waves the electric field component $E_z = 0$ (E_x, E_y, H_z), H_z-polarization (p-polarization). For TM waves the component of the magnetic field $H_z = 0$ (H_x, H_y, E_z), E_z-polarization (s-polarization). The Helmholtz equations with the field time dependence $\exp(-i\omega t)$ for these two types of waves may be given as:

$$\frac{\partial}{\partial x}\left(\frac{1}{\varepsilon_\perp(x)}\frac{\partial H_z}{\partial x}\right) + \frac{1}{\varepsilon_\perp(x)}\frac{\partial^2 H_z}{\partial y^2} + k^2\mu_\parallel H_z = 0, \tag{6.1}$$

$$\frac{\partial}{\partial x}\left(\frac{1}{\mu(x)}\frac{\partial E_z}{\partial x}\right) + \frac{1}{\mu(x)}\frac{\partial^2 E_z}{\partial y^2} + k^2\varepsilon_\parallel E_z = 0. \tag{6.2}$$

Here $\varepsilon_{\perp j}(x) = \varepsilon_j\left(1 - \frac{\varepsilon_{aj}^2}{\varepsilon_j^2}\right)$ is the effective value of dielectric permeability of MPhC layers media; $\mu(x) = \mu_j$ is the magnetic permeability of layers. The relationship of the tangential field components H_y and E_y through the components E_z and H_z (6.2), (6.3), is determined by the equations:

$$E_y = \left(\frac{1}{ik\varepsilon_\perp}\right)\left(\frac{\partial H_z}{\partial x} + i\frac{\varepsilon_a}{\varepsilon}\frac{\partial H_z}{\partial y}\right), \tag{6.3}$$

$$H_y = \left(\frac{1}{-ik\mu_j}\right)\frac{\partial E_z}{\partial x}. \tag{6.4}$$

From the presented (6.1), (6.2) for the both field components E_z, H_z and from the expressions of the tangential fields (6.3) and (6.4) it follows that the principle of permutation duality is fulfilled for the two types of TM and TE waves. Further, when the field component E_z in (6.1) is replaced by H_z and, at the same time, $\overleftrightarrow{\varepsilon} \leftrightarrow -\overleftrightarrow{\mu}$, it passes to (6.2). These facts allow us to simplify the consideration of a general electrodynamics problem and to limit for ourselves only one wave type (TE or TM) implementations for any type of medium. Equations (6.1) and (6.2) can be reduced by using the method of separation of variables to one type of equations of the Hill's equation with periodic coefficients [19–22], namely:

$$\frac{\partial}{\partial x}\left(p(x)\frac{\partial X}{\partial x}\right) + q(x)X_z = 0, \tag{6.5}$$

where $p(x)$ and $q(x)$ is the periodical coefficients which are determined by expressions $p(x) = \frac{1}{\varepsilon_{\perp j}(x)}$, $q(x) = p(x)\left(k^2\mu_{\parallel j}(x)\varepsilon_{\perp j}(x) - \beta^2\right)$ for TE waves and $p(x) = \frac{1}{\mu_{\perp j}(x)}$, $q(x) = p(x)\left(k^2\varepsilon_{\parallel j}(x)\mu_{\perp j}(x) - \beta^2\right)$ for TM waves. β is the wave propagation constant along the axis $Oy(exp(\pm i\beta y))$. Substitution $\frac{\partial^2}{\partial y^2} = -\beta^2$ is used when the variables are separated in (6.1).

The one-dimensional (6.5) is the Hill's equation [19–22] with periodic functions $p(x)$ and $q(x)$ such that $p(x+L) = p(x), q(x+L) = q(x)$. Equation (6.5) with the corresponding boundary conditions is the boundary Sturm-Liouville problem. The boundary conditions for finding solutions to (6.5) are related to the continuity of the tangential components of the magnetic

$$H_z(x, y) = X(x)e^{i\beta y}$$

and electric

$$E_y(x, y) = \left(\frac{1}{ik\varepsilon_\perp}\right)\left(\frac{\partial X(x)}{\partial x} - \beta\frac{\varepsilon_a}{\varepsilon}X(x)\right)$$

fields at the boundaries of the layers and reduced to the following equations:

$$X_1(a) = X_2(a),$$

$$\frac{1}{\varepsilon_{\perp 1}}\left(\frac{\partial X_1(a)}{\partial x} - \beta\frac{\varepsilon_{a1}}{\varepsilon_1}X_1(a)\right) = \frac{1}{\varepsilon_{\perp 2}}\left(\frac{\partial X_2(a)}{\partial x} - \beta\frac{\varepsilon_{a2}}{\varepsilon_2}X_2(a)\right). \tag{6.6}$$

In addition to these boundary conditions, we use the Floquet theorem to relate solutions on the MPhC period, namely:

$$\rho X_1(0) = X_2(0 + L),$$

$$\rho\frac{1}{\varepsilon_{\perp 1}}\left(\frac{\partial X_1(0)}{\partial x} - \beta\frac{\varepsilon_{a1}}{\varepsilon_1}X_1(0)\right)$$

$$= \frac{1}{\varepsilon_{\perp 2}}\left(\frac{\partial X_2(0+L)}{\partial x} - \beta\frac{\varepsilon_{a2}}{\varepsilon_2}X_2(0+L)\right). \tag{6.7}$$

Using the Hill's equation solution in the form $X(x) = A\psi_1(x) + B\psi_2(x)$ we obtain the following fundamental solutions for the functions $\psi_1(x)$ and $\psi_2(x)$:

$$\psi_1(x) = \begin{cases} \cos\xi_1 x + \beta\frac{\varepsilon_{a1}}{\varepsilon_1}\frac{\sin\xi_1 x}{\xi_1}, & 0 < x < a \\ A\cos\xi_2(x-a) + \\ \quad + B\frac{\sin\xi_2(x-a)}{\xi_2}, & a < x < L \end{cases}, \tag{6.8}$$

$$\psi_2(x) = \begin{cases} \varepsilon_{\perp 1}\frac{\sin\xi_1 x}{\xi_1}, & 0 < x < a \\ D\cos\xi_2(x-a) + \\ \quad + C\frac{\sin\xi_2(x-a)}{\xi_2}, & a < x < L \end{cases}, \tag{6.9}$$

where

$$A = \cos\xi_1 a + \beta\frac{\varepsilon_{a1}}{\varepsilon_1}\frac{\sin\xi_1 a}{\xi_1}, \quad B = \beta\frac{\varepsilon_{a2}}{\varepsilon_2}\cos\xi_1 a - \frac{\varepsilon_{\perp 2}}{\varepsilon_{\perp 1}}\xi_1\sin\xi_1 a$$
$$+ \beta\beta\frac{\varepsilon_{a1}}{\varepsilon_1}\left(\frac{\varepsilon_{a2}}{\varepsilon_2} - \frac{\varepsilon_{\perp 2}}{\varepsilon_{\perp 1}}\frac{\varepsilon_{a1}}{\varepsilon_1}\right)\frac{\sin\xi_1 a}{\xi_1}, \quad D = \varepsilon_{\perp 1}\frac{\sin\xi_1 a}{\xi_1},$$
$$C = \varepsilon_{\perp 2}\cos\xi_1 a + \beta\varepsilon_{\perp 2}\left(\frac{\varepsilon_{a2}}{\varepsilon_2}\frac{\varepsilon_{\perp 1}}{\varepsilon_{\perp 2}} - \frac{\varepsilon_{a1}}{\varepsilon_1}\right)\frac{\sin\xi_1 a}{\xi_1}.$$

Using the Floquet theorem and (6.7), we obtain in general the characteristic equation for the definition of a constant $\rho = e^{iKL}$, namely:

$$(\rho + \rho^*) = 2\cos KL = \frac{1}{\varepsilon_{\perp 2}}\left[\psi_2'(L) - \beta\frac{\varepsilon_{a2}}{\varepsilon_2}\psi_2(L)\right] + \psi_1(L),$$

which, taking into account fundamental solutions of the Hill's (6.8), (6.9), takes the following form for determining the Floquet wave number $K = K_{TE}$ in the MPhC TE waves:

$$\cos K_{TE}L = \cos\xi_1 a\cos\xi_2 b - \frac{1}{2}\left[\begin{array}{l}\frac{\varepsilon_{\perp 2}}{\varepsilon_{\perp 1}}\frac{\xi_1}{\xi_2} + \frac{\varepsilon_{\perp 1}}{\varepsilon_{\perp 2}}\frac{\xi_2}{\xi_1} \\ + \frac{\beta^2}{\xi_1\xi_2}\left(\frac{\varepsilon_{\perp 2}}{\varepsilon_{\perp 1}}\right)\left(\frac{\varepsilon_{a1}}{\varepsilon_1} - \frac{\varepsilon_{\perp 1}}{\varepsilon_{\perp 2}}\frac{\varepsilon_{a2}}{\varepsilon_2}\right)^2\end{array}\right]\sin\xi_1 a\sin\xi_2 b. \tag{6.10}$$

Note that the dispersion (6.10) exactly coincides with the equation which was obtained by the transfer matrix method [11].

The fundamental solutions of the Hill's (6.8), (6.9) make it possible analytically to determine the reflection and transmission coefficients for a plane wave that propagates through a bounded MPhC.

$$T_N = 2\left[\left(W_{22}^N + \frac{\varepsilon_{in}}{\xi_{in}}\frac{\xi_{ex}}{\varepsilon_{ex}}W_{11}^N\right) - \left(\frac{\varepsilon_{in}}{\xi_{in}}kW_{21}^N + \frac{\xi_{ex}}{\varepsilon_{ex}}\frac{1}{k}W_{12}^N\right)\right]^{-1},$$
$$R_N = 1 - T_N\left(\frac{\varepsilon_{in}}{\xi_{in}}\frac{\xi_{ex}}{\varepsilon_{ex}}W_{11}^N + i\frac{\varepsilon_{in}}{\xi_{in}}ikW_{21}^N\right), \quad (6.11)$$

where

$$\mathbf{W}^N = \mathbf{W}\frac{\sin(NKL)}{\sin Kl} - \vec{\mathbf{1}}\frac{\sin[(N-1)KL]}{\sin Kl},$$
$$\begin{pmatrix} W_{11} & W_{12} \\ W_{21} & W_{22} \end{pmatrix} = \begin{pmatrix} \psi_1(L) & ik\psi_2(L) \\ \frac{1}{ik\varepsilon_{\perp 2}}\begin{bmatrix}\psi_1'(L) \\ -\beta\frac{\varepsilon_{a2}}{\varepsilon}\psi_1(L)\end{bmatrix} & \frac{1}{\varepsilon_{\perp 2}}\begin{bmatrix}\psi_2'(L) \\ -\beta\frac{\varepsilon_{a2}}{\varepsilon}\psi_2(L)\end{bmatrix} \end{pmatrix}.$$

The minimum values of the $|T_N|^2_{\min}$ are determined by the expression

$$|T_N|^2_{\min} = \left[1 + \left(\left(\frac{1}{2\sin KL}\right)^2 \left|\left(\frac{\varepsilon_{in}}{\xi_{in}}kW_{21} + \frac{\xi_{ex}}{\varepsilon_{ex}}\frac{1}{k}W_{12}\right)\right|^2 - 1\right)\right]^{-1}$$

6.3 Analysis of Results

Let's turn to analysis of the waves propagation in MPhC that contains the semiconductor plasma and the dielectric layers. Two cases are considered for positive and negative values of the effective permittivity of semiconductor plasma for the single MPhC layer.

Figure 6.2 shows the dispersion characteristic as a dependence of the Floquet-Bloch wave number $K_{TE}L$ on the frequency parameter $\frac{\omega L}{2\pi c}$ (dotted blue curves). This dispersion characteristic was calculated with following parameters of the MPhC: $a = 0.2L$; $\varepsilon_1 = 18$; $\varepsilon_2 = 2.5$; $\varepsilon_{a1} = 10$. Forbidden zones are accented by shaded areas. The dependence of the limited MPhC reflection coefficient $|R|^2_N$ on $\frac{\omega L}{2\pi c}$ for $N = 6$ periods is represented on the same figure.

The same dependencies as in Fig. 6.2 only for negative values $\varepsilon_{1\perp} < 0$ and $\varepsilon_{a1} = 20$ are shown in Fig. 6.3. The transition from positive to negative magnitudes is accompanied not only by the transmission and forbidden zones location changes but also by their width.

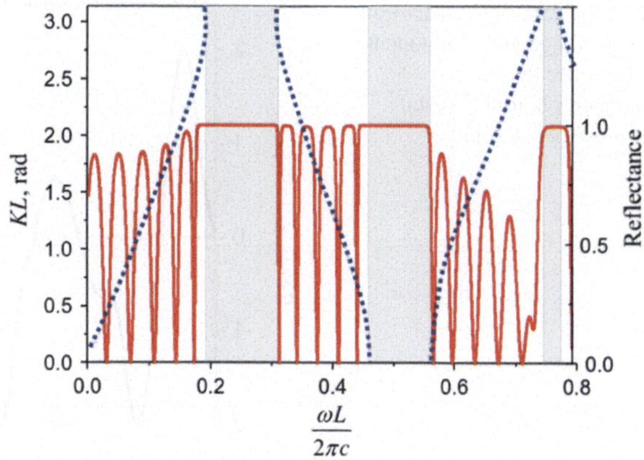

Fig. 6.2 Dispersion characteristics of the magnetophotonic crystal (dotted curves) and reflectance for positive effective value of permittivity (solid curve)

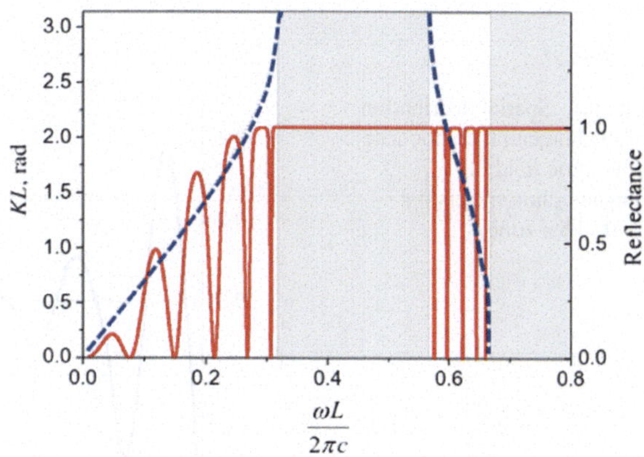

Fig. 6.3 Dispersion characteristics of the magnetophotonic crystal (dotted curves) and reflectance for negative effective value of permittivity (solid curve)

The number of the reflection (transmission) coefficient maximums in the bandwidth is equal $(N - 1)$, and their value is determined by the expression $|R_N|^2_{\max} = 1 - |T_N|^2_{\min}$.

Figures 6.4 and 6.5 show spatial distributions of fields $\mathrm{Re}\, E_y$ in layers for two cases: full passing (Fig. 6.4, $\frac{\omega L}{2\pi c} = 0.6231$) in the transmission zone and full reflection in the forbidden zone (Fig. 6.5, $\frac{\omega L}{2\pi c} = 0.7$).

6.4 Conclusions

An analytical theory of gyrotropic plasma MPhC for determining eigenfunctions, dispersion characteristics, reflection and transmission coefficients of TM modes with arbitrary plasma layers material parameters have been developed. The obtained characteristics allow to construct the theory of controlled waveguide structures, whose guided surfaces can be Bragg reflecting surfaces from the considered MPhCs.

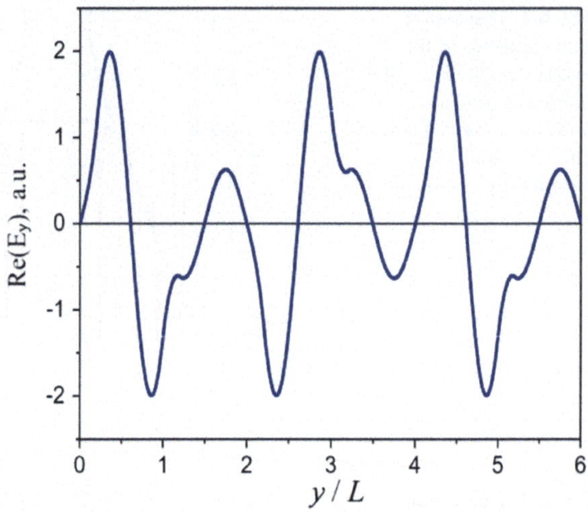

Fig. 6.4 Spatial distribution of the tangential component of electric field in magnetophotonic crystal (transmission zone)

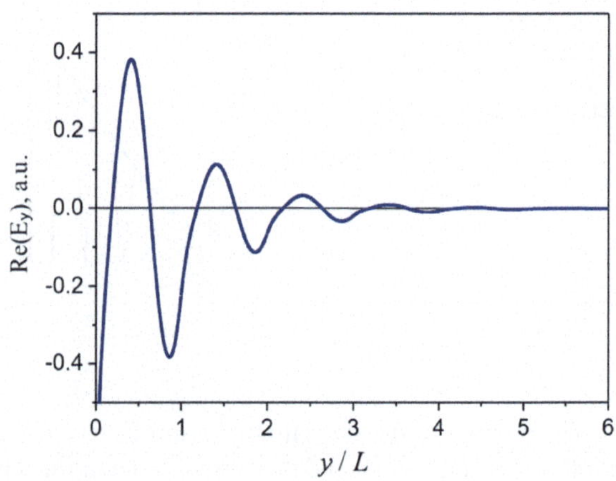

Fig. 6.5 Spatial distribution of the tangential component of electric field in magnetophotonic crystal (forbidden zone)

References

1. P. Yeh, A. Yariv, Chi-Shain Hong. J., Opt. Soc. Am. **67**, 4 (1977)
2. A. Yariv, P. Yeh, Photonics. *Optical Electronics in Modern Communications* (Oxford University press, New York, 2007)
3. F.G. Bass, A.A. Bulgakov, *Kinetic and Electrodynamic Phenomena in Classical and Quantum Semiconductor* (Nova Science Publishers, New York, 1997)
4. J. Lekner, Opt. Soc. Am. A **11**, 2156 (1994)
5. S. Sakaguchi, N. Sugimoto, J. Lightwave Technol. **17**, 6 (1999)
6. M. Inoue, K. Arai, T. Fujii, M. Abe, Appl. Phys. **85**, 5768 (1999)
7. I.L. Lyubchanskii, N.N. Dadoenkova, M.I. Lyubchanskii, E.A. Shapovalov, J. Phys. D: Appl. Phys. **36**, 277–287 (2003)
8. A.A. Shmat'ko et al., in *Proceedings of the 7th International Conference on Advanced Optoelectronics and Lasers (CAOL'2016)*, Odessa, Ukraine, September 2016, pp. 126–128
9. O.V. Shramkova, Prog. Electromagn. Res. **7**, 71 (2009)

10. J.-X. Fu, R.-J. Liu, Z.-Y. Li, Europhys. Lett. **89**, 64003 (2010)
11. A.A. Shmatko et al., in *Theoretical Foundations and Applications of Photonic Crystals*, ed. by A. Vakhrushev (InTech, 2018) p. 228
12. I.L. Lyubchanskii, N.N. Dadoenkova, M.I. Lyubchanskii, E.A. Shapovalov. J. Phys. D: Appl. Phys. **36**(18), R277 (2003)
13. E.N. Odarenko, A.A. Shmat'ko, in *12th IEEE International Conference on Modern Problems of Radio Engineering, Telecommucations, and Computer Science*, Lviv-Slavsko, February 2016 (TCSET, Lviv, 2016), p. 345
14. E.N. Odarenko, Y.V. Sashkova, A.A. Shmatko, N.G. Shevchenko, in *17th International Conference on Mathematical Methods in Electromagnetic Theory*, Kyiv, July 2018 (in recognition of the centenary of the National Academy of Sciences of Ukraine, Kyiv, 2018), p. 164
15. A.A. Shmat'ko, V.N. Mizernik, E.N. Odarenko, in *14th International Conference on Advanced Trends in Radioelectronics, Telecommunications and Computer Engineering*, Lviv-Slavsko, February 2018 (TCSET, Lviv, 2018), p. 436
16. E.N. Odarenko, Y.V. Sashkova, A.A. Shmat'ko in *IEEE Microwaves, Radar and Remote Sensing Symposium*, Kyiv, August 2017 (MRRS, Kyiv, 2017), p. 147
17. E.N. Odarenko, A.A. Shmat'ko in *Proceedings of the 13th International Conference on Laser and Fiber-Optical Networks Modeling (LFNM2016)*, ed. by O.V. Shulika, I.A. Sukhoivanov, Odessa, September 2016 (LFNM, Odessa, 2016), p. 53
18. D.W.L. Sprung, H. Wu, J. Martorell, Am. Phys. **61**, 1118 (1993)
19. J.J. Stoker, *Nonlinear Vibrations* (Wiley, New York, 1950)
20. V.A. Yakubovich, V.M. Starzhinskii, *Linear Differential Equations with Periodic Coefficients* (Wiley, 1975)
21. M.S.P. Eastham, *The Spectral Theory of Periodic Differential Equations* (Scottish Academic, 1975)
22. W. Magnus, S. Winkler, *Hill's Equation* (Dover, 2004)
23. J.K. Nurligareev, V.A. Sychugov, Quantum Electron. **38**, 452 (2008)
24. G.V. Morozov, D.W.L. Sprung, Europhys. Lett. **96**, 54005 (2011)
25. J.K. Nurligareev, Surf. Invest. **5**, 193 (2011)
26. G.V. Morozov, D.W.L. Sprung, Opt. Soc. Am. B **29**(12), 3231 (2012)
27. G.V. Morozov, D.W.L. Sprung, Opt. **17**, 035607 (2015)
28. D.J. Vezzetti, M.M. Cahay, Phys. D Appl. Phys. **12**(4), L53 (1986)
29. A.G. Gurevich, *Ferrites at Microwave Frequencies* (Consultans Bureau, New York, 1963)

Chapter 7
Investigation of the Process of Spatial Self-organization of Linear Defects in Nanocrystalline Materials

O. V. Yushchenko and A. S. Krivets

Abstract On the basis of the evolution equation for dislocation density, considering the properties of grain boundaries, which are the main sources, drains, and barriers for dislocations, the theoretical model of the process of spatial self-organization of linear defects in nanocrystalline materials was constructed. When solving the system of synergetic equations, we find the effective energy of the system, the minima of which correspond to stable states of nanocrystalline material. The analysis of these states determines the conditions for the formation of shear micro bands.

7.1 Introduction

Now we can talk about such a stage of development of nanotechnology, in which the next their application does not cause great surprise or excitement [1–3]. The nomenclature of nanomaterials and their applications has been expanding exponentially, covering almost all technical fields of human activity [1–5].

Thus, the nanostructure (nanosystem) includes an ultrathin structure, which can be characterized by an average size of phases or conglomerates, layers or threads of the order of or less than 100 nm [1–3]. It is precisely in this scale area the changes in the fundamental properties of a substance or material (mechanical, optical, electrical, and magnetic) occur, and the quantum-dimensional effects, that are not characteristic of macro-dimensional analogs, appear [1–5]. At the same time, often the change of these properties and parameters begins on a micrometric scale, which to some extent generalizes micro- and nanostructures or systems (polycrystalline fine-grained materials with an average grain size of 50–150 nm are sometimes called microcrystalline).

Also, technologies for creating metamaterials and zone engineering are already widely known. They are similar to nanostructures and systems in the manifestation of the unique properties of materials (left-side refraction, overcoming diffraction limitations), not inherent in the substances from which they are created. Moreover,

O. V. Yushchenko (✉) · A. S. Krivets
Sumy State University, Sumy, Ukraine
e-mail: o.yushchenko@phe.sumdu.edu.ua

A. D. Pogrebnjak and O. Bondar (eds.), *Microstructure and Properties of Micro- and Nanoscale Materials, Films, and Coatings (NAP 2019)*,
Springer Proceedings in Physics 240, https://doi.org/10.1007/978-981-15-1742-6_7

these effects are established (confirmed) and are used up to the centimeter range of wavelengths, and hence the size of individual elements of such structures.

If we talk about nanomaterials, it is impossible to bypass the issues of the ratio of volume and surface (the number of atoms in the volume to the number of atoms that create the surface). In a nanoparticle (nanocluster) this ratio is much smaller than in a macroparticle, and hence the surface effects become comparable or surpass the volume effects in influencing the particle properties and its interaction with other particles. In fact, there is a transformation of the surface properties of the crystal into its bulk properties. And by changing the size and shape of nanocrystals, one can predict the properties of a nanomaterial [1–3].

Thus, it can be argued that the manifestation of the unique properties of nanomaterials is determined by the properties of individual nano-sized elements, their spatial structure, and the nature of the interaction between nanoparticles. This last statement is widely used to modify the properties of nanocrystalline materials [1–5].

Historically, the first nanosystems were micro- and nanocrystalline (metal and ceramic) materials [3]. Particular attention was paid to the modification of their properties—hardening of materials at low temperatures and superplasticity at high. These effects are primarily associated with the Hall-Petch grain-boundary hardening and its disturbances or the dislocation mechanisms of nanopacity and nanoplasticity [4, 5]. According to experimental and theoretical studies, an important role in hardening and superplasticity is played by both the crystal size and the defects of the crystal structure. It is already known that defects in the crystal structure, which are typical for single crystals and large crystallites in polycrystals at nanomaterials, often cannot be kept inside the grain and come to the surface, creating a defect-free structure inside the grain. At the same time, in the array of nanoclusters (nanocomposites) the boundaries between the grains are characterized by high instability, which determines the high diffusion mobility of atoms (5–6 orders of magnitude greater than that of conventional polycrystalline materials).

Dislocation mechanisms of nanopacity and nanoplasticity also affect the creation of periodic structures during plastic deformation [4, 5]. Such modification of the surface of materials, in addition to strength, can determine their optical and hydrophobic properties. It is the problem (the dislocation mechanism of nanopacity and nanoplasticity) that this work is devoted to.

7.2 Basic Equations

The plastic deformation of nanocrystalline (d <100 nm) and submicrocrystalline (d <100 μm) metals is accompanied by heterogeneous microscale distribution by deformable material [4–6]. In the papers [7, 8], when the nanocrystalline iron was deformed, the authors observed the formation of the system of shear microbands, the w width of which increased with the growth of the average size of nanorods d and was in an order of magnitude larger, $w \approx 60d$. The formation of localized shear bands was observed in nanocrystalline palladium, nickel, copper and submicrocrystalline aluminum alloy. For large grain sizes, the deformation of the material proceeds homogeneously, evenly covering all grains without localizing in the form of microbands.

Similarly to [7], we will assume that the process of self-organization of dislocations dominates during the formation of shear microbands. Evolution equation of average density of the dislocations $\rho(x, y, t)$ in the nanometall we write in the form of the reduced Swift-Hohenberg equation [9, 10]

$$\frac{\partial \rho(x, y, t)}{\partial t} + \frac{\partial j_y}{\partial y} = (1 - \beta_{gb})nu + \frac{\beta}{d}u\rho - h_a u\rho^2. \quad (7.1)$$

Here, x is the coordinate in the direction of elongation of microbands (in the direction of the action of tangential stresses), y—coordinate in the direction of their expansion, t is the time, u is the velocity of the dislocations, j_y is the diffusion flow of dislocations, d is the grain size, h_a is the annihilation distance of screw dislocations in the microband by the cross slip mechanism [10]. In Eq. (7.1), the first term on the right-hand side describes the balance of bulk densities sources n and $\beta_{gb}n$ drains of dislocations in the grain boundaries, the second term reflects the accumulation of dislocations inside nanograins due to the restriction of the free path length of the grain ($\beta \approx 1$), the third term takes into account the annihilation of the screw components of dislocation loops. In this case, the parameter β_{gb} is equal to the relative efficiency of grain boundaries as sources and drains for dislocations. Further we will consider a simpler case of a constant diffusion flow of dislocations ($j_y = const$).

Assume that in the formation of shear microbands, the process of self-organization of dislocations is dominant. It is known that for the description of self-organization processes the simplest is the synergetic three-parameter model given by the equations for the rates of change of the order parameter, the conjugate field and the control parameter [11–15]. In the case of the process of spatial self-organization of dislocations and the formation of shear microbands, the main parameter that distinguishes the state of the system is the density of dislocations. Let's consider the relative density of dislocations, which in the usual state acquires a zero value, and for the state of formation of microbands the displacement is nonzero. From the experimental data, it is known that microbands are formed precisely at large values of the density of dislocations. The equation for the evolution of the average density of dislocations ρ in a nanomaterial is written in the form of a reduced Swift-Hohenberg equation which contains the dislocation processes characteristic of the nanosized structures on its right side. As a conjugate field and a control parameter, let us take the velocity of the dislocations and the density of the grain boundaries, respectively. Then the system of self-organization equations, describing the transition to the state of formation of shear microbands at a constant value of the diffusion flow, takes the form

$$\frac{\partial \rho}{\partial t} = -\frac{\rho}{\tau_\rho} + \frac{\beta}{d}u\rho - h_a u\rho^2 + (1 - \beta_{gb})nu,$$

$$\frac{\partial u}{\partial t} = -\frac{u}{\tau_u} + a_1 \rho n,$$
$$\frac{\partial n}{\partial t} = \frac{n_e - n}{\tau_n} - a_2 \rho u. \tag{7.2}$$

Here τ_i $(i = \rho, u, n)$ are the relaxation times for density and velocity of dislocations and density of grain boundaries, respectively; $a_1, a_2 > 0$ constants of nonlinear feedback.

In the first equation of system (7.2), the first term on the right side describes the relaxation process of the relative density of dislocations in the autonomous mode; the second term reflects the process of accumulation of dislocations inside the nanosized grains due to the restriction of the length of their free run by the grain size, the third term takes into account the annihilation the scroll component of the dislocation loops, the last term describes the balance of volumetric densities of sources and drains of dislocations in the grain boundaries.

In the second equation of the system (7.2), the rate of change of the conjugate field (acceleration of the dislocation motion) is given by the relaxation component with the corresponding relaxation time τ_u and the positive feedback relation of the density of dislocations with the density of grain boundaries (a_1 is a positive coupling constant). It is this positive feedback between the densities of dislocations and grain boundaries that is the cause of self-organization.

The third equation of the system (7.2) specifies the rate of change of grain boundary density (control parameter). The relaxation term on the right side describes, in this case, the change in the density of the grains boundaries not to zero, but to the equilibrium value of n_e, which is given by the external conditions. The second term on the right side is given by the negative feedback between the density and the velocity of the dislocations with the rate of change in the density of grain boundaries.

It is impossible to solve analytically the system (7.2). In addition, the analysis of these equations is hindered by a large number of constants. To reduce this amount, we will analyze the dimensionless system.

$$\frac{\partial \rho}{\partial t} = -\rho + u\rho - Au\rho^2 + B(1 - \beta_{gb})nu,$$
$$\delta \frac{\partial u}{\partial t} = -u + \rho n,$$
$$\sigma \frac{\partial n}{\partial t} = n_e - n - \rho u. \tag{7.3}$$

At the same time for the main parameters of the system (time, density and velocity of dislocations, and density of grain boundaries) the following scales were chosen

$$t_m = \tau_p,\ \rho_m = (\tau_u \tau_n a_1 a_2)^{1/2},\ u_m = \frac{d}{\beta \tau_\rho},\ n_m = \frac{d}{\beta \tau_\rho} \sqrt{\frac{\tau_n a_2}{\tau_u a_1}}.$$

Also, parameters and ratios of the relaxation times were introduced.

$$A \equiv h_a \tau_\rho \rho_m u_m,\ B \equiv \tau_\rho \frac{u_m n_m}{\rho_m};\ \delta \equiv \frac{\tau_u}{\tau_\rho},\ \sigma \equiv \frac{\tau_n}{\tau_\rho}.$$

7.3 Stationary Value of Dislocation Velocity

Since system (7.3) has no analytical solution, we will use approximation $\tau_\rho \gg \tau_u, \tau_n(\delta, \sigma \approx 0)$ which means that the velocity of the dislocation movement and the density of the grain boundaries change much faster than the density of dislocations.

As a result, the dependence of the dimensionless density of grain boundaries on the density of dislocations becomes the form

$$n = \frac{n_e}{1+\rho^2}. \tag{7.4}$$

Accordingly, the stationary value of the dimensionless velocity of the dislocation motion, depending on the density of the dislocation, has the form

$$u = \frac{n_e \rho}{1+\rho^2}. \tag{7.5}$$

As can be seen from Fig. 7.1 in stationary mode, the velocity of dislocations, depending on the dimensionless density of dislocations, increases from 0 to $n_e/2$ in the range from 0 to ρ_m. The growth of the dislocation velocity is the reason for the process of self-organization and the formation of shear microbands. As the dislocation density reaches ρ_m, the velocity of dislocations begins to decrease.

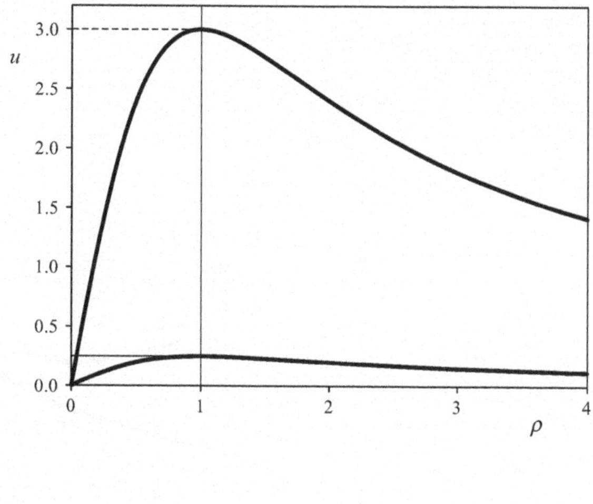

Fig. 7.1 The dimensionless velosity of dislocations, depending on the density of dislocations. The upper curve corresponds to $n_e = 6$, the lower $n_e = 0.5$

7.4 Effective Energy

Substituting the expressions for the dislocation velocity (7.5) and density of grain boundaries (7.4) into the first equation of system (7.3), we obtain the differential equation, which has the form of the Landau-Khalatnikov equation

$$\frac{\partial \rho}{\partial t} = -\frac{\partial V}{\partial \rho}, \tag{7.6}$$

where V is the effective energy of the system.

After integration, we obtain an explicit form of the effective energy of the system, depending on the density of dislocations

$$V = \rho^2/2 - n_e[\rho - \arctan \rho] + An_e\left[\rho^2 - \ln\left(\rho^2 + 1\right)\right]/2 + \frac{\left(1 - \beta_{gb}\right)Bn_e^2}{2\left(1 + \rho^2\right)}. \tag{7.7}$$

The corresponding dependence is shown in Fig. 7.2. The dependence of the effective energy on the density of the dislocations and the parameter of the grain boundary as sources and drains of dislocations is shown in Fig. 7.3.

As can be seen from Figs. 7.2 and 7.3, when the density of grain boundaries n_e, given by external conditions, exceeded the critical value ($n_c = n_m$), a non-zero minimum appears on the dependence of the effective energy. The latter means that with such parameters the process of the formation of shear microbands becomes

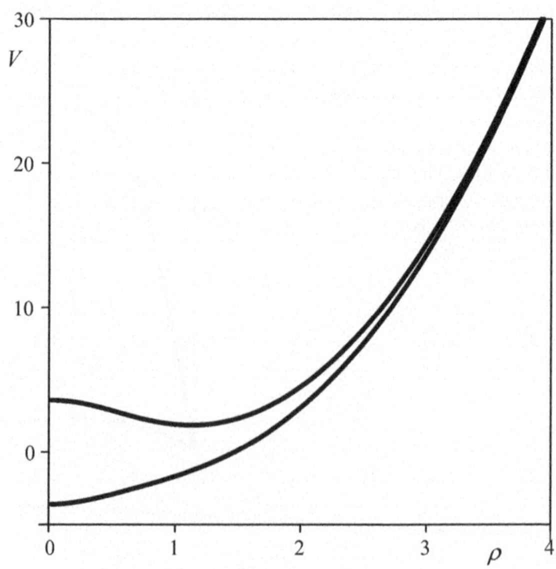

Fig. 7.2 The dependence of the effective energy of the system on the dislocation density for a set of parameters: $\beta_{gb} = 0.8$ (upper curve), $\beta_{gb} = 1.2$ (lower curve), $A = 1$; $B = 1$; $n_e = 6$

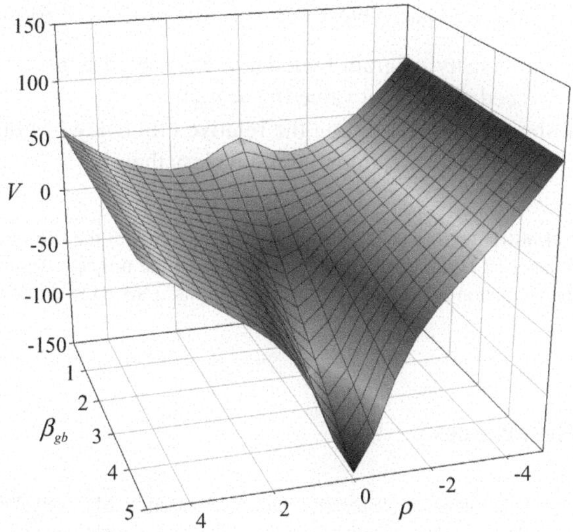

Fig. 7.3 The dependence of the effective energy of the system on the density of dislocations and the parameter of grain boundary efficiency as sources and drains of dislocations for a set of parameters: $A = 1$; $B = 2$; $n_e = 6$

possible. It should be noted that the effectiveness of the boundaries of the grains is given precisely as the source of dislocations. At values n_e, smaller than critical $n_c = n_m$, microbands are not formed regardless of the effectiveness of grain boundaries (as sources and drains of dislocations). An analysis of the influence of coefficients *A*, *B* in the form of effective energy of the system was also conducted. As can be seen from Fig. 7.4, this effect is only quantitatively reflected on the difference between the energies of the ground state (at a minimum at the point $\rho = 0$) and the state of the formation of shear microbands.

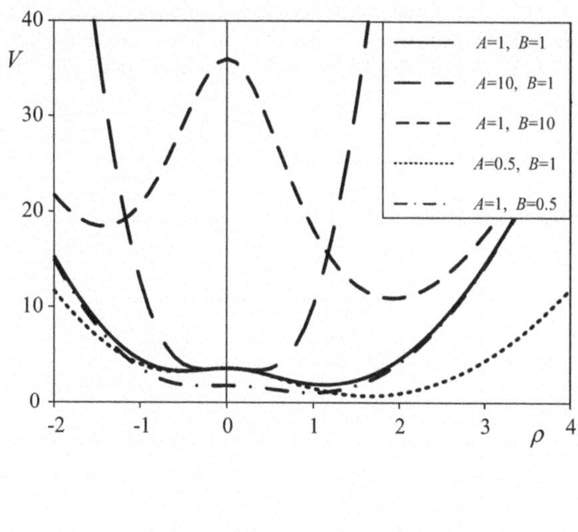

Fig. 7.4 The dependence of the effective energy on the density of dislocations for a set of parameters $\beta_{gb} = 0.8$, $n_e = 6$

Thus, two parameters influence the process of formation of shear microbands:

- the density of grain boundaries n_e, which is given by external conditions, must exceed the critical value ($n_c = n_m$);
- the parameter defining the relative efficiency of grain boundaries as sources and drains of dislocations should be less than 1.

Acknowledgements The work was executed under the support of the Ministry of Education and Science of Ukraine within the framework of the project "Atomic and statistical representation of the formation and friction nanoscale systems" (No. 0118U003584).

References

1. A.A. Shchuka, *Nanoelectronics* (Fizmatkniga, Moscow, 2007)
2. I.P. Suzdalev, *Nanotechnology: Physical Chemistry of Nanoclusters, Nanostructures and Nanomaterials* (Combook, Moscow, 2006)
3. N.A. Azarenkov, V.M. Beresnev, A.D. Pogrebnyak, *Nanomaterials, Nano Coatings, Nanotechnology* (V.N. Karazin Kharkiv National University, Kharkov, 2009)
4. G.A. Malygin, Phys. Usp. **54**, 1091 (2011)
5. G.A. Malygin, Phys. Usp. **42**, 887 (1999)
6. G.A. Malygin, Phys. Solid State **51**(9), 1814 (2009)
7. D. Jia, K.T. Ramesh, E. Ma, Acta Mater. **51**, 3495 (2003)
8. Q. Wei, D. Jia, K.T. Ramesh, E. Ma, Appl. Phys. Lett. **81**, 1240 (2002)
9. M.C. Cross, P.C. Hohenberg, Rev. Mod. Phys. **65**, 2492 (1993)
10. G.A. Malygin, Phys. Solid State **37**(1), 3 (1995)
11. A.I. Olemskoi, A.V. Khomenko, Rev. E. **63**, 036116 (2001)
12. L.S. Metlov, M.M. Myshlyaev, A.V. Khomenko, I.A. Lyashenko, Tech. Phys. Lett. **38**(11), 972 (2012)
13. O.V. Yushchenko, M.A. Rudenko, J. of Nano- Electron. Phys. **10**(6), 06019 (2018)
14. O.V. Yushchenko, AYu. Badalyan, J. Nano- Electron. Phys. **4**(3), 03009 (2012)
15. O.V. Yushchenko, T.I. Zhylenko, Condens. Matter Phys. **16**(1), 13605 (2013)

Chapter 8
Method for Analysis XPS Data of Nanolayered Samples

Yurij Kazarinov and Vasyl Gritsyna

Abstract A method has been developed for fitting the XPS spectra with a background of inelastic loss, which was realized analytically for Fourier calculated images. The use of the Tougaard method allowed us to obtain the contribution of the background of inelastic losses, which depends on the morphology of the sample. Instead of the universal inelastic cross section, an analytical expression was used for the cross section of the Shirley method to define the peaks in the form of Doniah-Sunjic. Due to computations for Fourier images, it became possible to carry out simultaneous fitting of several spectra obtained for different chemical elements of one sample and to consider the hardware effects when synthesizing the spectra. As a result of the analysis of the spectra, the structure and composition of the nanolayered chromium-carbide protective coating deposited by method of metal–organic chemical vapor deposition (MOCVD) were established.

8.1 Introduction

The information about the depth distribution of atoms in the surface nanolayers of materials is of great importance in many technological applications, from chemical catalysis and coatings to the parameters and properties of nanoparticles. The existing methods for analyzing XPS spectra have limitations associated with the used algorithms and do not allow to extract all the useful information contained in the measured spectra.

Y. Kazarinov (✉)
Institute of High-energy Physics and Nuclear Physics, NSC Kharkiv Institute of Physics and Technology, Kharkiv, Ukraine
e-mail: yu.kazarinov@karazin.ua

V. Gritsyna
Department of Physics and Technology, V. N. Karazin
Kharkiv National University, Kharkiv, Ukraine
e-mail: vtgritsyna@karazin.ua

A. D. Pogrebnjak and O. Bondar (eds.), *Microstructure and Properties of Micro- and Nanoscale Materials, Films, and Coatings (NAP 2019)*,
Springer Proceedings in Physics 240, https://doi.org/10.1007/978-981-15-1742-6_8

The determination of the chemical composition of homogeneous materials by measuring the intensity of the peaks in the XPS spectra requires careful consideration of the background of inelastic losses [1]. When analyzing nanostructured materials, it is often assumed that the distribution of atoms is uniform. This leads to significant errors since the intensity of the peaks depends not only on the concentration of atoms but also on the distribution of atoms in-depth [2]. The general method for calculating the background of the inelastic electron loss proposed by Tougaard in [3] makes it possible to reconstruct the surface morphology, but only for elements in a matrix of a different material.

In addition, after subtracting the background of inelastic losses by the Tougaard method, the tail of intrinsic losses remains [4] in the spectrum, and such a spectrum cannot be fitted using Gauss-Lorenz functions. On the other hand, the subtraction of the background by the method of Shirley solves this problem but is applicable only for homogeneous materials.

The development of the Shirley method in the active approach of fitting XPS lines with an individual background contribution was proposed in [5]. This approach effectively considers the background of complex shape in the synthesis of the spectrum, but the result of such processing is the removal of the inelastic background and the neglect of the surface morphology.

In this work, a technique has been developed for fitting the X-ray photoelectron spectra with peaks having an inelastic loss background. It considers the morphology of the material and the influence of the parameters of excitation source and electron analyzer. Due to the self-consistent fitting of the spectra of the three chemical elements, the reliability of determining the structure of the surface layer of the solids consisting of overlapping peaks of different chemical compounds is increased.

8.2 Method

To fit the XPS spectra containing the background component it is necessary to perform a convolution procedure of the spectrum without a background and a function describing the inelastic loss background. If we compute the convolution for the Fourier transforms, then instead of calculating the integral for each point of the spectrum, the images are simply multiplied. In [6], a method of XPS spectra fitting is described that is performed for the Fourier transforms of the spectra. In this case, the inverse Fourier transform is performed only for visual inspection and for recording the final result of the modeling spectrum.

Then each X-ray photoelectron line $Y(s)$ in the spectrum is described by the following expression:

$$Y(s) = X(s) \cdot A(s) \cdot C \cdot Sc \cdot L(s) \cdot P(s), \tag{8.1}$$

where $X(s)$ is the Fourier transform of the exciting X-ray spectrum (including satellites if necessary), $A(s)$ is the Fourier transform of the instrumental function of the electron analyzer in the form of a Gaussian profile, C is the concentration of atoms in the layer, Sc is the cross section of the photoelectron line excitation by Scofield [7]. $L(s)$ is the Fourier transform of the Doniach-Sunjic spectral line [8]

$$L(s) = (is)^{-\alpha} \cdot \exp((iE_0 - w)s),$$

where w is the peak width determined by the decay time of the hole state (lifetime states by electron capture), E_0 is the binding energy, $\alpha > 0$ is the shape parameter. The Fourier transform of $P(s)$ describes the background component of the spectrum in accordance with the formula for the layered material based on the method proposed by Tougaard [3]:

$$P(s) = \int_{x1}^{x2} \exp\left(-\frac{x}{\lambda \cdot \cos(\Theta)} \cdot (1 - \lambda \cdot K(s))\right) dx, \tag{8.2}$$

where $x1$, $x2$ are the layer boundaries, Θ is the emission angle between detector and surface normal. The λ is the electron mean free path to inelastic scattering, and $K(s)$ is the Fourier transform of the differential inelastic scattering cross section. Due to the fact that the universal cross section of the Tougaard does not consider intrinsic losses, it cannot be used to synthesize the model spectrum with the background.

The background in the Shirley method [9] is well combined with the further fitting with peaks, but to use it in the Tougaard algorithm, it is necessary to obtain the corresponding expression for $K(s)$.

Usually, the Shirley method is used as an iterative procedure, and the final corrected spectrum $F(E)$ is obtained from the original spectrum $J(E)$ as follows [8]:

$$F(E) = J(E) - k \int_{E}^{\infty} F(E')dE'. \tag{8.3}$$

where k is the parameter that determines the intensity of the background. To obtain an expression for cross section K, expression (8.3) should be transformed to the form for a corrected spectrum of a bulk material in the Tougaard method:

$$F(E) = J(E) - \lambda \int_{E}^{\infty} J(E')K(E' - E)dE', \tag{8.4}$$

To do this, we rewrite (8.3) in the form of a differential equation for the inelastic background part of spectrum $B(E) = k\int_E^\infty F(E')dE'$:

$$\frac{1}{k}\frac{d}{dE}B(E) + B(E) = J(E). \tag{8.5}$$

The solution of (8.5) gives us the background component of the spectrum, and the corrected spectrum is expressed through the measured spectrum:

$$F(E) = J(E) - \int_{E}^{\infty} J(E')ke^{k(E'-E)}dE'. \tag{8.6}$$

Comparing expressions (8.4) and (8.6) we get the desired expression:

$$K(E' - E) = \lambda^{-1} \cdot k \cdot \exp(-k(E' - E)) \tag{8.7}$$

After the Fourier transform, we obtain the required $K(s)$:

$$K(s) = \lambda^{-1} \cdot k/(k + is). \tag{8.8}$$

For a dielectric layer with a band gap E_g, expression (8.8) should be multiplied by $\exp(iE_g)$, which will lead to a shift of the background contribution and correspond to the inability to transfer energy to the dielectric less than E_g.

In this work, we used a flat model of a sample with layers of constant composition: outer layer #1 with thickness d, inner layer #2 with thickness D, bulk substrate #3. This allows reducing the number of parameters and describing frequently encountered surface variants. Using expression (8.8) in expression (8.3) and specifying the boundaries of the layers, we obtain the final expression for Fourier images of the background contribution of each layer:

$$\begin{aligned}
P_1(s) &= \frac{\cos(\Theta)}{\Sigma_1(s)} \cdot \left(1 - \exp\left[-d\frac{\Sigma_1(s)}{\cos(\Theta)}\right]\right),\\
P_2(s) &= \frac{\cos(\Theta)}{\Sigma_2(s)} \cdot \left(1 - \exp\left[-D\frac{\Sigma_2(s)}{\cos(\Theta)}\right]\right) \cdot \exp\left(-d\frac{\Sigma_1(s)}{\cos(\Theta)}\right),\\
P_3(s) &= \frac{\cos(\Theta)}{\Sigma_3(s)} \cdot \exp\left(-D\frac{\Sigma_2(s)}{\cos(\Theta)}\right) \cdot \exp\left(-d\frac{\Sigma_1(s)}{\cos(\Theta)}\right).
\end{aligned} \tag{8.9}$$

where $n = 1, 2, 3$ is the layer number, $\Sigma_n(s) = (1 - k/(k - is)) \cdot \exp(isE_{gn}))/\lambda_n$, where λ_n is the inelastic mean free path of electrons in the material of the layer. For each layer, it is calculated based on the actual composition $\lambda_n = a \cdot 0.72(aE)^{1/2}$, where a is the average atomic size, E is the electron energy. In the case of the elementary composition of the layer, $\lambda_n = a \cdot 0.41(aE)^{1/2}$ [10]. As a result of the summation of expressions in the form of (8.1), we obtain a model Fourier image. A similar procedure is performed for all spectra and the total difference between the model and the original Fourier images is calculated. The difference is minimized by varying the parameters of the peaks and layers.

When performing a discrete Fourier transform of the spectra with a background, a jump occurs due to the difference in the signal level at the edges of the spectrum. Some trick was used to avoid this problem: the cosine Fourier transform of the original spectrum is calculated over a half period. In the full period, the spectrum is mirrored and has no jumps at the edges, and its part from 0 to π corresponds to the original spectrum. In this case, the Fourier calculations of the images using formulas (8.1), (8.9) are carried out for complex harmonics, and for fitting and the inverse Fourier transform, only the real (cosine) part is used.

8.3 Experiment Details

The spectra were obtained using a homemade X-ray photoelectron spectrometer equipped with a non-monochromatized X-ray source with an aluminum anode with a Be window of 7 μm in thickness. The source operated at 15 kV, with an anode current of 20 mA. The electron analyzer of the type cylindrical mirror with a retarding system worked in the regime of constant transmission energy. The total energy resolution of the spectrometer was 1.0 eV. The residual pressure in the chamber is 5×10^{-8} Torr.

The experimental X-ray photoelectron spectra for processing by using the developed method were taken from two chromium-containing samples.

For calibration of an algorithm we have used bulk metal chromium with a natural oxide on the surface. Before measurement sample was etched by bombardment with 1.4 keV Ar^+ ions to the fluence of 2.1×10^{17} ions/cm^2.

The experimental spectra for analysis were taken from the sample which prepared by deposition of the chromium-carbide coating on a steel substrate by MOCVD. The coating has a layered structure with a total thickness of 10 μm. The spectra were obtained after sputtering by 1.4 keV Ar^+ ions to the fluence 4.7×10^{17} ions/cm^2. Before processing, the spectra were corrected for the transmission function of the electron energy analyzer.

8.4 Application

For the operation fitting algorithm, it is necessary to determine the values of two constants: the sensitivity coefficient of the spectrometer, which matches the amplitudes of the model and experimental spectra and the background intensity parameter k in expression (8.8). To do this, the spectra of metallic chromium, with a layer of natural oxide and a layer of hydrocarbons were taken and analyzed. The beforehand known chemical composition of the layers reduced the number of free parameters in the process of fitting. Thus, only the thicknesses of layers and parameter k need to be varied. After fitting the experimental spectrum, the values for the parameters were obtained: $k = 0.045$ eV^{-1} and the thickness of the hydrocarbons layer 2.3 nm, the thickness of the oxide layer 2.2 nm.

The developed technique was applied to fit the X-ray photoelectron spectrum of the layered chromium-carbide coating sputtered with argon ions to the condition with the maximum signal intensity of the chromium carbide. A survey XPS spectrum is

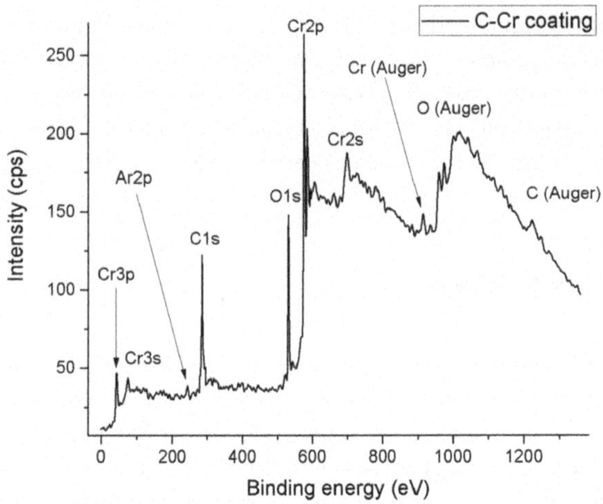

Fig. 8.1 The survey X-ray photoelectron spectrum of C–Cr coating deposited on a steel substrate by the MOCVD and subjected to sputtering with Ar^+ ions with an energy of 1.4 keV and a dose of 4.7×10^{17} ions/cm^2

shown in Fig. 8.1. As can be seen, in addition to carbon and chromium, the coating contains a significant amount of oxygen and argon implanted during the sputtering process.

The spectral ranges of three characteristic lines corresponding to the chemical elements were selected for processing: carbon—C1 s, oxygen—O1 s and chromium—Cr2p, shown with circles in Fig. 8.2. The lines show the spectra of the individual peaks obtained as a result of the fit. In the spectra, there are observed overlapping lines from various compounds of the elements with the background contribution, which is determined from the distribution of chemical compounds over the depth of the sample.

In the process of fitting, the arrangement of chemical compounds in the layers was selected. Since the spectra of carbon (Fig. 8.2a) and oxygen (Fig. 8.2b) do not have a significant background component observed in the chromium spectrum (Fig. 8.2c), the presence of chromium oxide and carbide in layer #3 was excluded. Due to the low background intensity in Fig. 8.2b (at binding energy above 535 eV) all oxygen compounds were placed in outer layer #1. Such a distribution of compounds in depth was unique for a reasonable fit of all three spectra.

As a result of the fit, it was found that the thickness of the outer layer #1 was 1.9 nm, the inner one #2 was 1.6 nm. The binding energies of atoms, their concentrations, and location in layers are listed in Table 8.1. The total computational time was about 30 s for a processor with a frequency of 2.1 GHz.

The obtained ratios of concentrations of elements showed that chromium oxide has the formula CrO_2, but the binding energies of this chromium compound correspond to CrOOH [11]. The C1s line with a binding energy of 282.8 eV corresponds

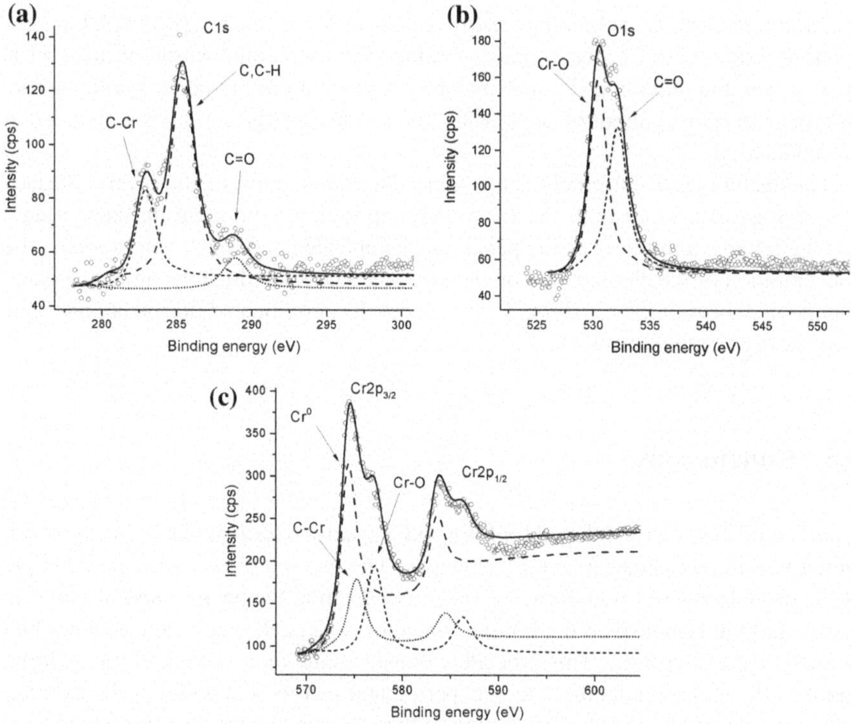

Fig. 8.2 XPS spectra of characteristic lines of atoms contained in chromium carbide coating (circles—experimental data) and modelled peaks with an inelastic loss background, obtained as a result of fitting (lines): **a** C1s—chromium carbide, carbon, carbon oxide, **b** O1s—chrome oxide and oxide carbon, **c** Cr2p-metallic chromium, chromium carbide and chromium oxide

Table 8.1 Types of the chemical bonds, binding energies, and relative concentration of constituent chemical compounds in layers. Data was obtained from analysis of photoelectron spectra of the chromium carbide sample

Lines	Parameters			
	Chemical bond	Binding energy, eV	Concentration, at.%	Number of layer
C1 s	C, C–H	285.3	40	1
	C–O	288.7	6	1
	C–Cr	282.8	46	2
O1 s	Cr–O	530.3	15	1
	C–O	532.0	13	1
$Cr2p_{3/2}$	Cr–O	577.0	8	1
	C–Cr	575.1	22	2
	Cr	574.2	100	3

to chromium carbide, as binding energy of 575.1 eV for line $Cr2p_{3/2}$ [12], but concentration ratio C to Cr has an unusual value. The total composition of layer #1 is 90 at.%, and the remaining 10 at.% may be hydrogen atoms associated with carbon, and for layer #2 it is about 68 at.% where the residual is free carbon not included in the calculations.

This technique is extremely sensitive to the composition of the layers. Surface morphology affects on both the form and amplitude of the inelastic background, and the relative amplitude of the peaks. Such dependence of peak amplitude on the concentration of the element and on the layer thickness in which the atom is located and the thickness of the overlying layers makes the complicated fitting, but provides a physically reliable result.

8.5 Conclusions

A procedure has been developed for fitting XPS spectra with an inelastic background, which was implemented for Fourier images. The method uses a special technique of the discrete Fourier transform for spectra containing the background of inelastic losses, and the synthesis of the Fourier image of each peak in the form of a product of analytical expressions. This procedure considers the morphology of the material and the effects of broadening due to experimental factors: the width of the exciting radiation line and the limited resolution of the electron energy analyzer. Due to the self-consistent fitting of the spectra of the three chemical elements, it was possible to determine the structure and composition of the surface layer of the chromium carbide coating in the presence of a strong overlapping of the photoelectron peaks of individual compounds.

Acknowledgements The authors are grateful to Dr. S.A. Krokhmal from NSC KIPT for providing samples of chromium-carbide coatings.

References

1. S. Tougaard, C. Jansson, Surf. Interface Anal. **20**, 1013 (1993)
2. S. Tougaard, Appl. Surf. Sci. **100/101**, 1 (1996)
3. S. Tougaard, J. Electron Spectros. Relat. Phenom. **52**, 243 (1990)
4. A.M. Salvia, J.E. Castle, Electron Spectros. Relat. Phenom. **95**, 45 (1998)
5. M. Herrera-Gomez, O. Bravo-Sanchez, M.O. Ceballos-Sanchez, Vazquez-Lepe. Surf. Interface Anal. **46**, 897 (2014)
6. Y.G. Kazarinov, V.T. Gritsyna, Y.V. Sidorenko, Funct. Mater. **11**, 131 (2004)
7. J.H. Scofield, J. Electron Spectros. Relat. Phenom. **8**, 129 (1976)
8. S. Doniach, M. Sunjic, Phys. C Solid State Phys. **3**, 285 (1970)
9. D.A. Shirley, Phys. Rev. B **5**, 4709 (1972)

10. M.P. Seah, W.A. Dench, Surf. Interface Anal. **1**, 2 (1979)
11. G.P. Halada, J. Electrochem, Electrochem. Soc. **138**, 2921 (1991)
12. K.H. Stern, D.R. Rolison, Electrochem. Soc. **137**, 178 (1990)

Chapter 9
Temperature Dependence of Conductivity in Composite Film of Single-Walled Carbon Nanotubes with Graphene Oxide

Nikita Kurnosov and Victor Karachevtsev

Abstract The results of low-temperature studies (5–291 K) of conductivity in the composite film of graphene oxide (GO) with single-walled nanotubes (SWNTs) are presented. The composite film was obtained by vacuum filtration of aqueous suspension containing both GO and SWNTs. It was shown that conductivity of composite is largely conditioned by the nanotubes, while graphene oxide film obtained similarly demonstrated no conductivity. Semiconductor behavior with negative temperature coefficient of conductivity was revealed for both composite and nanotube films. Further analysis confirmed that conductivity in the range of 5–240 K is well described within the framework of the 3D Mott model. The conductivity mechanism involves thermoactivated tunneling of electrons through barriers with a variable range hopping (VRH), which is common for disordered semiconductors. At higher temperatures (T > 240 K) the Arrhenius model was used. Such parameters as distance and energy of electron hopping as well as energy barriers were estimated.

9.1 Introduction

Graphene oxide (GO) and single-walled carbon nanotubes (SWNTs) are carbon nanomaterials that exhibit unique physical properties and are promising for applications in such fields as nanoelectronics, photonics, and sensing. GO is graphene with oxygen-containing groups at the edges (carboxyl-COOH and hydroxyl-OH groups) and on the plane (epoxy (C–O–C) and hydroxyl groups) [1, 2]. Due to these groups GO is dispersible in water, which is a major advantage necessary for biomedical purposes [3, 4]. GO sheets contain domains with both sp^2— and sp^3—hybridized electrons that govern electronic properties of this nanomaterial. Another important form is reduced graphene oxide (rGO) with partially removed oxygen groups. rGO

N. Kurnosov (✉) · V. Karachevtsev
Department of Molecular Biophysics, B. Verkin ILTPE of NASU, Kharkiv, Ukraine
e-mail: n.kurnosov@ilt.kharkov.ua

V. Karachevtsev
e-mail: karachevtsev@ilt.kharkov.ua

A. D. Pogrebnjak and O. Bondar (eds.), *Microstructure and Properties of Micro- and Nanoscale Materials, Films, and Coatings (NAP 2019)*,
Springer Proceedings in Physics 240, https://doi.org/10.1007/978-981-15-1742-6_9

contains mainly sp^2 domains so that electronic structure is more similar to that of graphene, but it remains soluble in water.

SWNTs have high electrical and thermal conductivity as well as very structure-dependent electronic and optical properties. Significant progress has been made in the production of hybrids formed by carbon nanotubes with rGO or GO [5]. These structures combine 2D GO and 1D SWNTs so that GO sheets are interleaved and connected with nanotubes. At that synergistic effect is possible and resulting characteristics are improved compared to those of individual components. Various methods are proposed to obtain such composites: deposition from suspension under pressure (spray method), vacuum filtration, layer-by-layer deposition, and chemical vapor deposition.

It is expected that GO-SWNTs composite material has a significant potential for use as transparent conductive electrodes in solar cells, active electrodes in supercapacitors or in lithium-ion batteries. The electrical conductivity of GO-SWNTs composites as well as of separate GO/rGO or SWNTs films and networks is a prominent topic of research [6–14]. Mechanism of conductivity can essentially depend on structural peculiarities such as film thickness and alignment of GO or SWNTs and also on the oxidation degree of GO and conductivity type of SWNTs. Variable-range hopping (VRH) [15] and fluctuation-induced tunneling (FIT) [16] conductivity models elaborated for disordered systems are used for approximation of GO-SWNTs and SWNTs conductivity temperature dependence.

The main task of the present work is the study of electronic transport in GO-SWNTs composite film in the temperature range of 5–291 K. The electrical resistance temperature dependence was analyzed using VRH model and also compared with that of SWNTs film. The films were obtained by vacuum filtration method and contained strongly oxidized GO and SWNTs with prevailing content of semiconducting species. The physical characteristics of the films were studied earlier in more detail [17].

9.2 Experimental Details

9.2.1 Materials

Graphene oxide synthesized by chemical oxidation of graphite using the modified Hummers' method was purchased from Graphenea (San Sebastian, Spain). In GO the content of carbon and oxygen was 49–56% and 41–50%, respectively (so the C/O ratio was 1.2–1.3). The SWNTs grown by chemical vapor deposition CoMoCAT method were purchased from SouthWest NanoTechnologies (USA). The SWNT sample contained predominantly semiconducting nanotubes (~95%), among them, nanotubes of chirality (6, .5) prevailed.

9.2.2 Methods

The preparation method of aqueous suspensions of GO-SWNTs hybrids is based on the ultrasound treatment (60 min) of the mixture of the aqueous suspension of GO (mainly monolayers [4]) with carbon nanotubes added to this suspension. The GO concentrations in suspension were ~0.2 mg/ml while the weight ratio of GO:SWNT was 1:1. The composite GO-SWNTs films were obtained from aqueous suspensions of GO-SWNTs (0.5–2 ml) deposited on the PTFE membrane (diameter 12.5 mm, pores 0.24 μm, Millipore, USA) by vacuum filtration. The thickness of GO-SWNTs film detached from the membrane was ~10 μm. SWNTs film was prepared similarly from SWNTs suspension in acetone and contained SWNT bundles. The low-temperature measurements of the film conductivity in the range of 5–291 K were carried out in a helium cryostat with temperature stabilization of the film kept in the gas helium atmosphere. The samples were film stripes 0.4–1.5 mm wide attached between contacts distanced at 3 mm, the average rate of temperature changes was about 1.5–2 K/min.

9.3 Results and Discussion

9.3.1 Temperature Dependence of GO-SWNTs Composite Film Conductivity

It was found that conductivity of composite GO-SWNTs films depends on the GO:SWNTs ratio. Conductivity was registered for the film with 1:1 ratio and was much smaller or absent at larger ratios. It was concluded that conductivity in obtained composites is largely conditioned by nanotubes, as the pure GO film has shown no conductivity.

The resistance of composite GO-SWNTs film (GO:SWNTs 1:1 ratio) and of SWNTs film was measured in the temperature range of 5–291 K. Data are shown in Fig. 9.1 as well as evaluated conductivities (see Fig. 9.1 inset). Both films demonstrate decrease in conductivity towards low temperatures.

While both films show a negative temperature coefficient of resistance ($dR/dT < 0$), there are certain distinctions in $\sigma(T)$ behavior. For the GO-SWNTs film conductivity decreased more than 150 times with larger and more uniform slope. So, despite that in composite conductivity occurs mainly through SWNTs, the GO also defines resulting electronic transport. This is because in composite film SWNTs contact not only between themselves but also with GO sheets containing sp^2 and sp^3 domains. Such contacts can create additional energy barriers for electronic transport. Also, most likely, conductive sp^2 domains in strongly oxidized GO sheets do not form continuous network and do not provide conductivity, but still can affect the conductivity of bulk film.

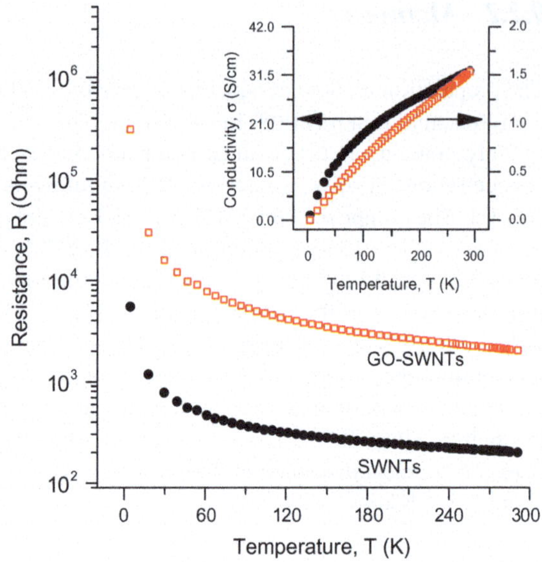

Fig. 9.1 Temperature dependence of resistance of GO-SWNTs (empty squares) and SWNTs (full circles). Inset shows the specific conductivity of both films (same symbols)

We should specifically note that extrapolation of σ(T) to zero temperature yields zero conductivity for both films studied. Such σ(T) dependence means that most conductive paths contain major disordered areas in which electronic transport is "frozen". The VRH model proposed by Mott for disordered semiconductors is more appropriate for an approximation of such conductivity temperature dependence [15].

9.3.2 Approximation of Temperature Dependencies of GO-SWNTs and SWNTs Films with Different Models

Resistance temperature dependence R(T) is often considered when experimental data are fitted with certain electronic transport model. This is partially due to the fact that resistance value changes greatly at low temperatures. In the present study we have tested such models as FIT, VRH and Arrhenius equation for approximation of R(T).

In the FIT model [16] R(T) is described as:

$$R = R_0 \exp[T_1/(T + T_0)], \quad (9.1)$$

where R_0 is a pre-exponential factor, T_1 depends on energy barrier height and width, T_0 is characteristic temperature so that if $T << T_0$ the resistance becomes constant.

In the VRH model [15] R(T) is described as:

$$R = R_0 \exp\left[\left(T_M/T\right)^{\frac{1}{1+d}}\right], \quad (9.2)$$

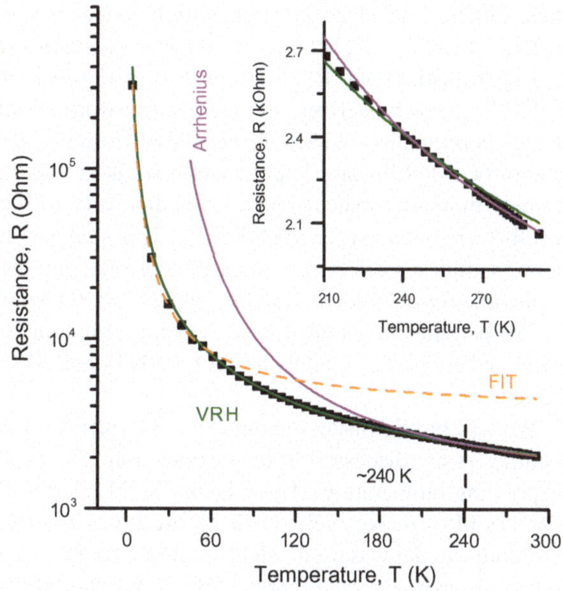

Fig. 9.2 Temperature dependence of GO-SWNTs film resistance fitted with curves according to models described in text. Inset shows high-temperature range where Arrhenius equation is applicable (at T > 240 K)

where R_0 is a pre-exponential factor, T_M is governed by the density of states in the vicinity of Fermi energy $N(\varepsilon_F)$ and localization length of electron wave function ξ, d depends on the transport dimensionality.

The Arrhenius equation for the R(T) is as follows:

$$R = R_0 \exp(E_G/k_B T), \tag{9.3}$$

where E_G is the energy barrier height, R_0 is a pre-exponential factor.

The experimental data for GO-SWNTs film fitted with these three models in different temperature ranges are shown in Fig. 9.2.

It was found that the FIT model (9.1) fits experimental data only in the low-temperature range (5–60 K) and then deviates sharply. In addition, the proper use of FIT model implies that there should be saturation in R(T) growth below certain temperature T_0, which was not observed in our experiment (in [11] FIT model was used at temperatures below 2 K). On the contrary, the VRH model (9.2) yields good fit in much wider temperature range, small deviation starts only from 240 K (see Fig. 9.2 insert), where approximation with Arrhenius equation (9.3) is more precise. The results for the SWNTs film were qualitatively similar. On the basis of comparative analysis performed we have chosen VRH (9.2) and Arrhenius (9.3) models for approximation of our experimental data.

Another important stage of approximation was definition of transport dimensionality in VRH model (parameter d in (9.2)) for our samples. We have considered two values, d = 2 and d = 3 for composite GO-SWNTs, which correspond to two- and three-dimensional electronic transport. It should be noted that for SWNTs films d = 3 [10, 11] and d = 2 [12] were used earlier at studies of electronic transport.

Experimental data, R(T), are plotted in Fig. 9.3 in new coordinates, ln(R) and $T^{-1/(1+d)}$. It can be derived that (9.2) is transformed into equation of a straight line if such coordinates are used. As it follows from Fig. 9.3, at d = 2 experimental data cannot be fitted linearly, the deviation starts at T < 40 K. On the contrary, at d = 3 linear fit approximates experimental data very well up to T ≈ 240 K, and similar results were obtained for SWNTs film. This analysis shows that electronic transport in our samples is effectively three-dimensional and corresponds to 3D VRH model applicable for disordered systems (also called 3D Mott model [15]).

The parameters obtained from the approximation of experimental R(T) data with VRH and Arrhenius models for GO-SWNTs and SWNTs films are summarized in Table 9.1.

We will briefly discuss parameters obtained from Arrhenius model, while VRH parameters are discussed in the next section. The evaluated energies E_G yield corresponding temperatures $T_G = E_G/k_B$ as 212.3 and 130.6 K for GO-SWNTs and SWNTs films respectively. Both T_G are lower than that temperature range, where Arrhenius model was used, which confirms Arrhenius mechanism as thermal activation of electrons over the energy barriers. We should note that similar energy barriers for rGO (10–50 meV) were reported earlier [7].

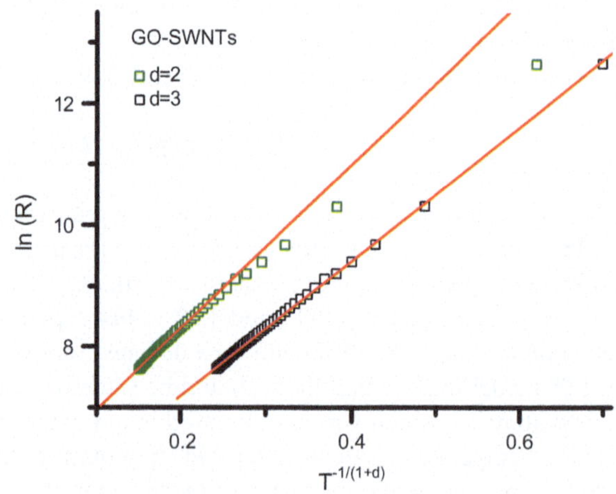

Fig. 9.3 Temperature dependence of GO-SWNTs film resistance in coordinates $T^{-1/(1+d)}$ and ln(R) at d = 2 and d = 3 in VRH model (2)

Table 9.1 The values of parameters obtained in VRH (R_0 and T_M) and Arrhenius (R_0 and E_G) models for GO-SWNTs and SWNTs films

GO-SWNTs	SWNTs		
VRH	Arrhenius	VRH	Arrhenius
R_0 = 130.1 Ohm	R_0 = 999.4 Ohm	R_0 = 34.6 Ohm	R_0 = 128.8 Ohm
T_M = 17449 K	E_G = 18.3 meV	T_M = 2857.4 K	E_G = 11.3 meV

9.3.3 Parameters of Electronic Transport in GO-SWNTs and SWNTs Films Within Framework of VRH Model

VRH model, according to Mott, describes electronic transport that takes place due to tunnelling of electrons between localized states in the potential wells divided by energy barriers [15]. It is also important that additional energy provided by interaction with thermoactivated phonons is needed for tunneling. Probability of tunneling, and consequently conductivity, is defined by the localization length of electron wave function ξ, electron hopping distance r and electron hopping energy W. In this section it is shown that these physical parameters can be derived from fitting parameters obtained from approximation of experimental R(T) data.

As was mentioned before, parameter T_M in (9.2) depends on density of states $N(\varepsilon_F)$ and localization length of electron wave function ξ. It was shown that in case of 3D VRH (d = 3), this relation is [11]:

$$T_M \sim \frac{18.2}{k_B N(\varepsilon_F)\xi^3} \tag{9.4}$$

In order to estimate ξ for our samples we need to know the density of state value. The analysis of literature data was performed for this task. For example, $N(\varepsilon_F)$ of two SWNTs samples [11] were evaluated using (9.4) and yielded 8.5×10^{19} and 7.1×10^{20} eV^{-1} cm^{-3}. As for GO and rGO, data from [18, 19] gives $N(\varepsilon_F)$ as 6×10^{18} and 10^{20} eV^{-1} cm^{-3}. Therefore, we can conclude that $N(\varepsilon_F)$ for different GO and SWNTs samples can vary from ~10^{19} to 7×10^{20} eV^{-1} cm^{-3}. For the ξ estimations we have used three $N(\varepsilon_F)$ values: 5×10^{20}, 10^{20} and 10^{19}, which yield ξ as ~5, 9, 17 nm for the SWNTs film and~ 3, 5, 11 nm for the composite GO-SWNTs film. The first two values for our SWNTs sample roughly correlate with data in [11].

Electron hopping distance r can be defined from ξ [11]:

$$r = 0.38(T_M/T)^{1/4}\xi \tag{9.5}$$

Note that according to (9.5) r should increase towards low temperatures. The ξ and r (calculated at 5 and 240 K) are presented in Table 9.2.

We can conclude from values in Table 9.2 that ξ in case of composite GO-SWNTs film is almost twice lower than that of SWNTs film. This, in addition to higher energy barriers (see data from Arrhenius model), partially explains weaker conductivity in GO-SWNTs film. Most likely, SWNTs contacts with GO sheets create additional and more pronounced barriers that decrease ξ and hamper the electronic transport in composite.

Electron hopping energy was evaluated as [18]:

$$W = \frac{3}{4\pi N(\varepsilon_F) r^3} \tag{9.6}$$

Table 9.2 The estimated values of localization length ξ and hoping distance r obtained in VRH model at different densities of states N(εF) for GO-SWNTS and SWNTs films

Sample	$N(\varepsilon_F)$ (eV^{-1} cm^{-3})	ξ (nm)	r at 5 K (nm)	r at 240 K (nm)
SWNTs	5×10^{20}	5.4	9.9	3.6
	10^{20}	9.2	16.9	6.2
	10^{19}	17	34.7	12.7
GO-SWNTs	5×10^{20}	2.9	8.4	3.1
	10^{20}	5	14.5	5.3
	10^{19}	10.7	31.1	11.4

As the r increases at low temperature, W should decrease which is a consequence of smaller energy of thermoactivated phonons. Decreased W is a reason which makes necessary longer-range tunneling at low temperatures and results in decreased conductivity.

We have plotted the hopping range and hopping energy temperature dependencies for both GO-SWNTs and SWNTs films (using the density of states value $N(\varepsilon_F) = 10^{20}$ eV^{-1} cm^{-3} which gave rather good correlation of calculated ξ with literature data). r(T) and W(T) are presented in Fig. 9.4a and b respectively. It can be seen that in the whole temperature range of 5–240 K r is lower for GO-SWNTs film. The situation is opposite for W. We have noted that at 240 K the value of hopping energy W reached 9 and 14 meV for the SWNTs and GO-SWNTs films respectively, which is only slightly lower energy barriers height evaluated from Arrhenius model (11 and 18 meV). This means that roughly near 240 K VRH tunneling transport mechanism is gradually changed to over-the-barrier electron activation described by Arrhenius equation.

We should also note that for the SWNTs film the condition $W(T) < k_BT$ is fulfilled in the whole temperature range. This is also the case for GO-SWNTs film, except for the range 5–60 K in which $W(T) \approx k_BT$. This observation implies the viability of VRH model as the hopping energy should be of the same value order or less than thermal energy. The W(T) dependencies calculated using other $N(\varepsilon_F)$ values, namely, 5×10^{20} and 10^{19} eV^{-1} cm^{-3}, have shown that for our samples the $N(\varepsilon_F)$ most likely lies in the interval $10^{20} - 5\times10^{20}$ eV^{-1} cm^{-3}.

9.4 Conclusions

The analysis of resistance of composite film GO-SWNTs (components ratio 1:1) containing strongly oxidized graphene oxide and nanotubes with prevailing content of semiconducting species was performed in the temperature range of 5–291 K. It was shown that conductivity of composite film is due to nanotubes present, as the GO film had no conductivity and SWNTs film displayed rather large conductivity. The

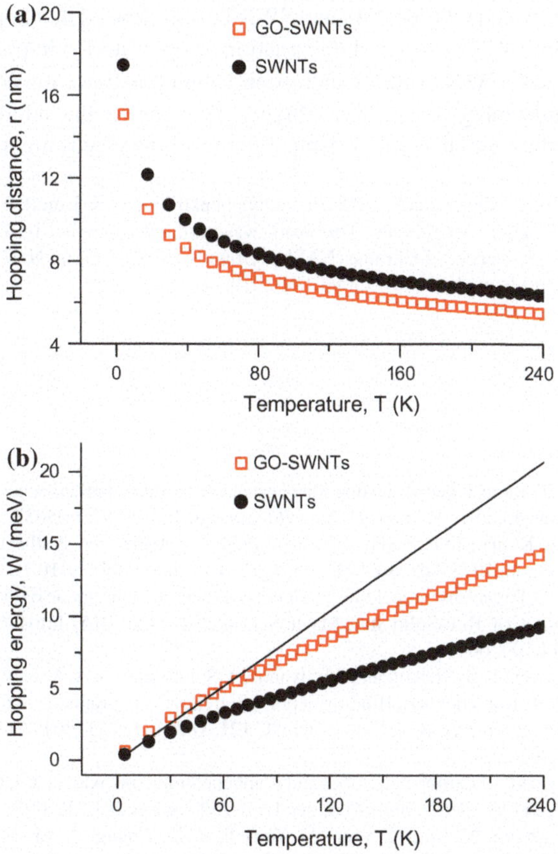

Fig. 9.4 Temperature dependencies of hopping distance r (**a**) and hopping energy (**b**) calculated for the GO-SWNTs (empty squares) and SWNTs (full circles) films. The straight line in Fig. 9.4b denotes the thermal energy k_bT

decrease of conductivity towards low temperature is similar to conductivity behavior in disordered semiconductor systems.

Three-dimensional variable-range hopping model of electronic transport (3D Mott model) based on tunneling of electrons between localized states near Fermi energy was proved applicable for the resistance of both GO-SWNTs and SWNTs films in the low-temperature range of 5–240 K. The Arrhenius model was used for the higher temperatures (T > 240 K). Approximation of experimental data using these models resulted in evaluation of such parameters of electron transport as localization length ξ, hopping distance r, hopping energy W and energy barriers E_G. The density of states near Fermi energy $N(\varepsilon_F)$ was estimated to be in the interval $10^{20} - 5\times10^{20}$ eV^{-1} cm^{-3}.

It was shown that at temperature increase from 5 to 240 K hopping distance r is decreased in three times for both GO-SWNTs and SWNTs films. The comparison of r(T) and W(T) between GO-SWNTs and SWNTs film showed that hopping distance r is larger for the SWNTs film, while situation is opposite for hopping energy W. Also, in case of GO-SWNTs film localization length ξ is twice lower comparing to SWNTs film while energy barrier E_G is higher. This implies that additional contacts between GO surface and nanotubes hamper the resulting electronic transport.

Acknowledgements Authors thank A.M. Plokhotnichenko and A.S. Linnik for the help provided in low-temperature experiments. This work was partially supported by funding from the National Academy of Sciences of Ukraine (NASU) (Grant N 15/19-H, Grant N 07-01-18/19). N.K. acknowledges support from the NASU Grant 1/H-2019.

References

1. C.K. Chua, M. Pumera, Chemical reduction of graphene oxide: a synthetic chemistry viewpoint. Chem. Soc. Rev. **43**, 291–312 (2014). https://doi.org/10.1039/C3CS60303B
2. R.K. Singh, R. Kumar, D.P. Singh, Graphene oxide: strategies for synthesis, reduction and frontier applications. RCS Adv. **6**, 64993–65011 (2016). https://doi.org/10.1039/C6RA07626B
3. S.S. Nanda, G.C. Papaefthymiou, D.K. Yi, Functionalization of graphene oxide and its biomedical applications. Crit. Rev. Solid State Mater. Sci. **40**, 291–315 (2015). https://doi.org/10.1080/10408436.2014.1002604
4. M.V. Karachevtsev, S.G. Stepanian, AYu. Ivanov, V.S. Leontiev, V.A. Valeev, O.S. Lytvyn, L. Adamowicz, V.A. Karachevtsev, Binding of polycitydylic acid to graphene oxide: spectroscopic study and computer modeling. J. Phys. Chem. C **121**, 18221–18233 (2017). https://doi.org/10.1021/acs.jpcc.7b04806
5. H.X. Kong, Hybrids of carbon nanotubes and graphene/graphene oxide. Curr. Opin. Solid State Mater. Sci. **17**, 31–37 (2013). https://doi.org/10.1016/j.cossms.2012.12.002
6. Q. Zheng, B. Zhang, X. Lin, X. Shen, N. Yousefi, Z.-D. Huang, Z. Li, J.-K. Kim, Highly transparent and conducting ultralarge graphene oxide/single-walled carbon nanotube hybrid films produced by Langmuir-Blodgett assembly. J. Mater. Chem. **22**, 25072–25082 (2012). https://doi.org/10.1039/C2JM34870E
7. V. Skakalova, A.B. Kaiser, Y.-S. Woo, S. Roth, Electronic transport in carbon nanotubes: From individual nanotubes to thin and thick networks. Phys. Rev. B **74**, 085403–1–10 (2006). https://doi.org/10.1103/physrevb.74.085403
8. G. Eda, C. Mattevi, H. Yamaguchi, H. Kim, M. Chhowalla, Insulator to semimetal transition in graphene oxide. J. Phys. Chem. C **113**, 15768–15771 (2009). https://doi.org/10.1021/jp9051402
9. V. Skakalova, V. Vretenar, L. Kopera, P. Kotrusz, C. Mangler, M. Mesko, J.C. Meyer, M. Hulman, Electronic transport in composites of graphite oxide with carbon nanotubes. Carbon **72**, 224–232 (2014). https://doi.org/10.1016/j.carbon.2014.02.006
10. C. Morgan, Z. Alemipour, M. Baxendale, Variable range hopping in oxygen-exposed single-wall carbon nanotube networks. Phys. Stat. Sol. A **205**, 1394–1398 (2008). https://doi.org/10.1002/pssa.200778113
11. M. Salvato, M. Lucci, I. Ottaviani, M. Cirillo, S. Orlanducci, E. Tamburri, V. Guglielmotti, F. Toschi, M.L. Terranova, M. Pasquali, Low temperature conductivity of carbon nanotube aggregates. J. Phys.: Condens. Matter **23**, 475302–1–7 (2011). https://doi.org/10.1088/0953-8984/23/47/475302
12. S. Ravi, A.B. Kaiser, C. Bumby, Charge transport in surfactant-free single walled carbon nanotube networks. Phys. Stat. Sol. B **250**, 1463–1467 (2013). https://doi.org/10.1002/pssb.201300033

13. H.-J. Kim, D. Kim, S. Jung, S.N. Yi, Y.J. Yun, S.K. Chang, D.H. Ha, Charge transport in thick reduced graphene oxide film. J. Phys. Chem. C **119**, 28685–28690 (2015). https://doi.org/10.1021/acs.jpcc.5b10734
14. R. Cheruku, D.S. Bhaskaram, G. Govindaraj, Variable range hopping and relaxation mechanism in graphene oxide sheets containing sp3 hybridization induced localization. J. Mater. Sci.: Mater. Electron. **29**, 9663–9673 (2018)
15. N.F. Mott, E.A. Davis, *Electronic Processes in Non-Crystalline Materials* (Clarendon, Oxford, 1979), pp. 1–608
16. P. Sheng, Fluctuation-induced tunneling conduction in disordered materials. Phys. Rev. B **21**, 2180–2195 (1980). https://doi.org/10.1103/PhysRevB.21.2180
17. V.A. Karachevtsev, A.M. Plokhotnichenko, M.V. Karachevtsev, A.S. Linnik, N.V. Kurnosov, Composite films of single-walled carbon nanotubes with strong oxidized graphene: characterization with spectroscopy, microscopy, conductivity measurements (5–291 K) and computer modeling. Low Temp. Phys. **45**(7), 881–891 (2019). https://doi.org/10.1063/1.5111303
18. B. Muchharla, T.N. Narayanan, K. Balakrishnan, P.M. Ajayan, S. Talapatra, Temperature dependent electrical transport of disordered reduced graphene oxide. 2D Materials **1**, 011008-1–011008-10 (2014). https://doi.org/10.1088/2053-1583/1/1/011008
19. R. Kumar, A. Kaur, I.E.T. Circ, Device Syst. **9**, 392 (2015)

Chapter 10
Structure and Properties of Nitride Coatings Based on Ti and Cr Synthesized by PIII&D Technique

V. V. Vasyliev, A. I. Kalinichenko, F. F. Komarov, S. V. Konstantinov, A. A. Luchaninov, E. N. Reshetnyak and V. E. Strel'nitskij

Abstract The results of studies of the structure, stress, and hardness of Ti–N, Cr–N and Ti–Cr–N coatings deposited by the PIII&D method using the rectilinear vacuum-arc plasma filter are presented. X-ray structural analysis was used to study the effect of the pulse bias potential amplitude in the range from 0.5 to 1.5 kV on the characteristics of the coatings. In accordance with the composition of the cathode, the coatings have the nanocrystalline cubic structure TiN, CrN or solid solution (Ti, Cr)N (structural type NaCl) with axial texture [110]. The content of Cr in (Ti, Cr)N coatings is 1.5 times higher than in the cathode. It was found that the change of the pulse bias potential amplitude practically does not affect the elemental composition of the coatings. The level of compression stress and the size of crystallites in (Ti, Cr)N coatings are less than in TiN coatings obtained under similar conditions. All TiN and (Ti, Cr)N coatings have a high hardness of 29–36 GPa.

10.1 Introduction

Vacuum arc coatings based on transition metal nitrides are widely used to protect and harden the surface of materials. TiN is a very common coating that can provide a high degree of wear resistance and is widely used as a coating for tools. CrN is characterized by a lower coefficient of friction and increased resistance to oxidation, therefore, it has found application in automotive components and dies for the molding of metals [1, 2]. However, the need for nitride coatings that can be used in more severe and difficult conditions, especially at high temperatures, when exposed to corrosion and ionizing radiation led to the development of multi-element nanostructured nitride coatings [3–5]. A continuous series of solid solutions are formed in

V. V. Vasyliev · A. I. Kalinichenko · A. A. Luchaninov · E. N. Reshetnyak · V. E. Strel'nitskij (✉)
National Science Center "Kharkov Institute of Physics and Technology", Kharkiv, Ukraine
e-mail: strelnitskij@kipt.kharkov.ua

F. F. Komarov · S. V. Konstantinov
A. N. Sevchenko Institute of Applied Physics Problems, Belarusian State University, Minsk, Belarus

A. D. Pogrebnjak and O. Bondar (eds.), *Microstructure and Properties of Micro- and Nanoscale Materials, Films, and Coatings (NAP 2019)*,
Springer Proceedings in Physics 240, https://doi.org/10.1007/978-981-15-1742-6_10

nanocrystalline coatings by combining solid metal nitrides with unlimited solubility (TiN, TaN, ZrN, NbN, CrN, etc.). Usually, there is no significant grain refinement or abrupt hardening of the coating. However, the resulting compositions combine the advantages of individual systems and provide increased resistance to different types of wear, heat resistance, corrosion resistance, etc. [6–10]. For example, the addition of Cr to TiN coating reduces friction and increases oxidation resistance [8].

The peculiarity of the arc discharge is the presence in the plasma stream of drops of cathode material—macroparticles which, fall into the coating, break its uniformity and lead to a decrease in hardness, an increase of surface roughness, deterioration of adhesion, which can adversely affect the performance characteristics of coatings. The size and number of macroparticles depend on the material of the cathode, its temperature, the type of discharge current (stationary or pulsed) and its value, the speed of movement of the cathode spot, the gas pressure, etc. Typically, the size of the macroparticles in the coatings is in the range of 0.1–10 μm [1, 2].

The presence of macroparticles is permissible for many industrial applications of vacuum-arc technology, so most commercial and experimental vacuum-arc coating installations for nitride coatings are not equipped with special filters that allow purification of plasma from macroparticles. A fundamentally different situation occurs when obtaining high-quality nanostructured coatings, the creation of which is impossible in the presence of macroparticles. To solve this problem magnetoelectric filters have been developed, the principle of which is based on the separation in the plasma-optical systems of charged plasma components (ions, electrons) and neutral (droplets, neutral atoms). The acceleration and change of the direction of ion motion are carried out in a self-consistent radial electric field arising in a "magnetized" plasma. Under the action of electromagnetic fields, plasma ions move along curvilinear trajectories along the magnetic field lines (the plasma guide axis) from the cathode to the deposition surface. The droplets do not feel the fields, they move in a straight line, they are captured by the walls and screens of the plasma guide and do not fall on the substrate [1].

To further improve the characteristics of nitride coatings one can to supply high-voltage pulsed bias potential to substrate. That is the basis of the plasma immersion ion implantation and deposition (PIII&D) method [11]. With this approach intense ion bombardment of the surface occurs during the coating deposition which leads to the synthesis of more dense nanostructured layers and allows to adjust the level of residual stress in the coatings. Pulse parameters are key characteristics that allow controlling the structure and properties of the coating. The role of the amplitude, frequency and duration of the pulse bias potential during PIII&D synthesis was studied for TiN and Ti–Al–N based coatings [12–15], but there are only a few studies in the literature concerning Ti–Cr–N coatings.

The purpose of this study is to study the composition, structure, residual stress and hardness of Ti–Cr–N coatings obtained by the PIII&D method using various values of pulse bias voltages, and comparing them with the characteristics of TiN and CrN coatings.

10.2 Experiment Details

Nitride coatings were deposited by the vacuum-arc installation "Bulat-6" equipped with a rectilinear type plasma filter, a high-voltage pulse generator to supply bias potential to the substrate; an inverter power source for the vacuum arc, which provides high stability of the arc discharge in the conditions of wide variation of the voltage drop on the arc. The absence of direct visibility between the plasma source and the substrate in the design of the rectilinear type filter is achieved by placing the screen on the axis of the rectilinear cylindrical plasma line [16, 17]. The main advantages of this source are the following: high productivity; stability of parameters regardless of the degree of the cathode burnup; ensuring uniform thickness of coating over a large area at the high quality of plasma purification from macroparticles; structural simplicity, low cost of manufacture compared to known sources with curvilinear filters. The main characteristics of the filter are as follows: cathode diameter is 60 mm; arc current is 50–100 A; magnetic field strength is 200 E; voltage drop on the arc is 35–60 V; ionic current at the outlet of the filter is up to 5 A. Diameter of the region of uniform deposition of TiN-based coatings at the distance of 150–180 mm is 180 mm at the deposition rate up to 20 μm/h. Relative deviation in thickness and composition does not exceed ±5%. Improved vacuum-arc evaporator with a cylindrical cathode, special construction of screens that capture microparticles, and the original configuration of the magnetic field that transports the plasma stream, provides the filter transmittance up to 50%.

The photo and the scheme of installation of the experimental deposition of the coating are presented in Fig. 10.1. The rectilinear macroparticle filter consists of two cylindrical sections: the anode and the plasma guide, both with external magnetic coils. The so-called "magnetic island" is inside the anode section. This element

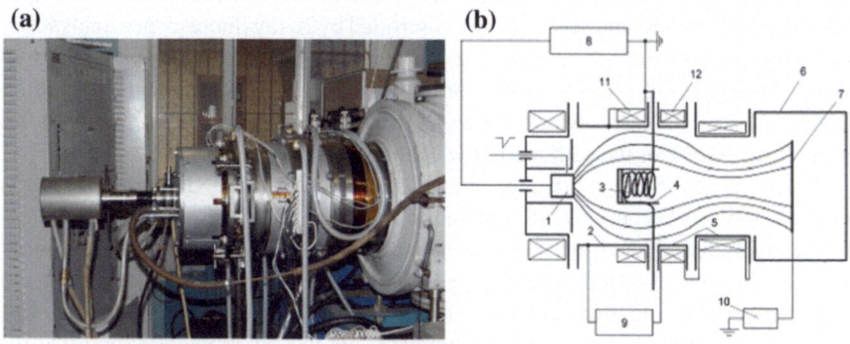

Fig. 10.1 Photo (**a**) and scheme (**b**) of the vacuum arc unit for coatings deposition with plasma filter of rectilinear type: 1—cathode; 2—anode; 3—additional coil of "magnetic island"; 4—"magnetic island" casing; 5—input and output parts of the plasma guide; 6—vacuum chamber; 7—substrate holder; 8—power supply of the vacuum arc; 9—power supply of the plasma guide; 10—generator of high-voltage pulses; 11—additional electromagnetic anode coil; 12—additional electromagnetic coil of the plasma guide

has the additional magnetic coil, the body of which prevents the direct passage of macroparticles from the cathode surface to the filter exit. Additional coils included in the power supply of the arc discharge area at the anode and inside the "magnetic island".

The instantaneous distribution of magnetic field lines is shown in Fig. 10.1. The direction of the magnetic field created by the outer coils is opposite to that created by the coil of the magnetic island, which leads to the curvature of the field lines. The magnetic field of the additional coils improves the conditions for the transport of charged plasma components through the filter. As the plasma jet approaches the anode or magnetic island, the current through the coil is increased and an additional magnetic field is created that pushes the plasma jet away from the surface of the anode or magnetic island. As a result, plasma losses are significantly reduced.

For the synthesis of coatings, the cathodes of the technically pure Ti, Cr, alloys $Ti_{0.8}Cr_{0.2}$ and $Ti_{0.6}Cr_{0.4}$ were used. The deposition of nitride coatings was carried out in the presence of the reaction gas—nitrogen at a pressure of 0.1 Pa. The arc current was 100 A. The coatings with thickness in the range of 4–10 μm were deposited onto polished substrates made of zirconium alloy (Zr with 3% Nb) and stainless steel (12X18H10T). The substrate was under the negative pulse bias potential with amplitude U in the range of 0–1.5 kV. Under the TiN coating synthesis, the pulse duration t_p and their frequency f was of 5 μs and of 24 kHz, respectively. In the intervals between pulses, the substrate was under the "floating" potential not exceeding 30 V. During the CrN and TiCrN coatings synthesis t_p and f were 6 μs and 12 kHz, respectively. The surface temperature of the coating deposition was not forcibly regulated and is controlled by the Raynger MX4 optical infrared pyrometer. Radiation heating of the surface did not exceed 300 °C.

The structure and elemental composition of the coating surface were investigated by the JEOL JSM-6360 LA scanning electron microscope with JED 2200 X-ray Spectral Energy Inspection Microanalysis System. In addition, the elemental composition of multi-component coatings was controlled by X-ray fluorescence analysis on a vacuum spectrometer SPRUT-VM. According to the intensity of the corresponding characteristic lines, the values of the concentration of metal components (excluding nitrogen) were calculated, which allows us to estimate the degree of reproducibility of the composition of the cathode in the coatings.

Structural investigations of the samples are conducted via XRD measurements by diffractometer DRON-3. CuKα radiation is used in all investigations. XRD measurements are conducted in Bragg–Brentano (θ–2θ) configuration. The crystallite size (i.e., the zone of coherent scattering) in the nitride films was calculated from the (220) peak broadening using the Scherer relation. Intrinsic stress in the coatings was measured by the X-ray tensometry technique using the $\sin^2\psi$ method modified for textured films [1, 18].

Hardness and Young's modulus of the TiN coatings are determined by using a nanoindenter MTS G200. To exclude the influence of the substrate on the results of the analysis, the value of the mechanical characteristics was determined by indentation to the depth of 300 nm that did not exceed 10% of the thickness of the coatings.

10.3 Results and Discussion

Experiments on the coating deposition show that the filtered vacuum-arc plasma source of rectilinear type works stably with alloy cathodes of various compositions and provides high-quality filtration from macroparticles of the plasma stream. The principal differences depending on the composition of the cathodes were not detected. Figure 10.2 shows the microscopic images of the original surface of the polished steel substrate (Fig. 10.2a) and the surface of TiN coatings. The relief of the surface of the coating deposited without plasma filtration is defective with a large number of droplets of various sizes (Fig. 10.2b). The surface of the coating deposited using a filter (Fig. 10.2c) is smooth and reproduces the relief and defects of the substrate.

A typical electron microscopic image of a transverse fracture of a specimen (Fig. 10.3) indicates that there is no droplet not only on the surface but also in the volume of the coating. It can be seen that the coatings are uniform in thickness and consist of nanoscale structural elements. Electron microscopic images of the coating cross-section indicate that all coatings consist of narrow ($\leq$100 nm) columnar grains. The structure of the coating cross-section is dense and has a staircase appearance.

It was found that with identical parameters of the synthesis process the deposition rate of CrN coatings was 2 times lower than the others and did not exceed 4 μm/h. It

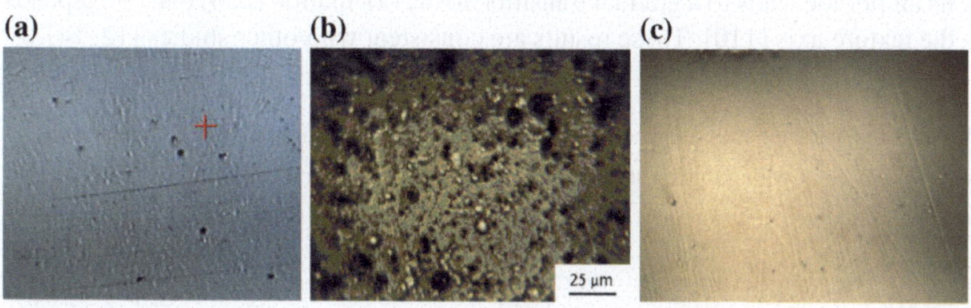

Fig. 10.2 Microscopic images of the initial surface of the tool steel substrate (**a**) and the surface of TiN coatings deposited on such substrate of unfiltered (**b**) and filtered plasma (**c**)

Fig. 10.3 Electron microscopic image of the cross section of TiN coating

may be due to the low rate of the cathode spot movement on the surface of the chromic cathode. At the same time, the change in the parameters of the substrate pulsed bias potential had an insignificant effect on the coatings deposition rate. For example, the Ti–Cr–N coating deposition rate decreased by only 12% when the amplitude of the pulsed potential increased from 0.5 to 1.5 kV.

Table 10.1 shows the parameters of deposition and characteristics of nitride coatings deposited using cathodes of various compositions. The samples #1–4 were deposited from the titanium cathode at different values of the amplitude of the high-voltage pulsed bias potential U in the range of 0–1.5 kV. The only crystalline phase in all coatings deposited from the titanium cathode was TiN with a cubic structure of the NaCl type and texture of the axial type. The content of nitrogen in the coatings was close to the stoichiometric ratio (the ratio of atoms N/Ti is 1 ± 0.1).

Figure 10.4a displays X-ray diffraction patterns of samples with TiN coatings. It can be seen that an increase in the amplitude leads to a change in the ratio of the intensity of the diffraction lines indicating the change in the preferential orientation of the nitride crystallites in the coatings. At the floating bias potential ($U = 0$) in the coating, the crystallite orientation is dominated by (111) planes parallel to the coating surface that corresponds to the texture axis [111]. With the increase in pulse amplitude, the intensity of reflection (111) decreases whereas the intensity of reflection (220) increases. Thus, the supply of the pulse bias potential and the growth of its amplitude leads to a gradual transition to the orientation (220) that corresponds to the texture axis [110]. These results are consistent with other studies [12, 14].

Table 10.1 Characteristics of nitride coatings deposited from cathodes of various composition

#	U (kV)	Coating thickness (μm)	Coating composition	H (GPa)	E (GPa)	H/E
Cathode Ti						
1	0	7.1	TiN	36.1	401	0.09
2	0.5	0.5	TiN	36.2	396	0.09
3	1.0	1.0	TiN	32.0	404	0.08
4	1.5	1.5	TiN	32.1	375	0.09
Cathode Cr						
5	0.5	3.5	CrN	19.3	195	0.10
Cathode $Ti_{0.8}Cr_{0.2}$						
6	0.5	5.6	$Ti_{0.71}Cr_{0.29}N$	30.4	331	0.09
7	1.0	5.1	$Ti_{0.65}Cr_{0.35}N$	32.2	392	0.08
8	1.5	3.9	$Ti_{0.70}Cr_{0.30}N$	28.6	286	0.10
Cathode $Ti_{0.6}Cr_{0.4}$						
9	0.5	7.6	$Ti_{0.45}Cr_{0.55}N$	31.6	359	0.09
10	1.0	7.4	$Ti_{0.45}Cr_{0.55}N$	29.0	333	0.09
11	1.5	6.8	$Ti_{0.45}Cr_{0.55}N$	29.5	352	0.07

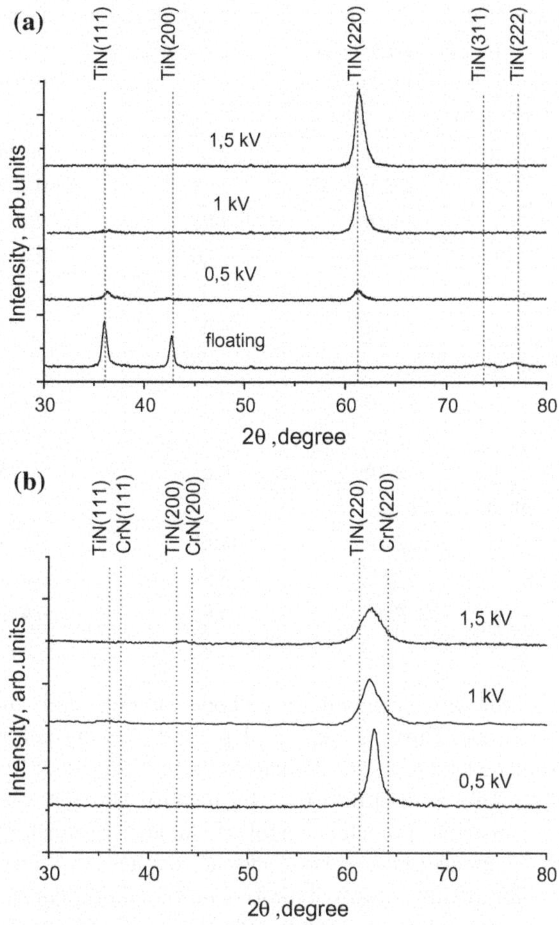

Fig. 10.4 The diffractograms of the coatings obtained at various values of the amplitude of the pulse bias potential on the substrate: **a**—TiN coating; **b**—(Ti, Cr)N coating

The results of determining the parameters of the crystalline structure in the coatings are given in Table 10.2. For the TiN coating deposited at the floating bias potential, the crystallite size is 27 nm. The supply of the pulse bias potential leads to the decrease of crystallite size to 12–13 nm. The parameter of the TiN crystal lattice normal to the surface of the film is in the range of 0.425–0.430 nm.

The resulting values of the crystal lattice parameter exceed the value of 0.4243 nm which is characteristic of the massive TiN stoichiometric composition. This feature is associated with the stress formation in the coatings due to the ion bombardment of the growth surface. The results of determining the residual stress in coatings show that with increasing the amplitude of the pulsed bias potential *U* in the range 0–1.5 kV the level of compression stress in TiN coatings increases initially from 5.2 to 9.9 GPa and then (at U > 500 V) decreases to 5.7 GPa.

The alloy cathodes $Ti_{0.8}Cr_{0.2}$ and $Ti_{0.6}Cr_{0.4}$ were used for deposition of Ti–Cr–N coating (samples #6–11), but the chromium content in the synthesized coatings was

Table 10.2 Results of X-ray structure analysis of nitride coatings

#	U (kV)	Preferential orientation	Crystallite size (nm)	Crystal lattice parameter (nm)	Compression stress (GPa)
Cathode Ti					
1	0	(111)	27.2	0.4264	5.2
2	0.5	(220)	12.3	0.4277	9.7
3	1	(220)	12.1	0.4255	7.9
4	1.5	(220)	13.0	0.4252	5.7
Cathode Cr					
5	0.5	(220)	5.0	0.4240	
Cathode $Ti_{0.8}Cr_{0.2}$					
6	0.5	(220)	7.3	0.4258	
7	1	(220)	5.5	0.4265	
8	1.5	(220)	4.8	0.4252	
Cathode $Ti_{0.6}Cr_{0.4}$					
9	0.5	(220)	10.5	0.4194	3.0
10	1	(220)	5.5	0.4223	6.1
11	1.5	(220)	4.8	0.4224	6.1

1.5 times higher than in the cathode and practically did not change with increasing amplitude. Thus, for coatings deposited from the cathode $Ti_{0.8}Cr_{0.2}$ the Cr content in Ti made up about 30 at.% and in coatings deposited from the cathode $Ti_{0.6}Cr_{0.4}$ – 55 at.%. This result differs from the data for Ti–Al–N system coatings for which when the pulse bias potential is applied, the good correspondence between the content of the elements in the cathode and the coatings is achieved [13, 15].

The diffractograms of the Ti–Cr–N coatings deposited from the cathode $Ti_{0.6}Cr_{0.4}$ at various values of the amplitude of the pulsed bias potential are shown in Fig. 10.4b. When using the $Ti_{0.8}Cr_{0.2}$ cathode the diffractograms are similar to the above ones. Chromium has an unlimited solubility in a crystalline lattice TiN and forms a continuous series of solid solutions. The lines of nitride on the diffractogram are located between the corresponding positions of the lines for TiN and CrN. The values of the crystal lattice parameters for cubic nitrides of the stoichiometric composition of TiN, CrN are 0.424 nm, 0.415 nm, respectively. Depending on the composition of the multicomponent coatings the value of the period of the crystalline lattice of nitride is changed almost in accordance with the Wegard rule which indicates the formation of (Ti, Cr)N solid solutions in the coatings. Similar to the TiN coating the application of the pulsed potential leads to the formation of strong axial texture in the coating [110].

Thus, irrespective of the composition of the cathodes used the only crystalline phase in the coatings was nitride with the cubic lattice of the type NaCl and preferred orientation (220). In Fig. 10.5 shows sections of diffractograms of coatings of the various composition containing the line (220). In the diffractogram of CrN coating,

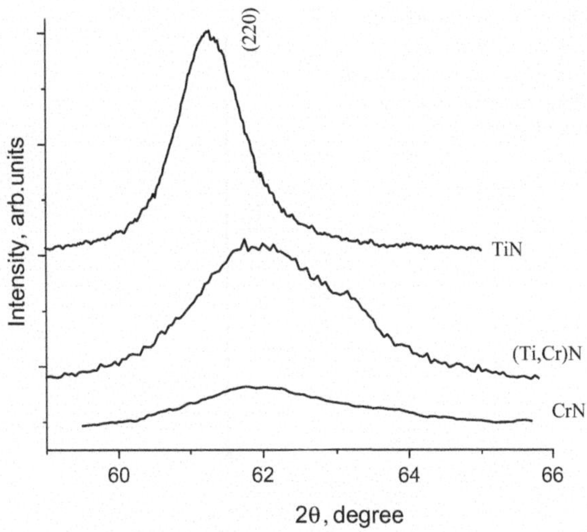

Fig. 10.5 Part of X-ray diffraction patterns of vacuum-arc coatings of various elemental composition, deposited with pulse bias potential amplitude 0.5 kV

the nitride line is very weak and wide, in comparison with other coatings. This indicates the presence of a significant number of an amorphous phase. Coating CrN has an amorphous-crystalline structure with a crystalline size of 5–7 nm. The addition of Ti to CrN contributes to the crystallization of coatings, the growth of crystalline size up to 10 nm and stabilizes the nanocrystalline state of a solid solution. This result is similar to the effect of adding yttrium to amorphous-crystalline coatings (Ti, Al)N [13].

The change in amplitude does not cause fundamental differences in the phase composition and microstructure of the (Ti, Cr)N coatings. However, unlike the TiN coatings, an increase in amplitude results in a decrease in the intensity and expansion of the line (220). The results of determining the residual stress and the crystallite size in coatings (Ti, Cr) N are shown in Table 10.2. The crystallite size of nitride does not exceed 10 nm and decreases to 5 nm with increasing amplitude. The level of compression stress in the coatings increases from 3 to 6 GPa. Unlike TiN coatings, there is no decrease in stress at $U > 0.5$ kV.

Some coatings were deposited in one experiment simultaneously on the substrates of steel and zirconium alloy. The substrate type did not affect the general appearance of the diffraction pattern. However, the diffraction lines of nitride were shifted. Figure 10.6 shows fragments of diffractograms containing line (220) for coatings (Ti, Cr) N deposited on steel and zirconium substrates. It can be seen that for the specimen on the steel substrate the line (220) is shifted to smaller angles. That is, for this coating the value of the crystal lattice parameter in the direction normal to the surface is greater. A similar difference was also found for CrN coatings.

The differences are due to the difference in the level of residual stress in the coatings. The general level of stress in coatings is considered to be the sum of two components: structural and thermal [18]. The structural compression stress σ_s arises from ion bombardment during deposition. The thermal component of the stress σ_t

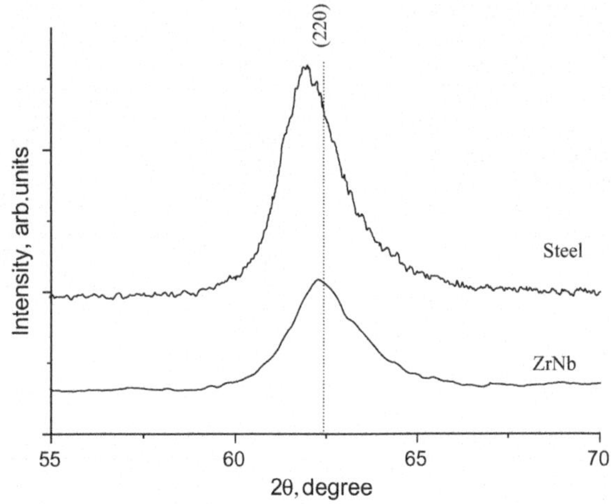

Fig. 10.6 Diffractogram fragments of coatings $(Ti_{0.45}Cr_{0.55})$ N obtained on different substrates at a bias potential of 1 kV

is due to the difference in the coefficients of thermal expansion of the film (α_1) and the substrate (α_2):

$$\sigma_t = \frac{E}{1 - \upsilon}(\alpha_1 - \alpha_2)\Delta T, \tag{10.1}$$

where E and υ are Young's modulus and the Poisson's ratio of coating material; ΔT is the difference between the temperature of the deposited coating and temperature of measurement. Estimation of the residual stress of thermal origin for coating TiN deposited at a temperature of 300 °C on the steel substrate gives −0.4 GPa. That is, the thermal compression stress is added to the intrinsic one after cooling of the sample to room temperature. In the case of the zirconium substrate the value of the stress is +0.6 GPa. The tensile stress of thermal origin reduces the overall level of residual stress. Thus, the stresses in the samples on the steel substrates can be significantly higher (up to 1 GPa) than on the substrates of zirconium alloy. This feature should be taken into account in the manufacture of nitride coatings on zirconium parts working at elevated temperatures because, when heated, the overall level of stress will increase which may lead to the destruction of the coating.

It is known that the pulse parameters (f—frequency, t_p—duration, U—amplitude) are the key characteristics that allow controlling the structure of the coating, the level of residual stress, and hence their properties [11]. This approach is commonly used to reduce the high compression stress that is characteristic of vacuum arc coatings and can cause low operating properties and even their destruction [19, 20]. It was found that supply to the substrate of 20 μs pulses leads to the stress decrease in the TiN and AlN coatings at ion energies exceeding 500 eV. The reduction can be achieved both by changing the bias potential and the pulse frequency [13]. In order to compare the effect of the potential on the characteristics of coatings it is advisable to consider them depending on the parameter Uft_p which characterizes average value of the displacement potential during deposition, that is, the power of ion bombardment and the radiative heating of the surface. Figure 10.7 shows the crystallite size and the residual stress in some nitride coatings. It is seen that the crystallite size and stress in the coating $Ti_{0.45}Cr_{0.55}N$ are lower than in TiN. The level of compression stresses in the TiN coatings varies not monotonically in the range from 5 to 10 GPa depending on the parameter $U f t_p$. The level of compression stresses in the $Ti_{0.45}Cr_{0.55}N$ coatings increases from 3 to 6 GPa with the growth of the parameter $U f t_p$. Unlike TiN coatings there is no stress decrease at $U f t_p > 100$ V.

To explain the patterns of stress formation in nitride coatings that deposited under ion implantation the approach based on the concept of nonlocal thermoelastic peaks (NTPs) that are generated by ions bombarding the surface of deposition is effectively used. To determine the intrinsic stress the modified Davis model is used which assumes that intrinsic stress is formed as a result of two simultaneous, but differently oriented processes: (1) the generation of stress due to the implantation of the incident beam of ions; and (2) stress relaxation due to thermoactivated migration processes of defects (interstitial atoms) in NTP of ions [21, 22]. Calculations within this model indicate that the stress dependence on the pulse amplitude for the TiN and CrN coatings is similar, and the stress levels correlate with the elastic modulus

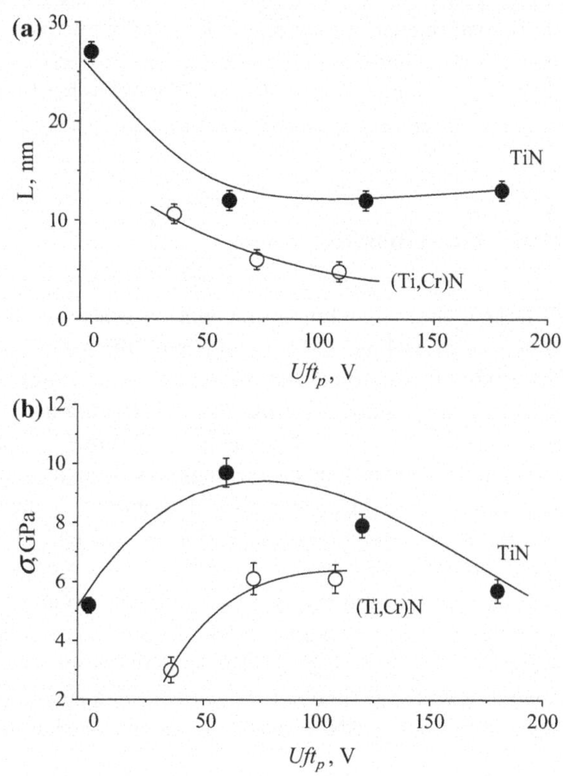

Fig. 10.7 The effect of changing the parameter ($U f{\cdot}t_p$) on the size of crystallites (**a**) and compressive stresses (**b**) in TiN and $(Ti_{0,45}Cr_{0,55})N$ coatings on steel

of the coating material. Within this approach, it is impossible to explain the experimental difference during the curves in Fig. 10.7. The discrepancy can be eliminated taking into account the radiation-acoustic mechanism for accelerating the migration of defects in nanocrystalline materials when interacting with low-energy ions (1–10 keV) which is considered in the paper [23]. The rapid heating of substance in the NTP of a heavy low energy ion during ion-ion relaxation results in the generation of a powerful thermoelastic pulse of subpicosecond duration which has both static and wave components. Acoustic pulses from the Ti^+ ion bombarding the nanocrystalline TiN coating can "sweep out" the defects to the boundary of the nanocrystallites with the size of $d \leq 30$ nm and accelerate their migration in larger nanocrystallites with the size of $d \leq 100$ nm. In materials with relatively low coefficients of thermal expansion, as is the case with CrN coatings, the mechanism of thermoacoustic activation of defect migration is ineffective. It is possible that the lower migration activity of interstitial atoms also causes a smaller crystallite size in the coatings (Ti, Cr)N as compared with TiN.

According to the data of nanoindentation, all TiN coatings were characterized by the sufficiently high hardness of 32–36 GPa regardless of the amplitude of the bias potential which significantly exceeds the values for microcrystalline TiN coatings deposited by the vacuum-arc method without plasma filtration (20–24 GPa). The ratio of H/E was within the range of 0.08–0.09. The CrN coating has the least hardness of 20 GPa. Indeed, the hardness of solid CrN is only 11 GPa which is 2 times less than TiN solid hardness. The addition of Ti to CrN leads to an increase of hardness of (Ti, Cr)N coatings up to 29−31 GPa due to hard soldering. The ratio of H/E for all (Ti, Cr)N coatings is within the range of 0.07–0.1.

10.4 Conclusions

PIII&D method using filtered vacuum-arc plasma of Ti and Cr cathodes was applied to deposit of TiN and CrN coatings with the nanocrystalline cubic structure of the NaCl type. It was found that in the nitride coatings obtained using $Ti_{0.8}Cr_{0.2}$ and $Ti_{0.6}Cr_{0.4}$ alloy cathodes forms cubic solid solution (Ti, Cr)N. It was established that in the coatings (Ti, Cr)N the content of Cr is in 1.5 times higher than in the cathode. The content of nitrogen in the coatings was close to the stoichiometric ratio. Using the high-voltage pulse bias potential with amplitude in the range of 0.5–1.5 kV provides the formation of the strong axial texture of coating with the axis [110] in the direction normal to the surface.

It has been determined that CrN coatings contain a significant amount of X-ray amorphous phase in contrast to TiN. Adding Ti to CrN contributes to better crystallization of (Ti, Cr)N coatings. The level of compressive stress in the coatings varies depending on the parameter Uft_p which characterizes the average bias potential during deposition. For TiN coatings this dependence has the maximum at $Uft_p \approx 70$ V.

Further growth of this parameter leads to a decrease in stress due to relaxation processes. For coatings (Ti, Cr)N the overall stress level is lower, however, even with $Uft_p \approx 110\,\text{V}$ stress reduction is not observed.

Differences in structure and stresses in nitride coatings of TiN and (Ti, Cr)N can be associated with the suppression of the mechanism of thermoacoustic activation of defect migration due to the low thermal expansion coefficient of CrN.

All TiN and (Ti, Cr)N coatings have a high hardness of 29–36 GPa.

Acknowledgements This research was carried out within the framework of the joint scientific project of the National Academy of Sciences of Ukraine and the National Academy of Sciences of Belarus "Creation of radiation-resistant nanostructured metal and nitride coatings based on titanium, aluminum, and chromium by vacuum-arc deposition from filtered plasma", 2018 (Reg. No. of Ukrainian projects: 02-03-18).

References

1. Cathodic Arcs Anders, *From Fractal Spots to Energetic Condensation* (Springer, New York, 2008)
2. R.L. Boxman, D.M. Sanders, P.J. Martin (eds.), *Handbook of Vacuum Arc Science and Technology* (Fundamentals and Applications (Noyes Publications, New York, 1995)
3. A. Cavaleiro, J.T.M. De Hosson (eds.), *Nanostructured Coatings* (Springer, New York, 2006)
4. A.D. Pogrebnjak, A.A. Bagdasaryan, A. Pshyk, K. Dyadyura, Phys.-Uspekhi **60**(6), 586 (2017)
5. K.V. Smyrnova, A.D. Pogrebnjak, V.M. Beresnev, S.V. Litovchenko, S.O. Borba-Pogrebnjak, A.S. Manokhin, S.A. Klimenko, B. Zhollybekov, A.I. Kupchishin, Y.O. Kravchenko, O.V. Bondar, Met. Mater. Int. **24**(5), 1024 (2018)
6. G. Abadias, Surf. Coat. Tech. **202**, 2223 (2008)
7. G. Abadias, L.E. Koutsokeras, S.N. Dub, G.N. Tolmachova, A. Debelle, T. Sauvage, P. Villechaise, J. Vac. Sci. Technol. **28**(4), 541 (2010)
8. D.H. Jung, H.S. Park, H.S. Na, J.W. Lim, J.J. Lee, J.H. Joo, Surf. Coat. Tech. **169–170**, 424 (2003)
9. V.V. Uglov, V.M. Anishchik, S.V. Zlotski, G. Abadias, S.N. Dub, Surf. Coat. Tech. **200**, 178 (2005)
10. V.V. Uglov, V.M. Anishchik, S.V. Zlotski, G. Abadias, S.N. Dub, Surf. Coat. Tech. **202**(11), 2394 (2008)
11. A. Anders, *Handbook of Plasma Immersion Ion Implantation and Deposition* (Wiley, New York, 2000)
12. S.S. Akkaya, V.V. Vasyliev, E.N. Reshetnyak, K. Kazmanlı, N. Solak, V.E. Strel'nitskij, M. Ürgen, Surf. Coat. Tech. **236**, 332 (2013)
13. V.A. Belous, V.V. Vasyliev, V.S. Goltvyanytsya, S.K. Goltvyanytsya, A.A. Luchaninov, E.N. Reshetnyak, V.E. Strel'nitski, G.N. Tolmacheva, O. Danylina, Surf. Coat. Tech. **206**, 1720 (2011)
14. S. Mukherjee, F. Prokert, E. Richter, W. Möller, Surf. Coat. Tech. **200**, 2459 (2005)
15. V. Vasyliev, A. Luchaninov, E. Reshetnyak, V.E. Strel'nitskij, B. Lorentz, S. Reichert, V. Zavaleyev, J. Walkowicz, M. Sawczak, Probl. Atom. Sci. Tech. **106**(22), 244 (2016)
16. I.I. Aksenov, V.E. Strelnytskiy, V.V. Vasylyev, A. Voevodin, J. Jones, J. Zabinski, U.S. Patent 20070187229A1, 3 June 2008
17. V.V. Vasylyev, A.A. Luchaninov, V.E. Strel'nitskij, Probl. Atom. Sci. Tech. **1**, 97 (2014)
18. U. Welzel, J. Ligot, P. Lamparter, A.C. Vermeulenb, E.J. Mittemeijer, J. Appl. Crystallogr. **38**, 1 (2005)

19. M.M.M. Bilek, D.R. McKenzie, R.N. Tarant, S.H.M. Lim, D.G. McCulloch, Surf. Coat. Tech. **156**, 136 (2002)
20. M. Bilek, D. Mc Kenzie, W. Moeller, Surf. Coat. Tech.**186**, 21 (2004)
21. A.I. Kalinichenko, S.A. Kozionov, S.S. Perepelkin, V.E. Strel'nitskij, East Eur. J. Phys. **1**(4), 58 (2014)
22. A.I. Kalinichenko, S.S. Perepelkin, E.N. Reshetnyak, V.E. Strel'nitskij, Probl. Atom. Sci. Tech. **1**, 114 (2018)
23. A.I. Kalinichenko, V.E. Strel'nitskij, in *Advances in Thin Films, Nanostructured Materials, and Coatings*, ed. by A. Pogrebnjak, V. Novosad. 8th IEEE International Conference on "Nanomaterials: Applications & Properties", Odessa, September 2018. Lecture Notes in Mechanical Engineering (Springer, Singapore, 2019), pp. 207–214

Chapter 11
Features of the Microstructure of Multilayered (TiAlSiY)N/MoN Coatings Prepared by CA-PVD and Their Influence on Mechanical Properties

Olga Maksakova, Alexander D. Pogrebnjak, Oleksandr Bondar, Vyacheslav Stolbovoy, Yaroslav Kravchenko, Vyacheslav Beresnev and Pawel Zukowski

Abstract Thin nanolayered coatings composed of the consequent alternation of multi-element (TiAlSiY)N and binary MoN layers were deposited by the cathodic arc deposition. The elemental composition, phase structure, microstructure and mechanical properties of the coatings were studied by well-established experimental methods: SEM, EDS, XRD, TEM, and microindentation. It was found that (TiAlSiY)N/MoN coatings had a complex chemical composition, which preferably consisted of a mixture of Ti, Al and Si nitrides. The preferential crystallographic orientation along (200) plane was found for all samples. TEM results showed that investigated coatings composed of ununiform nano-scale multilayered structures with modulation periods ranged from 20 to 32 nm. The maximum microhardness of the deposited coatings reached 1087H_V0.1.

O. Maksakova (✉) · A. D. Pogrebnjak · O. Bondar · Y. Kravchenko
Sumy State University, Sumy, Ukraine
e-mail: maksakova.tereshenko@gmail.com

V. Stolbovoy
National Science Center Kharkiv Institute of Physics and Technology, Kharkiv, Ukraine

V. Beresnev
V. N. Karazin Kharkiv National University, Kharkiv, Ukraine

P. Zukowski
Lublin University of Technology, Lublin, Poland

A. D. Pogrebnjak and O. Bondar (eds.), *Microstructure and Properties of Micro- and Nanoscale Materials, Films, and Coatings (NAP 2019)*,
Springer Proceedings in Physics 240, https://doi.org/10.1007/978-981-15-1742-6_11

11.1 Introduction

Designing of the multilayered films that improving the properties of the coated materials is a difficult task but in case of its successful fulfillment obtained material gained improved hardness, wear resistance, temperature stability and better performance [1–6]. Synthesis of alternating nanoscale layers consisting of various materials is a particularly promising strategy through the design of atomic-scale architectures. The results reported in works [7–23] prove that multilayered coatings of simple-composition, such as TiN/ZrN, CrN/AlN, ZrN/CrN, TiN/NbN, CrN/MoN, TiN/MoN, TiN/CrN demonstrate the significant improvement of mechanical and tribological properties, as well as better thermal and corrosion resistance compared to the monolayer films.

Nowadays, a novel tendency, in which the multielement and multilayered strategies for the synthesis of functional coatings based on transition metal nitrides, metalloids, and refractory elements are combined, starts to develop widely [24–33]. Chen et al. [34] investigated TiAlSiN-based monolayers and multilayered coatings fabricated by the cathodic arc evaporation. It was shown that the deposition method exceedingly enhanced adhesive strength and toughness. According to the results of milling tests, the lifetime for TiAlSiN coated substrates increased by approximately 172%. Tribology experiments did by Çalışkan's [35] indicated that the nanocomposite TiAlSiN/TiSiN/TiAlN multilayered coating had a higher value of load (L_{C3}) at almost 44% and longer functioning time than single layer TiN and TiAlN coatings. The work-team of the manuscript [36] investigated the TiAlN, TiAlSiN and TiAlN/TiAlSiN multilayered coatings deposited by magnetron sputtering. Obtained results pointed out that multilayered composite demonstrated enhanced adhesion strength compared with TiAlSiN. Therefore, a multilayered concept of TiAlSiN-based coatings arises much interest and can be used for the protection of cutting tools and working parts used in different industrial processes, where the high hardness, wear and corrosion resistance, and thermal stability at over 1100 K are required.

It is suggested that the addition of a low amount of Yttrium to the TiAlSiN alloy will improve the high-temperature oxidation resistance. The nitride of TiAlSiY will maintain high hardness as the formation of YO_2 phase at the grain boundaries is possible. This phase blocks inward diffusion of oxygen and outward diffusion of metal components of the coating because Yttrium has a high affinity towards Oxygen [37–42]. Considering all these advantages, the study of TiAlSiYN-based multilayered coatings presents great scientific interest.

The main focus of this work is concentrated on the detailed microstructural research of recently developed multilayered (TiAlSiY)N/MoN coatings and the establishment of the effect of their structural features on mechanical properties.

11.2 Methods and Experiments

11.2.1 Deposition Technology

Cathodic arc deposition (CA-PVD) is a widely applied method for the deposition of nitride-based coatings. The primary advantage of this technique lies in its ability to produce highly ionized plasma. Ion energies of plasma produced by CA-PVD can be further increased or tuned due to a negative potential applied to the substrate. Application of a direct current substrate bias during deposition leads to the formation of coatings with dense structure and well adhesion. However, this technological parameter increases residual compressive stresses of the coatings that influence on mechanical and tribological properties of the coated material [43–45].

Multi-purpose cathodic arc evaporation system was utilized to deposit multilayered coatings using circular TiAlSiY and Mo targets operated at the cathode currents of 100 and 150 A, respectively.

A composite multi-element cathode had the following elemental ratio Ti—58 at. %; Al—38 at. %; Si—3 at. %, Y—1 at. %. The composite cathode was sintered using the spark plasma sintering unit. The purity of Mo cathode was 99.8%. The nitrogen with a purity of 99.95% was fed into the chamber as reaction gas through the control facility. A constant nitrogen pressure of 0.53 Pa was constantly kept during the deposition process. The coatings were deposited onto a 321S51 steel substrates of 18 × 20 × 2 mm size. The substrates stopped in front of each cathode for 1 min for the deposition of alternating layers. The deposition time for the films was 1 h.

11.2.2 Investigation Methods

The surface of the deposited samples was examined by scanning electron microscope (JEOL) (JEM-7001TTLS). The cross-section of the coatings was prepared and studied by the focused ion beam (FIB) (JEOL JEM-9320). ImageJ program was used to calculate the total and bilayer thickness of experimental composites [46]. The TEM and EDS investigations were conducted using a JEOL ARM 200F operated at 200 keV. The X-ray diffraction (XRD) investigations were carried out using a PANalytical diffractometer equipped with a $CuK\alpha$ X-ray source with PIXcel 3D detector. XRD measurements were carried out in the θ–2θ mode and obtained results were used to determine the phase state, preferred orientation and crystallite size. The hardness tests were carried out using the statistical microindentation method taking into account the area of the tip imprint. The Shimadzu HMV-G Micro Vickers Hardness Tester equipped with a tetrahedral diamond pyramidal tip with an angle of 136° between the opposite faces was used. The delay time after reaching the specified load was 10 s.

11.3 Results and Discussion

11.3.1 Surface Morphology and Elemental Composition

The morphology of the surface of experimental coatings has a rather rough texture due to a large number of drop constituents that can be seen in Fig. 11.1a. This phenomenon is typically observed for CA-PVD products and attributed to the technological process when active gas in the vacuum chamber extensively undergoes a reaction with the evaporated material and forms solid refractory compounds [47–50]. The EDS spectrum of the surface elemental composition of investigated coatings is shown in Fig. 11.1b.

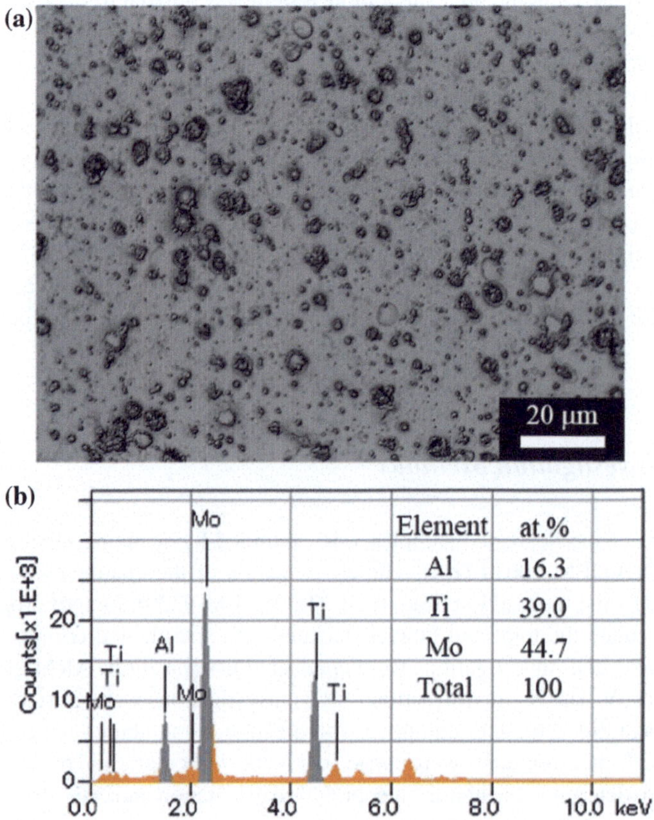

Fig. 11.1 Results of SEM with EDS analysis of multilayered (TiAlSiY)N/MoN coatings: surface image (**a**); EDS spectrum and elemental composition (**b**)

The surface topography of the (TiAlSiY)N/MoN coatings, studied using 3D model, ensured the estimation of some surface peculiarities, which include: the average roughness (0.27 μm), the texture aspect ratio (0.86) and the maximum height of droplet constituents (8 μm).

11.3.2 Phase State and Microstructure

The diffraction spectra of the multilayered coatings show a strong (200) preferential orientation and low contribution from (111), (220) and (311) planes of fcc-TiN phase (Fm3 m space group) (ICCD: 04-001-2272) (see Fig. 11.2). The (200) diffraction peak has been shifted toward lower angles comparing with bulk values. It indicates a decrease of the inter-planar distance for (200) plane ($d = 2.1043$ Å) and related with the incorporation of Al to the coating. The broad shoulder of the (200) peak ranged from 45.15° to 50.5° can be assigned to the formation of the solid solutions of (Ti, Si)N and (Ti, Al)N that have generated in a result of the substitution of Si and Al for Ti in TiN lattice [51–53]. It is because the ionic radiuses of Si^{4+} (0.041 nm) and Al^{3+} (0.053) ions are smaller than that of Ti^{3+} (0.075 nm) ion. Additionally, it possibly indicates the formation of highly disordered or even amorphous-like phase in the coatings [37]. The presence of the hexagonal δ-MoN phase (P63mc space group) is submitted by (112) and (211) peaks (ICCD: 00-064-0129).

The average crystallite size calculated using the Scherrer method from (200) reflection is approximately 12 nm. The intensities of other reflections are too low for a correct estimation of the average crystallite size.

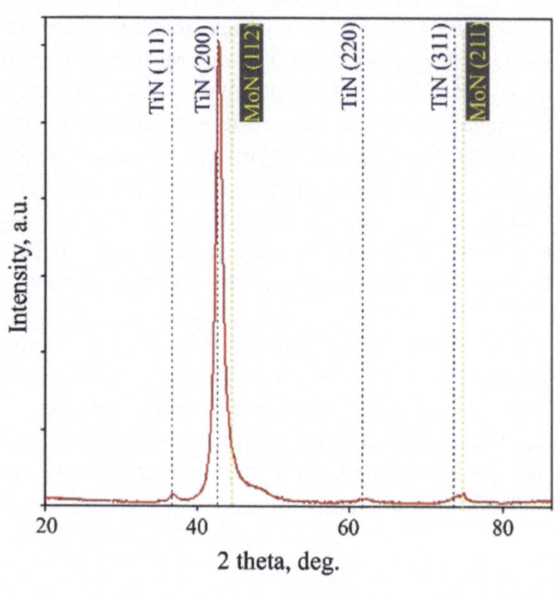

Fig. 11.2 Typical XRD pattern of multilayered (TiAlSiY)N/MoN coatings

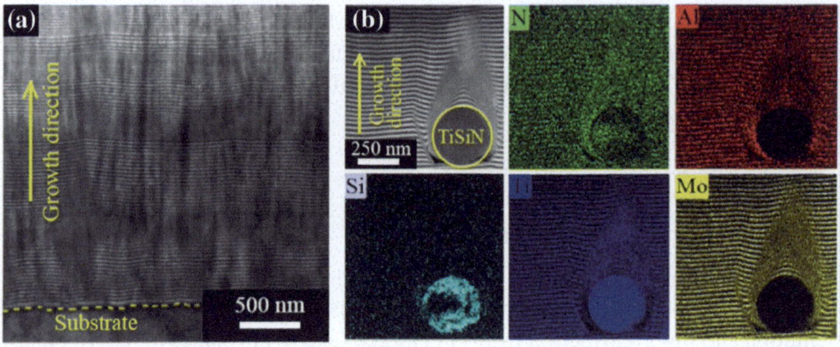

Fig. 11.3 TEM images of the cross-sectional view of multilayered (TiAlSiY)N/MoN coatings: coating-substrate zone (**a**); TEM-EDS elemental mapping of the surface zone (**b**)

The microstructure of the layers and interfaces, as well as elemental composition of (TiAlSiY)N/MoN coatings, have been analyzed from cross-section analysis by means of TEM and EDS. As shown in Fig. 11.3a distinct interface can be seen at the boundary between the steel substrate and the first nitride layer being deposited. The micrographs suggest that the appearance of the MoN layers is darker compared with the lighter contrast of the multi-element (TiAlSiY)N layers. This is because of the higher atomic number of MoN layers [37, 54, 55]. The total thickness of the experimental coatings is approximately 7 μm.

The microstructure of (TiAlSiY)N/MoN composites consists of columnar grains with an average of width ranges from 40 to 120 nm. They are oriented in such a way that the longer axes of the grains are parallel to the growth direction of the coating. The columnar microstructure is typical of the coatings deposited at low temperature and low gas pressure in CA-PVD process. Evident straight and sharp interfaces between (TiAlSiY)N and MoN layers are identified. It is pronounced due to the immiscibility of (TiAlSiY)N and MoN layers. The samples do not display any inter-lamella cracking indicating good adhesion. The thin layers of both deposited condensates have minor disparity expressed as the insignificant imperfection of the thickness of the layers in cross-sectional images. The modulation period extracted from the TEM image is ranged from 20 to 32 nm. Spherical defects start to appear in the bulk of the coating close to the surface. Their average size is approximately 350 nm and elemental composition mostly consists of Ti, Si and N elements (see Fig. 11.3b).

The SAED patterns indicate that nanocomposite coatings have polycrystalline structures with a preferred δ-TiN phase (NaCl-type structure). The separation of some diffraction rings becomes extremely difficult to fulfill in electron microscopy, due to the small differences in the inter-atomic spacing. There is no indication of rings corresponding to the appearance of Si_3N_4 that could indicate the formation of solid (Ti, Si)N in the investigated area.

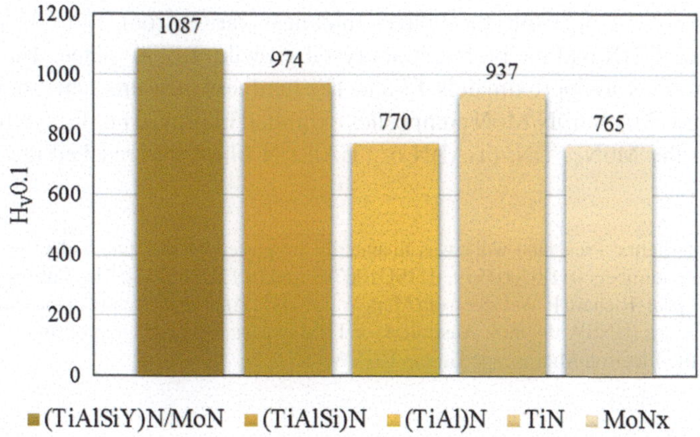

Fig. 11.4 Values of Vickers microhardness of experimental multilayered (TiAlSiY)N/MoN coatings and monolayer coatings based on constituent elements [60–63]

11.3.3 Microhardness

It is known that multilayered hard coating formed on the steel surface definitely increases the hardness of the coated material. In particular, multilayered (TiAlSiY)N/MoN coatings show considerably enhanced hardness (see Fig. 11.4) due to the several factors, which are: multilayered strengthening, which promotes impeding dislocation motion across the interfaces and the difference in elastic modulus of the layers [56, 57]; Hall-Petch strengthening, based on the increasing of volume fraction of grain boundaries with high interfacial energies [58]; Orowan strengthening, which acts in structures with nanometer modulation wavelength [59].

For all experimental coatings, it was observed a strong tendency of decreasing intensity and broadening of the width of the TiN (200) peak, which is in the result of the diminution of the grain size or the residual stress induced in the crystal lattice [64–68]. It is also suggested that the Si incorporation reduced the crystallites size and the residual stress and, hence, has ensured the hardening of nanolayered (TiAlSiY)N/MoN coatings.

11.4 Conclusions

(TiAlSiY)N/MoN multilayered coatings were successfully fabricated by the cathodic arc deposition onto steel substrates under the following deposition condition: arc currents applied to the evaporators were 100 A for the multi-component TiAlSiY cathode and 150 A for molybdenum one; constant substrate bias was-200 V and working gas pressure was 0.53 Pa. The obtained composites had fairly linear layers and well-defined interfaces between layers. The general coatings thickness was approximately 7 μm, while the bilayer thickness varied from 20 to 32 nm. The main phase of TiN had the preferential crystal growth of (200) plane. The average crystallites size was approximately 12 nm. The hardness measurements showed that multilayered (TiAlSiY)N/MoN composites exhibited improved hardness when compared with the MoN_x, TiN, (TiAl)N or (TiAlSi)N films, that reached the value of $1087H_V0.1$.

Acknowledgements Presented work was financially supported by the state budget programs of Ukraine [grant numbers 0118U003579, 0119U100787 and 0117U003923]. The authors gratefully acknowledge prof. Richard F. Webster from Mark Wainwright Analytical Centre of the University of New South Wales (UNSW, Sydney, Australia) for TEM experiments and analysis and Microscopy Australia at the Electron Microscope Unit at UNSW.

References

1. V.I. Lavrentiev, A.D. Pogrebnjak, High-dose ion implantation into metals. Surf. Coatings Technol. **99**, 24–32 (1998). https://doi.org/10.1016/S0257-8972(97)00122-9
2. H.A. Jehn, Multicomponent and multiphase hard coatings for tribological applications. Surf. Coatings Technol. **131**, 433–440 (2000)
3. H.C. Barshilia, B. Deepthi, K.S. Rajam, in *Nanostructured Thin Films and Coatings: Mechanical Properties*, ed. by S. Zhang (CRC Press, 2010), pp. 427–480
4. M. Khadem, O.V. Penkov, H.-K. Yang, D.-E. Kim, Tribology of multilayer coatings for wear reduction: a review. Friction **5**, 248–262 (2017). https://doi.org/10.1007/s40544-017-0181-7
5. A.A. Bagdasaryan, A.V. Pshyk, L.E. Coy, P. Konarski, M. Misnik, V.I. Ivashchenko, M. Kempi´nski, N.R. Mediukh, A.D. Pogrebnjak, V.M. Beresnev, S. Jurga, A new type of (TiZrNbTaHf)N/MoN nanocomposite coating: microstructure and properties depending on energy of incident ions. Compos. Part B Eng. **146**, 132–144 (2018). https://doi.org/10.1016/j.compositesb.2018.04.015
6. V. Vereschaka, S. Tabakov, N. Grigoriev, G. Sitnikov, N. Oganyan, F. Andreev, Milovich, Investigation of wear dynamics for cutting tools with multilayer composite nanostructured coatings in turning constructional steel. Wear **420–421**, 17–37 (2019). https://doi.org/10.1016/j.wear.2018.12.033
7. X.M. Xu, J. Wang, J. An, Y. Zhao, Q.Y. Zhang, Effect of modulation structure on the growth behavior and mechanical properties of TiN/ZrN multilayers. Surf. Coatings Technol. **201**, 5582–5586 (2007). https://doi.org/10.1016/j.surfcoat.2006.07.132
8. V. Pogrebnjak, O. Ivashchenko, V. Bondar, O. Beresnev, K. Sobol, S. Załęski, E. Jurga, P. Coy, B. Konarski, Postolnyi, Multilayered vacuum-arc nanocomposite TiN/ZrN coatings before and after annealing: Structure, properties, first-principles calculations. Mater. Charact. **134**, 55–63 (2017). https://doi.org/10.1016/j.matchar.2017.10.016
9. J. Lin, J.J. Moore, B. Mishra, M. Pinkas, W.D. Sproul, Nano-structured CrN/AlN multilayer coatings synthesized by pulsed closed field unbalanced magnetron sputtering. Surf. Coatings Technol. **204**, 936–940 (2009). https://doi.org/10.1016/j.surfcoat.2009.04.013
10. V. Chawla, D. Holec, P.H. Mayrhofer, Stabilization criteria for cubic AlN in TiN/AlN and CrN/AlN bi-layer systems. J. Phys. D Appl. Phys. **46**, 045305 (2013). https://doi.org/10.1088/0022-3727/46/4/045305
11. N. Bagcivan, K. Bobzin, S. Theiß, Synthesis of nano-structured HPPMS CrN/AlN coatings. J. Phys. D Appl. Phys. **46**, 084001 (2013). https://doi.org/10.1088/0022-3727/46/8/084001
12. N. Bagcivan, K. Bobzin, A. Ludwig, D. Grochla, R.H. Brugnara, CrN/AlN nanolaminate coatings deposited via high power pulsed and middle frequency pulsed magnetron sputtering. Thin Solid Films **572**, 153–160 (2014). https://doi.org/10.1016/j.tsf.2014.06.058

13. O. Maksakova, S. Simoẽs, A. Pogrebnjak, O. Bondar, Y. Kravchenko, V. Beresnev, N. Erdybaeva, The influence of deposition conditions and bilayer thickness on physical-mechanical properties of CA-PVD multilayer ZrN/CrN coatings. Mater. Charact. **140**, 189–196 (2018). https://doi.org/10.1016/j.matchar.2018.03.048
14. Z.G. Zhang, O. Rapaud, N. Allain, D. Mercs, M. Baraket, C. Dong, C. Coddet, Microstructures and tribological properties of CrN/ZrN nanoscale multilayer coatings. Appl. Surf. Sci. **255**, 4020–4026 (2009). https://doi.org/10.1016/j.apsusc.2008.10.075
15. H.C. Barshilia, K.S. Rajam, Structure and properties of reactive DC magnetron sputtered TiN/NbN hard superlattices. Surf. Coatings Technol. **183**, 174–183 (2004). https://doi.org/10.1016/j.surfcoat.2003.09.070
16. M. Nordin, M. Larsson, S. Hogmark, Mechanical and tribological properties of multilayered PVD TiN/CrN, TiN/MoN, TiN/NbN and TiN/TaN coatings on cemented carbide. Surf. Coatings Technol. **106**, 234–241 (1998). https://doi.org/10.1016/S0257-8972(98)00544-1
17. A.D. Pogrebnjak, V.M. Beresnev, O.V. Bondar, B.O. Postolnyi, K. Zaleski, E. Coy, S. Jurga, M.O. Lisovenko, P. Konarski, L. Rebouta, J.P. Araujo, Superhard CrN/MoN coatings with multilayer architecture. Mater. Design **153**, 47–59 (2018). https://doi.org/10.1016/j.matdes.2018.05.001
18. B. Gilewicz, Warcholinski, Deposition and characterisation of Mo_2N/CrN multilayer coatings prepared by cathodic arc evaporation. Surf. Coatings Technol. **279**, 126–133 (2015). https://doi.org/10.1016/j.surfcoat.2015.08.042
19. B. Han, Z. Wang, N. Devi, K.K. Kondamareddy, Z. Wang, N. Li, W. Zuo, D. Fu, C. Liu, RBS Depth Profiling Analysis of (Ti, Al)N/MoN and CrN/MoN Multilayers. Nanoscale Res. Lett. **12**, 161 (2017). https://doi.org/10.1186/s11671-017-1921-3
20. O.V. Bondar, B.A. Postol'nyi, V.M. Beresnev, G. Abadias, P. Chartier, O.V. Sobol, D.A. Kolesnikov, F.F. Komarov, M.O. Lisovenko, A.A. Andreev, Composition, structure and tribotechnical properties of TiN, MoN single-layer and TiN/MoN multilayer coatings. J. Superhard Mater. **37**, 27–38 (2015). https://doi.org/10.3103/s1063457615010050
21. M. Nordin, M. Larsson, Deposition and characterisation of multilayered PVD TiN/CrN coatings on cemented carbide. Surf. Coatings Technol. **116–119**, 108–115 (1999). https://doi.org/10.1016/S0257-8972(99)00263-7
22. S.Y. Lee, G.S. Kim, J.H. Hahn, Effect of the Cr content on the mechanical properties of nanostructured TiN/CrN coatings. Surf. Coatings Technol. **177–178**, 426–433 (2004). https://doi.org/10.1016/j.surfcoat.2003.09.020
23. A.D. Pogrebnjak, K. Smyrnova, O.V. Bondar, Nanocomposite multilayer binary nitride coatings based on transition and refractory metals: structure and properties. Coatings **9**, 155 (2019). https://doi.org/10.3390/coatings9030155
24. A.D. Pogrebnjak, V.M. Beresnev, D.A. Kolesnikov, O.V. Bondar, Y. Takeda, K. Oyoshi, M.V. Kaverin, O.V. Sobol, R. Krause-Rehberg, C. Karwat, Multicomponent (Ti-Zr-Hf-V-Nb)N nanostructure coatings fabrication, high hardness and wear resistance. Acta Phys. Polon. A **123**, 816–818 (2013). https://doi.org/10.12693/APhysPolA.123.816
25. Y.O. Kravchenko, L.E. Coy, B. Pepli´nska, I. Iatsunskyi, K. Załeski, M. Kempinski, V.M. Beresnev, P. Konarski, S. Jurga, A.D. Pogrebnjak, Nano-multilayered coatings of (TiAlSiY)N/MeN (Me = Mo, Cr and Zr): influence of composition of the alternating layer on their structural and mechanical properties. J. Alloy. Compd. **767**, 483–495 (2018). https://doi.org/10.1016/j.jallcom.2018.07.090
26. D.D. Kumar, N. Kumar, S. Kalaiselvam, S. Dash, R. Jayavel, Wear resistant super-hard multilayer transition metal-nitride coatings. Surfaces Interfaces **7**, 74–82 (2017). https://doi.org/10.1016/j.surfin.2017.03.001
27. Y.Y. Chang, H. Chang, L.J. Jhao, C.C. Chuang, Tribological and mechanical properties of multilayered TiVN/TiSiN coatings synthesized by cathodic arc evaporation. Surf. Coatings Technol. **350**, 1071–1079 (2018). https://doi.org/10.1016/j.surfcoat.2018.02.040
28. B. Navinsek, S. Seal, Transition metal nitride functional coatings. JOM **53**, 51–54 (2001). https://doi.org/10.1007/s11837-001-0072-1

29. Y.-Y. Chang, W.-T. Chiu, J.-P. Hung, Mechanical properties and high temperature oxidation of CrAlSiN/TiVN hard coatings synthesized by cathodic arc evaporation. Surf Coating Technol. **303**, 18–24 (2016). https://doi.org/10.1016/j.surfcoat.2016.02.047
30. J.-W. Yeh, S.-K. Chen, S.-J. Lin, J.-Y. Gan, T.-S. Chin, T.-T. Shun, C.-H. Tsau, S.-Y. Chang, Nanostructured high-entropy alloys with multiple principal elements: novel alloy design concepts and outcomes. Adv. Eng. Mater. **6**, 299–303 (2004). https://doi.org/10.1002/adem.200300567
31. A.D. Pogrebnjak, O.V. Sobol, V.M. Beresnev, P.V. Turbin, G.V. Kirik, N.A. Makhmudov, M.V. Il'yashenko, A.P. Shypylenko, M.V. Kaverin, M.Yu. Tashmetov, A.V. Pshyk, in *Nanostructured Materials and Nanotechnology IV*, ed. by S. Mathur, S.S. Ray, T. Ohji (Hoboken, John Wiley & Sons Inc., 2010), pp. 127–138
32. A.D. Pogrebnjak, A.P. Shpak, N.A. Azarenkov, V.M. Beresnev, Structures and properties of hard and superhard nanocomposite coatings. Phys.-Usp. **52**, 29–54 (2009). https://doi.org/10.3367/UFNe.0179.200901b.0035
33. A.D. Pogrebnjak, A.A. Bagdasaryan, I.V. Yakushchenko, V.M. Beresnev, The structure and properties of high-entropy alloys and nitride coatings based on them. Rus. Chem. Rev. **83**, 1027–1061 (2014). https://doi.org/10.1070/RCR4407
34. L. Chen, S.Q. Wang, Y. Du, S.Z. Zhou, T. Gang, J.C. Fen, K.K. Chang, Y.W. Li, X.A. Xiong, Machining performance of Ti-Al-Si-N coated inserts. Surf. Coatings Technol. **205**, 582–586 (2010). https://doi.org/10.1016/j.surfcoat.2010.07.043
35. H. Çalışkan, C. Kurbanoğlu, P. Panjan, M. Čekada, D. Kramar, Wear behavior and cutting performance of nanostructured hard coatings on cemented carbide cutting tools in hard milling. Tribol. Int. **62**, 215–222 (2013). https://doi.org/10.1016/j.triboint.2013.02.035
36. X.D. Sui, G.J. Li, X.S. Qin, H.D. Yu, X.K. Zhou, K. Wang, Q. Wang, Relationship of microstructure, mechanical properties and titanium cutting performance of TiAlN/TiAlSiN composite coated tool. Ceram. Int. **42**, 7524–7532 (2016). https://doi.org/10.1016/j.ceramint.2016.01.159
37. Y. Pshyk, E. Kravchenko, M. Coy, I. Kempiński, K. Iatsunskyi, A. Załęski, S. Pogrebnjak, Jurga, Microstructure, phase composition and mechanical properties of novel nanocomposite (TiAlSiY)N and nano-scale (TiAlSiY)N/MoN multifunctional heterostructures. Surf. Coatings Technol. **350**, 376–390 (2018). https://doi.org/10.1016/j.surfcoat.2018.07.010
38. E. Pflüger, A. Schröer, P. Voumard, L. Donohue, W.-D. Münz, Influence of incorporation of Cr and Y on the wear performance of TiAlN coatings at elevated temperatures. Surf. Coatings Technol. **115**, 17–23 (1999). https://doi.org/10.1016/S0257-8972(99)00059-6
39. T. Moria, M. Noborisaka, T. Watanabe, T. Suzuki, Oxidation resistance and hardness of TiAlSiN/CrAlYN multilayer films deposited by the arc ion plating method. Surf. Coatings Technol. **213**, 216–220 (2012). https://doi.org/10.1016/j.surfcoat.2012.10.050
40. M. Moser, P.H. Mayrhofer, Yttrium-induced structural changes in sputtered Ti1-xAlxN thin films. Scripta Mater. **57**, 357–360 (2007). https://doi.org/10.1016/j.scriptamat.2007.04.019
41. H. Riedl, D. Holec, R. Rachbauer, P. Polcik, R. Hollerweger, J. Paulitsch, P.H. Mayrhofer, Phase stability, mechanical properties and thermal stability of Y alloyed Ti-Al-N coatings. Surf. Coatings Technol. **235**, 174–180 (2013). https://doi.org/10.1016/j.surfcoat.2013.07.030
42. M. Moser, D. Kiener, C. Scheu, P.H. Mayrhofer, Influence of yttrium on the thermal stability of Ti-Al-N thin films. Materials (Basel) **3**, 1573–1592 (2010). https://doi.org/10.3390/ma3031573
43. L.C. Hernández, L. Ponce, A. Fundora, E. Lopez, E. Perez, Nanohardness and residual stress in TiN coatings. Materials (Basel) **4**, 929–940 (2011). https://doi.org/10.3390/ma4050929
44. Cavaleiro, J.Th.M. De Hosson, *Nanostructured Coatings* (Springer-Verlag, New York, 2006)
45. M. Ohring, *Materials Science of Thin Films* (Academic Press, London, Deposition and Structure, 2002)
46. C.A. Schneider, W.S. Rasband, K.W. Eliceiri, Nat. Methods **9**, 671–675 (2012)
47. V. Beresnev, O. Sobol, S. Grankin, U. Nemchenko, V. Novikov, O. Bondar, K. Belovol, O. Maksakova, D. Eskermesov, Physical and mechanical properties of (Ti–Zr–Nb)N coatings fabricated by vacuum-arc deposition. Inorg. Mater. Appl. Res. **7**, 388–394 (2016). https://doi.org/10.1134/S2075113316030047

48. O.V. Maksakova, S.S. Grankin, O.V. Bondar, Ya.O. Kravchenko, D.K. Yeskermesov, A.V. Prokopenko, N.K. Erdybaeva, B. Zhollybekov, J. Nano- Electron. Phys. **7**, 04098 (2015)
49. K.V. Smyrnova, A.D. Pogrebnjak, V.M. Beresnev, S.V. Litovchenko, S.O. Borba-Pogrebnjak, A.S. Manokhin, S.A. Klimenko, B. Zhollybekov, A.I. Kupchishin, Y.O. Kravchenko, O.V. Bondar, Microstructure and physical–mechanical properties of (TiAlSiY)N nanostructured coatings under different energy conditions. Met. Mater. Int. **24**, 1024–1035 (2018). https://doi.org/10.1007/s12540-018-0110-y
50. A.D. Pogrebnjak, A.A. Bagdasaryan, A. Pshyk, K. Dyadyura, Adaptive multicomponent nanocomposite coatings in surface engineering. Phys.-Usp. **60**, 586–607 (2017). https://doi.org/10.3367/UFNe.2016.12.038018
51. S. Balasubramanian, A. Ramadoss, A. Kobayashi, J. Muthirulandi, Nanocomposite Ti–Si–N coatings deposited by reactive dc magnetron sputtering for biomedical applications. J. Am. Ceram. Soc. **95**, 2746–2752 (2012). https://doi.org/10.1111/j.1551-2916.2011.05029.x
52. A.D. Pogrebnjak, ShM Ruzimov, Increased microhardness and positron annihilation in Al exposed to a high-power ion beam. Phys. Lett. A **120**, 259–261 (1987). https://doi.org/10.1016/0375-9601(87)90221-0
53. M. Braic, V. Braic, A. Vladescu, C.N. Zoita, M. Balaceanu, Solid solution or amorphous phase formation in TiZr-based ternary to quinternary multi-principal-element films. Prog. Nat. Sci: Mater. Int. **24**, 305–312 (2014). https://doi.org/10.1016/j.pnsc.2014.06.001
54. M.K. Kazmanli, M. Ürgen, A.F. Çakir, Effect of nitrogen pressure, bias voltage and substrate temperature on the phase structure of Mo–N coatings produced by cathodic arc PVD. Surf. Coatings Technol. **167**, 77–82 (2003). https://doi.org/10.1016/S0257-8972(02)00866-6
55. M. Ürgen, O.L. Eryilmaz, A.F. Çakir, E.S. Kayali, B. Nilüfer, Y. Işik, Characterization of molybdenum nitride coatings produced by arc-PVD technique. Surf. Coatings Technol. **94–95**, 501–506 (1997). https://doi.org/10.1016/S0257-8972(97)00432-5
56. M. Stueber, H. Holleck, H. Leiste, K. Seemann, S. Ulrich, C. Ziebert, Concepts for the design of advanced nanoscale PVD multilayer protective thin films. J. Alloys Compd. **483**, 321–333 (2009). https://doi.org/10.1016/j.jallcom.2008.08.133
57. A. Leyland, A. Matthews, On the significance of the H/E ratio in wear control: a nanocomposite coating approach to optimised tribological behavior. Wear **246**, 1–11 (2000). https://doi.org/10.1016/S0043-1648(00)00488-9
58. P.M. Anderson, C. Li, Hall-Petch relations for multilayered materials. Nanostruct. Mater. **5**, 349–362 (1995). https://doi.org/10.1016/0965-9773(95)00250-I
59. H. Söderberg, M. Odén, J.M. Molina-Aldareguia, L. Hultman, Nanostructure formation during deposition of TiN/SiNx nanomultilayer films by reactive dual magnetron sputtering. J. Appl. Phys. **97**, 114327 (2005). https://doi.org/10.1063/1.1935135
60. L. Jakab-Farkas, S. Papp, G. Strnad, G. Safran, I. Vida-Simiti, D. Biro, Preparation and study of nanostructured TiAlSiN thin films. Paper presented at 5th Engineering International Conference, "Petru Maior" University of Tîrgu Mureş, Romania, 2011
61. M. Wang, S. Miyake, in *Lubrication—Tribology, Lubricants and Additives*, ed. by D.W. Johnson (Intech Open, 2018), pp. 77–100
62. A.K. Sahu, S.S. Mahapatra, in *Additive Manufacturing of Emerging Materials*, ed. by B. AlMangour (Springer, 2019), pp. 29–53
63. L. Gmelin, Gmelin *Handbook of Inorganic Chemistry*, 8th edn. (Springer, 1990)
64. M. Diserens, J. Patscheider, F. Levy, Improving the properties of titanium nitride by incorporation of silicon. Surf. Coatings Technol. **108–109**, 241–246 (1998). https://doi.org/10.1016/S0257-8972(98)00560-X
65. X. Chu, S.A. Barnett, Model of superlattice yield stress and hardness enhancements. J. Appl. Phys. **77**, 4403–4411 (1995). https://doi.org/10.1063/1.359467
66. A.D. Pogrebnjak, V.I. Ivashchenko, P.L. Skrynskyy, O.V. Bondar, P. Konarski, K. Załęski, S. Jurga, E. Coy, Experimental and theoretical studies of the physicochemical and mechanical properties of multi-layered TiN/SiC films: temperature effects on the nanocomposite structure. Compos. Part B Eng. **142**, 85–94 (2018). https://doi.org/10.1016/j.compositesb.2018.01.004

67. V. Vladescu, M. Braic, M. Braic, Balaceanu, Arc plasma deposition of TiSiN/Ni nanoscale multilayered coatings. Mater. Chem. Phys. **138**, 500–506 (2013). https://doi.org/10.1016/j.matchemphys.2012.12.010
68. A. Vladescu, M. Kiss, C.M. Braic, P. Cotrut, M. Drob, C. Balaceanu, V. Vasilescu, Braic, Vacuum arc deposition of nanostructured multilayer coatings for biomedical applications. J. Nanosci. Nanotechnol. **8**, 733–738 (2008). https://doi.org/10.1166/jnn.2008.d211

Chapter 12
Changing Cohesive Energy Between Atoms in Metal-to-Metal Transition Layer for Fe–Sn and Fe–Cu–Sn Compounds in the Course of Spark Alloying Process

Ie. Konoplianchenko, V. Tarelnyk, V. Martsynkovskyy, A. Belous, V. Gerasimenko, G. Smolyarov, A. Tolbatov, V. Tolbatov and M. Chuprina

Abstract There have been analyzed design and technological methods to improve quality of Babbit sliding bearings (SB). A process for forming intermediate layer of copper by the electrospark alloying (ESA) method is described. The process provides strong adhesion of steel substrate to antifriction babbit layer, as well as more intense heat removal from the friction zone. The numerical values of the cohesive energy between atoms in the metal-to-metal transition layer with the use traditional (Fe–Sn) and proposed (Fe–Cu–Sn) technologies have been determined. The feasibility of using copper to increase the cohesive strength of the babbit layer to the steel substrate has been proved. It has been stated that in the case of combining the elements according to the compound system scheme of Fe–Cu–Sn, the cohesive energy in Sn increases, and with the elements being in the sequence Fe–Sn, it decreases. When combining metals according to the scheme of Fe–Sn, the weakest link of the compound system, namely Sn, results in decreasing the strength of Sn and strengthening Fe, and when doing it according to the scheme of Fe–Cu–Sn, the weakest link of Sn is strengthened.

12.1 Introduction

One of the ways to improve the reliability and durability of machines and mechanisms is to solve the problem of increasing wear resistance of their friction units. Usually the solution of the problem is achieved by applying traditional methods of strengthening

Ie. Konoplianchenko (✉) · V. Tarelnyk · V. Martsynkovskyy · A. Belous · V. Gerasimenko · G. Smolyarov · A. Tolbatov · V. Tolbatov
Sumy National Agrarian University, Sumy, Ukraine
e-mail: konoplyanchenko@ukr.net

M. Chuprina
National Technical University of Ukraine "Igor Sikorsky KPI", Kyiv, Ukraine

A. D. Pogrebnjak and O. Bondar (eds.), *Microstructure and Properties of Micro- and Nanoscale Materials, Films, and Coatings (NAP 2019)*, Springer Proceedings in Physics 240, https://doi.org/10.1007/978-981-15-1742-6_12

their contacting surfaces, for example, centrifugal reinforcement [1], alloying [2], electrochemical chromium plating [3]. Effective methods are the formation of special coatings with the use of concentrated energy and substance flows [4–10], two-layer oxide–aluminum coating [11, 12], including optimization of the shape of surfaces of the rubbing parts [13].

A large number of works [14–20] are devoted to implementing new and modernized units of sliding bearings (SB), which meet the current level of development and ensure high reliability of equipment operation both at the transitional start and stop modes and also at technological modes of any kind.

Bearing alloys formed on the basis of lead, tin, zinc or aluminum and the so-called babbits, which are usually deposited on a steel substrate with the use of various methods, have special properties, namely, low friction coefficient, increased wear resistance, and they withstand high specific pressures [21–27]. Thus, studying the processes for the formation of babbit coatings, which are aimed at improving their quality, is timely and relevant, and creating the surface layers with special properties for sliding friction units, in this case, sliding bearings (SB), is an urgent problem to solve.

12.2 Analysis of Major Achievements and Publications

Operations at increased speeds, the use of flexible rotors, cantilever and aerodynamic loads, self-oscillations and other factors have complicated the problem of controlling the rotor-bearing system characteristics. Currently, to minimize the instability and vibration activity of this system, various SB designs have been put in place or are being developed [28–33]. At the same time, the problems of improving the SB quality are solved not only by developing proper designs [34] but also by applying new technological methods [35–41].

Traditionally, babbit alloys are applied to the base of SB liners in various ways, those are manual casting, centrifugal casting, under pressure and gas-thermal spraying [42]. As a rule, in all the cases, the basis for spraying is a steel of Ст.20 grade (ГОСТ (GOST) 1050-88). Although the other materials having good adhesion with tin, such as bronze, brass and the steels of Ст.10 and Ст.15 grades can be used. Along with the traditional methods of manufacturing and repairing sliding bearings (SB), there is also observed the appearance of new ones, such as galvanic build-up (electroplating) and stamping within a temperature range [43, 44]. Recently, to apply antifriction coatings, the electrospark alloying (ESA) method has been increasingly used. This method consists in transferring a required material to the surface being treated with a spark electric discharge [45]. The regularities of the processes for forming the surface layers in the course of the ESA method are sufficiently reflected in [46–49].

In work [50], it is proposed a process for treating SB liners (Pat.UA64663A), which, due to forming an intermediate layer of copper by the ESA method, increases the cohesive strength of the babbit layer to the steel substrate by 35%. The formation of the intermediate copper layer being firmly bonded, on the one hand, with a steel substrate, and, on the other hand, with a tin layer, by forming the substitution solid solutions and babbit, will provide the stronger bond between the steel substrate and the babbit, as well as more intensive heat removal from the friction zone.

The preliminary studies have confirmed the fact that, as compared with the methods of surfacing, electroplating, performing metallization and ion bombardment, the most promising method for this purpose is an electroerosive method, which provides minimal penetration with high stability of the quality for the applied coatings [45].

In paper [51], there are described the results of the preliminary studies of the transition layers of copper, formed by the ESA method on the specimens made of the steel of Ст.20 grade. The metallographic studies were performed on the optical microscope of «Неофот-2» (Neofot-2) model and the raster electron microscope of JOEL JSM-540 type. There was carried out the assessment of the continuity of the layer and the thickness and structure of the sublayer, namely, the diffusion zone and the heat-affected one. Under conditions of applying different loads, the microhardness meter of ПМТ-3 (PMT-3) model was used to analyze the microhardness distribution in the surface layer over the depth thereof.

To study the distribution of the elements over the depth of the layer, there was provided the X-ray microanalysis based on recording the characteristic X-ray radiation excited by the electron beam of the chemical elements which were present in the excitation microvolume. In this case, there was used the electron microscope-microanalyzer provided by the ISIS 300 Oxford Instruments.

It was set that the best quality of the coating (continuity, roughness, uniformity) had been achieved when applying the coatings in argon at discharge energy of 0.2–0.4 J, which provided 100% continuity and roughness Rz equal to of 8 to 10 mμ. At performing the ESA method in air, the thickness of the formed layers was not uniform and, depending on the mode, it was of 30 to 70 mμ for copper. The continuity of the layer was at the level of ~90%. In argon, the uniformity of the layer significantly increased, and its thickness at the ESA method on the same modes was of 20 to 30 mμ for copper. The continuity of the layer for the alloying modes used was at the level of 100%.

The investigation of the electrospark coating structure showed that the surface layer of the steel of Ст.20 grade with the copper coating, both in air and in argon, consisted of two zones, namely, the so-called white layer arranged closer to the surface, and the transition layer located below the white layer. The presence and depth of the third layer having the increased microhardness, namely, the heat-affected zone (HAZ), depends on the spark discharge energy value.

Durametric analysis shows that in case of alloying with copper in various environments, the microhardness on the surface of the layer is, respectively, of 850–900 MPa. As deepening, in the HAZ zone, it has been gradually increasing to the value of 2500–3000 MPa and then decreasing to the microhardness of the steel that is of 1750–1800 MPa. At alloying on the ninth mode, the thickness of the HAZ zone, having a microhardness value higher than the microhardness of the base metal, reaches 50 μm in air and 40 μm in argon. As compared to alloying in air, the decrease in the HAZ zone thickness at alloying in argon medium is explained by cooling of the ESA zone with a jet of the supplied gas.

In Fig. 12.1, there is shown the microstructure (a) and the microhardness distribution in the boundary zone (b) between the steel of Cт.20 grade and tin applied using the traditional technology of tinning.

The transition layer between the tin and the substrate (the steel of Cт.20 grade) is missed. The microhardness values are sharply varying in the magnitude of 310 to

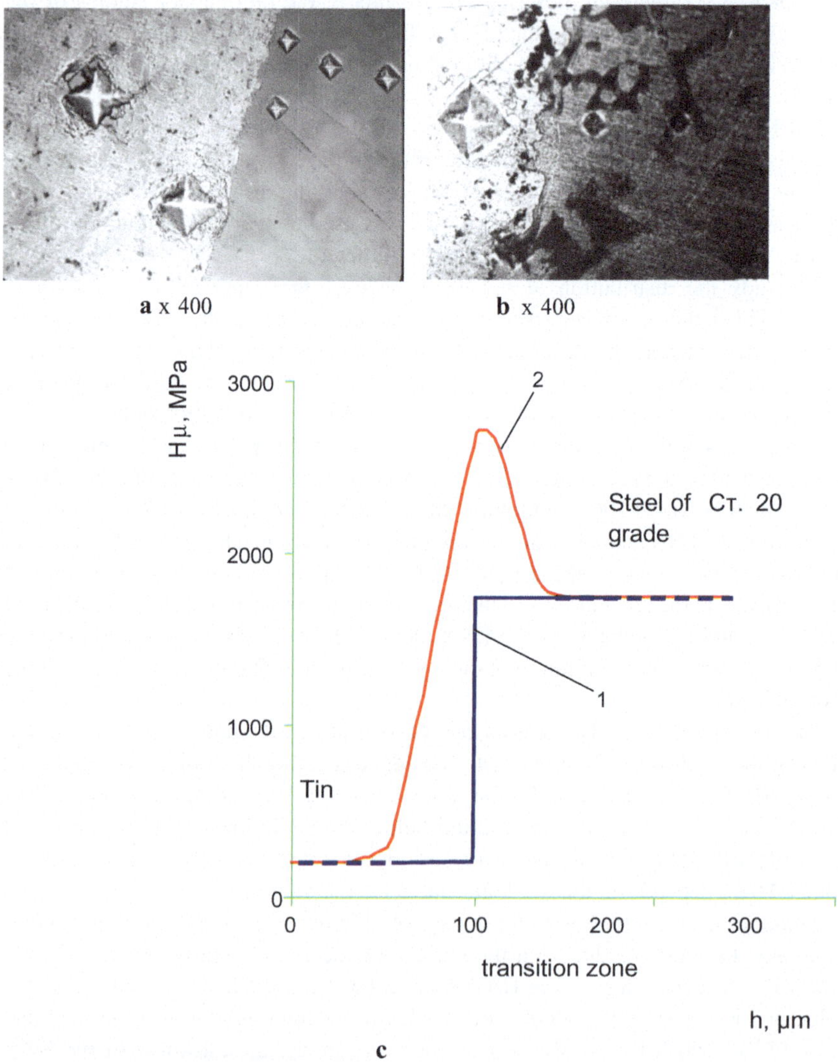

Fig. 12.1 The microstructures after tinning the steel of Cт.20 grade (**a**) and with a sublayer of copper (ESA) (**b**); microhardness distribution (**c**): 1—steel-tin, 2—steel-copper (ESA)

340 MPa (tin) to 1750 to 1800 MPa (the steel of Ст.20 grade). In this case, the tin is retained on the steel substrate only by adhesion.

At processing the steel of Ст.20 grade by the ESA method with copper, in the diffusion zone between tin and copper, there are formed substitution solid solutions providing a stronger bond. Initially, the microhardness in the transition zone smoothly rises from 210 to 230 MPa (tin) to of 2700 to 2800 MPa (the HAZ zone), and then it gradually decreases to the microhardness of the substrate (Fig. 12.1b, c).

In Fig. 12.2, there is shown the most characteristic area of the surface layer of the steel of Ст.20 grade after processing it by the ESA method with copper and tin in a plane perpendicular to the coating. The formed surface layer consists of two zones. On the surface, there is a layer of tin having a thickness of 10–15 mμ. Below, there is a layer mainly consisting of copper with a thickness of 15–20 mμ. Both in the first and second layers, there are closed pores of sizes from a moiety (small fraction) up

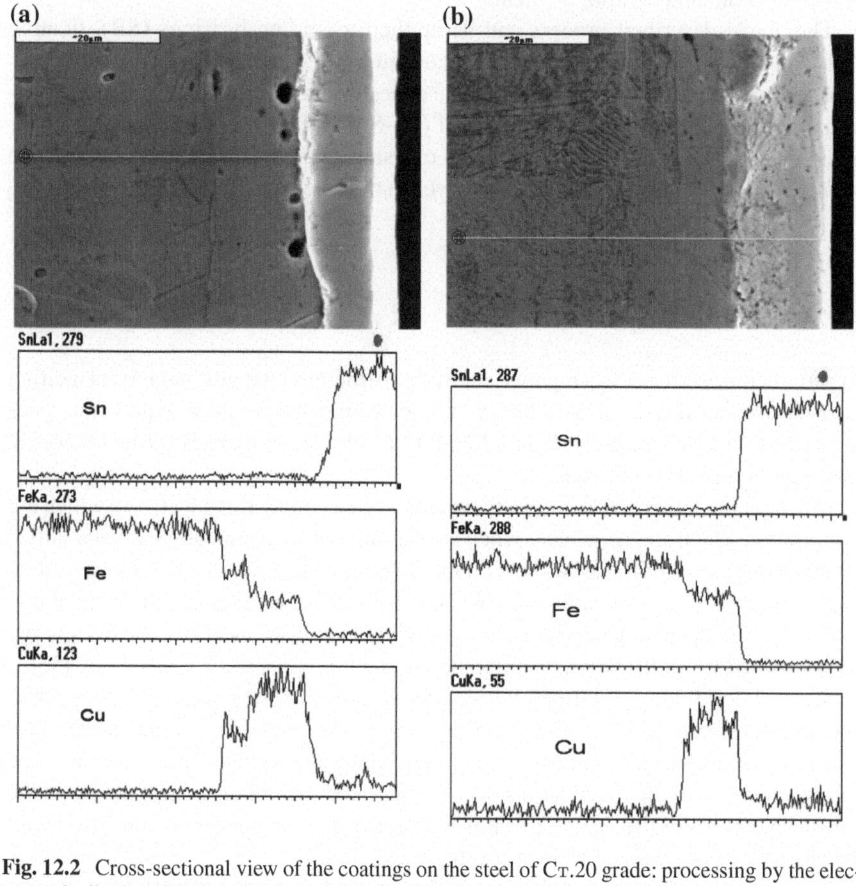

Fig. 12.2 Cross-sectional view of the coatings on the steel of Ст.20 grade: processing by the electrospark alloying (ESA) method successively with copper and then tin (**a**) and by the electroerosive alloying (EEA) method with copper and subsequent tinning (**b**) and the distribution of Sn, Fe and Cu over the depth of the coatings, x 2000

to 3 mμ, with the larger pores located in the copper layer. The study of the coating composition over the depth testifies about the fact that the boundary areas between tin and copper and the steel of Cт.20 grade are composed of the elements of the adjacent layers, which event indicates to their diffusion redistribution in the coating. As deepening, the concentration of copper in the bottom layers of tin increases. At the border between copper and steel, the concentration of copper decreases, and the same of iron increases.

In the case of processing the steel by the ESA method with copper followed by subsequent tinning with tin (Fig. 12.2b), there were not observed significant differences in the structure of the formed surface layer. The study of the distribution of the elements in the coating indicates the absence of a transition layer between copper and tin. In the transition zone beginning from the copper coating to the iron base, the copper concentration decreases. In the same area, the concentration of iron gradually increases. This indicates to intense mixing and providing mutual diffusion of the substrate and coating elements.

The above-described process can be applied to sliding bearings (SB), floating seals, supporting fingers (rods) for planetary multipliers, etc. [51, 52].

The aim of the work is to determine the cohesive energy between atoms in metal-to-metal transition layer using traditional (Fe–Sn) and proposed (Fe–Cu–Sn) technologies and, as a result, the feasibility of using copper to increase the cohesive strength (binding power) of the babbit layer to the steel substrate.

12.3 Research Results

At certain temperatures, aggregative state of pure metals, changes, namely, at melting temperature, solid state goes to liquid one, at boiling temperature, liquid state goes to gaseous one. When going from a liquid to a solid state, there is formed a crystal lattice and created the crystals.

In the crystal lattice of the metal, there appear free conduction electrons, which are a degenerate electron gas, which, when interacting with the ions of the crystal lattice of the metal, localizes the atoms in a certain region of space and holds them together. Such a bond is called metallic. The work that needs to be spent to melt metal and turn it into gas determines the cohesive energy between atoms. The cohesive energy is determined by sublimation heat of a certain metal with sufficient accuracy [53].

When a solid body is formed by individual atoms as they approach each other, due to interaction between them, the energy levels of the individual atoms turn into energy bands (zones) of the closely located energy levels of the electrons. The cohesive energy of the metal atoms occurs due to the shift in the energy levels of the valence electrons of the crystal towards the decrease in the energy as compared with the energy of the valence electrons of the free atoms.

If we take an electron energy on the surface of the crystal $E_s = 0$, then we can present a diagram of the cohesive energy in the metal by indicating the energy levels of the electrons (Fig. 12.3), where E_s is a zero energy of the electron; $E_{free\,atom}$ is

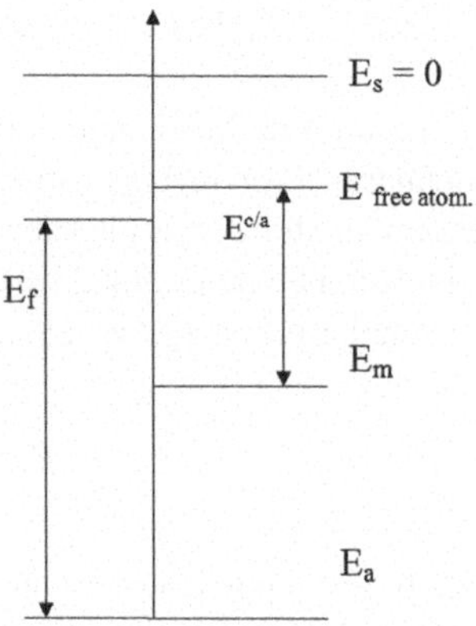

Fig. 12.3 Location of electrons energy levels in metal

the energy of the ground state of the valence electron in a free atom (atom ionization energy); E_m is an average translational energy of an electron in a metal; E_f is a Fermi level; E_a is the lowest state of the valence electron at $\vec{k} = 0$, where $\vec{k}$ is a wave vector, the absolute value $k = 2\pi\, p/\hbar$, where p is a particle momentum, $\hbar$ is Planck's constant.

Considering the designations given in Fig. 12.3, we can write the expression for the cohesive energy per atom ($E^{c/a}$) [53].

$$E^{c/a} = E_{\text{free atom.}} - E_a + E_m \tag{12.1}$$

The state of the electrons is denoted by the energy $E_{free\,atom}$ when the atoms are at large distances (vapor state), and the interaction between the electrons of the metal atoms is equal to zero.

When the metal crystals form, the metal atoms approach each other, and the electrons, due to their interaction, form the energy bands having widths of E_f to E_a. The formation of the energy bands is due to the Pauli principle, which prohibits two electrons (more precisely, particles), which obey the Fermi-Dirac statistics, to be situated on the same energy level.

Thus, the cohesive energy in a crystalline metal is considered the energy that must be expended in order to transfer all the electrons from the energy bands of E_f to E_a to the energy levels which they occupied before the formation of the metal crystals, i.e., to $E_{free\,atom}$.

To analyze the cohesive energy of the atoms in the metal-to-metal transition layers that occur in Fe–Sn and Fe–Cu–Sn compounds obtained in the course of fusion by the EEA and other methods, it is necessary, for convenience, to obtain diagrams of the energy levels of the metals.

The energy of a metal atom in a free state is determined by the ionization energy of the atom [54], i.e., $E_{free\,atom} = E_i$.

Ionization potentials are determined for all the chemical elements and they are given in electron volts (eV). So for Fe, Cu and Sn we can find from [55] as follows:

$$E_i(\text{Fe}) = 7.83\ \text{eV};\ E_i(\text{Cu}) = 7.72\ \text{eV};\ E_i(\text{Sn}) = 7.30\ \text{eV}.$$

To analyze the change in the cohesive energy between the atoms in the metal-to-metal transition layers formed by Fe–Sn and Fe–Cu–Sn metals, having the ionization energy of atoms, it is also necessary to calculate the average translational energy E_m, which can be determined by calculating the Fermi energy E_f.

The Fermi energy is calculated for the chemical elements from the ratio [54]:

$$E_f = \frac{\hbar^2}{2m_e}\sqrt[3]{(3\hbar^2)^2}\cdot\sqrt[3]{(N/V)^2} \quad (12.2)$$

where $\hbar$ is Planck's constant, m_e is an electron rest mass, (N/V) is a concentration of free electrons in the crystalline metal.

The concentration of free electrons (N/V) is determined by the product of the metal valence and the number of atoms in 1 cm^3 [54, 56].

The concentration of electrons for Fe, Cu, Sn is given in [52], and it is equal to: $(N/V)\,(\text{Fe}) = 17.0 \times 10^{22}\ 1/\text{cm}^3$; $(N/V)\,(\text{Cu}) = 8.45 \times 10^{22}\ 1/\text{cm}^3$; $(N/V)\,(\text{Sn}) = 14.48 \times 10^{22}\ 1/\text{cm}^3$.

The Fermi energy for the chemical elements of Fe, Cu, Sn is given in [54, 56], and it is equal to: $E_f(\text{Fe}) = 11.16733$ eV; $E_f(\text{Cu}) = 7.00$ eV; $E_f(\text{Sn}) = 10.03$ eV.

In the isolated state, that is, in the absence of the contact of the metals), the electron gas of the metals is characterized by the work function W (electron releasing work), which for Fe, Cu, Sn is, respectively, equal to [48]:

$$W(\text{Fe}) = 4.31\ \text{eV};\ W(\text{Cu}) = 4.4\ \text{eV};\ W(\text{Sn}) = 4.38\ \text{eV}.$$

The work function (electron releasing work) is an amount of energy required to remove an electron from the Fermi level, arranged in a potential well of a metal, to a point located outside the metal near its surface, i.e.

$$W = -E_f. \quad (12.3)$$

At combining metals, namely, when Fe having the work function W (Fe) and Sn having the work function W (Sn) are combined in such a way that electrons can transfer from one of them to the other, after the equilibrium had been established, the electron gas will have a common chemical potential. In metals, the relationship between the Fermi energy and the chemical potential is expressed by the ratio [50]:

$$\mu = 1 - ((\pi K_B T)/2E_f))^2/3, \quad (12.4)$$

where μ—chemical potential; T—absolute temperature; K_B—Boltzmann constant. At room temperatures, there can be taken $\mu = E_f$.

If one tries to schematically depict the process of filling the energy levels with electrons, the work function, and Fermi energy, with positioning them relative to their overall zero level at $T = 0$ (Fig. 12.4a), then one could see the Fe electrons

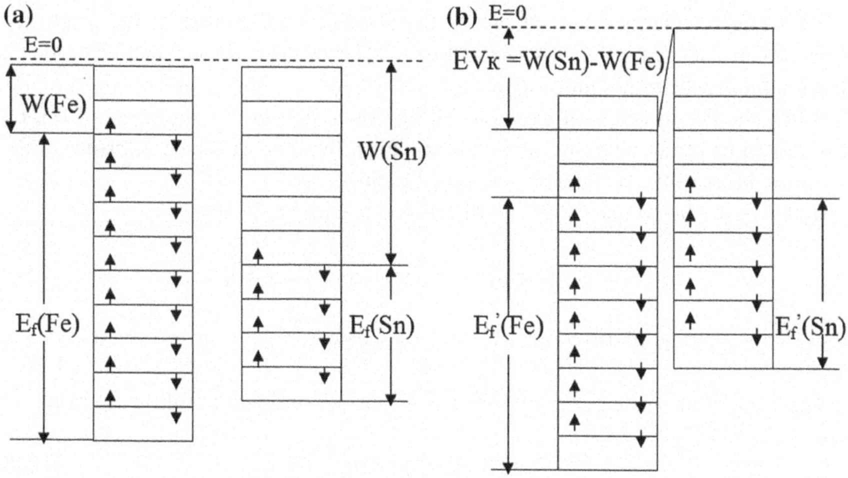

Fig. 12.4 The diagram of the conditional arrangement of the electrons, work function and Fermi energies in Fe and Sn metals: **a** before combining, **b** after combining

being arranged at higher levels than the Sn ones. While those metals being combined, the electrons of iron will have the opportunity for moving to the lower energy levels located in tin.

Transferring of the electrons from iron to tin occurs before the Fermi energy being equalized in both metals (Fig. 12.4b).

The alignment of the Fermi energy levels occurs as follows. An electron, being at a higher energy level of Fe, on moving to a lower level of Sn, transfers a negative charge. Consequently, Sn, when acquiring the electron, is negatively charged, and Fe, when losing the electron, is positively charged. With the transition of the next electron from the Fe level to the Sn one, the electron takes on itself the action of a repulsive force, since the Sn metal is negatively charged. Overcoming the repulsive force, the electron transfers from the Fe metal to the free level in the Sn metal with increasing the potential energy in Sn by eV, where e is the electron charge; V is the potential difference at distance d, where the electric field $\vec{E}$ was formed as caused by the uncompensated charge.

If denoting the distance d whereat the voltage V acts, then, when the electron passes this distance, the acquired kinetic energy while turning into the potential energy, increases the potential energy in the metal Sn by eV, which fact can be seen from the following calculation:

$$A = e\int_0^d \vec{E}d\vec{r} = e\int_0^d \frac{V}{d}dr = \frac{\text{eV}}{d}r\Big|_0^d = \frac{\text{eV}}{d}d = \text{eV}, \quad (12.5)$$

where A is a work for moving the electron.

Thus, all the energy levels of the Sn metal rise by eV. Similarly, the electrons, while passing the distance *d* where the positive charge acts, are accelerated in the Fe metal reducing the potential energy.

When the Fermi levels of iron, on falling down, and the Fermi levels of tin, on rising up, turn out to be at the same energy level, there is set the equilibrium in the transition layer of the metal compound of Fe–Sn.

In this case, the contact potential difference is equal to [53]:

$$V_K = (W(\text{Sn}) - W(\text{Fe}))/e = V_1 - V_2, \tag{12.6}$$

where V_{κ} is a potential difference; V_1 and V_2 are potentials, respectively, for Sn and Fe.

The alignment of the Fermi energy levels is expressed by the following ratio:

$$E'_f(\text{Fe}) = E'_f(\text{Sn}), \tag{12.7}$$

and, from ratio (12.2) it follows that the Fermi energy depends on the electron density values in the metals. Therefore, the density of the electrons in iron decreases after their transition into tin, and in tin, their density increases.

The magnitude of the decrease in the Fermi energy in iron, as well as the magnitude of the increase in the Fermi energy in tin, is determined by the ratio:

$$(W(\text{Sn}) - W(\text{Fe}))/2 = \text{eV}_K/2. \tag{12.8}$$

Consequently, after setting the thermodynamic equilibrium, the reduced Fermi energy for iron is determined by the ratio:

$$E'_f(\text{Fe}) = E_f(Fe) - (W(\text{Sn}) - W(\text{Fe}))/2. \tag{12.9}$$

The increased Fermi energy for tin is:

$$E'_f(\text{Sn}) = E_f(\text{Sn}) + (W(\text{Sn}) - W(\text{Fe}))/2. \tag{12.10}$$

Since the Fermi energy depends on the electron density values (12.2), we can find the electron density by calculating E'_f (Fe) using formula (12.9), having written the ratio (12.2) as follows:

$$(N/V) = 4.542291 \times 10^{21} \times \sqrt{(E'_f(Fe))^3}, \tag{12.11}$$

here (*N*/*V*) is the electron density in Fe after combining.

Having entered the numerical value of E'_f (Fe) into (12.1), one can obtain the electron density value in the iron after combining: (*N*/*V*) (Fe) $= 1.687151 \times 10^{23}$ $1/\text{cm}^3$.

The number of the electrons transferred from Fe to Sn is found by calculating the difference between the density values in Fe before combining and the density values in Fe after combining with Sn:

$$\Delta n = (N/V)(\text{Fe}) - (N/V)'(\text{Fe}) = 1.7 \times 10^{23}$$
$$- 1.687151 \times 10^{23} = 7.996 \times 10^{20}\, 1/\text{cm}^3.$$

Decreasing the electron density value in Fe after combining with Sn is about 0.5%.

The value of increasing density in Sn after combining with Fe is equal to: $(N/V)'\,(\text{Sn}) = 14.48 \times 10^{22} + 7.996 \times 10^{20} = 1.455996 \times 10^{23}\ 1/\text{cm}^3$.

The Fermi energy values calculated by formulas (12.9) and (12.10) after setting the thermodynamic equilibrium are equal for metals Fe and Sn, respectively, as it follows:

$$E'_f(\text{Fe}) = E_f(\text{Fe}) - (W(\text{Sn}) - W(\text{Fe}))/2 = 11.16733$$
$$-(4.38 - 4.31)/2 = 11.13233\ \text{eV};$$

$$E'_f(\text{Sn}) = E_f(\text{Sn}) + (W(\text{Sn}) - W(\text{Fe}))/2 = 10.03$$
$$+ (4.38 - 4.31)/2 = 10.065\ \text{eV}.$$

With the data of E'_f (Fe) and E'_f (Sn), one can calculate the average values of the kinetic translational energy for iron and tin, which, according to [47], is determined by the ratio:

$$E_f m = \frac{3}{5} E_f, \qquad (12.12)$$

then:

$$E_f m(\text{Fe}) = \frac{3}{5} E_f(\text{Fe}) = 6.679398 eV;\ E_f m(\text{Sn}) = \frac{3}{5} E_f(\text{Sn}) = 6.039\ \text{eV}.$$

Using the ratio (12.12), one can calculate the values of the translational kinetic energy of the electrons in Fe and Sn before combining, which are respectively equal to:

$$E_f m(\text{Fe}) = 0.6 \cdot 11.16733 = 6.700398\ \text{eV},\ E_f m(\text{Sn}) = 0.6 \cdot 10.03 = 6.018\ \text{eV}$$

To analyze the fact of decreasing the cohesive energy for Sn when combined with Fe, it is necessary to enter the values $E_f m(\text{Sn})$ and E^{cla} into ratio (12.1) and then obtain the ratio for the cohesive energy in Sn before combining it with Fe.

The values of the cohesive energy for the chemical elements of Fe, Cu, Sn, according to [54] are, respectively, equal to:

$$E^{c/a}(\text{Fe}) = 4.29\ \text{eV};\ E^{c/a}(\text{Cu}) = 3.50\ \text{eV};\ E^{c/a}(\text{Sn}) = 3.12\ \text{eV}.$$

The cohesive ratio (12.1) for Sn before combining is:

$$3.12 = E_{free\,atom} - E_a + 6.018. \tag{12.13}$$

While entering the data into the ratio (12.1) after having combined Sn with Fe, it is necessary to take into account that after the electrons migrations from Fe to Sn, in tin, there appears an uncompensated charge (negative charge), which uniformly raises all the energy levels along with the Fermi level by the value of

$$(W(\text{Sn}) - W(\text{Fe}))/2 = 0.35\,\text{eV}.$$

In addition, due to increasing the electron density, the average kinetic translational energy increases by the value of

$$E'_f m(\text{Sn}) - E_f m(\text{Sn}) = 6.039 - 6.018 = 0.21\ \text{eV}.$$

Therefore, after combining Fe and Sn, the ratio (12.1) takes the form:

$$E^{c/a} = E_{free\,atom} - E_a + 0.021 + 6.021 + 6.018 + 0.035. \tag{12.14}$$

Having rewritten the ratio (12.14), one can get:

$$E^{c/a} = 3.12–0.056 = E_{free\,atom.} - E_a + 6.018. \tag{12.15}$$

Thus, the cohesive energy decreases and is equal to $E^{c/a} = 3.064$ eV.

Similarly, it is possible to analyze the event of increasing the cohesive energy for Fe after combining it with Sn, taking into account that all the energy levels of the electrons decrease, including the Fermi level E_f, due to reducing the potential energy by $(W(\text{Sn})–W(\text{Fe}))/2 = 0.35$ eV and decreasing the kinetic translational energy by $6.\,700398 - 6.679398 = 0.021$ eV.

On analyzing the event, similarly to the analysis of the ratios (12.14) and (12.15), one can conclude that the cohesive energy in the metal-to-metal transition layer in Fe increases by 0.056 eV and becomes equal to:

$$E^{c/a} = 4.29 + 0.056 = 4.346\ \text{eV}.$$

The remaining calculations of the cohesive energy for the metal-to-metal compounds in the order of Fe–Cu–Sn have been carried out and represented without detailed explanations.

First, it is determined the cohesive energy in the Cu–Sn transition layer, and then it is determined the cohesive energy in the Fe–Cu transition layer.

Using formula (12.9), having set the equilibrium in the transition layer, it is determined the Fermi energy:

$$E'_j(\text{Sn}) = E_f(\text{Sn}) - (W(\text{Cu}) - W(\text{Sn}))/2 = 10.03 - E^{c/a} = 10.02\,\text{eV}.$$

From the ratio (12.1), there is found the electron density after setting the thermodynamic equilibrium:

$$(N/V)'(\text{Sn}) = 4.542291 \times 10^{21} \times \sqrt{(E'_f(\text{Sn}))^3}$$
$$= 4.542291 \times 10^{21} \times \sqrt{(10.02)^3} = 1.440709 \times 10^{23}\ 1/\text{cm}^3.$$

Then it is found the number of the electrons transferred from Sn to Cu:

$$\Delta n = (N/V)(\text{Sn}) - (N/V)'(\text{Sn}) = 14.48 \times 10^{22}$$
$$- 1.440709 \times 10^{23} = 7.2902 \times 10^{20}\,1/\text{cm}^3.$$

Further, from the ratio (12.2), there is determined the average translational kinetic energy before and after combining with Cu:

$$E_f m(\text{Sn}) = 0.6 \cdot 10.03 = 6.018\ \text{eV};\ E'_f m(\text{Sn}) = 0.6 \cdot 10.02 = 6.012\ \text{eV}.$$

Consequently, the cohesive energy in the Sn layer increases due to a decrease in the energy levels of the electrons on account of the alignment of the work functions of Cu and Sn by 0.01 eV and at the expense of reducing the translational kinetic energy by 0.006 eV.

Thus, the cohesive energy in the Sn layer will become equal to:

$$E^{c/a} = 3.12 + 0.016 = 3.136\ \text{eV}.$$

The cohesive energy in the Cu transition layer decreases by 0.016 eV and becomes equal to:

$$E^{c/a} = 3.50–0.016 = 3.484\ \text{eV}.$$

At the Fe–Cu compound, due to transferring the electrons from Fe to Cu, the Fermi energy level rises at the expense of increasing the potential energy, and according to (12.10), it is equal to:

$$E'_j(\text{Cu}) = E_f(\text{Cu}) + (W(\text{Cu}) - W(\text{Fe}))/2 = 7.00 + (4.4 - 4.31)/2 = 7.045\ \text{eV}.$$

The average kinetic translational energy values in the Cu transition layer before and after combining events, according to (12.2), are equal to:

$$E_f m(\text{Cu}) = 0.6 \times 7.00 = 4.200\ \text{eV};\ E'_f m(\text{Cu}) = 0.6 \times 7.045 = 4.227\,\text{eV}.$$

Consequently, due to increasing the kinetic translational energy, the cohesive energy in Cu decreases by 4.227 − 4.200 = 0.027 eV.

Thus, the cohesive energy in Cu decreases by 0.027 + 0.045 = 0.072 eV, and it is equal to: $E^{c/a}(E) = 3.50 - 0.072 = 3.428\,\text{eV}$,

And the cohesive energy in the transition layer of iron will increase by 0.072 eV. Then:

$$E^{c/a}(\text{Fe}) = 4.29 + 0.072 = 4.362\,\text{eV}.$$

If compared the values of the cohesive energies per atom for the chemical elements of Fe, Cu, and Sn, which, according to [54], are equal:

$$E^{c/a}(\text{Fe}) = 4.29\ \text{eV};\ E^{c/a}(\text{Cu}) = 3.50\,\text{eV};\ E^{c/a}(\text{Sn}) = 3.12\ \text{eV},$$

then the smallest cohesive energy per atom is for the chemical element Sn.

After the elements of Fe and Sn have been combined, the cohesive energy in the Sn layer will decrease by 0.056 eV, and it will become

$$E^{c/a}(\text{Sn}) = 3.064\,\text{eV}.$$

When the elements are connected according to the Fe–Cu–Sn scheme, the cohesive energy for the Sn element of the transition layer will increase by 0.016 eV, and it will become

$$E^{c/a}(\text{Sn}) = 3.136\,\text{eV}.$$

It should be noted that in this case, the cohesive energy for Cu will decrease by 0.016 eV, and it will become $E^{c/a}(\text{Cu}) = 3.8484$ eV, but even after such a decrease, the cohesive energy per atom for Cu will be much more significant than the cohesive energy for Sn.

12.4 Conclusions

As a result of determining the cohesive energy between atoms in the metal-to-metal transition layer using the traditional (Fe–Sn) and proposed (Fe–Cu–Sn) technologies, it has been proved that copper should be used to increase the cohesive strength of the babbit layer to the steel substrate.

In the case of combining the elements according to the scheme of Fe–Cu–Sn, the cohesive energy for Sn increases, and in the sequence according to the scheme of Fe–Sn, it decreases.

The weakest link in the system of the compound, namely Sn, when combining metals according to the Fe–Sn scheme, results in decreasing the Sn strength and increasing the Fe strength, and when combined after the Fe–Cu–Sn scheme, the weakest link, namely Sn, is strengthened.

References

1. L. Ropyak, I. Schuliar, O. Bohachenko, Influence of technological parameters of centrifugal reinforcement upon quality indicators of parts. East.-Europ. J. Enterp. Tech. **1**(5), 53 (2016). https://doi.org/10.15587/1729-4061.2016.59850
2. T.M. Radchenko, V.A. Tatarenko, H. Zapolsky, Statistical thermodynamics and ordering kinetics of $D0_{19}$-type phase: application of the models for H.C.P.-Ti–Al Alloy. Solid State Phenom. **138**, 283 (2008). https://doi.org/10.4028/www.scientific.net/SSP.138.283
3. L. Ropyak, V. Ostapovych, Optimization of process parameters of chrome plating for providing quality indicators of reciprocating pumps parts. East.-Europ. J. Enterp. Tech. **2**(5), 50 (2016). https://doi.org/10.15587/1729-4061.2016.65719
4. A.A. Bagdasaryan, A.V. Pshyk, L.E. Coy, P. Konarski, M. Misnik, V.I. Ivashchenko, M. Kempinski, N.R. Mediukh, A.D. Pogrebnjak, V.M. Beresnev, S. Jurga, Compos. Part B Eng. **146**, 132 (2018). https://doi.org/10.1016/j.compositesb.2018.04.015
5. A.D. Pogrebnjak, V.I. Ivashchenko, P.L. Skrynskyy, O.V. Bondar, P. Konarski, K. Załęski, S. Jurga, E. Coy, Experimental and theoretical studies of the physicochemical and mechanical properties of multi-layered TiN/SiC films: Temperature effects on the nanocomposite structure. Compos. Part B Eng. **142**, 85 (2018). https://doi.org/10.1016/j.compositesb.2018.01.004
6. O. Maksakova, S. Simoẽs, A. Pogrebnjak, O. Bondar, Y. Kravchenko, V. Beresnev, N. Erdybaeva, The influence of deposition conditions and bilayer thickness on physical-mechanical properties of CA-PVD multilayer ZrN/CrN coatings. Mater. Charact. **140**, 189 (2018). https://doi.org/10.1016/j.matchar.2018.03.048
7. A.D. Pogrebnjak, V.M. Beresnev, O.V. Bondar, B.O. Postolnyi, K. Zaleski, E. Coy, S. Jurga, M.O. Lisovenko, P. Konarski, L. Rebouta, J.P. Araujo, Superhard CrN/MoN coatings with multilayer architecture. Mater. Des. **153**, 47 (2018). https://doi.org/10.1016/j.matdes.2018.05.001
8. A.D. Pogrebnjak, V.M. Beresnev, K.V. Smyrnova, Y.O. Kravchenko, P.V. Zukowski, G.G. Bondarenko, The influence of nitrogen pressure on the fabrication of the two-phase superhard nanocomposite (TiZrNbAlYCr)N coatings. Mater. Lett. **211**, 316 (2018). https://doi.org/10.1016/j.matlet.2017.09.121
9. S.Y. Donets, V.F. Klepikov, V.V. Lytvynenko, Yu.F. Lonin, A.G. Ponomarev, R.I. Starovoytov, O.A. Startsev, V.T. Uvarov, Aluminum surface coating of copper using high-current electron beam. Probl. Atom. Sci. Tech. **4**(98), 302 (2015)
10. A.B. Batracov, M.I. Bazaleev, S.E. Donets, V.F. Klepikov, V.V. Lytvynenko, Y.F. Lonin, A.G. Ponomarev, V.V. Uvarov, V.T. Uvarov, The particularities of the high current relativistic electron beams influence on construction materials targets. Probl. Atom. Sci. Tech. **6**(88), 225 (2013)
11. L.Y. Ropyak, I.P. Shatskyi, M.V. Makoviichuk, Influence of the oxide-layer thickness on the ceramic–aluminium coating resistance to indentation. Metallofiz. Noveishie Tekhnol. **39**(4), 517 (2017). https://doi.org/10.15407/mfint.39.04.0517
12. L.Y. Ropyak, I.P. Shatskyi, M.V. Makoviichuk, Analysis of interaction of thin coating with an abrasive using one-dimensional model. Metallofiz. Noveishie Tekhnol. **41**(5), 647 (2019). https://doi.org/10.15407/mfint.41.05.0647
13. O. Vlasiy, V. Mazurenko, L. Ropyak, O. Rogal, Improving the aluminum drill pipes stability by optimizing the shape of protector thickening. East.-Europ. J. Enterprise Technol. **7**(85), 25 (2017). https://doi.org/10.15587/1729-4061.2017.65718
14. V.B. Wrublewski, A.B. Nevzorov, V.A. Dovgyalo, *Wood-Based Sliding Bearings: Design and Interchangeability Thereof: Manual* (BelSUT, Gomel, 2001)

15. G.S. Batkis, V.A. Maksimov, V.K. Khaisanov, in *Abstracts from Report. XI International Conference on Compressor Technology*, St. Petersburg, 1998
16. V.V. Kramin, G.A. Pospelov, Trenie i Iznos **3**(4), 691 (1982)
17. V.K. Krivonos, A.I. Poddubny, Res. Hydrostat. Brg. Seal. Aircraft Engines **2**, 79 (1986)
18. V.A. Maksimov et al., Calculation and some results of study of turbocompressor conical bearings. Chem. Petrol. Eng. **11**, 16 (1991)
19. G.S. Batkis, V.F. Maksimov, V.K. Khaysanov, *Compilation. Design and Study of Compressor Machines* (JSC NIIturbocompressor, Kazan, 1997), p. 124
20. I. Pavlenko et al., in *DSMIE'18: Application of Artificial Neural Network for Identification of Bearing Stiffness Characteristics in Rotor Dynamics Analysis*, ed. by V. Ivanov et al., 1st Conference Advances in Design, Simulation and Manufacturing, Sumy, June 2018. Lecture Notes in Mechanical Engineering, (Springer, Cham, 2019), p. 325
21. N.P. Barykin et al., As for the problem of babbit B83 structure. Mater. Eng. **8**, 24 (2001)
22. F.A. Sadykov, N.P. Barykin, I.S. Valeev, Effects of temperature and strain rate on mechanical properties of babbit B83 having different structures. Probl. Strength **2**, 121 (2002)
23. F.A. Sadykov, N.P. Barykin, I.S. Valeev, Influence of structural state on babbit B83 mechanical properties. Phys. Chem. Mater. Process. **2**, 86 (2001)
24. B.N. Arzamasov, G.G. Mukhin (eds.), *Materials Engineering: Textbook for Universities*, 5th edn. (Bauman MSTU, Moscow, 2003)
25. B.A. Potekhin, A.N. Glushchenko, V.V. Ilyushin, Structural dependence of babbite B83 properties. Metal. Technol. **1**, 28 (2006)
26. A.D. Pogrebnjak, A.A. Bagdasaryan, A. Pshyk, K. Dyadyura, Adaptive multicomponent nanocomposite coatings in surface engineering. Usp. Phys. Nauk **60**, 586 (2017)
27. A. Pogrebnjak, K. Smyrnova, O. Bondar, Nanocomposite multilayer binary nitride coatings based on transition and refractory metals: structure and properties. Coatings **9**(3), 155 (2019)
28. I.V. Pavlenko, V.I. Simonovskiy, M.M. Demianenko, Dynamic analysis of centrifugal machines rotors supported on ball bearings by combined application of 3D and beam finite element models. I.O.P. Conf, Ser. Mater. Sci. Eng. **233**, 012053 (2017). https://doi.org/10.1088/1757-899X/233/1/012053
29. Glacier Metal Co. Ltd., *Alperton High Speed Equipment Bearings* (Wembley, England, 1984)
30. N. Wada, E. Hiraishi, Recent Technology for Gas Field Compressors. Paper presented at 7th international compressor symposium, St. Petersburg Institute of technology, St. Petersburg, May 2001
31. D.E. Bently, Pursuit of better bearings. Orbit J. **21**(2), 33 (2000)
32. D. Nelson, L. Hollingsworth, Problemy Treniya i Smazki **1**(F), 127 (1973)
33. V.A. Voskresensky, V.I. Dyakov, A.Z. Zile, *Calculation and Design of Film Lubrication Bearings* (Machine Building, Moscow, 1983)
34. I. Pavlenko, J. Trojanowska, V. Ivanov, O. Liaposhehenko, Scientific and Methodological Approach for the Identification of Mathematical Models of Mechanical Systems by Using Artificial Neural Networks, in *Lecture Notes in Electrical Engineering*, ed. by J. Machado, F. Soares, G. Veiga (Springer, Cham, 2019), p. 299
35. N.P. Barykin, R.F. Fazlyakhmetov, A.K. Valeeva, Influence of babbit B83 structure on wear intensity of triboconjugations. Met. Sci. Heat Treat. **2**(608), 44 (2006)
36. V. Martsinkovsky, V. Yurko, V. Tarelnik, Yu. Filonenko, Designing radial sliding bearing equipped with hydrostatically suspended pads. Procedia Eng. **39**, 157 (2012). https://doi.org/10.1016/j.proeng.2012.07.020
37. V. Martsinkovsky, V. Yurko, V. Tarelnik, Y. Filonenko, Designing thrust sliding bearings of high bearing capacity. Procedia Eng. **39**, 148 (2012). https://doi.org/10.1016/j.proeng.2012.07.019
38. V.Y. Stetsenko, A.I. Rivkin, Effect of carbon nanotubes on structure and frictional wear resistance of cast babbitts. Foundry Prod. **2**, 9 (2011)
39. B.A. Potekhin, V.V. Ilyushin, A.S. Khristolyubov, Influence of casting methods on structure and properties of tin babbit. Metall. Heat Treat. Metal. **8**, 16 (2009)

40. A.P. Perekrestov, V.A. Chanchikov, Ways to improve performance of bearings. A.S.T.U. Vestnik, Mar. Eng. Technol. **1**, 147 (2010)
41. V. Tarelnyk, V. Martsynkovskyy, Upgrading of pump and compressor rotor shafts using combined technology of electroerosive alloying. Appl. Mech. Mater. **630**, 397 (2014). https://doi.org/10.4028/www.scientific.net/AMM.630.397
42. I.G. Galiakhmetov, *Construction Materials of Centrifugal and Screw Compressors. Choice and Technology of Their Application* (FEN, Kazan, 2009)
43. A.P. Lebedeva, T.N. Pogorelov, *Restoration of Machine Parts* (Machine Building, Moscow, 2003)
44. N.P. Barykin, R.F. Fazlyakhmetov, Forging and stamping production. Press. Process. Mater. **9**, 27 (2006)
45. V. Tarelnyk, V. Martsynkovskyy, A. Dziuba, New method of friction assemblies reliability and endurance improvement. Appl. Mech. Mater. **630**, 388 (2014). https://doi.org/10.4028/www.scientific.net/AMM.630.388
46. V.B. Tarelnyk, O.P. Gaponova, I.V. Konoplianchenko, M.Y. Dovzhyk, Investigation of regularities of the processes of formation of surface layers with electroerosive alloying. Part II. Metallofiz. Noveishie Tekhnol. **39**, 363 (2017). https://doi.org/10.15407/mfint.39.03.0363
47. V.B. Tarelnik, O.P. Gaponova, Y.V. Konoplyanchenko, M.Y. Dovzhyk, Investigation of regularities of the processes of formation of surface layers with electroerosive alloying. Part I. Metallofiz. Noveishie Tekhnol. **38**, 1611 (2016). https://doi.org/10.15407/mfint.38.12.1611
48. V.B. Tarelnik, O.P. Gaponova, I.V. Konoplianchenko, N.S. Evtushenko, V.A. Herasymenko, The analysis of a structural state of surface layer after electroerosive alloying. I. Features of Formation of Electroerosive Coatings on Steel 45. Metallofiz. Noveishie Tekhnol. **40**, 235 (2018). https://doi.org/10.15407/mfint.40.02.0235
49. V.B. Tarelnik, O.P. Gaponova, I.V. Konoplianchenko, N.S. Evtushenko, V.A. Herasymenko, The analysis of a structural state of surface layer after electroerosive alloying. II. Features of the Formation of Electroerosive Coatings on Special Steels and Alloys by Hard Wear-Resistant and Soft Antifriction Materials. Metallofiz. Noveishie Tekhnol. **40**, 795 (2018). https://doi.org/10.15407/mfint.40.06.0795
50. V.B. Tarelnik, V.S. Martsinkovsky, E.V. Konoplyanchenko, A.V. Belous, O.P. Gaponova, Improvement in babbit sliding bearing quality with electrospark alloying. Chem. Petrol. Eng. **54**, 598 (2018). https://doi.org/10.1007/s10556-018-0521-0
51. V.B. Tarelnik, A.V. Paustovskii, Y.G. Tkachenko, V.S. Martsinkovskii, E.V. Konoplyanchenko, K. Antoshevskii, Electric-spark coatings on a steel base and contact surface for optimizing the working characteristics of babbitt friction bearings. Surf. Eng. Appl. Electrochem. **53**, 285 (2017). https://doi.org/10.3103/S1068375517030140
52. V.B. Tarelnik, V.S. Martsinkovskii, V.I. Yurko, Improvement of compressor seal assembly elements. Part I. Chem. Petrol. Eng. **51**, 328 (2015). https://doi.org/10.1007/s10556-015-0047-7
53. C. Worth, R. Thomson, *Physics of Solid* (Mir, Moscow, 1968)
54. H. Kittel, *Introduction to Physics of Solids* (Nauka, Moscow, 1978)
55. M. Born, *Atomic Physics* (Mir, Moscow, 1970)
56. N. Ashcroft, N. Mermin, *Physics of Solids* (Mir, Moscow, 1979)

Chapter 13
New Process for Forming Multicomponent Wear-Resistant Nanostructures by Electrospark Alloying Method

V. Martsynkovskyy, V. Tarelnyk, Ie. Konoplianchenko, O. Gaponova, B. Antoszewski, Cz. Kundera, K. Dyadyura, N. Tarelnyk, B. Sarzhanov, M. Mikulina, O. Gapon and O. Semernya

Abstract The work is intended for discussing the problem of improving wear resistance of sliding friction pair parts. There are considered the processes of strengthening surfaces by highly concentrated flows of energy and substance, i.e. those being provided by the method of electrospark alloying (ESA). It is shown the ESA method allows creating macro-micro- and nanostructures to enhance operational properties of friction surfaces. There are represented the results of research and development of a process for protecting steel products from abrasive and other types of wear by applying wear-resistant coatings, made of refractory and wear-resistant metals, to the wear surfaces with the use of the ESA method, and enlarging thickness of increased hardness areas. The coatings formed by the ESA method in the sequence of C → Al → T15K6 have the largest area of increased hardness (of 320–360 μm), the maximum microhardness of the surface layer (10,000 MPa) and the smallest surface roughness (7.5 μm).

13.1 Introduction

Improving reliability and durability of dynamic equipment operating at high speeds, loads and temperatures, as well as under conditions of corrosive, abrasive and other

V. Martsynkovskyy · V. Tarelnyk · Ie. Konoplianchenko (✉) · N. Tarelnyk · B. Sarzhanov · M. Mikulina · O. Semernya
Sumy National Agrarian University, Sumy, Ukraine
e-mail: konoplyanchenko@ukr.net

O. Gaponova · K. Dyadyura
Sumy State University, Sumy, Ukraine

B. Antoszewski · Cz. Kundera
Kielce University of Technology, Kielce, Poland

O. Gapon
G.K. Parts Group LTD, Kyiv, Ukraine

A. D. Pogrebnjak and O. Bondar (eds.), *Microstructure and Properties of Micro- and Nanoscale Materials, Films, and Coatings (NAP 2019)*, Springer Proceedings in Physics 240, https://doi.org/10.1007/978-981-15-1742-6_13

types of working environment exposure remains an urgent problem and requires an integrated approach.

Modern strengthening technologies provide numerous methods for improving structures and properties of part surface layers, and each of those methods has optimal areas of application, advantages, and disadvantages.

The promising methods of strengthening surfaces and modifying thereof are the methods based on processing materials by concentrated flows of energy and substance (CFES).

The results of analysis of the CFES influence on the formation of the microstructures and the surface properties, as well as on the course of processes occurring on conditions of friction and wear of metals are represented in the works of Pogrebnjak [1–5], Litvinenko [6], Kostyuk [7, 8] and others. Strengthening metallic materials by the concentrated flows of energy and substance (CFES) in their various combinations makes it possible to intensify existing technological processes of surface treatment and to obtain results unattainable with traditional technologies such as alloying, centrifugal reinforcement and metal plating [9–11].

As for implementation issues and challenges, recently, multilayer multicomponent and nanocomposite (nanostructured) coatings have become the basis for creating surfaces having various functional purposes and improved characteristics depended thereof, such as hardness, wear, corrosion and heat resistance, fatigue strength, etc. [12–15], as well as medicinal properties [16]. For complex protection of steel against hydrogen saturation and wear, two-layer coatings of ceramics - aluminum are used, with the justified ratio of layer thicknesses when indenting and interacting with abrasive particles [17, 18].

The development of technological civilization initiates both the development of new high-energy technologies and the optimization of already known ones intended for processing materials. Among the modern methods of surface treatment by concentrated flows of energy and substance (CFES), there is electrospark alloying (ESA) method, which allows obtaining surface structures with unique physical, mechanical and tribological properties at nano level.

Despite the undeniable advantages of the ESA method, such as high adhesion strength of alloyed layer and base material, the possibility of applying any conductive materials to a surface to be strengthened, the low energy intensity of the process, environmental friendliness, ease of performing the technological operation, etc., the method has some drawbacks. They concern such problems as increasing surface roughness, reducing fatigue strength and limiting the thickness of the applied layer, which often restrains the application of this method for a wider range of machine parts and tools.

The numerous papers of Ukrainian [19–21] and foreign [22–26] scientists are devoted to improving the technology of electrospark alloying. Improving the reliability and durability of machine and equipment parts operating under difficult conditions (abrasive wear, high temperature, aggressive environment, etc.) is an actual technical problem of great importance.

13.2 Problem Formulation. Analysis of Major Achievements and Publications

Possessing wide possibilities for creating certain structures, phase and chemical compositions in part surfaces, the electrospark alloying method allows improving their operational properties. Formation of surface structures and properties thereof, as well as investigation of laws of surface friction and wear after processing surfaces by the ESA method, are research subjects by such scientists as Zolotykh [27], Lazarenko and Lazarenko [28], Samsonov [29], Verkhoturov [30], Gitlevich [31], Podchernyayeva [32], Mikhailova et al. [33].

Having summarized the studies of the above mentioned authors, it is possible to single out the following processes occurring during the electrospark alloying: the process of transferring material to the surface of a part from the alloying electrode occurs through the formation of mechanical mixtures, solid solutions, and chemical compounds; while performing the process of the surface enrichment with the elements of the alloying electrode, there is an abnormally high diffusion of the transferred material under the action of highly concentrated energy flows; the process of ultrafast quenching at short-term heating to a high temperature with a discharge of electric current is followed by instantaneous cooling; the process of plastic deformation occurs with a local pulse pressure impact on the material; the process is accompanied with creating extremely non-equilibrium structures with fine grains, high heterogeneity in composition, structure, and the process occurs when the material is subjected to localized pulse pressures and temperatures, as well as thermal stresses; the processes of nitration, carburization and oxidation take place under environment conditions [34].

Today the investigations conducted by the specialists are primarily focused on the development and improvement of new equipment for the ESA method, the creation of new electrode materials and the study of the structures and properties of coatings. But now, there is lack of information on applied technologies for the use of the ESA method. Therefore, it should be noted there is a problem that is rather relevant in modern conditions.

In paper [35], to attract attention to the use of the electrospark alloying method for repair purposes, there was considered the fact of applying a new process [36] for creating thick-layer coatings by the ESA method to restore worn parts operating under sliding friction conditions. The above said application is based on investigating the well-known publications devoted to the causes of electrospark coating limited thickness [37] and the ways to increase thereof [38]. Testing the coatings was carried out on the example of restoring the worn parts of the power cylinders friction pairs with wear and local defects of more than 300 microns depth. There has been well established the advantage of БрКМц3-1 (BrKMts3-1) bronze on applying so-called "hilly" relief coatings. There was determined the range of the optimal electrical modes for the ESA method with the pulse energy of 0.6–1.3 J, the pulse repetition rate of 250 Hz and the specific time for the ESA method of 42–53 s/cm^2.

It should be noted that in the course of performing the ESA method, a strengthened surface layer having a modified structure and properties can be formed both as a result of diffusion provided, for example, with the use of a graphite electrode as a tool to saturate the surface with carbon and to perform a kind of carburization with the help of the ESA method (CESA) [39], and also at applying coatings of refractory wear-resistant metals or their combination with soft antifriction metals [40].

In addition, while performing the ESA method, for example, with aluminum, the surface layer of the product, depending on the discharge energy (W_u), can be modified not only by diffusion (under 'soft' alloying conditions), but also owing to the occurrence of joint effect, namely, because of forming a diffusion layer and also creating a coating by transferring aluminum to the surface. It has been found that with increasing the amount of carbon in the alloyed surface and also with enhancing the discharge energy in the course of the process of aluminization by the ESA method, there is raised microhardness, continuity, the thickness of increased hardness zone and roughness of the obtained surface layer [41].

In papers [42–44], there are described the processes for increasing the thickness of strengthened surface layers of 12X18H10T steel and ХН58МБЮД nickel alloy at applying combined electrospark coatings (CEC) consisting of ВК8 hard alloy and copper, which were formed in the sequences of ВК8 → Cu → ВК8 and ВК8 → ВК8 → Cu, respectively. The methods are used to increase the wear resistance of the working surfaces of the pulse end seal rings for the turbopump units of the liquid rocket engines, and those consist in that before applying the CEC, there is produced the process using the CESA method. At the same time, the increased hardness zone is enlarged by the thickness of the CESA layer.

Thus, it should be noted that the thickness of the increased hardness zone of the surface layer is essential for parts of friction pairs at occurring abrasive and other types of wear, and the problem of increasing such a thickness remains relevant.

The aim of the work is to increase the efficiency of the electrospark alloying method to protect steel products from abrasive and other types of wear by applying wear-resistant coatings to the wear surfaces and enlarging the thickness of the increased hardness zone, due to preliminarily performing the CESA method and applying aluminum by the ESA method.

13.3 Research Methods

To study the structure and perform measurements of surface layer microhardness, the thin sections made of 12X18H10T steel and having dimensions of 10 × 10 × 8 were used after processing them for 1.0 min. by the CESA method. The process of alloying occurred on different conditions at the Элитрон-52А (Elitron-52A) unit equipped with a manual vibrator and at the Эил-9 (EIL-9) mechanized unit at the discharge energy range (W_p) of 0.5–6.8 J.

The surface of a thin section was oriented perpendicular to the surface being alloyed. To exclude edge effect in the course of alloying, the end face of a specimen had been milled by the depth of at least 2 mm before preparing the thin section thereof. To prevent the crimpling of the layer, the edge of the specimen was fastened with a counterbody in a clamp. To identify the structure, the thin section was subjected to chemical etching in a reagent.

After manufacturing, the thin sections were examined with the use of the Neofot-2 optical microscope, where the layer quality and continuity and also the thickness and the structure of the underlayer zones, namely the diffuse zone and the heat-affected zone, were evaluated. At the same time, there was carried out a durametric analysis to determine microhardness distribution in the surface layer along with the depth of the thin section beginning from the surface thereof. The microhardness measurement was performed using the PMT-3 microhardness tester by means of indenting a diamond pyramid under the load of 0.05 N. The roughness was measured by reading and processing profilograms with the use of the profilograph-profilometer measuring device of 201 model, "Kalibr" plant production. To study the distribution of the elements and carbon along the depth of the layer, the local X-ray spectroscopic analysis was performed using the Joel JSM-5400 scanning electron microscope equipped with the microanalyzer of the ISIS 300 Oxford Instruments.

13.4 Research Results

Figure 13.1 shows the microstructure (a, b) and the distribution of the microhardness in the surface layer (c, d) of 12X18H10T steel at processing by the CESA method with the discharge energy (W_u) of 3.4 J. The layer is not uniform, and it is of 160 μm (a, c) to 90 μm (b, d). On average, the depth of the increased hardness layer is ~100 μm.

Figure 13.2 represents the results of the electron microscopic studies, that is the area of the surface layer of 12X18H10T steel after processing it by the CESA method with $W_u = 3.4$ J, as well as the distribution of carbon and other elements, that is chromium, iron, nickel, and titanium, which are the parts of the steel, along the depth of the layer.

From the figure, it can be seen that the increased carbon content in the surface layer is maintained up to 100 μm, which fact is consistent with the data of durametric studies (Fig. 13.1).

Figure 13.3 shows the microstructure (a) and the distribution of the surface layer microhardness (b) of 12X18H10T steel after processing it by the CESA method and doing it with aluminum by the ESA method with $W_u = 3.4$ J. The layer is rather massive, but not uniform. The thickness of the layer having a modified structure is ~170 μm. At the same time, on the surface, there is an uneven soft, crumbly and porous layer, the thickness of which is of 70–100 μm in various areas, and the microhardness is of 1000–2500 MPa. Under this layer, there is a layer of a higher microhardness (up to 7500 MPa) having a depth of 80–120 μm.

Figure 13.4 shows the area of the surface layer of 12X18H10T steel after processing it by the CESA method and doing it with aluminum by the ESA method

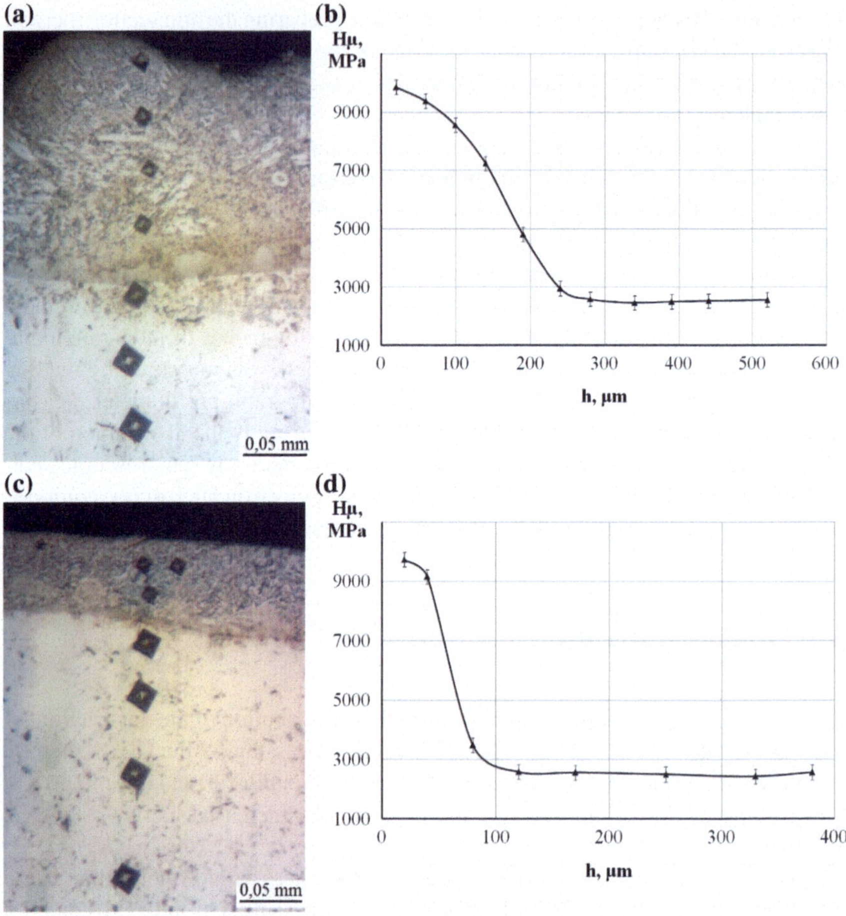

Fig. 13.1 Microstructure (**a**, **b**) and microhardness distribution (**c**, **d**) of the surface layer of steel 12X18H10T at processing it by the CESA method with $W_u = 3.4$ J

with $W_u = 3.4$ J, as well as the distribution of aluminum, iron, chromium, nickel, carbon, and titanium. The results of the local X-ray and microscopy microanalysis indicate that in the course of processing by the ESA method, aluminum diffuses into the steel substrate by the depth of 100 μm. The content of the elements (Fe, Cr, Ni, Ti) comprised into 12X18H10T steel in the surface layer is minimal, and at a distance of 80–100 μm starting measurement from the surface, it corresponds to the steel grade composition.

With sequential alloying of the steel according to the scheme of C → Al → T15K6 at $W_u = 3.4$ J, the formation of a massive layer having a thickness of 320–360 μm and a maximum microhardness on the surface of 9000–10,000 MPa is observed. The results of metallographic studies indicate that a significant amount of carbides and

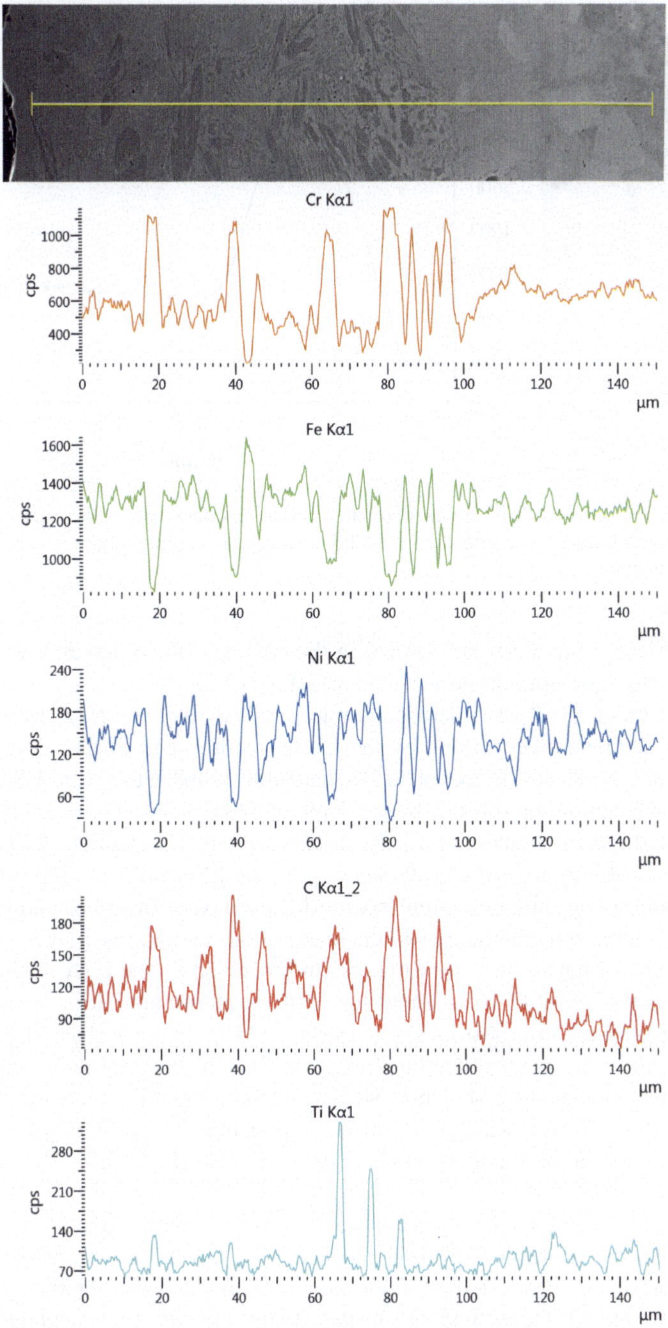

Fig. 13.2 Distribution of the elements in the surface layer of 12X18H10T steel after processing it by the CESA method

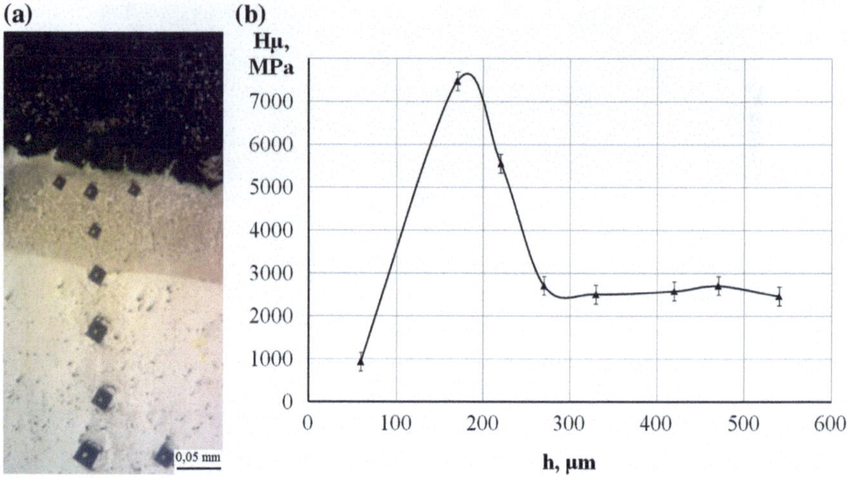

Fig. 13.3 Microstructure (**a**) and distribution of microhardness (**b**) in the surface layer of 12X18H10T steel after processing it by the CESA method and further doing it with aluminum by the ESA method

other chemical compounds are formed in the coating, which apparently results in significant strengthening of the surface layer (Fig. 13.5).

The results of the X-ray microscopy microanalysis indicate (Fig. 13.6) that, as a result of sequential processing by the CESA method and further doing it with aluminum and the electrode-tool of T15K6 hard alloy with the use of the ESA method at $W_u = 3.4$ J, aluminum diffuses by the depth of 320 μm, but its content decreases, tungsten as the main element of T15K6 hard alloy—by the depth of 270 μm, and carbon does it by up to 270 μm. Processing by the ESA method according to the described technology allows us to enlarge the diffusion zones of carbon and aluminum and also to increase the hardness and thickness of the strengthened layer.

The results of the studies show that, when processed by the ESA method under the scheme of C → T15K6 (without the ESA by aluminum), it is impossible to achieve the required microhardness values (Fig. 13.7). On the surface, it is equal to 9000 MPa, and as deepening, it gradually decreases to the microhardness value of the substrate, which is about 2500–2600 MPa. The thickness of the increased hardness layer is of 200–220 μm, which is 100 μm less than that as a result of processing by the ESA method in the sequence of C → Al → T15K6 (Fig. 13.5). The continuity of the layer is ~100%.

Figure 13.8 represents the results of the study for the distribution of the elements included into the substrate (12X18H10T steel) and also those comprised into the composition of the electrode-tool along the depth of the layer obtained after processing it by the CESA method and further doing it by the ESA method with the electrode-tool of T15K6 hard alloy at $W_u = 3.4$ J. Tungsten diffuses by the depth of 125 μm, and the layer with a higher carbon content is ~120 μm.

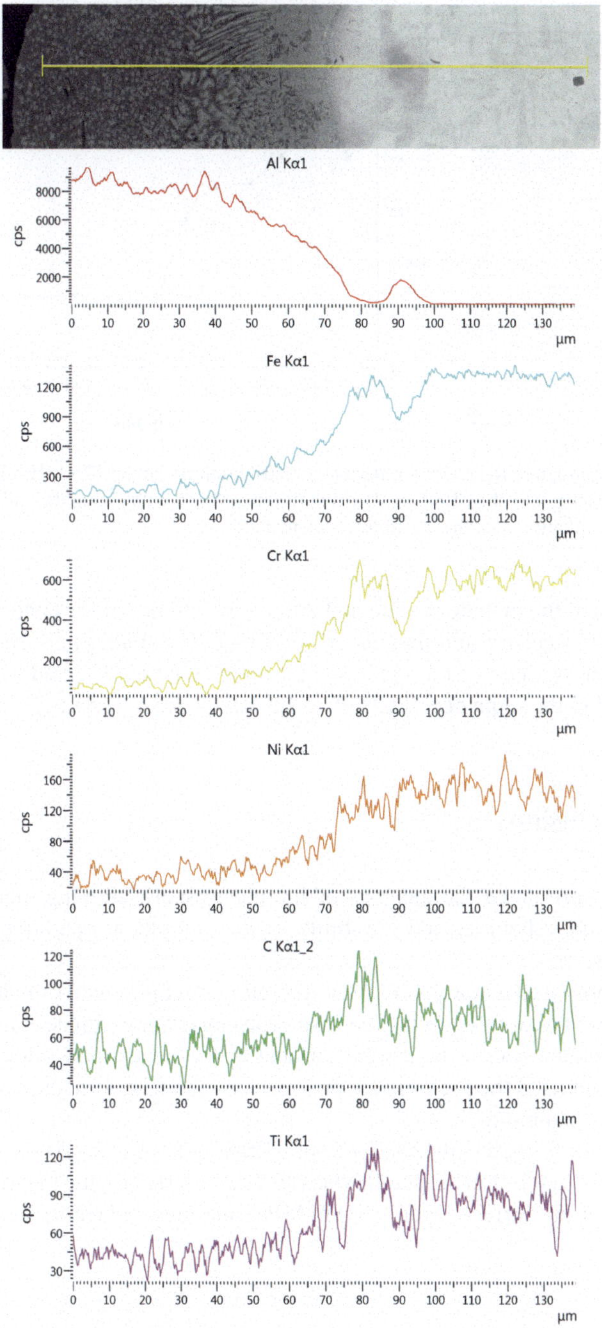

Fig. 13.4 Distribution of the elements in the surface layer of 12X18H10T steel after processing it by the CESA method and further doing it with aluminum by the ESA method

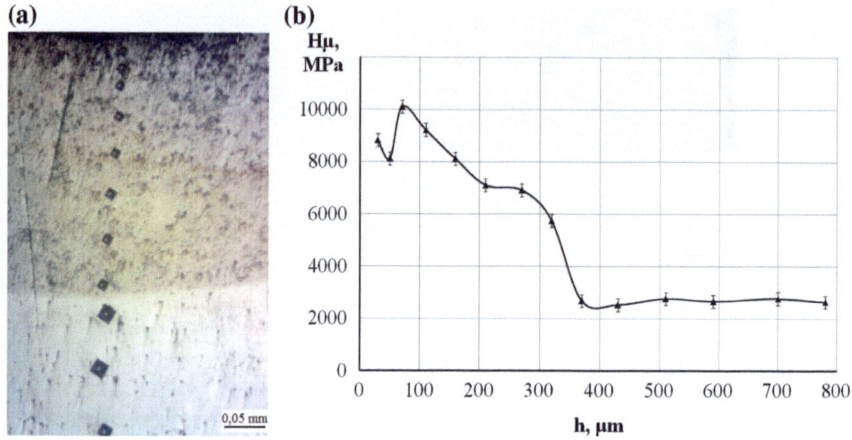

Fig. 13.5 Microstructure (**a**) and microhardness distribution (**b**) in the 12X18H10T steel surface layer after processing it by the CESA method and further doing it with aluminum and the electrode-tool of T15K6 hard alloy with the use of the ESA method

The results of the metallographic and durametric studies of the coatings obtained on 12X18H10T steel are summarized in Table 13.1. The studies have shown that the greatest thickness value of the increased hardness zone is observed in the surface layers formed in the sequence of CESA + ESA Al + ESA T15K6.

13.5 Conclusions

1. There is represented the analysis of the electrospark alloying method, which increases the reliability and durability of the equipment operating in difficult conditions.
2. There is proposed a new process for protecting steel products from abrasive and other types of wear by applying the wear-resistant coatings, made of the refractory wear-resistant metals, to their wear surfaces by the electrospark alloying method, having conducted the preliminary processes by the CESA method and the ESA method with aluminum.
3. The coatings formed in the sequence of CESA → ESA C → Al → T15K6 have the greatest zones of increased hardness (of 320–360 μm), the maximum micro-hardness of the surface layer (10,000 MPa) and the smallest surface roughness (7.5 μm).

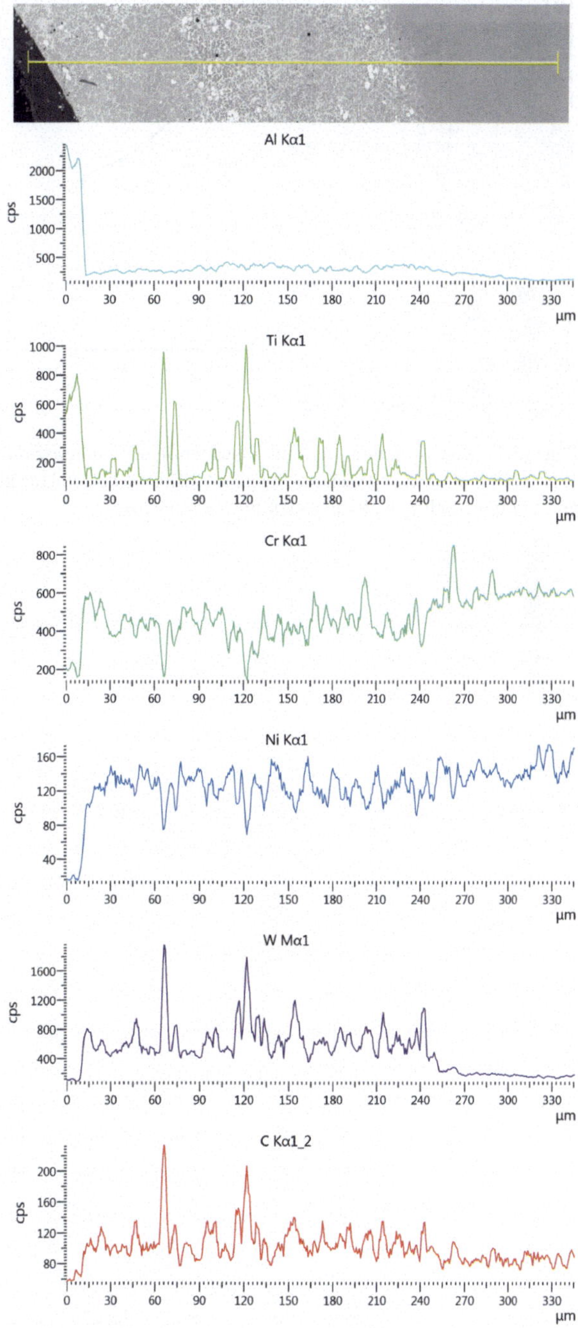

Fig. 13.6 Distribution of the elements in the surface layer of 12X18H10T steel after processing it by the CESA method and further doing it with aluminum and the electrode-tool of T15K6 hard alloy with the use of the ESA method

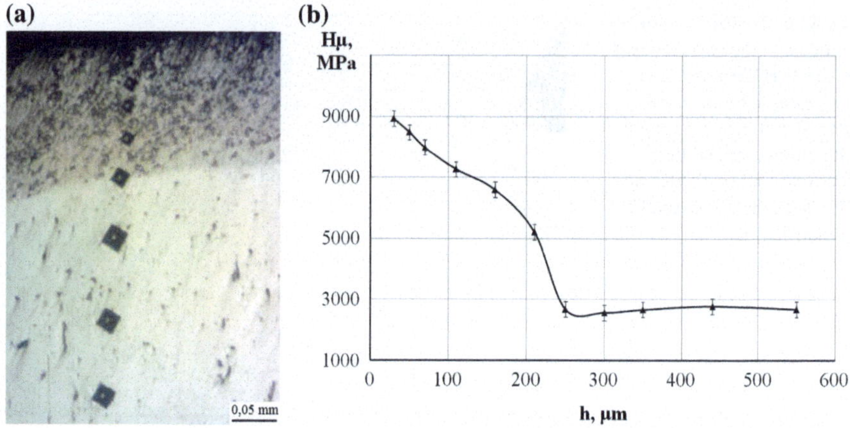

Fig. 13.7 Microstructure (**a**) and distribution of microhardness (**b**) of the surface layer of 12X18H10T steel after processing it by the CESA method and further doing it with the electrode-tool of T15K6 hard alloy with the use of the ESA method

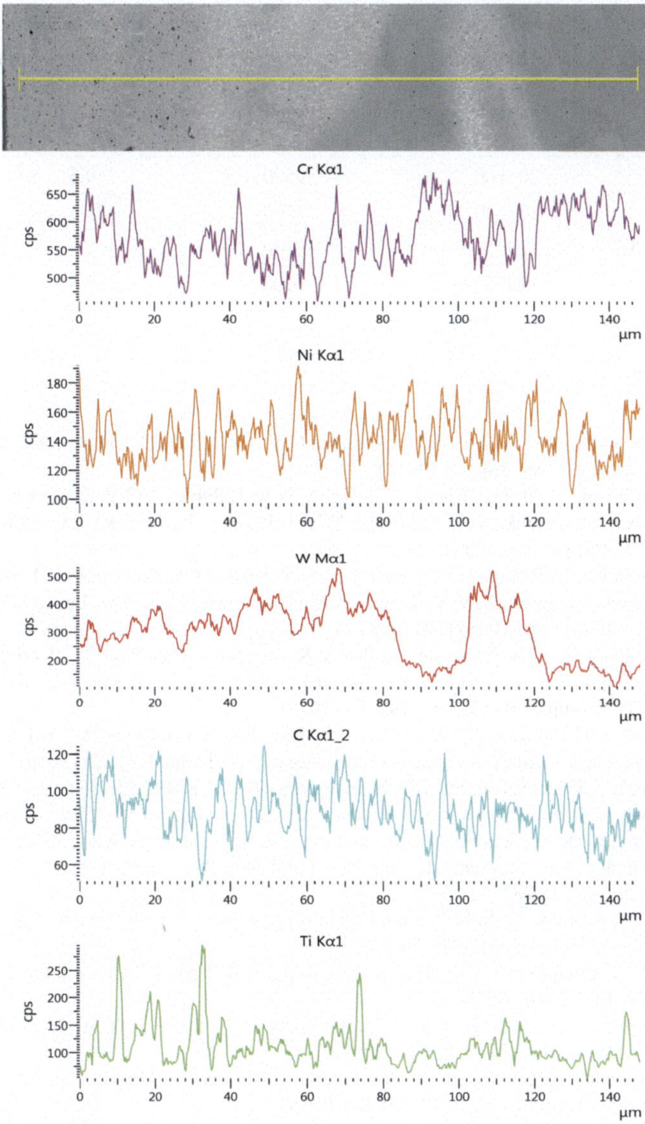

Fig. 13.8 Distribution of elements in the surface layer of steel 12X18H10T after processing it by the CESA method and further doing it by the ESA method with the electrode-tool of T15K6 hard alloy

Table 13.1 Results of metallographic studies of the specimens made of 12X18H10T steel

Electrode-tool material	Thickness of surface layer increased hardness zone (μm)	Maximum microhardness of strengthened surface layer (MPa)	Surfaces roughness, Ra (μm)
C	90–160	9500	9.5
C + Al	80–120	7500	25.5
C + Al + T15K6	320–360	10,000	7.5
C + T15K6	140–220	9000	8.5

References

1. A.D. Pogrebnyak, A.M. Mahmud, I.T. Karasha, G.V. Kirik, R.Y. Tkachenko, A.P. Sypylenko, J. Nano- Electron. Phys. **3**(4), 73 (2011)
2. A.D. Pogrebnyak, M.M. Danilionok, V.V. Uglov, N.K. Erdybaeva, G.V. Kirik, S.N. Dub, V.S. Rusakov, A.P. Shypylenko, P.V. Zukovski, YZh Tuleushev, Vacuum **83**, 235 (2009). https://doi.org/10.1016/j.vacuum.2009.01.071
3. A.D. Pogrebnyak, A.P. Shpak, V.M. Beresnev, G.V. Kirik, D.A. Kolesnikov, F.F. Komarov, P. Konarski, N.A. Makhmudov, M.V. Kaverin, V.V. Grudnitskii, Tech. Phys. Lett. **37**, 636 (2011). https://doi.org/10.1134/S1063785011070145
4. A.D. Pogrebnjak, A.A. Bagdasaryan, A. Pshyk, K. Dyadyura, Usp. Phys. Nauk **60**, 586 (2017)
5. A.D. Pogrebnjak, V.I. Ivashchenko, P.L. Skrynskyy, O.V. Bondar, P. Konarski, K. Załęski, S. Jurga, E. Coy, Compos. Part B Eng. **142**, 85 (2018)
6. A.B. Batracov, M.I. Bazaleev, S.E. Donets, V.F. Klepikov, V.V. Lytvynenko, YuF Lonin, A.G. Ponomarev, V.V. Uvarov, V.T. Uvarov, Probl. Atomic Sci. Technol. **6**, 225 (2013)
7. G.I. Kostyuk, S.S. Dobrotvorsky, V.A. Fadeev, News NTU "KhPI" **53**(959), 116 (2012)
8. G.I. Kostyuk, *Effective Coatings and Modified Strengthened Layers on Cutting Tools* (House of the International Academy of Science and Innovation Technology, Kiev, 2012)
9. T.M. Radchenko, V.A. Tatarenko, H. Zapolsky, Solid State Phenomena **138**, 283 (2008). www.scientific.net/SSP.138.283
10. L. Ropyak, I. Schuliar, O. Bohachenko, East.-Europ. J. Enterp. Tech. **1**(5), 53 (2016). https://doi.org/10.15587/1729-4061.2016.59850
11. L. Ropyak, V. Ostapovych, East.-Europ. J. Enterp. Tech. **2**(5), 50 (2016). https://doi.org/10.15587/1729-4061.2016.65719
12. A.D. Korotaev, V.Y. Moshkov, S.V. Ovchinnikov, Y.P. Pinzhin, V.M. Savostikov, A.N. Tyumentsev, Phys. Mesomech. **8**, 103 (2005)
13. A.D. Pogrebnjak, A.P. Shpak, N.A. Azarenkov, V.M. Beresnev, Phys. Usp. **52**, 29 (2009). https://doi.org/10.3367/UFNe.0179.200901b.0035
14. A.D. Pogrebnjak, A.G. Ponomarev, A.P. Shpak, YuA Kunitskii, Phys. Usp. **55**, 270 (2012). https://doi.org/10.3367/UFNe.0182.201203d.0287
15. A.D. Pogrebnyak, V.M. Beresnev, A.A. Demyanenko, V.S. Baidak, F.F. Komarov, M.V. Kaverin, N.A. Makhmudov, D.A. Kolesnikov, Solid State Phys. **54**(9), 1764 (2012)
16. O.Y. Popadyuk, O.S. Malyshevska, L.Y. Ropyak, V.S. Vytvytskyi, M.M. Droniak, Novosti Khirurgii **27**(1), 16 (2019). https://doi.org/10.18484/2305-0047.2019.1.16 (in Russian)
17. L.Ya. Ropyak, I.P. Shatskyi, M.V. Makoviichuk, Metallofiz. Noveishie Tekhnol. **39**(4), 517 (2017). https://doi.org/10.15407/mfint.39.04.0517
18. L.Ya. Ropyak, I.P. Shatskyi, M.V. Makoviichuk, Metallofiz. Noveishie Tekhnol. **41**(5), 647 (2019)
19. E.V. Korbut, B.A. Lyashenko, I.I. Podchernyaeva, D.V. Yurchenko, Proc. Mech. Proc. in Machine Build. **12**, 67 (2012) (in Russian)

20. O.S. Zavoiko, S.N. Novikov, Explor. Develop. Oil Gas Deposits **48**, 119 (2013). (in Russian)
21. A.V. Paustovskii, Y.G. Tkachenko, R.A. Alfintseva, S.N. Kirilenko, D.Z. Yurchenko, Surf. Eng. App. Electrochem. **49**, 4–12 (2013). https://doi.org/10.3103/S1068375513010109
22. V.V. Mikhailov, A.E. Gitlevich, A.D. Verkhoturov, A.I. Mikhailyuk, A.V. Belyakov, L.A. Konevtsov, Surf. Eng. Appl. Electrochem. **49**, 373 (2013). https://doi.org/10.3103/S1068375513050074
23. I. Pliszka, N. Radek, Procedia Eng. **192**, 707 (2017). https://doi.org/10.1016/j.proeng.2017.06.122
24. X. Hong, K. Feng, Y.-F. Tan, X.-L. Wang, H. Tan, Trans. Nonferrous Met. Soc. China **27**, 1767 (2017). https://doi.org/10.1016/S1003-6326(17)60199-7
25. P.D. Enrique, Z. Jiao, N.Y. Zhou, E. Toyserkani, J. Mater. Process. Technol. **258**, 138 (2018). https://doi.org/10.1016/j.jmatprotec.2018.03.023
26. M. Salmaliyan, F. Malek Ghaeni, M. Ebrahimnia, Surf. Coat. Tech. **321**, 81 (2017). https://doi.org/10.1016/j.surfcoat.2017.04.040
27. B.N. Zolotykh, Compil. TsNIIL-Elektrom. **38** (1957)
28. B.R. Lazarenko, N.I. Lazarenko, Electronic Process. Mater. **3**, 12 (1977)
29. G.V. Samsonov, A.D. Verkhoturov, G.A. Bovkun, V.S. Sychev, *Electrospark Alloying of Metal Surfaces* (Naukova Dumka, Kiev, 1976). (in Russian)
30. A.D. Verkhoturov, T.B. Ershova, L.A. Konevtsov, Theor. Found. Chem. Eng. **41**, 624 (2007)
31. A.E. Gitlevich, A.I. Mikhailyuk, V.V. Mikhailov, Electr. Proc. Mater. **3**, 12 (1995)
32. I.A. Podchernyaeva, O.N. Grigor'ev, V.I. Subbotin, Powder Metall. Met. Ceram. **43**, 391 (2004). https://doi.org/10.1023/B:PMMC.0000048133.84178.97
33. L.P. Kornienko, G.P. Chernova, V.V. Mikhailov, A.E. Gitlevich, Surf. Eng. Appl. Electrochem. **47**, 9 (2011). https://doi.org/10.3103/S106837551101011X
34. D.N. Korotayev, *Technological Capabilities for Forming Wear-Resistant Nanostructures by Electrospark Alloying* (SibADI, Omsk, 2009). (in Russian)
35. S.A. Velichko, P.V. Senin, V.I. Ivanov, P.V. Chumakov, Surf. Eng. Appl. Electrochem. **53**, 116 (2017). https://doi.org/10.3103/S1068375517020119
36. V.I. Ivanov, F.Kh. Burumkulov, V.A. Denisov, EA Patent 017066B1, 28 Sept 2011
37. V.I. Ivanov, Works GOSNITI **113**, 429 (2013)
38. V.I. Ivanov, Works GOSNITI **113**, 450 (2013)
39. V.B. Tarel'nik, A.V. Paustovskii, Y.G. Tkachenko, Surf. Eng. Appl. Electrochem. **54**, 147 (2018). https://doi.org/10.3103/s106837551802014x
40. V.B. Tarelnyk, E.V. Konoplianchenko, V.S. Martsynkovskyi, A.V. Paustovskii, Y.G. Tkachenko, B. Antoszewski, Powder Metall. Met. Ceram. **55**, 585 (2017). https://doi.org/10.1007/s11106-017-9843-2
41. G.V. Kirik, O.P. Gaponova, V.B. Tarelnyk, O.M. Myslyvchenko, B. Antoszewski, Powder Metall. Met. Ceram. **56**, 688 (2018). https://doi.org/10.1007/s11106-018-9944-6
42. V.B. Tarel'nik, V.S. Martsinkovskii, A.N. Zhukov, Chem. Petrol. Eng. **53**, 266 (2017). https://doi.org/10.1007/s10556-017-0333-7
43. V.B. Tarel'nik, V.S. Martsinkovskii, A.N. Zhukov, Chem. Petrol. Eng. **53**, 114 (2017). https://doi.org/10.1007/s10556-017-0305-y
44. V.B. Tarel'nik, V.S. Martsinkovskii, A.N. Zhukov, Chem. Petrol. Eng. **53**, 385 (2017). https://doi.org/10.1007/s10556-017-0351-5

Chapter 14
Effect of Structure on the Properties of Composite Cr_3C_2 + NiCr Coatings

Ye. Titkov, Olena Berdnikova, Yu. Tyurin, O. Kolisnichenko, Ye. Polovetskiy and O. Kushnaryova

Abstract Increase of the reliability and durability of products, the operating properties of which are determined by the properties of their working surfaces and make it possible to use them in extreme conditions: high temperature and pressure, intense wear by friction, alternating loads, etc. There are various technological solutions that provide the necessary properties of functional surface layers depending on the requirements for the products. The proposed technology for the creation of surface layers using a detonation gun sprayer of the original design, developed in the E. O. Paton Electric Welding Institute. This method allows meeting the increasing demands on the quality of the coatings obtained, energy and material efficiency, etc. The goal of the work is to study the phase composition and microstructure and their influence on the properties of cermet coatings of the Cr_3C_2–NiCr system sprayed in various modes using a multi-chamber detonation device. An integrated research approach was carried out, which made it possible to obtain the data on the structure and phase composition of the coatings under study by the methods of light, scanning and transmission electron microscopy. The obtained data on structural parameters made it possible to carry out analytical evaluation of the level of strength and crack resistance of the coatings obtained. With an increase in heat input per unit volume of sprayed powder the level of decarbonization of carbide particles increases. Evaluation of the local inner stresses of the coatings allowed us to determine that it leads to formation of the critical zones with high tendency of crack formation.

14.1 Introduction

Cermet coatings such as Cr_3C_2–NiCr are widely used for wear applications at high temperatures. The ductile metal matrix of NiCr acts as a binder, supporting the brittle reinforcing phase of Cr_3C_2 [1]. There are various technological solutions that provide

Ye. Titkov (✉) · O. Berdnikova · Yu. Tyurin · O. Kolisnichenko · Ye. Polovetskiy · O. Kushnaryova
E. O. Paton Electric Welding Institute of the NAS of Ukraine, Kiev, Ukraine
e-mail: ye.titkov@gmail.com

A. D. Pogrebnjak and O. Bondar (eds.), *Microstructure and Properties of Micro- and Nanoscale Materials, Films, and Coatings (NAP 2019)*,
Springer Proceedings in Physics 240, https://doi.org/10.1007/978-981-15-1742-6_14

the necessary properties of functional surface layers depending on the requirements for the products [2, 3]. The proposed technology for the creation of surface layers using a detonation gun sprayer of the original design, developed in the E. O. Paton Electric Welding Institute. This method allows meeting the increasing demands on the quality of the coatings obtained, energy and material efficiency, etc. [4, 5].

14.2 General Information

14.2.1 Objectives

The goal of the work is to study the phase composition and microstructure and their influence on the properties of cermet coatings of the Cr_3C_2–NiCr system sprayed in various modes using a multi-chamber detonation device.

14.2.2 Materials and Conditions for Obtaining the Coatings

For detonation sprayingб Amperit 584.054 composite powder (75% Cr_3C_2 + 25% NiCr) of 10–45 μm fraction was used. The powder used for spraying was obtained by agglomeration and sintering (Fig. 14.1).

The parameters of the technological modes of multi-chamber detonation spraying are presented in Table 14.1. Changing the specified process parameters leads to a change in the heat input to the sprayed powder.

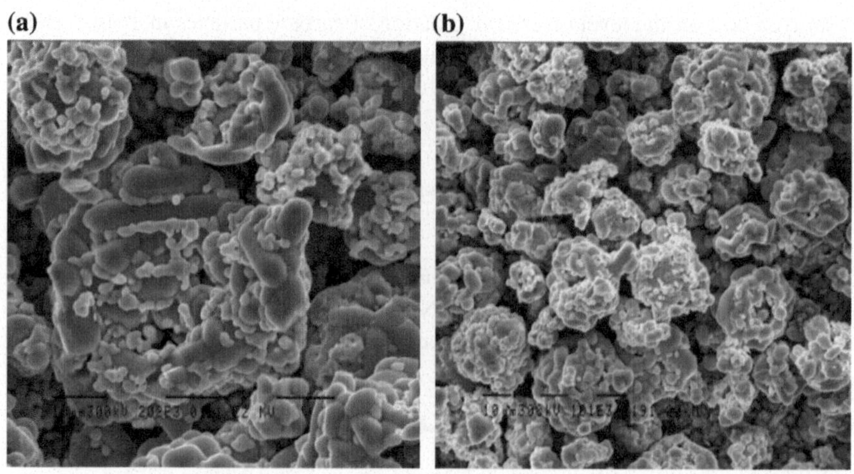

Fig. 14.1 Appearance of sprayed powders: **a** ×1010, **b** ×2020

Table 14.1 Spraying modes

Spraying mode	Gas consumption (m^3/h)	The ratio of oxidant to combustible gas		
I	Chamber 1	O_2	3.08	4.96
		Air	1.50	
		C_3H_8	0.68	
	Chamber 2	O_2	2.54	4.85
		Air	1.24	
		C_3H_8	0.57	
	Feeder	Air	0.95	
II	Chamber 1	O_2	2.35	4.95
		Air	2.32	
		C_3H_8	0.57	
	Chamber 2	O_2	1.96	4.97
		Air	1.95	
		C_3H_8	0.47	
	Feeder	Air	0.95	
III	Chamber 1	O_2	3.60	5.07
		Air	1.09	
		C_3H_8	0.75	
	Chamber 2	O_2	3.04	4.94
		Air	0.84	
		C_3H_8	0.65	
	Feeder	Air	0.95	

14.3 Results and Discussion

14.3.1 Microstructure Studies

Three variants of coatings up to 400 microns thick were obtained (Fig. 14.2). The highest porosities are observed in coatings group II and the lowest—in group I. Coatings also differ in the degree of lamellization: group I of coatings has an average volume fraction of lamellas in the matrix—18–22%, group II—minimum content—15–18% and group III with a maximum of 22–24%. It should be noted that the quality of the coating is better, the higher the volume fraction of lamellae formed by the molten and/or plasticized powder particles. The high content of melted and deformed particles determines the good cohesive strength of the resulting coating. At the same time, increasing the volume fraction of unmelted particles leads to disruption of the interaction processes of the particles with each other and with the surface of the substrate, significantly deteriorating the quality of the resulting coating.

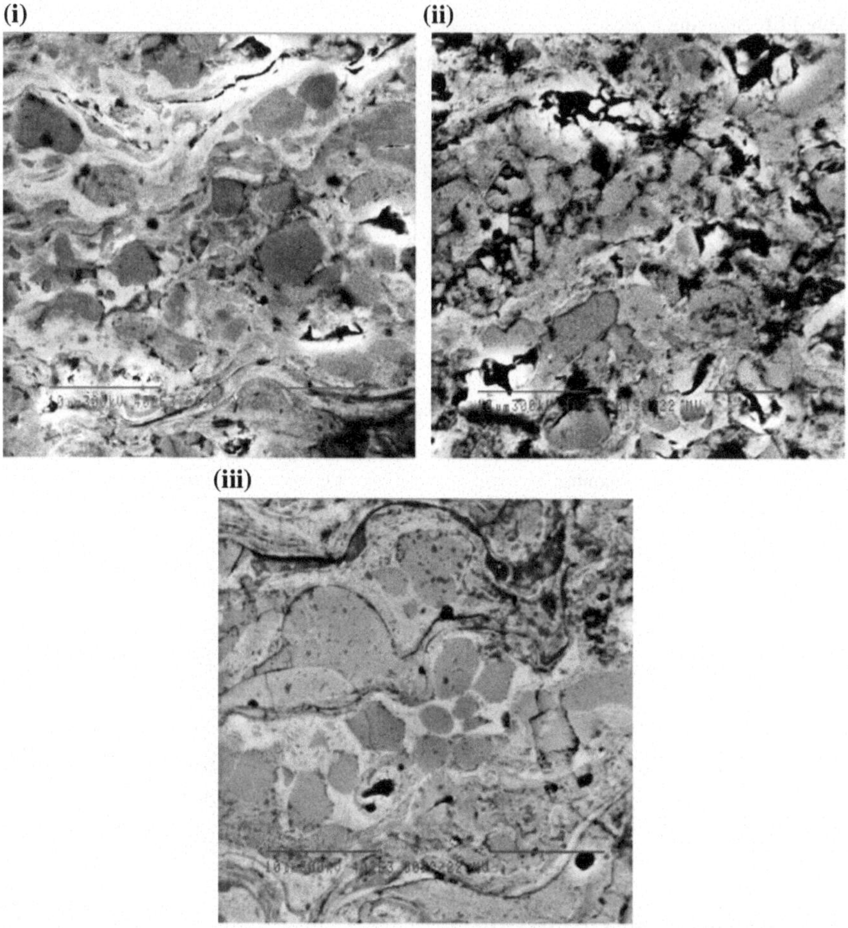

Fig. 14.2 Appearance of the coatings, ×4020

The results of the microstructure and microhardness study of the obtained coatings are presented in Table 14.2. X-ray diffraction studies of the coatings showed that the decarbonization of carbide particles was the strongest in the coatings obtained using mode III, the content of the Cr_3C_2 phase was the lowest during the formation

Table 14.2 Results of microstructure and microhardness studies

Spraying mode	Maximum thickness (μm)	Porosity (%)	Lamellae volume (%)	Microhardness $HV_{0.3}$ (GPa)
I	400	0.7–1.3	18–22	7–13.2
II	375	9–13	15–18	5.6–10.5
III	375	1–4	22–24	7.3–9.2

of 40% (Cr_7C_3 + Cr_2O_3). The least decomposition of Cr_3C_2 carbide particles is observed in group II of coatings. However, these coatings are characterized by the greatest porosity and gradients of microhardness, which can reduce their strength and crack resistance. Studies of coatings in group I showed an average degree of decarbonization of carbide particles with a content of about 50% Cr_3C_2 and 26% (Cr_7C_3 + Cr_2O_3). In this case, as noted above, their porosity is minimal with the most uniform distribution of microhardness.

Thus, the results of studies have shown that the use of mode I in the detonation spraying of Cr_3C_2–NiCr metal ceramics is the most optimal. Such coatings are distinguished by an average lamella content, an average degree of dissociation of carbide particles, and a minimum porosity compared with modes II and III. In turn, studies of coatings sprayed on mode II show a low degree of decomposition of higher carbides, but at the same time an insufficient degree of lamellization. In coatings sprayed using mode III, a higher volume fraction of lamellae is noted, but also a higher level of decarbonization of carbide particles.

14.3.2 Transmission Electron Microscopy

The change of the dislocation density (ρ) in the structural components, the substructure formations, the size of particles, their distribution, etc. were studied in 2 groups of coatings: I and III.

In both cases, the structure with large Cr_3C_2 carbide particles of 1–2 μm in size is clearly visible (Fig. 14.3a, 14.4b), around which dispersed Cr_7C_3 and Cr_2O_3 phases are formed (Fig. 14.3, 14.4). In coatings sprayed on mode I (Fig. 14.3), the particle size ($Cr_7C_3 + Cr_2O_3$) is 10–50 nm (Fig. 14.3b, 14.4a) with their uniform distribution in the NiCr matrix. The dislocation density in the volume of the NiCr lamellae of the matrix was $\rho = 2 \times 10^{10}$ cm^{-2}, at the lamella–lamella border $\rho = (4–5) \times 10^{10}$ cm^{-2}, at the “carbide–lamella” border $\rho = (5–6) \times 10^{10}$ cm^{-2}. It was established that the lamella thickness is 1.5–3.5 μm, and the size of the NiCr substructure of the matrix is 0.1–0.3 μm.

In coating sprayed on mode III (Fig. 14.4), in the NiCr matrix, larger particles of phase precipitates (15–80 nm) are found. The size of the subgrain is 0.2–0.3 μm. The formation of gradients in dislocation density is also noted. If the dislocation density in the NiCr lamella volume was $\rho = 8–9 \times 10^9$ cm^{-2}, and at the lamella – lamella boundary $\rho = 8 \times 10^{10}$ cm^{-2}, then at the carbide – lamella boundaries due to the formation of more massive interlayers of Cr_7C_3 and Cr_2O_3 the dislocation density increases to $\rho = 10^{11}$ cm^{-2}. In addition to the increase in the scalar dislocation density in the structure of the coating, the formation of extended dislocation clusters—potential sources of crack formation—was noted. In addition, in the case of a coating sprayed on mode III, the grinding of the structure and substructure is noted. The lamella thickness is Ds = 1.25 μm, and the average size of the NiCr substructure of the matrix is no more than 200 nm.

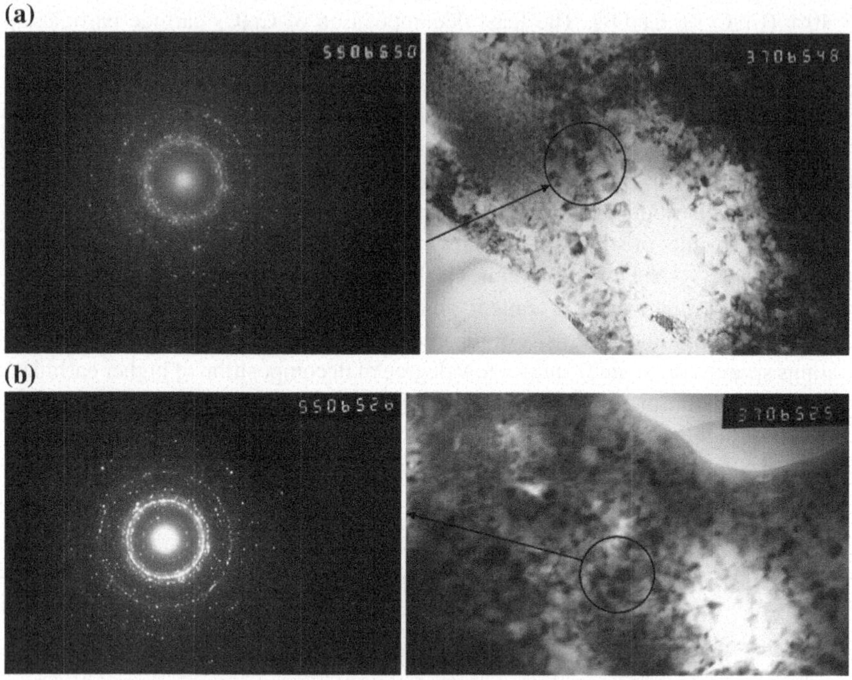

Fig. 14.3 Transmission electron microscopy studies of coatings in group

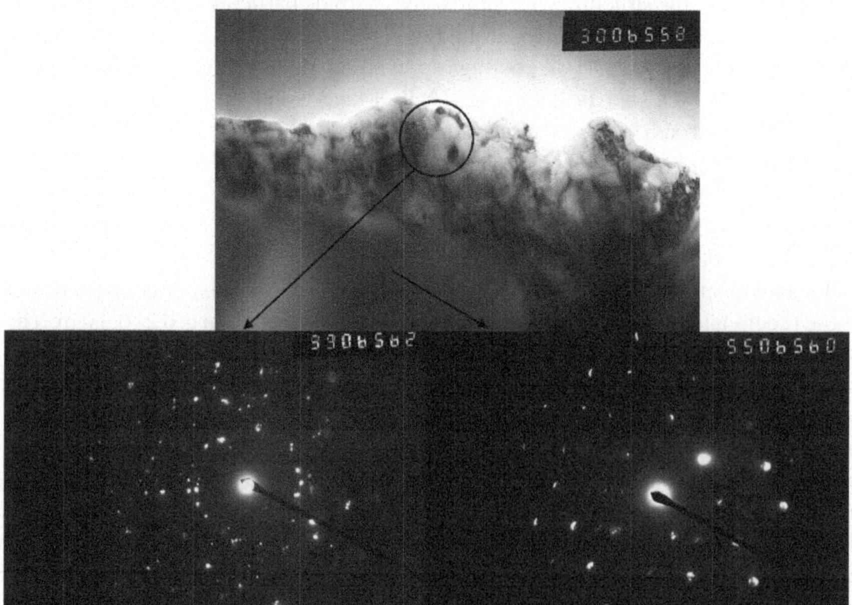

Fig. 14.4 Transmission electron microscopy studies of coatings in group III

Studies of the coatings structure revealed that during the transition from mode I to mode III: the volume fraction of carbide Cr_3C_2 decreases from 49 to 32% (1.5 times) with an increase of component ($Cr_7C_3 + Cr_2O_3$) from 26 to 40% (1.54 times) and enlargement of the substructure by 1.2 times. Coatings obtained using mode III are also characterized by the formation of extended dislocation clusters at the carbide–lamella boundaries due to the formation of massive interlayers of Cr_7C_3 and Cr_2O_3 with an increase of dislocation (from 5–6 × 10^{10} cm^{-2} to 9 × 10^{11} cm^{-2}) compared with coatings obtained using mode I. At the same time, the gradients in the dislocation density are from (8–9) × 10^9 cm^{-2} (in the volume of NiCr lamellae) to 8 × 10^{10} cm^{-2} (along their boundaries) and 9 × 10^{11} cm^{-2} (along the carbide/lamella borders)—maximum, which should lead to an increase in the level of local internal stresses—potential sources of cracking.

Thus, the use of mode I leads to more uniform structural changes, which should ensure high indicators of strength and crack resistance properties.

14.3.3 Analytical Evaluations of the Operational Properties

The obtained data made it possible to estimate the differentiated contribution of each of the structural parameters to the integral value of the yield strength using classical dependencies [6]. It was established that the total value of the yield strength for samples of coatings sprayed using mode I is 1745 MPa. The greatest contribution to the total value makes the hardening of the matrix by the particles of primary and secondary carbides—up to 35%. In the case of the use of mode III, the total value of the yield strength is 2105 MPa. As in the first case, the maximum contribution to the integral value is made by the particles of primary and secondary carbides—up to 40%.

The analytical method has shown that with an increase in heat input per volume unit of the sprayed powder, we can observe the increase of integral values of yield strength due to the refinement of the structure and substructure, increase in the density of dislocations, as well as increase in the volume fraction of secondary carbide particles in the matrix. A computational and analytical evaluation of the level of local internal stresses was made. It was found that the maximum level of local internal stresses for the first group of coatings does not exceed 1020 MPa or 0.082 from the level of the theoretical material shear strength. At the same time, the level of local internal stresses of the second sample was almost equal to the theoretical shear strength of this material (0.97 of the theoretical strength) and reached 12,100 MPa. Thus, the formation of extended dislocation clusters around significantly degraded carbide particles leads to a catastrophic increase in the level of local internal stresses. Such localization is likely to lead to the appearance of microcracks and crumbling of particles of primary carbides from the matrix during practical operation. Therefore, spraying modes that lead to a comparatively higher hardness of the matrix due to the mechanisms mentioned above and at the same time to the appearance of significant localized internal stresses are unacceptable. At the same time, a moderate increase in the strength properties and hardness of the matrix without significant degradation of the primary carbide particles and the formation of low values of local internal stresses should lead to an increase in the performance properties of the final products.

14.4 Conclusions

The use of a multi-chamber detonation spraying at optimum conditions ensures the production of high-quality Cr_3C_2–NiCr coatings.

In the process of spraying Cr_3C_2–NiCr powder undergoes changes due to the influence of high temperatures and the atmosphere of detonation products. Its phase composition changes with the formation of the complex heterogeneous state in the coatings. A layered structure of the regions of carbide particles and the matrix metal is found next to the "carbide–matrix" boundary with precipitates in the matrix of dispersed secondary carbides.

It is shown that the change in the parameters of the detonation spraying process has a significant impact on the final structural and phase state of the coatings obtained. The increase in heat input per mass unit of the powder, on the one hand, leads to an increase in the volume fraction of NiCr lamellae matrix, grinding the structure and reducing the porosity of coatings. On the other hand, it contributes to a significant intensification of the degradation processes of primary carbide particles and an increase in the volume fraction of secondary carbide particles in the matrix.

The analytical calculation method has established that the change in the technological parameters of detonation spraying has a significant effect on the change in such structurally dependent properties of the final coatings as the yield strength and the level of local internal stresses.

Increasing the heat input per mass unit of the powder of Cr_3C_2–NiCr promotes hardening of the coating material by grinding the structure and substructure, increasing the density of dislocations, as well as increasing the volume fraction of secondary carbide particles in the matrix. However, at the same time, it leads to the formation of extended dislocation clusters around carbide particles that degraded during the deposition process to a more significant extent, and, as a result, to a catastrophic increase in the level of local internal stresses up to the level of the theoretical matrix strength.

Technological modes of spraying, contributing to a moderate increase in the strength properties and hardness of the matrix without significant degradation of the primary carbide particles and the formation of high values of local internal stresses should lead to an increase in the performance properties of the final products. And spraying modes that contribute to the substantial hardening of the matrix and the appearance of significant localized internal stresses are unacceptable.

References

1. J.R. Davis, *Handbook of Thermal Spray Technology* (ASM International, 2004)
2. A.A. Tracton (ed.), *Coatings technology: Fundamentals, Testing, and Processing Techniques* (CRC Press, Taylor & Francis Group, 2007)
3. L. Pawlowski, *The Science and Engineering of Thermal Spray Coatings*, 2nd edn. (Wiley, New York, 2008)
4. Yu. Tyurin, A. Pogrebnyak, O. Kolisnichenko, I. Duda, The Hardening Technology and Coatings **5**, 27–33 (2009)
5. L.I. Markashova, Yu. Tyurin, O. Berdnikova, O. Kolisnichenko, I. Polovetskyi, Ye. Titkov, in *Advances in Thin Films, Nanostructured Materials, and Coatings*, ed. by A. Pogrebnjak, V. Novosad. 8th IEEE International Conference on "Nanomaterials: Applications & Properties", Odessa, September 2018. Lecture Notes in Mechanical Engineering (Springer, Singapore, 2019), pp. 109–117
6. L.I. Markashova, V.D. Poznyakov, A.A., Gaivoronskii, E.N. Berdnikova, T.A. Alekseenko, Mater. Sci. **47**(6), 799 (2012)

Chapter 15
Combined Magnetron-Ion-Source System for Reactive Synthesis of Complex Nanostructured Coatings

Alexander Zykov, Stanislav Dudin, Stanislav Yakovin, Nina Yefymenko, Artem Shchibrya and Alexander Dahov

Abstract Design and characteristics of the new combined magnetron-ion-source sputtering system (MISSS) are described. The system includes a magnetron sputtering system and Hall-type ion source with a common magnetic system allowing variation of both the topology and strength of magnetic field. The system provides independent control of ion energies (in the range of 10–500 eV) and current densities (0.1–10 mA/cm^2) of ion flow bombarding the growing film with the deposition rate of up to 3 nm/s at gas pressure 0.5–5 mTorr with no significant restrictions on the type of the gas. The system possesses the unique ability of surface potential management of dielectric coatings. Control of ion and electron flows to the surface allows setting the dielectric surface potential to a predefined value, thus providing energy control of ions bombarding the growing film.

15.1 Introduction

Recent studies have shown that plastic super-hard coatings without internal stress may be formed in T-zone of Thornton diagram, which is defined by the T/T_m ratio of the sample temperature T and the melting point of the material T_m as well as by ion bombardment of the coating during deposition. The critical parameter of influence on physical properties of the coatings is the specific energy input to the substrate per one condensed atom [1–4]. The energy flow consists of: (1) equilibrium heating of the sample, (2) non-equilibrium heating of the surface due to relaxation of the kinetic energy of ions, electrons and fast neutrals, (3) exothermic energy of chemical reactions, and (4) radiation energy flux from deposition source and plasma. It is the nonequilibrium heating of the surface by ion-atomic flows that can shift the zone T to the region of lower sample temperatures [4]. At the same time, the problem of determining the degree of influence of various components of the nonequilibrium surface heating on the structure and properties of condensate remains unsolved. Existing

A. Zykov · S. Dudin (✉) · S. Yakovin · N. Yefymenko · A. Shchibrya · A. Dahov
V. N. Karazin Kharkiv National University, Kharkiv, Ukraine
e-mail: dudin@karazin.ua

A. D. Pogrebnjak and O. Bondar (eds.), *Microstructure and Properties of Micro- and Nanoscale Materials, Films, and Coatings (NAP 2019)*,
Springer Proceedings in Physics 240, https://doi.org/10.1007/978-981-15-1742-6_15

technological equipment, in particular magnetron sputtering systems (MSS), doesn't allow variation of energy and density of ion flow to the deposited coating in wide range, independently from the neutral atom flow. Thus, the task of developing special research equipment allowing independent variation of parameters of the sputtered atom flow and ion bombardment of the growing film is actual. Additional problem appears in the case of non-conductive coating. The usual method of substrate biasing allowing ion energy control for conductive coatings is not efficient for dielectrics.

To solve these problems the authors of the present paper have developed the new combined magnetron-ion-source sputtering system (MISSS). In the paper [5], the design and characteristics of MISSS are described including ignition curves, current-voltage characteristics in dependence on gas pressure, magnitude and topology of a magnetic field.

The present paper describes the new optimized design of MISSS and gives its physical and technological characterization. The study is focused on the control of ion and electron flows to the processed surface, especially on energy control of ions bombarding the growing film. Special attention is paid to the specific case of non-conductive coatings.

15.2 MISSS Design

The schematic diagram of the developed device with diagnostic equipment is presented in Fig. 15.1. MISSS consists of a conical Hall-type ion source (IS) with closed near-anode electron drift in crossed EH fields [6] and a planar magnetron [7, 8], integrated into a joint structure. The ion source forms a conical coinciding beam of gas ions with a convergence angle of 60°, the initial diameter of 100 mm, and a thickness of about 5 mm.

The magnetron body 11 of 76 mm diameter and 45 mm in length is made of soft magnetic steel and is mounted on a dielectric flange in the center of the ion source. Permanent magnets with surface magnetic field strength up to 3.5 kOe, are located on the axis of the system inside the magnetron body, which is cooled by water. A target 12 of 76 mm diameter is mounted on the magnetron body with a vacuum seal.

Both the IS and the MSS possess the common magnetic system. Using the solenoid 8, it is possible to change the configuration of the magnetic field lines (MFL) in the magnetron and to realize balanced or unbalanced magnetron types. The MFL topologies at various solenoid currents are given in the paper [5].

The magnetic field strength at the center of the magnetron target can be adjusted in the range of 300–600 Oe, while the radial magnetic field strength in the discharge gap of the ion source varies linearly in the range of 0–4000 Oe with the solenoid current of $I_s = (0\text{–}4)$ A.

Gas dynamics in MISSS significantly affects the characteristics of both IS and MSS. The working gases can be input through two different channels separately or simultaneously (see Fig. 15.1): through the annular gap in the anode cavity of the ion source (Gas inlet 1); through the annular gap between the inner cathode of the

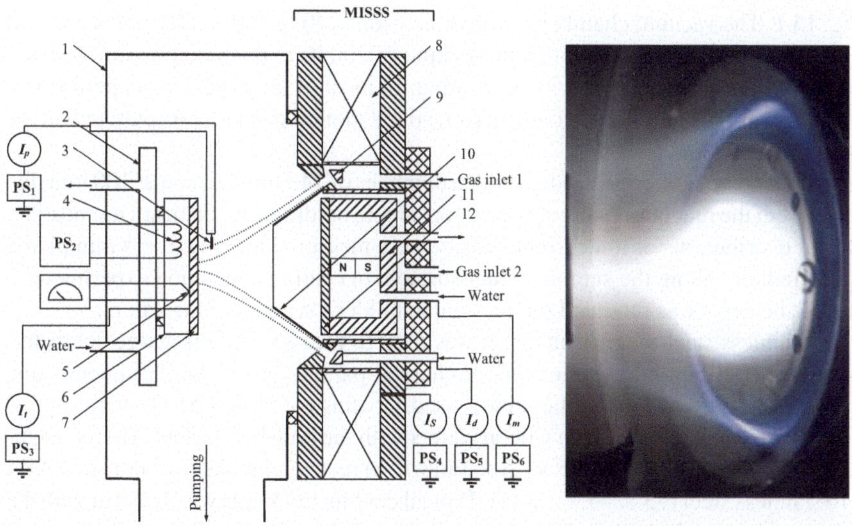

Fig. 15.1 Schematic diagram of experimental set-up and photograph of MISSS at simultaneous operation of IS and MSS. 1—vacuum chamber, 2—water-cooled sample holder, 3—flat probe for ion current density measurement, 4—infrared heater, 5—thermocouple, 6—heater screen, 7—sample, 8—solenoid, 9—anode of ion source, 10—conical anode of magnetron, 11—magnetron, 12—magnetron target. The ion beam is shown by a dotted line

ion source and the magnetron body 11 (Gas inlet 2). The Gas inlet 1 was used for injection of Ar or reactive gases O_2, N_2, while the Gas inlet 2 served as Argon input only.

A new component in the MISSS design in comparison with one presented in [5] is the conical anode 10 of the magnetron, which is 30 mm long and is made of non-magnetic material (copper). The conical anode performs several functions:

- provides a pressure drop between the magnetron discharge and the deposition chamber at gas input to Gas inlet 2;
- restricts reactive gas flow into the magnetron discharge when the gas is input into the ion source (Gas inlet 1) reducing poisoning of the magnetron target;
- focuses the ion flow from the magnetron plasma to the treated surface;
- limits the flow of sputtered atoms to the chamber walls.

15.3 Experimental Setup and Diagnostic Equipment

The experimental setup was built for MISSS characterization and coating deposition experiments (including conductive and dielectric coatings) by magnetron sputtering with variable ion bombardment in a wide range of ion energy (10–1000 eV) and current density (0–10 mA/cm^2). The schematic diagram of the setup is presented in

Fig. 15.1. The vacuum chamber 1 with dimensions 240 × 240 × 120 mm is pumped by a turbomolecular pump with a throughput of 700 l/s to the base pressure below 1 × 10^{-5} Torr within 10 min without heating. In most of the experiments performed, the chamber pressure was below 1 mTorr, so the motion of ions and sputtered atoms is mostly collisionless.

The flat conductive electrode 2 with a diameter of 140 mm located at 100 mm distance from the magnetron target is used for sample holding and ion beam monitoring. Radial distributions of ion current density were measured moving 5 × 5 mm planar probe radially along the sample holder surface. In coating deposition experiments a sample heater 7 was mounted on the sample holder via ceramic insulators.

A stainless steel magnetron target was used to measure the discharge characteristics of MISSS and the parameters of charged particle fluxes, while in deposition experiments pure Aluminum target was used. Coatings of Al and Al_2O_3 were applied on various samples: (a) 3-inch silicon wafers with the polished surface; (b) rectangular plates of polished glass (75 × 25 × 1 mm); (c) rectangular plates of polished AISI 316 stainless steel (75 × 25 × 1 mm). Typical coating thickness was 1–2 μm with the deposition time of 10–20 min. The hardness of the deposited coatings was measured using Micron-Gamma microhardness tester in different points of each sample, taking into account the significant variation of deposition conditions across the sample. The crystalline phase composition of the deposited coatings was estimated by XRD spectra taken using used X-ray diffractometer XRD–6100 (Shimadzu).

15.4 Physical Characterization of MISSS

For ease of comparison, the current and voltage dependencies on chamber pressure are presented in Fig. 15.2 for magnetron (Fig. 15.2a) and ion source (Fig. 15.2b) during their autonomous operation. Magnetron voltage decreases monotonously with the pressure growth while the ion current to the substrate holder shows a maximum at a pressure of about 1 mTorr.

The same pressure dependence for the ion source is completely different. Below 1 mTorr the discharge current is growing in direct proportion to the pressure. In this region ion source efficiently generates focused ion beam, and the ion current accepted by the sample holder is more than 80% of the discharge current.

In contrast, at the pressures above 3 mTorr the source operates in plasma mode, which is characterized by high current, low voltage, and absence of focused ion beam. In this regime, the whole volume of the deposition chamber is filled with glowing plasma [9, 10].

From the comparison of the plots for I_t in Fig. 15.2b, it turns out that the pressure ratio between the ion source cavity and the deposition chamber is the about $p_i/ p_k \approx 7$, which is defined by the pumping speed and cross section of the ion source channel.

Figure 15.3 allows comparison of the magnetron discharge and the discharge in the ion source in plasma mode where currents and voltages are comparable to the magnetron. It is seen that in the plasma mode the ion source current is distributed

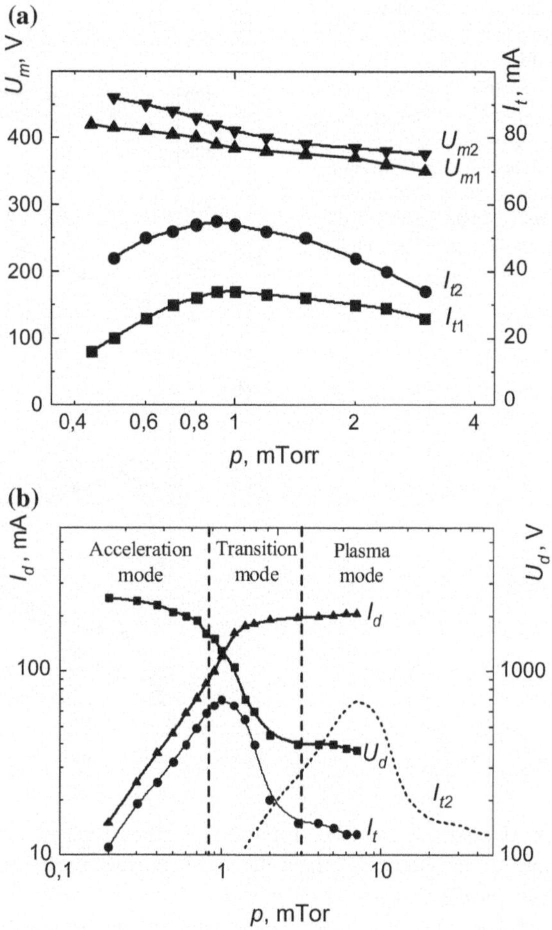

Fig. 15.2 **a** Magnetron voltage U_m and sample holder current I_t ($U_t = -75$ V) vs. pressure p (Argon is fed through Gas inlet 2) at constant discharge current: (1) $I_m = 300$ mA, (2) $I_m = 500$ mA; **b** IS anode voltage U_d, IS anode current I_d, and sample holder current I_t ($U_t = 0$) versus pressure p (Argon is fed through Gas inlet 1). Sample holder current I_{t2} at Argon feeding through Gas inlet 2 is shown with dashed line

between all the surfaces bounding plasma in the deposition chamber, particularly to the sample holder and to the magnetron target.

The impact of the conical anode on the magnetron discharge is illustrated by Fig. 15.4. One can see from the ignition curves that the limitation of the magnetron discharge volume by the conical anode when gas is injected into the ion source leads to increase of the minimum pressure $p_{\min}$ of magnetron discharge operation by factor of 1.5 and to increase of the solenoid current corresponding to p_{min} from 0.7 A up to 1.5 A (curves 1, 2). The physical reason for this phenomenon is investigated in detail in [11–13]. When the gas is injected into the magnetron cavity, the discharge distinction pressure decreases by a factor of 2 (curves 2, 3) as a result of the appearance of a pressure drop between the cavity and the deposition chamber.

Ion source operation is also affected by the gas feeding direction. Figure 15.5 allows a comparison of current-voltage characteristics at different gas pressures in

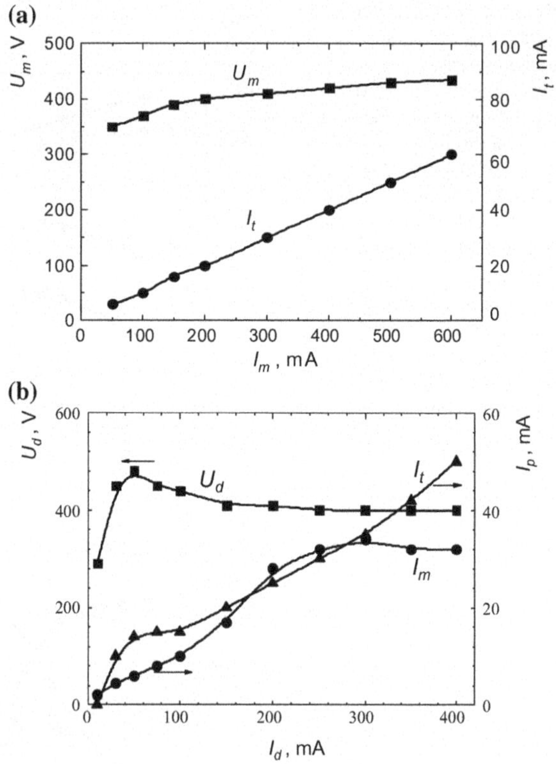

Fig. 15.3 **a** Magnetron voltage U_m and sample holder current I_t versus magnetron discharge current I_m ($p = 0.8$ mTorr); **b** IS anode voltage U_d, magnetron target current I_m and sample holder current I_t versus anode current of the ion source in plasma mode ($p = 3$ mTorr)

the deposition chamber for the two directions of gas injection. When the gas is injected into the Gas inlet 1 the anode current is about the order of magnitude higher than for the Gas inlet 2.

It should be noted that, in fact, the sample holder 2 is an essential part of MISSS. Depending on its potential U_t, the sample holder performs different functions and U_t value significantly affects the parameters of the discharges. Figure 15.6a shows typical current-voltage characteristics of the sample holder at different solenoid currents; Fig. 15.6b presents dependencies of electron and ion current to the sample holder and its floating potential on the solenoid current. As it follows from the plots in Fig. 15.6a, at $U_t > 0$, the sample holder serves as anode gathering the electron current, which is equal to the ion current to the magnetron target I_m. In this case, the plasma potential increases and U_m decreases, respectively. At negative potentials, $U_t < -60$ V, the ion current to the sample holder reaches saturation (Fig. 15.6a), and its value is determined by the parameters of the secondary plasma of the magnetron discharge.

The floating potential U_{fl} of the sample holder during magnetron operation is negative and depends on the magnitude and topology of the magnetic field determined by I_s and varies in the range (−50–0) V (Fig. 15.6b). In this case, the bulk of electrons go to the grounded conical anode and the chamber walls. For the ion source, the

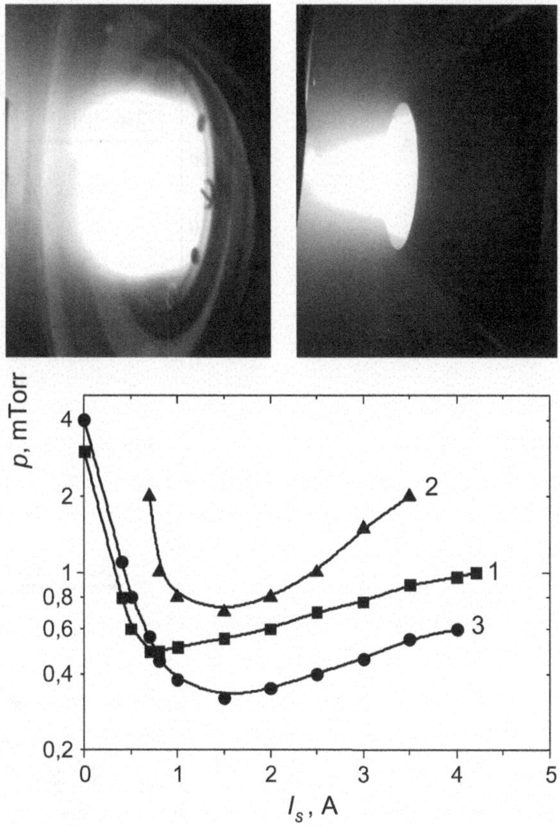

Fig. 15.4 Photograph of magnetron discharge without and with the conical anode and ignition curves of magnetron discharge at various gas inlets: 1—without conical anode, gas inlet 1; 2—with the conical anode, gas inlet 2; 3—with the conical anode, gas inlet 2

sample holder is a collector for the ion beam with the current I_i, and its floating potential is positive and reaches approximately 20% of anode voltage of the ion source.

We discussed above the integral value of charged particle current to the substrate holder, but the particle flows are highly non-homogeneous over the sample holder surface. In order to measure radial distributions of the charged particle flows we used a movable planar probe located near the sample holder surface. Figure 15.7a shows the current-voltage characteristic of the probe located on the axis of the system at the distance of 10 mm from the sample holder. Depending on the probe potential it can gather ions or electrons coming from magnetron plasma. Figure 15.7b shows the radial distribution of the probe current at various potentials of the probe. The measurements were carried out during the operation of the magnetron discharge with a current of 100 mA. As for the sample holder as a whole (Fig. 15.6a), Fig. 15.7 confirms that we can extract much more electron current from the magnetron plasma then ion current.

Figure 15.8 shows the comparison of radial profiles of ion fluxes to the flat probe in ion saturation mode for the magnetron discharge and the ion source, depending

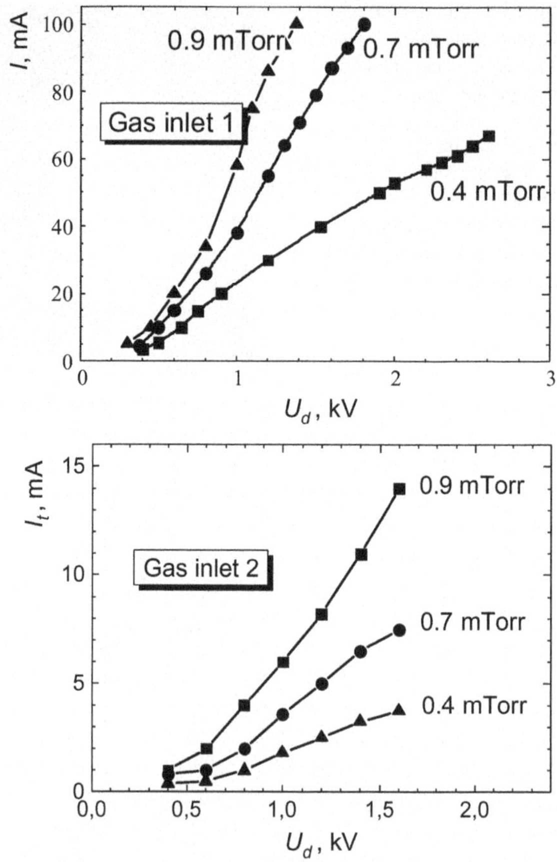

Fig. 15.5 Current-voltage characteristics of the ion source in acceleration mode under different Argon pressures and different gas channels. $I_s = 2$ A

on the solenoid current. One can see that the ion current from magnetron discharge has a maximum on the axis with a rapid decrease of ion current density outside the central region. Solenoid current changes only the amplitude of the peak, but not the shape. In contrast, the ion source produces ring-like ion current distribution with the void in the center. The solenoid current controls the level of the beam defocusing that allows radial scanning of the beam.

15.5 Technological Characterization of MISSS

The technological testing was carried out with the aim of MISSS characterization. Alumina was chosen as the main material for the test coatings since its properties are well studied [14]. Over the past decades, numerous papers on the deposition of stoichiometric Al_2O_3 films with different methods have been published. Therefore,

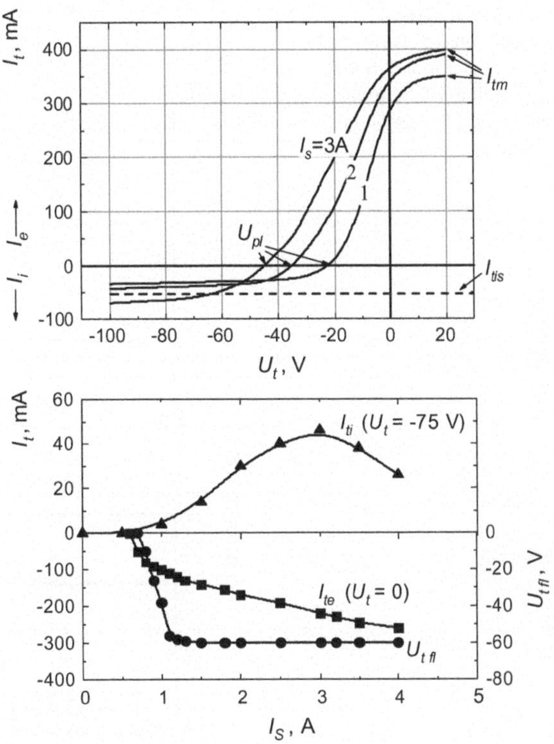

Fig. 15.6 Current-voltage characteristics of the sample holder 2 and dependencies of current to the substrate holder at $U_t = 0$, $U_t = -75$ V, and it's floating potential versus solenoid current at the separate operation of magnetron and ion source. $I_m = 400$ mA, $U_m = -400$ V, $I_d = 70$ mA, $U_d = 1.6$ kV, $p = 0.8$ mTor (Argon)

our task was to identify the advantages and disadvantages of MISSS in comparison with other devices.

The coating process included initial sample cleaning in an ultrasonic bath with alcohol, pumping to the base pressure of 1×10^{-5} Torr, ion cleaning of the sample using the ion source with Argon feeding (the ion beam was scanned over the sample surface by changing the solenoid current, see Fig. 15.8b). After the preliminary steps, magnetron was switched on, and Oxygen was added to the Gas inlet 1.

The important feature of our system is illustrated by the graphs shown in Fig. 15.9. It shows hysteretic dependencies (that is typical for reactive magnetron sputtering) of magnetron voltage and oxygen partial pressure in the deposition chamber for two cases: with or without the conical anode. As can be seen from the graph for U_1, without the conical anode, discharge jump to the target poisoning mode occurs at the low vapor pressure of oxygen and at the gas flow of 12.5 sccm. Moreover, the obtained Al_2O_3 coatings in all the modes were non-stoichiometric (non-transparent, black or brown). With the conical anode the discharge jump to poisoning mode is shifted to the region of higher partial oxygen pressures that provide a possibility to obtain transparent stoichiometric coatings over the entire surface of the sample at $q_{O2} > 3$ sccm.

Fig. 15.7 **a** Typical current-voltage characteristic of the planar probe; **b** radial distributions of the probe current at different probe potentials

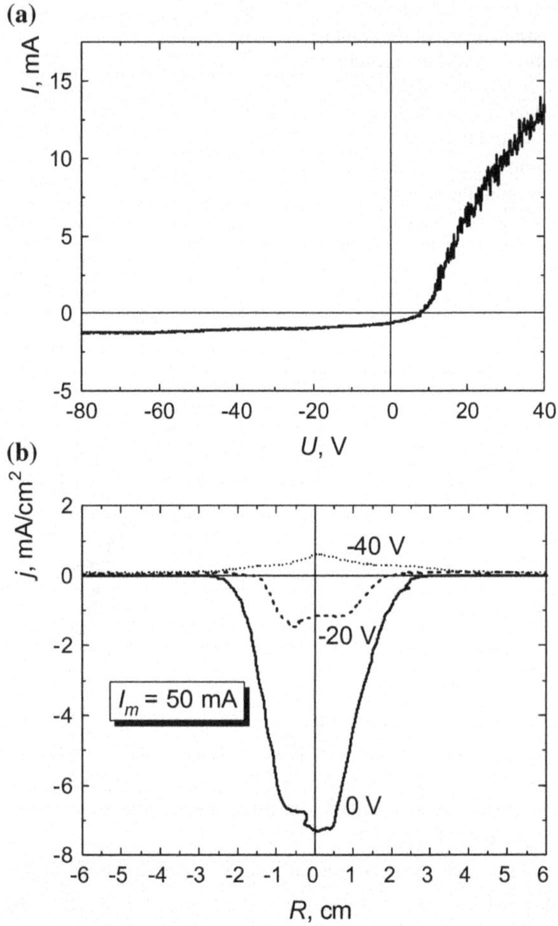

Figure 15.9b shows the thickness profiles of the pure aluminum film along the sample radius at different distances from the magnetron target with and without conical anode. It is noteworthy that the anode surface limits the coating area, and uniformity of about 90% is achieved at a diameter of 50 mm. The coating rate was 16 nm/s at a magnetron discharge current of 500 mA.

To study the possibility of crystalline phase control by means of ion bombardment, a series of experiments were carried out on the deposition of alumina with different sample temperatures and different parameters of the ion flux. The results of the influence of the sample temperature and ion bombardment on the properties of the stoichiometric Al_2O_3 coatings are presented in Fig. 15.10.

At the temperature 300 °C and below the alumina films deposited by the magnetron without the ion source assistance are always amorphous which is confirmed by XRD analysis and relatively low hardness values. With the temperature growth,

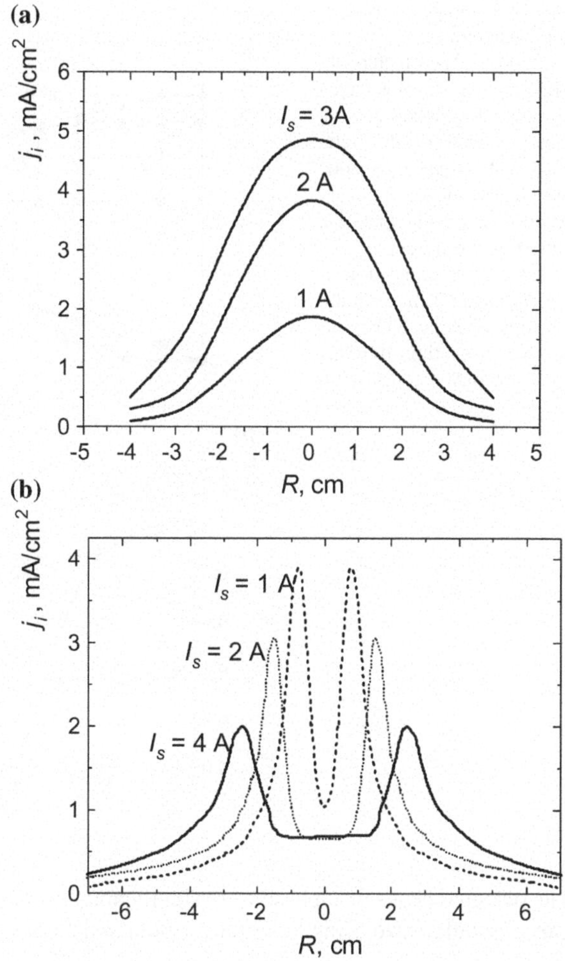

Fig. 15.8 Radial profiles of ion current to sample holder measured by the planar probe in ion saturation mode for magnetron discharge (**a**) and ion source (**b**) in dependence on solenoid current

the increasing level of crystallinity appears in the films. At the temperatures above 500 °C γ-phase of Al_2O_3 became dominant in the coating that is the reason for the hardness increase. It is important that addition of bombardment by the ions emitted by the ion source at the same sample temperature changes the picture dramatically. The film's hardness further increases reaching the peak values of about 25 GPa that allows us to identify this as a sign of α-phase formation. It should be noted that 500 °C is the maximum temperature reachable with our sample heating system, but at this temperature the maximum hardness can be found only within narrow ring corresponding to the maximum ion beam current density. The rest of the sample is filled with Alumina in γ-phase. Probably the integral amount of α-phase within the XRD measurement area was quite low, so we could not see any α-phase peak.

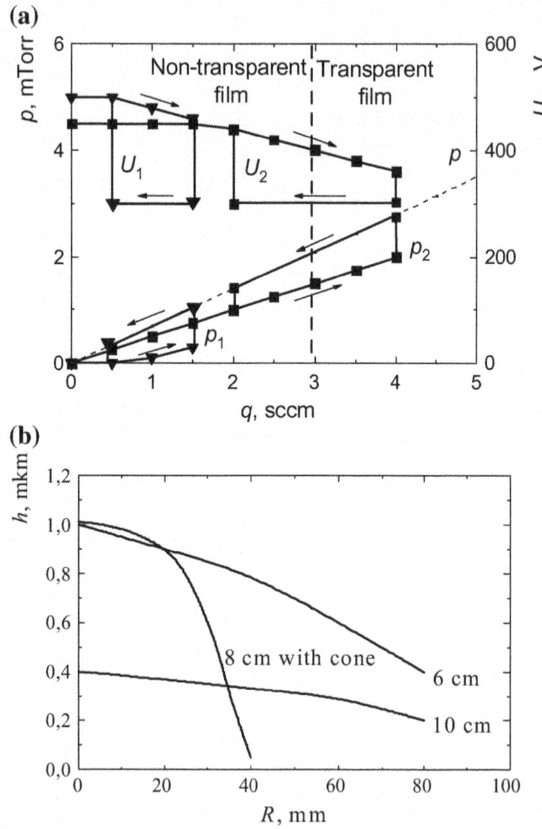

Fig. 15.9 Impact of the conical anode on technological parameters of MISS: **a** magnetron voltage U_m and Oxygen partial pressure versus O_2 flow rate. $p_{Ar} = 0.8$ мTopp, $I_m = 500$ mA; **b** dependence of Aluminum coating thickness on radius at different distances from magnetron target (6 and 10 cm without the conical anode, 8 cm with it). Deposition time 10 min, $I_m = 500$ mA, $p = 0.8$ mTorr

The detected peaks of γ-phase were significantly broadened that allowed to estimate nanocrystallite size using Scherrer formula as 12 nm.

15.6 Discussion

Summarizing the obtained experimental data, we can distinguish the following main properties of MISSS.

1. The regions of stable operation by the pressure in the deposition chamber of IS and MSS overlap in the range of 0.5 mTorr < p < 3 mTorr.
2. In this pressure range, depending on the gas inlet channel, the ion source in acceleration mode generates a high-energy ion beam with average energy $\varepsilon_i =$ 300–1000 eV, or ion flow with an average energy $\varepsilon_i =$ 100–300 eV in the plasma mode.

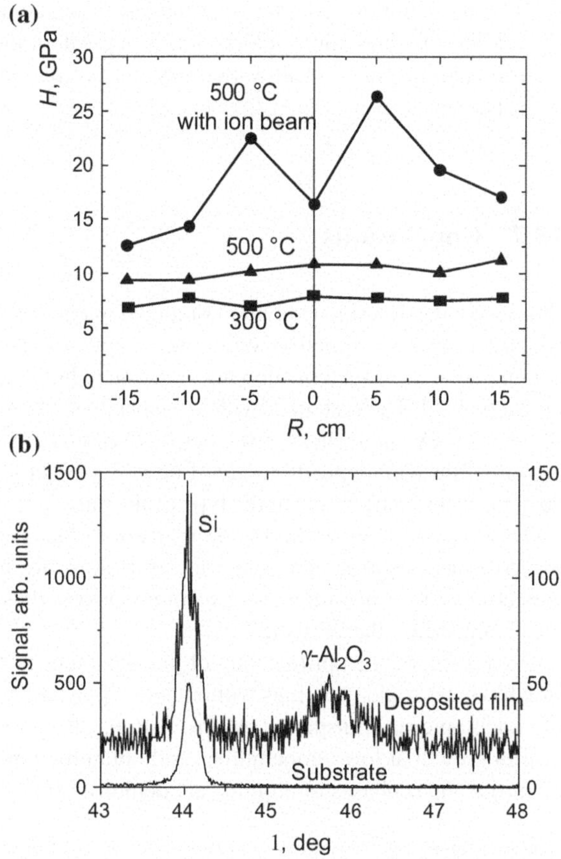

Fig. 15.10 **a** The Hardness of alumina coatings deposited at different sample temperatures and ion bombardment conditions; **b** XRD spectrum of the coating deposited at 500 °C with ion bombardment

3. The magnetron discharge also forms a flow of low-energy ions on the sample holder with energy $\varepsilon_i = eU_t = 10$–$100$ eV.
4. The magnitude of ion current density j_{im} from magnetron plasma depends on the strength and configuration of the magnetic field and can be controlled over a wide range: $j_{im} = 0$–7 mA/cm^2.
5. By varying the magnetic field strength, it is possible to control the radial profile of the ion source beam (defocusing effect) on the processed surface.

Investigation of discharge characteristics at the simultaneous operation of magnetron and ion source has shown a number of positive features of this approach.

1. The ion source operation (even with a low discharge current of about 10 mA) leads to a significant (1.5–2 fold) decrease in the minimum pressure of the existence of the magnetron discharge.
2. The ion beam presence provides a decrease in the magnetron discharge voltage by 50–100 V and stabilizes its operation at gas pressures less than 1 mTorr.

3. The electron flow from magnetron plasma can compensate for the current of the ion beam to the sample holder. This is of great importance at deposition of dielectric coatings since it allows energy control of ion bombardment of insulating surface.

15.7 Conclusions

Thus, we have described the new combined magnetron-ion-beam sputtering system, which includes a magnetron sputtering system and the Hall-type ion source with a common magnetic system allowing variation of both the topology and strength of the magnetic field. The system provides independent control of ion energies (in the range of 10–1000 eV) and current densities (0.1–10 mA/cm^2) of ion flow bombarding the growing film with deposition rate of up to 3 nm/s at gas pressure 0.5–5 mTorr with no significant restrictions on the type of the gas.

The system possesses the unique ability of surface potential management of dielectric coatings. Control of ion and electron flows to the surface allows setting the dielectric surface potential to a predefined value, thus providing energy control of ions bombarding the growing film.

As a result of the optimisation of gas dynamics in MISSS, the ability to deposit stoichiometric oxide coatings without target poisoning was shown. For the synthesis of stoichiometric transparent Al_2O_3 coating, the possibility of control of kinetics of film growth at low temperatures, and obtaining amorphous nanocrystalline (10–12 nm) γ and α phases of alumina are demonstrated.

Acknowledgements This work was supported by the Ministry of Education and Science of Ukraine.

References

1. S. Yick, I. Levchenko, S. Kumar, Z.J. Han, M.M.A. Yajadda, K. Ostrikov, I.E.E.E. Trans, Pl. Sci. **40**(4), 1094 (2012)
2. J. Musil, Surf. Coat. Technol. **207**, 50 (2012)
3. A. Anders, M. Panjan, R. Franz, J. Andersson, P. Ni, Appl. Phys. Lett. **103**, 44103 (2013)
4. J. Musil, RSC Adv. **5**(74), 60482 (2015)
5. S. Dudin, O. Tkachenko, A. Shchybria, S. Yakovin, A. Zykov, N. Yefymenko, Probl. Atom. Sci. Tech. **118**(6), 263 (2018)
6. YuP Maishev, Vacuum Tech. Technol. **2**(3–4), 53 (1992)
7. J. Musil, Vacuum **50**, 363 (1998)
8. R.D. Arnell, P.J. Kelly, Surf. Coat. Technol. **112**, 170 (1999)
9. I. Safi, Surf. Coat. Technol. **127**, 203 (2000)
10. J. Musil, J. Vlček, P. Baroch, in *Materials Surface Processing by Directed Energy Techniques*, ed. by Y. Pauleau (Elsevier Science Publisher B.V, Oxford, UK, 2006), pp. 67–106

11. J. Walkowicz, A. Zykov, S. Dudin, S. Yakovin, R. Brudnias, Tribologia **6**, 163 (2006)
12. A.V. Zykov, S.D. Yakovin, S.V. Dudin, Phys. Surf. Eng. **7**(3), 195 (2009)
13. S. Yakovin, S. Dudin, A. Zykov, V. Farenik, Probl. Atom. Sci. Tech. **71**(1), 152 (2011)
14. J. Andersson, Dissertation No. 987, Linköping (2005)

Chapter 16
$AlCoNiFeCrTiV_x$ High-Entropy Coatings Prepared by Electron-Beam Cladding

Alexandra Yurkova, Vadym Chernyavsky, Alexandr Matveev and Maxim Sysoyev

Abstract This study reports the investigation of high-entropy coatings obtained by electron-beam cladding in a vacuum of Al-Co-Ni-Fe-Cr-Ti-V_x powder blend on a steel substrate. V was added to the Al-Co-Ni-Fe-Cr-Ti equiatomic system and the effects of this added element on structure, phase composition and microhardness of $AlCoNiFeCrTiV_x$ high entropy coatings resulted from electron beam cladding were studied. The $AlCoNiFeCrTiV_0$ coatings consist of two solid solutions with BCC1 and BCC structure with different lattice parameters and a small volume fraction of σ-phase. It was shown that with an increase in V content from $x = 0$ to $x = 1.5$, the phase composition of the coatings transforms from two solid solutions to single BCC solid solution and σ-phases of different compositions. The σ-phase volume fraction increased with an increase in the V content. The addition of V to AlCoNiFe-CrTi shows the strengthening effect of the $AlCoNiFeCrTiV_{0.5-1.5}$ coatings and the Vickers hardness increased from 8.4 to 11 GPa. Microhardness of the coatings was affected by the sigma phase. The hardness enhancement can be likely attributed to the effect of solid solution strengthening and to the presence of σ-phase particles in the coating structure.

16.1 Introduction

In many situations, only the contact surface properties are important in determining the performance of the component in practical applications. To increase the life of mechanical parts, their surface is treated to increase hardness and wear resistance. Therefore, the use of a coating from new materials with high physical and mechanical characteristics, such as high-entropy alloys (HEA) [1, 2], has several attractive advantages. HEAs have been found to have novel microstructures and unique properties.

A. Yurkova · V. Chernyavsky · A. Matveev (✉) · M. Sysoyev
Igor Sikorsky Kyiv Polytechnic Institute, Kiev, Ukraine
e-mail: alexmatveev1994@gmail.com

A. D. Pogrebnjak and O. Bondar (eds.), *Microstructure and Properties of Micro- and Nanoscale Materials, Films, and Coatings (NAP 2019)*,
Springer Proceedings in Physics 240, https://doi.org/10.1007/978-981-15-1742-6_16

As a new focus on materials research and a novel alloy design concept [3], high entropy alloys (HEAs) have been achieved successfully by changing or designing the component elements of metallic material [1, 2]. HEAs are new class alloys and are quite different from traditional alloys, which are based on one or two elements. HEAs are typically defined as multicomponent alloys whose principal elements are at least five in equiatomic or near equiatomic ratio, and the concentration of each constituent element is between 5 and 35 at.% [1–3].

HEAs have been studied extensively in recent years, and they are expected to be abundant in academic research and industrial applications. Some core effects in HEAs would be more pronounced than those in conventional alloys [1, 2, 4–6]: high entropy—it enhances the formation of simple solid solution phases, such as bcc and/or fcc structures, and thus simplifies the microstructures; severe lattice distortion—it arises from the atomic size difference among different elements in solid solution phases and might markedly influence mechanical, physical and chemical properties; sluggish diffusion—it is due to the inefficient cooperative diffusion of various species and thus slow down phase transformations; and cocktail—it comes from the ideal mixing and inevitably excess interactions among the multi-principal elements in solid solution phases, and thus is a composite effect on properties.

HEAs have more excellent properties than conventional alloys and exhibit promising future for engineering applications due to their good thermal stability, high hardness and high strength, excellent wear resistance, superior oxidization, and corrosion resistance, as well as many other outstanding properties [1, 2]. The promising properties of HEAs offer the potential to be used in many applications, such as tools, molds, dies, mechanical parts and furnace parts which require high strength, thermal stability, and wear and oxidation resistance. These alloys can be used for various applications that demand high-temperature strength, oxidation, corrosion and wear resistance [1–4].

The main method of manufacturing HEAs is vacuum arc remelting for bulk cast ingot, and surface coating is also possible [1, 2]. Electron-beam cladding materials processing is a promising route, where a high-power electron beam is used as a source of heat for the processing of materials. Faster processing speed, refinement, and homogenization of microstructure, and heat-affected zone are the notable advantages associated with electron-beam heating materials processing. Thanks to the rapid heating, only a thin surface layer of the material is heated. Electron-beam surface cladding involves melting of clad material coating using electron beam and applying it on the surface with a minimum dilution at the interface. In the electron beam cladding process, the cooling rate can reach 103–106 K/s and this process is in situ synthesis. More importantly, electron beam cladding process has the capability of achieving metallurgical bonding between the coating and the substrate, small thermal deformation, and nonequilibrium reaction.

Until now this novel method for preparing high-entropy (HE) coatings has not just been reported by any organizations and the main properties of electron-beam cladding made HE coatings are still blank in the scientific research. Taking into account HEA's tendency to form simple structures and nano-crystallines, fabricating high-entropy coatings by electron-beam cladding is of great significance and potential for extensive use. In this study, a novel Al–Co–Ni–Fe–Cr–Ti–Vx high-entropy coating system was manufactured as the coating of the steel substrate by electron-beam cladding; also structure, phase and chemical composition, the hardness of the AlCoNiFeCrTiVx high-entropy coatings were tested under the different ratio of V element. References [7–9] have reported that V has great influence on the structure and properties of HEAs.

16.2 Experiment Details

16.2.1 Materials and Processing

In order to investigate the effect of V on the microstructure and the mechanical properties of the HE coatings, the coatings with different V contents are fabricated. Al, Co, Ni, Fe, Cr, Ti and V powders with commercial purity (99.7 wt%) and particle size of $\leq$50 μm was chosen for experimentations, prepared and well mechanically mixed as the raw materials. All the elements except V are equiatomic. Then the mixed powders were preplaced onto the surface of St3 steel substrate (C: 0.14–0.22; Ni: $\leq$0.3; Cr: $\leq$0.3; Mn: 0.4–0.65, Si: 0.15–0.3, Cu: $\leq$0.3; S: $\leq$0.05, P: $\leq$0.04; Fe: 97 balance in mass percentage) to form powder bed with the thickness of 1.7–2.0 mm. In order to evaluate the influence of V content in AlCoNiFeCrTiV$_x$ coatings, the x factor was set as variable quantity and the experiments were divided into 4 groups: x = 0, 0.25, 0.5, and 1.0. Powders of the raw components were mixed in a planetary ball mill for 5 min. With ELA-6 electron-beam equipment, in vacuum (10^{-2} Pa), the HEA coatings were synthesized on the surface of the steel substrate. The performance of the cladding coatings is controlled by the electron-beam power and scanning speed. The electron beam scanning speed along the surface of the powder pressing was 4 mm/s.

16.2.2 Structural Characterization

The effects of vanadium content on the structure and phase transformation of electron-beam coatings were investigated by means of Rigaku Ultima IV X-ray diffractometer (XRD) with Cu Kα radiation and the working voltage and current were 30 kV and 20 mA, respectively. The range of scanning angle (2θ) was from 20° to 120°. The scanning step was 0.04° with an angular speed of the goniometer rotation of 2°/min. The lattice parameters of solid solutions were calculated according to the positions of the gravitation centers of the diffraction profiles by the linear extrapolation method.

A scanning electron microscope (SEM) equipped with an energy dispersive spectrometer (EDS) was used to obtain basic information concerning microstructural features and measure the chemical composition of coatings. Prior to SEM observation the specimens with coatings were sectioned perpendicular to the scanning track for preparing a slice by standard metallographic technique with polishing by diamond slurry.

16.2.3 Mechanical Testing

Microhardness measurements were performed using a conventional microhardness machine Digital Microhardness Tester MHV-1000 equipped by standard Vickers' pyramid. Microhardness numbers were determined under indentation loads not higher than 1.5 N and a duration time of 15 s. Each coating was tested for multipoint from the interface to the surface with an equal interval of about 0.3 mm. 10–20 measurements were made on each coating/sample for an average.

16.3 Results and Discussion

16.3.1 Structure Evolution of Elemental Powders to Coatings by Electron-Beam Cladding

Figure 16.1 presents the X-ray diffraction patterns of the initial Al–Co–Ni–Fe–Cr–Ti–M_x powder blend and $AlCoNiFeCrTiV_0$, $AlCoNiFeCrTiV_{0.5}$, $AlCoNiFeCrTiV_{1.0}$, $AlCoNiFeCrTiV_{1.5}$ coatings, resulted from electron-beam cladding. The XRD pattern of powder blend exhibits peaks of all the pure elements in the initial powder mix. It can be seen that the phase composition of the V-free $AlCoNiFeCrTiV_0$ as-clad coating showed the presence of two body-centered-cubic (BCC) solid solution phases with different lattice parameters: BCC and BCC1 (β- and β1-phases, respectively) and the small volume fraction of σ-phase which can be identified as Co_3Ti type.

Thus Al, Co, Ni, Fe, Cr, and Ti elements are expected to distribute in two multi-element supersaturated solid solutions of BCC structure and increase the distortion of the crystal lattices. It is apparent that one BCC1 solid solution is with a larger lattice parameter of 0.2931 nm and the other BCC with smaller one of 0.2877 nm from the strongest (110) peak as marked by β_1-phase and β-phase in Fig. 16.1, respectively, the larger BCC1 (β_1-phase) has a higher volume percent.

Peak broadening and their small intensity can be observed and a lot of peaks become nearly invisible. The intensity decrement and peak broadening in the diffraction can be attributed to the formation of high lattice strain, different atomic diffraction factor proportional to the atomic number, crystallinity decrement, and formation of solid solutions.

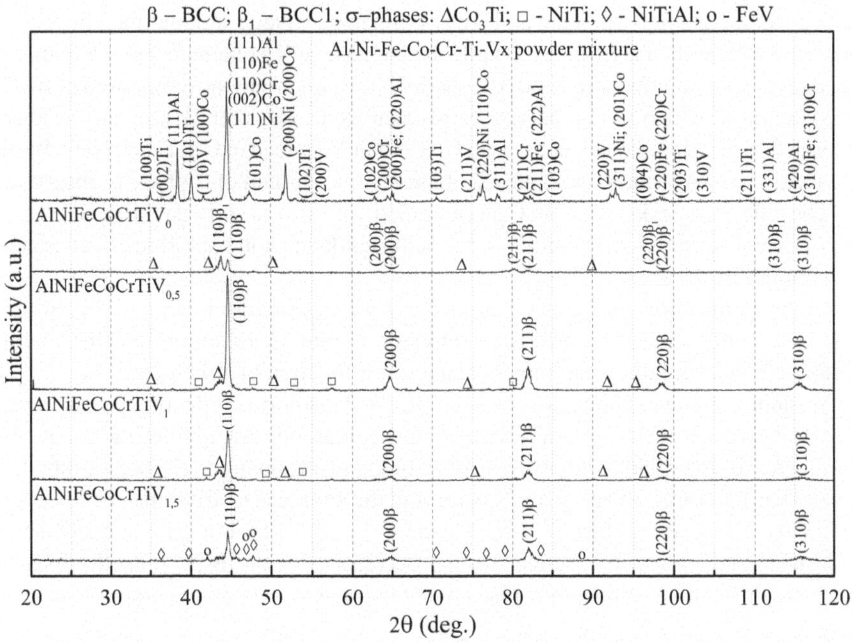

Fig. 16.1 XRD patterns of Al–Co–Ni–Fe–Cr–Ti–V$_x$ powder blend and AlCoNiFeCrTi$_0$, AlCoNiFeCrTiV$_{0.5}$, AlCoNiFeCrTiV$_{1.0}$, and AlCoNiFeCrTiV$_{1.5}$ coatings, resulted from electron-beam cladding

For the XRD peak intensity, the distorted atomic planes increase X-ray diffuse scattering effect and give smaller peak intensity [10]. Moreover, size mismatch effect between the constituents and increasing dislocation density leads to the increment of lattice strain [10].

According to the XRD analysis it may be reasonable to consider that the addition of V significantly influences the structure of HE coatings. Really, the addition of V element to the HE coatings composition promotes the transition of two BCC1 and BCC solid solutions and one type Co_3Ti of σ-phase to single solid solution of BCC structure and intermetallic σ-phases of different types as it can compositions in Fig. 16.1. Furthermore, the relative intensity of the σ-phase reflections is evidently lower than that of the BCC solid solution phase, revealing that the main phase still is the latter in this coating system. This may be ascribed to the high-entropy effect that greatly lowers the Gibbs free energy of the system, which more easily yields the formation of solid solutions during solidification, rather than ordered compounds especially at the high temperature and leads to the total number of phases well below the maximum equilibrium number allowed by the Gibbs phase rule [1, 2, 5, 6].

Intermetallic σ-phases of different compositions are formed, probably because of the decomposition of the BCC solid solution under V addition. The as-cladded AlCoNiFeCrTiV$_{0.5–1.0}$ HE coatings are composed of BCC solid solutions with the lattice parameters of 0.2878 nm and 0.2874 nm, respectively, and intermetallic compounds Co_3Ti and NiTi (σ-phase), while the increase of V content to x = 1.5 induces changes in phase composition and $AlCoNiFeCrTiV_{1.5}$ coatings are composed of BCC solid solution, which lattice parameter is calculated to be 0.2871 nm, and another intermetallic compounds (σ-phase): NiTiAl and FeV instead of Co_3Ti and NiTi. With the increase of V content, the increase of the relative amount of σ-phase is observed.

The reflection peaks of the BCC phases practically are not shifted with the increase of V content from 0.5 to 1.5 mol. due to the insignificant solid solution effect as can be seen in Fig. 16.2a.

In this connection, the lattice parameter of BCC structure is around a constant and varies only within the measurement error. Figure 16.2b shows the BCC solid solution (β-phase) lattice parameter changes as a function of V content.

V addition has an evident effect on the BCC solid solution reflection peaks intensity. The addition of 0.5 mol. of V to the V-free equiatomic AlCoNiFeCrTi composition increases the relative intensity of BCC phase peaks owing to the decreasing the crystal lattice distortion because of decreasing the atomic size difference from δ=7.2 to δ=6.94. However the further V increasing to 1 and 1.5 mol in spite of decreasing in δ parameter to δ = 6.46 effects in the decreasing of the BCC reflection peaks intensity through increment of σ-phase volume fraction in coating composition.

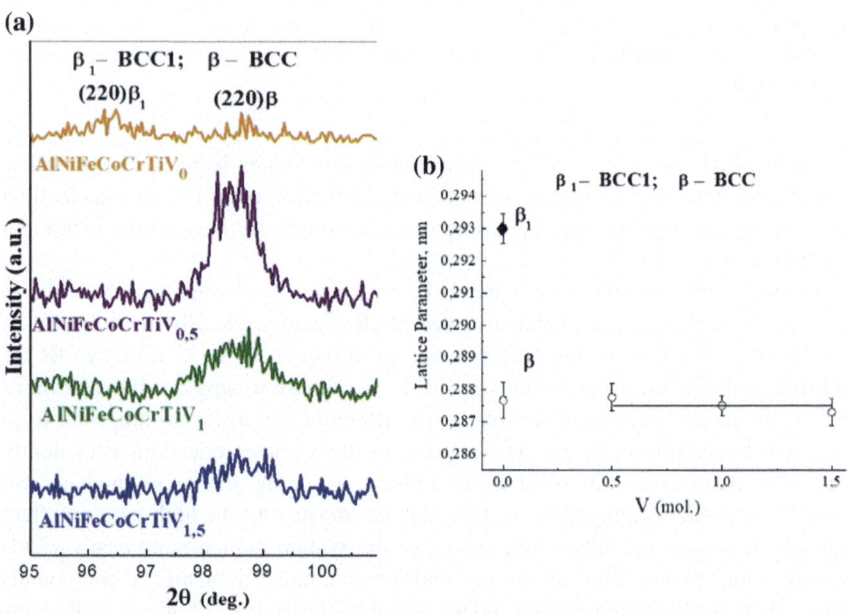

Fig. 16.2 The shift of the XRD (220)β peaks in the XRD of $AlCoNiFeCrTiV_x$, coatings with different V content (**a**); and Lattice constant of $AlCoNiFeCrTiV_x$, coatings as functions of V content (**b**)

16.3.2 Microstructure Characterization of AlCoNiFeCrTiV$_x$ High-Entropy Coatings Resulted from Electron-Beam Cladding

Figure 16.3 shows the microstructure of the as-cladded AlCoNiFeCrTiV$_x$ HE coatings. A dendrite structure was observed in the solidified AlCoNiFeCrTiV$_x$ coatings.

The chemical compositions of the dendrite area and interdendrite area in Fig. 16.3 characterized by EDS. It is shown that the dendrite area is rich in Ni, Fe, Co, V, and interdendrite area is rich in Ni. Al and Ti elements are slightly rich in dendrite area.

16.3.3 Mechanical Properties of AlCoNiFeCrTiV$_x$ High-Entropy Coatings Resulted from Electron-Beam Cladding

Figure 16.4 shows the Vickers microhardness distribution from the surface to the interface of the AlCoNiFeCrTiV$_x$ HE coatings. The thickness of the coatings is about 1.5 mm. The microhardness of V-free AlCoNiFeCrTiV$_0$ coating is 8.4 ± 0.2 GPa. As can be seen, with V addition from 0.5 to 1.5 mol the microhardness of the AlCoNiFeCrTiV$_x$ coatings increases significantly and reaches about 11 ± 0.3 GPa, which is much higher than most of the HEAs prepared by other methods with the hardness value of 3–6 GPa (300–600 HV) [1, 2, 5, 8, 9, 11, 12] and is much higher than that of the similar alloys prepared by laser cladding technique with the value of 4–7.6 GPa [13, 14] and almost four times higher than the steel substrate with the value of about 2.3 GPa. The high hardness could be attributed to the strong super-saturation hardening solid solution strengthening and the presence of σ-phase precipitations [1, 2, 5, 6]. The very high hardness of 10–11 GPa proves the promising future of the HE coating. These AlCoNiFeCrTiV$_x$ HE coatings would also have good wear resistance according to the conclusion, that is, the wear resistance of materials is in general proportional to their Vickers hardness.

16.4 Conclusions

The evolution of multicomponent powder Al–Co–Ni–Fe–Cr–Ti–V$_x$ (x = 0; 0.5; 1.0; and 1.5) mixture from elemental components to high-entropy coatings during electron-beam cladding was shown.

The addition of V element to the V-free AlCoNiFeCrTi HE coating changes the initial phase composition from two solid solutions with BCC1 and BCC crystal structure with larger lattice parameter of 0.2931 nm and smaller one of 0.2877 nm, respectively, and a small volume fraction of intermetallic σ-phase (Co_3Ti) to single

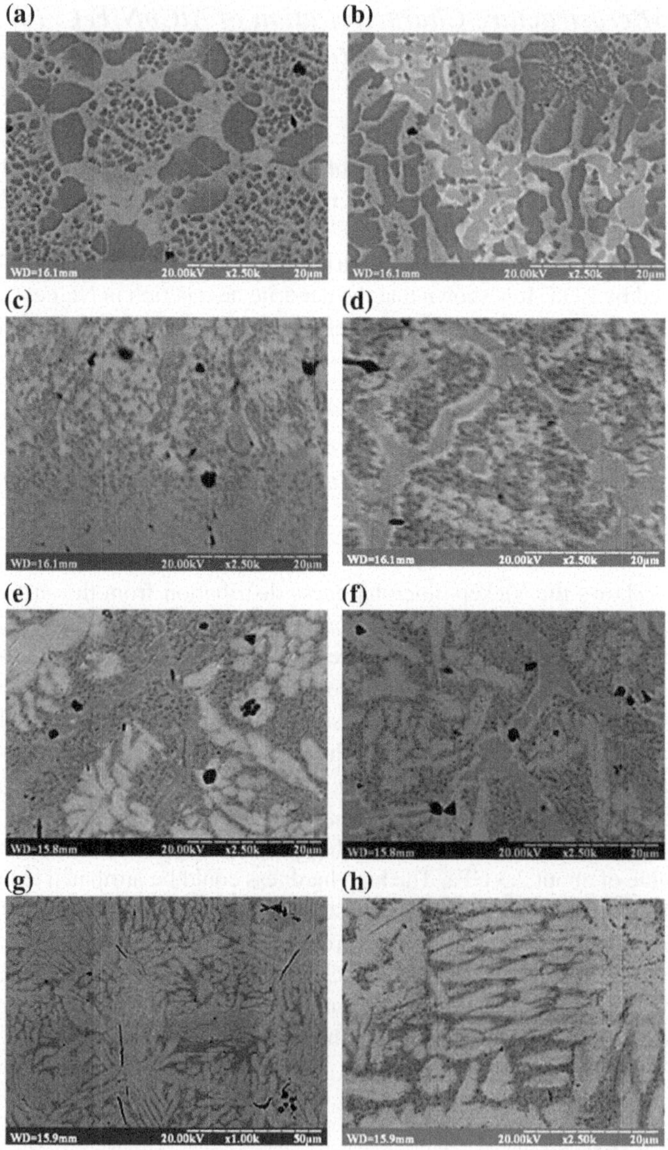

Fig. 16.3 SEM images of the AlCoNiFeCrTiV$_x$ coatings, x = 0, 0.5, 1.0, and 1.5 corresponding to (**a**, **b**), (**c**, **d**), (**e**, **f**) and (**g**, **h**), respectively; (**a**, **c**, **e**, **g**) near the substrate and (**b**, **d**, **f**, **h**) near the surface

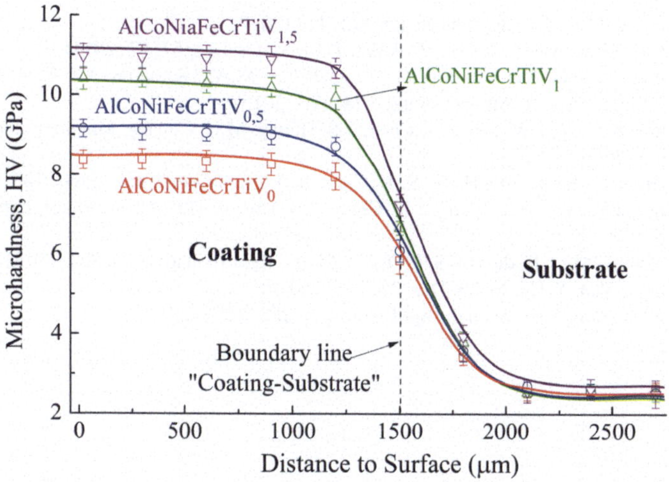

Fig. 16.4 Microhardness distribution of the AlCoNiFeCrTiV$_0$, AlCoNiFeCrTiV$_{0.5}$, AlCoNiFeCrTiV$_1$, and AlCoNiFeCrTiV$_{1.5}$ high-entropy coatings resulted from electron-beam cladding

BCC solid solution in AlCoNiFeCrTiV$_{0.5–1.5}$ coatings with increasing in the volume fraction of intermetallic σ-phases of different type (Co_3Ti, NiTi, NiTiAl, and FeV) while increasing vanadium content from x = 0.5 to x = 1.5. Moreover, the lattice parameter of AlCoNiFeCrTiV$_{0.5–1.5}$ coatings BCC solid solutions is around a constant and varies only within the measurement error.

This research indicates that with the addition of V element the AlCoNiFeCrTiV$_{0–1.5}$ coatings prepared by electron-beam cladding exhibit a monotonic trend of increasing microhardness. Higher vanadium coatings exhibit higher hardness. With the increase of V content from x = 0 to x = 1.5 the values of Vickers hardness increased from 8.4 to 11 GPa because of increase in the sigma-phase volume fraction resulted in continuous strengthening.

AlCoNiFeCrTiV$_x$ HE coatings show a solid solution hardening effect because different-sized atoms occupying the BCC crystal lattice sites cause lattice distortion and σ-phase precipitations harden effect serve as harden mechanism too.

References

1. B.S. Murty, J.W. Yeh, S. Ranganathan, *High-Entropy Alloys* (Elsevier, Amsterdam, 2014), p. 218
2. M.C. Gao, J.-W. Yeh, P.K. Liaw, Y. Zhang (eds.), *High-Entropy Alloys. Fundamentals and Applications* (Elsevier, Amsterdam, 2015), p. 516
3. J.W. Yeh, Y.L. Chen, S.J. Lin, Mater. Sci. Forum **560**, 1 (2007)
4. Y.F. Ye, Q. Wang, J. Lu, C.T. Liu, Y. Yang, Mater. Today **19**(6), 349 (2016)
5. M.-H. Tsai, J.-W. Yeh, Mater. Res. Lett. **2**(3), 107 (2014)

6. D.B. Miracle, O.N. Senkov, Acta Mater. **122**, 448 (2017)
7. Y. Dong, K. Zhou, Y. Lu, X. Gao, T. Wang, T. Li, Mater. Des. **57**, 67 (2014)
8. N.D. Stepanov, D.G. Shaysultanov, G.A. Salishchev, M.A. Tikhonovsky, E.E. Oleynik, A.S. Tortika, O.N. Senkov, J. Alloys Compd. **628**, 170 (2015)
9. M.R. Chen, S.J. Lin, J.W. Yeh, S.K. Chen, Y.S. Huang, M.H. Chuang, Metall. Mater. Trans. A **37A**, 1363 (2006)
10. J.-W. Yeh, S.Y. Chang, Y.D. Hong, S.-K. Chen, S.-J. Lin, Mater. Chem. Phys. **103**, 41 (2007)
11. C.C. Tung, J.W. Yeh, T.T. Shun, S.K. Chon, Y.S. Huang, H.C. Chen, Mater. Lett. **61**(1), 1 (2007)
12. K.K. Alaneme, M.O. Bodunrin, S.R. Oke, J. Mater. Res. Technol. **5**(4), 384 (2016)
13. H. Zhang, Y. Pan, Y. He, Mater. Des. **32**, 1910 (2010)
14. H. Zhang, Y. Pan, Y. He, H. Jiao, Appl. Surf. Sci. **257**, 2259 (2011)

Chapter 17
A Novel Method for the Formation of Silver-Containing Nanocomposites—Thermochemical Reduction of Ag^+ Ions in Polymer Films

V. Demchenko, S. Riabov, M. Iurzhenko and N. Rybalchenko

Abstract Structure of nanocomposites based on interpolyelectrolyte complex, biodegradable polymers and Ag nanoparticles formed by the thermochemical reduction of silver ions in polymer films have been investigated.

17.1 Introduction

During the last decades, the interest in studying nanosize particles of different metals and their oxides has grown up [1–4]. Primarily, it is due to their unique characteristics, which dramatically differ from their analogs—micro-scale objects. Thus, this fact opens new possibilities for diverse applications of nanomaterials that have advanced properties.

Hybrid materials containing silver nanoparticles show promising features for the design of catalytic systems and they are currently used in optoelectronics and nanophotonics [5–10]. In its turn, nanocomposite materials [10–14] with silver nanoparticles have found a wide application in medicine due to their antibacterial and antiviral properties. Within the recent years, the emergence of microorganisms resistant to known antimicrobial agents has caused a demand in designing new high-effective substances for saturation of dressing materials in medical facilities, for the formation of antimicrobial coatings for providing the sterile conditions in biological and medical laboratories, for food packaging, and for the antimicrobial air filters etc.

V. Demchenko (✉) · S. Riabov · M. Iurzhenko
Institute of Macromolecular Chemistry of the NAS of Ukraine, Kharkivske shosse 48, Kiev 02160, Ukraine
e-mail: dvaleriyl@ukr.net

V. Demchenko · M. Iurzhenko
E. O. Paton Electric Welding Institute, of the NAS of Ukraine, Kazymyra Malevycha 11, Kiev 03680, Ukraine

N. Rybalchenko
D. K. Zabolotny Institute of Microbiology and Virology of the NAS of Ukraine, Zabolotnoho 154, Kiev 03143, Ukraine

A. D. Pogrebnjak and O. Bondar (eds.), *Microstructure and Properties of Micro- and Nanoscale Materials, Films, and Coatings (NAP 2019)*,
Springer Proceedings in Physics 240, https://doi.org/10.1007/978-981-15-1742-6_17

Metal nanoparticles of copper, silver, and zinc are of great interest for researchers, involved into development of effective antimicrobial agents, due to their biocidal activity and high stability in severe conditions.

The development of such materials is impossible without fundamental researches and studying their structure, physical-chemical and mechanical properties.

The current review of scientific sources revealed that data concerning investigations of structural features, physical-mechanical, and antimicrobial characteristics of Ag-containing nanocomposites obtained by thermochemical reduction of Ag^+ ions in the interpolyelectrolyte–metal complexes (IMC) have not been published yet.

Therefore, the aim of this work is to study the structure of nanocomposites prepared involving natural and synthetic polymers and Ag nanoparticles, synthesized by the thermochemical reduction of silver ions in the polymer films.

17.2 Experiment Details

The thermochemical reduction of Ag^+ ions in the polymer films was performed by heating up to 100–160 °C for 5–30 min. Samples were heated in the oven using precise thermal regulator VRT-3. The precision of the temperature regulation was about ±0.5 °C.

The structure of silver-containing nanocomposites was studied via the wide-angle X-ray scattering (WAXS) method using DRON-4-07 diffractometer with the X-ray optic scheme set to "transmission" of the primary beam through samples. All the XRD measurements were performed under the following conditions: CuKα radiation, a Ni filter, T =20 ± 2 °C.

The size of Ag nanoparticles and their distribution in the polymer matrix were examined by the Jeol JEM-1230 transmission electron microscope (Japan) with a resolution of about 0.2 nm.

17.3 Results and Discussion

The analysis of X-ray patterns has shown that interpolyelectrolyte complexes (IPEC) formed from pectin and polyethylenimine (PEI) is characterized by a short-range ordering during translation of fragments of oppositely charged polyelectrolyte macromolecular chains in space.

This circumstance is indicated by the appearance of one diffuse diffraction maximum with $2\theta_m$ ~ 20.8° on the X-ray pattern of IPEC sample (Fig. 17.1, curve 1) obtained by the Wide-angle X-ray scattering (WAXS). The average value of the period of the short-range ordering of fragments of complementary macromolecular chains of oppositely charged polyelectrolytes in the IPEC (the Bragg distance between the macromolecule chains of anionic and cationic polyelectrolytes in the IPEC) is 4.3 Å, i.e., slightly less than that in the cationic polyelectrolyte. However,

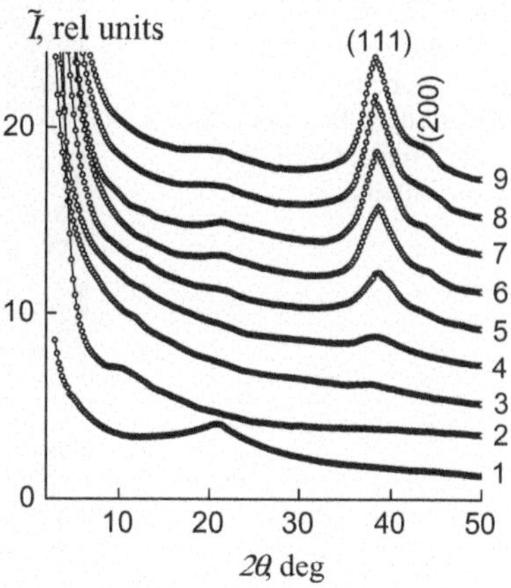

Fig. 17.1 Wide-angle X-ray patterns of (1) IPEC, (2) IMC, and (3–9) nanocomposites obtained by the thermal reduction method from IMC at temperature (3) 100, (4) 110, (5) 120, (6) 130, (7) 140, (8) 150, and (9) 160 °C for 30 min

the adsorption of $AgNO_3$ by the IPEC sample and formation of the IPEC–Ag^+ IMC is accompanied by change in the diffraction pattern. This result is proved by the appearance of an intense diffuse diffraction maximum at $2\theta_m \sim 11.2°$, featuring the structure polyelectrolyte–metal complex pectin–Ag^+–PEI [15–17] (curve *2*). However, the amorphous halo disappears at $2\theta_m \sim 20.8°$, which is related to the structural peculiarity of pectin–PEI IPEC. This indicates the full transfer of polyelectrolyte complexes into polyelectrolyte–metal complexes.

Analysis of diffraction patterns of silver-containing nanocomposites prepared by the thermochemical reduction of Ag^+ ions in IMC in the wide range of temperatures (100–160 °C) has shown that the content of silver nanoparticles is growing up when the temperature gradually increases to $T = 150$ °C. Therefore, the increase in the intensity of two peaks at $2\theta_m \sim 38.2°$ and 43.8° confirms the presence of silver in the system (Fig. 17.1, curves 3–8). Further increase in temperature up to 160 °C did not lead to the rise in the intensity of the diffraction maxima characterizing silver structure (Fig. 17.1, curves 8–9). Thus, according to the WAXS data we can conclude that the optimal temperature for the reduction of Ag^+ ions in IMC with further formation of nanocomposite is 150 °C. The decrease of silver ions at elevated temperatures is most likely due to the transfer of electrons from the nitrogen atoms of PEI to Ag^+ ions [1].

Microphotography analysis (Fig. 17.2) demonstrated that silver nanoparticles were statistically distributed in a polymer matrix and had an average size of 5.5 nm.

The silver-containing nanocomposites based on polyelectrolyte complexes from polymers of natural origin were obtained by the thermochemical reduction of silver ions (Fig. 17.3).

From the analysis of a microphotographs (Fig. 17.4) it is clear that polymeric system of $pectin^-$-$chitosan^+$-Ag has the best statistical distribution of silver nanoparticles with the smallest size.

The method of X-ray diffraction analysis showed that the reduction of silver ions in a biodegradable polymer-polylactide occurs at the temperature of 160 °C when the

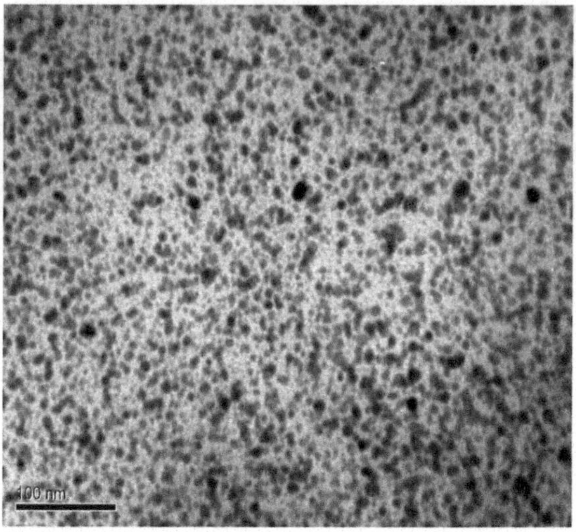

Fig. 17.2 TEM images of nanocomposites obtained by the thermochemical reduction method from IMC at the temperature 150 °C during 30 min

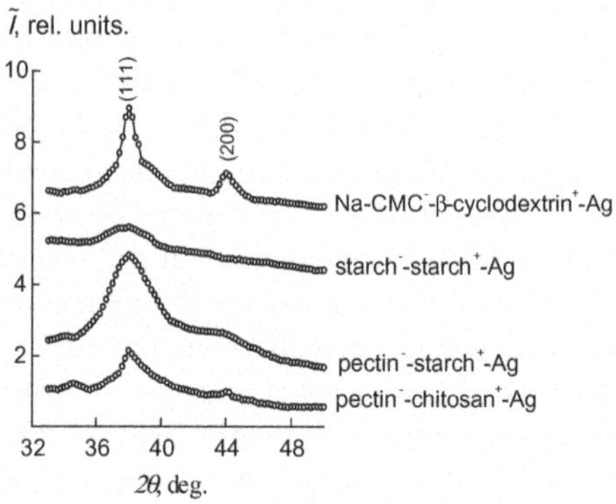

Fig. 17.3 Wide-angle X-ray diffractograms of IPEC–Ag nanocomposites

film is heated for 5 min. The diluents were pre-introduced into polylactide polymers containing nitrogen—polyethyleneimine (Fig. 17.5) or chitosan (Fig. 17.6).

Analysis of microphotographs of silver-based nanocomposites based on polylactide (Fig. 17.7) showed that silver nanoparticles are worse distributed in a polymer matrix than nanocomposites based on polyelectrolyte complexes.

17.4 Conclusions

X-ray diffraction analysis revealed that for the polymeric system pectin-polyethyleneimine-Ag, the optimum temperature of thermochemical reduction at which the structural presence of the metallic silver is fully realized is 150 °C. Different

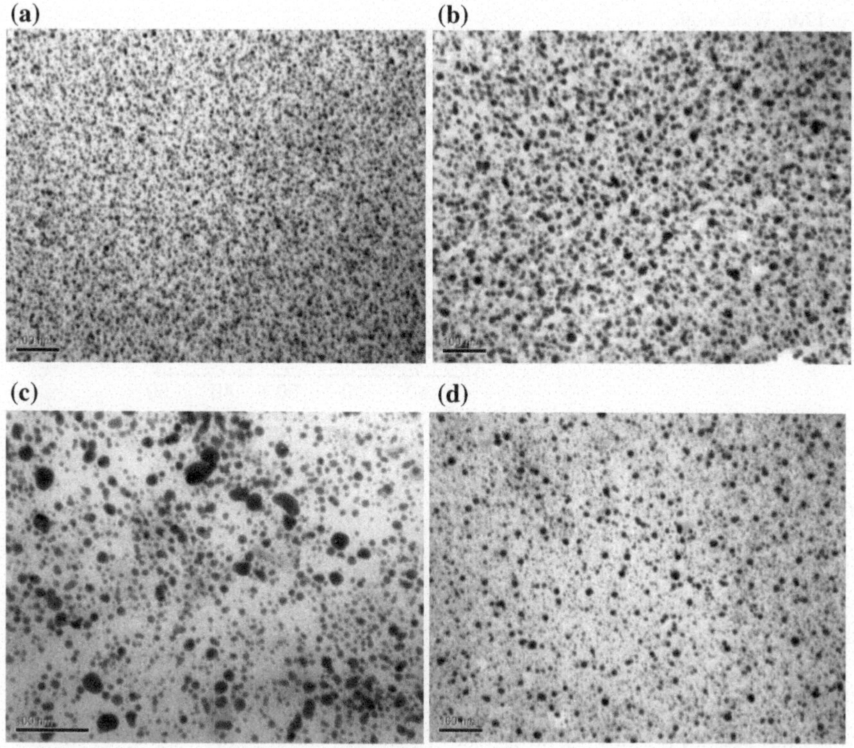

Fig. 17.4 TEM images of nanocomposites obtained by the thermochemical reduction method from IMC at the temperature 150 °C during 30 min: **a** Na-CMC^{-}-β-cyclodextrin^{+}-Ag; **b** starch^{-}-starch^{+}-Ag; **c** pectin^{-}-starch^{+}-Ag; **d** pectin^{-}-chitosan^{+}-Ag

Fig. 17.5 Wide-angle X-ray diffractograms of the PLA-filament; (2) PLA-film are annealed 100 °C; and the nanocomposites obtained by the thermochemical reduction method from films PLA-PEI at the temperature 60–160 °C for 5 min

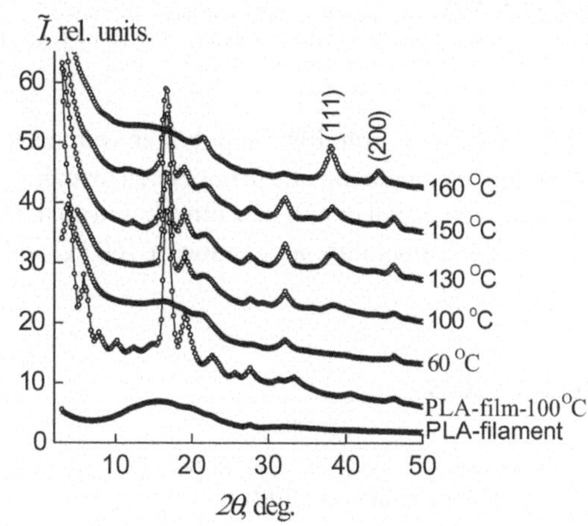

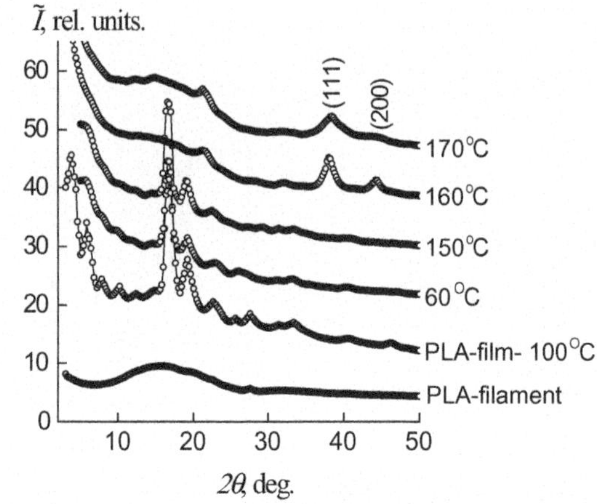

Fig. 17.6 Wide-angle X-ray diffractograms of the PLA-filament; (2) PLA-film are annealed 100 °C; and the nanocomposites obtained by the thermochemical reduction method from films PLA-chitosan at the temperature 60–170 °C for 5 min

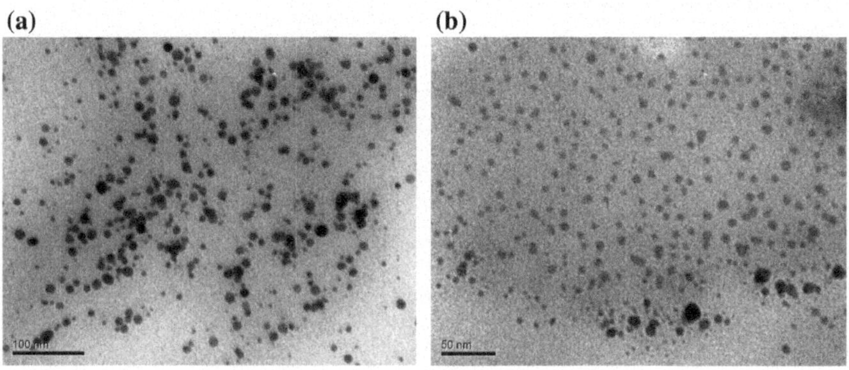

Fig. 17.7 TEM images of nanocomposites obtained by the thermochemical reduction method at the temperature of 150 °C during 5 min: **a** PLA-PEI-Ag; **b** PLA-chitosan-Ag

structures and morphology in polymer systems based on polyelectrolyte complexes of natural and synthetic origin have been revealed. It has been established that thermochemical reduction in polymer films based on the biodegradable polymer-polylactide occurs at a temperature and a time of 160 °C and 5 min, respectively.

References

1. V. Demchenko, S. Riabov, N. Rybalchenko, L. Goncharenko, S. Kobylinskyi, V. Shtompel', Eur. Polym. J. **96**, 326 (2017)
2. V. Demchenko, V. Shtompel, S. Riabov, Eur. Polym. J. **75**, 310 (2016)
3. V.L. Demchenko, V.I. Shtompel, S.V. Riabov, Polym. Sci. A **57**, 635 (2015)
4. V.L. Demchenko, V.I. Shtompel, Structuring, Polym. Sci. B **56**, 927 (2014)

5. A.D. Pomogailo, V.N. Kestelman, *Metallopolymer Nanocomposites* (Springer, New York, 2005)
6. N.L. Rosi, C.A. Mirkin, Chem. Rev. **105**, 1547 (2005)
7. M. Ballauff, Y. Lu, Polymer **48**, 1815 (2007)
8. P. Ruiz, J. Macanas, M. Munoz, D.N. Muraviev, Nanoscale Res. Lett. **6**, 343 (2011)
9. M.L. Bruening, D.M. Dotzauer, P. Jain, L. Ouyang, G.L. Baker, Langmuir **24**, 7663 (2008)
10. Z. Deng, H. Zhu, B. Peng, H. Chen, Y.F. Sun, X.D. Gang, P.J. Jin, J.L. Wang, ACS Appl. Mater. Interfaces **4**, 5625 (2012)
11. G.F. Prozorova, A.S. Pozdnyakov, N.P. Kuznetsova, S.A. Korzhova, A.I. Emel'yanov, T.G. Ermakova, T.V. Fadeeva, L.M. Sosedova, Int. J. Nanomed. **9**, 1883 (2014)
12. H.S. Barud, T. Regiani, R.F.C. Marques, W.R. Lustri, Y. Messaddeq, S.J.L., Ribeiro, J. Nanomater. **2011**, 721631 (2011)
13. A.L. Stepanov, V.V. Vorobev, V.I. Nuzhdin, V.F. Valeev, A.M. Rogov, Y.N. Osin, in *Ion Implantation. Synthesis, Applications and Technology*, ed. by A. Pogrebnjak (Nova Science Publishers, New York, 2018), p. 77
14. K. Sivaiah, K.N. Kumar, V. Naresh, S. Buddhudu, Mater. Sci. Appl. **2**(11), 1688 (2011)
15. V.I. Shtompel, Yu. Yu. Kercha, in *Structure of Linear Polyurethanes* (Naukova dumka, Kiev, 2008) (in Russian)
16. V. Demchenko, V. Shtompel, S. Riabov, E. Lysenkov, Nanoscale Res. Lett. **10**, 479 (2015)
17. V. Demchenko, S. Riabov, V. Shtompel, Nanoscale Res. Lett. **12**, 235 (2017)

Chapter 18
Application of Multicomponent Wear-Resistant Nanostructures Formed by Electrospark Allowing for Protecting Surfaces of Compression Joints Parts

V. Tarelnyk, Ie. Konoplianchenko, O. Gaponova, B. Antoszewski, Cz. Kundera, V. Martsynkovskyy, M. Dovzhyk, M. Dumanchuk and O. Vasilenko

Abstract The paper discusses the specific features of the process for forming multicomponent wear-resistant nanostructures on the structural steel of 38X2MЮA grade using the method of electrospark alloying (ESA) in the course of simultaneously saturating the surface layers with carbon (carbonizing), sulfur (sulfidizing) and aluminum (aluminizing), which can be used to improve microhardness and wear resistance, prevent frictional seizure, increase resistance to atmospheric corrosion and, thus, protect the surfaces of the parts for compression joints against fretting corrosion. At processing steel using the ESA method by graphite electrode with the discharge energy of Wp = 0.13; 0.52 and 4.9 J and the productivity of 0.5–2.5 cm^2/min, a consistent matter containing sulfur and an aluminum powder is applied to the surface of the part to be strengthened, and then, without waiting for the matter to dry, the alloying process is carried out, while the consistent matter with the content of the aluminum powder of not more than 56% is applied. There were conducted metallographic and durametric analyses of the surface layers of the structural steel after simultaneously aluminizing, sulfidizing, and carbonizing by the ESA method. It was shown that the layer structure consisted of three zones, namely, the white layer, the diffuse zone, and the base metal. While increasing the discharge energy, such surface layer qualitative parameters as thickness, microhardness, and continuity of the white layer as well as surface roughness thereof had been increasing. With increasing the discharge energy,

V. Tarelnyk (✉) · Ie. Konoplianchenko · V. Martsynkovskyy · M. Dovzhyk · M. Dumanchuk · O. Vasilenko
Sumy National Agrarian University, Sumy, Ukraine
e-mail: tarelnik@i.ua

Ie. Konoplianchenko
e-mail: konoplyanchenko@ukr.net

O. Gaponova · Cz. Kundera
Sumy State University, Sumy, Ukraine

B. Antoszewski
Kielce University of Technology, Kielce, Poland

A. D. Pogrebnjak and O. Bondar (eds.), *Microstructure and Properties of Micro- and Nanoscale Materials, Films, and Coatings (NAP 2019)*,
Springer Proceedings in Physics 240, https://doi.org/10.1007/978-981-15-1742-6_18

the presence of the enhanced sulfur content in the coating had been increasing from 40 to 100 μm.

18.1 Introduction

The most characteristic cases of compression and compression-keyed joints failures are represented by violations of mating strength of their component parts, fretting fatigue damages and breakages caused by fatigue thereof.

Relative micro displacements of connected surfaces observed in the course of wear process occur because of the deformation of parts under loading and vibration conditions accompanying the operation of machines and equipment. The wear intensity increases at operating the parts in aggressive environments. In this case, damaging of the mating surfaces occurs under conditions of fretting corrosion, which, as a rule, takes place with their slight oscillating relative displacements. The significant intensifications of oxidation and frictional seizure are caused by the dynamic nature of loading, when, as at contact, temperature gradient and deformation gradient sharply increase.

Fretting corrosion (FC) is observed at various press fits on rotating shafts, in points of fitting turbine blades, in splined, keyed, bolted and riveted joints. As a result of FC, there is reduced fatigue strength of the parts, which fact can cause serious accidents. The technical result consisting in improving the efficiency of the above said joints, to a great extent, depends on a lot of factors, namely, studying the problems of fretting process, investigating the causes of such a type of wear, analyzing the nature of the processes of interaction and damage of the surfaces of the bodies interacting in a frictional contact, etc. The achievement of this result is one of the most important tasks in ensuring the reliability and durability of machine parts.

18.2 Actual Scientific Researches and Issues Analysis

At the FC, wear occurs at slight oscillatory, cyclic, reciprocating movements with small amplitudes.

The minimum value of the relative micro displacements between the mating surfaces, which is sufficient for fretting occurrence, is extremely small. According to G. Tomlinson, one of the first researchers of the fretting wear phenomenon, its value makes up several nanometers.

It is known that with fretting, quality of a part surface is significantly deteriorated. The roughness parameters are significantly worsened. Under such a condition, there is possible occurrence of deep cavities reaching 200 μm and more deep into the surface layers [1].

Assembling fixed surfaces can be provided, for example, by pressing a shaft into a hole, heating a part having the hole and enveloping it, or by cooling the shaft [2].

For obtaining the teeth of drill bits the method of centrifugal reinforcement with hard alloy is used [3], which provides both increased their wear resistance and the quality of their press joint with the steel body.

The analysis of the results of operating composite mill rolls of a number of unit sizes shows that there are a large number of cases of low reliability of the tire fixation on the roll axis in the course of assembling process by thermal action [4]. This contributed to the development of a wide range of additional structural, technological and other types of fastening methods and means [5].

Damages caused by the FC phenomenon depend on the great majority of factors, namely, relative slip amplitude, contact pressure, a number of cycles, oscillation frequency, material, and environment.

A typical example to study different types of causes for the FC occurrence is an elastic coupling (EC). The main advantage of the couplings with the elastic metal elements, in comparison with gear and bushing-finger couplings, is their high compensating ability, or the ability to operate with offset axes of the shafts without creating significant additional loads on the shafts and their supporting elements (bearings). Low reactive forces have a positive effect on the rotor system, which service life here is less dependent on the accuracy of the shaft alignment. The couplings simultaneously have a torsion stiffness and flexibility in axial and angular directions to compensate for significant shaft decentering values, including misalignment [6].

It should be noted that among all the mating surfaces of the EC parts, a special danger is represented by the mating of 'half-coupling-shaft' type, wherein the outer cylindrical surface of the shaft contacts the inner cylindrical surface of the half-coupling, and the parts here form a preloaded joint [7–10].

Also it should be noted that such critical parts of pump and compressor equipment as half-couplings, which form compression joints with shafts, are commonly made of the heat-resistant and relaxation-resistant structural steel of 38Х2МЮА grade (another designation is a 38ХМЮА grade), which may be replaced by the steels of 38Х2ЮА, 38Х3МВФ, 38ХВФЮ, 38Х2Ю grades.

As a rule, in the course of operation, between the mating cylindrical surfaces, usually in the vicinity of the ends of the half-coupling, there occurs fretting wear (FW), which can lead to loosening fit, increasing vibration resulted in the joint failure and accident. In addition, the FW leads to decreasing fatigue strength of the parts, which event can also cause serious accidents.

To significantly enhance the carrying capacity of compression joints, there has recently been widely developed the scientific and technical direction associated with introducing intermediate layers made of soft and hard materials into the joints contact zones [11–19].

The effectiveness of a particular coating depends on its thickness. There had been carried out the experiments which showed that fretting damage increased when the thickness of the electroplated silver decreased from 125 to 12.5 μm. According to [20], to provide the practical use for the most parts, it has been recommended to apply the coatings in the range of 75–125 μm thick, although in some cases, there were recommended thicknesses up to 300 μm.

In [21], the author cited the results of the work by A. Tuma and F. Wunderlich, wherein, there was discussed a significant effect of increasing the endurance limit of the shafts with the pressed-on parts by carbonizing thereof. It had been found that the endurance limit of the specimens having a diameter of 12 mm with the pressed-in bushings was influenced by their distortions during the process of quenching. After taking action against the distortions, the endurance limit increased from 137.3 to 412.0 MPa. According to the data represented by E. Lehr, the endurance limit of the carbonized specimens having a diameter of 60 mm was more than doubled as in pressure mounting assembly unit.

Today, the chemical and heat treatment (CHT) is one of the most effective methods for strengthening the surfaces of parts to increase their durability. Despite the fact that as a result of the CHT, the quality of the surface layers of machine parts significantly increases, the method has several disadvantages. Those are as follows: a part volumetric heating, resulting in changing its structure and initial geometrical parameters (deformations and warpage); cumbersome and expensive technological equipment; prolonged process duration, use of energy-intensive equipment, etc.

Recently, to improve the quality of the surface layers of machine parts, the method of electrospark alloying (ESA) has become increasingly important, namely, the process of transferring material to the surface of a product by an electric spark discharge [22–24]. Its specific features are environmental safety, locality of action, low energy consumption, firm connection of the applied material with the base, etc. Using different electrode materials, the electroerosion alloying (EEA) method can be used for conducting the processes alternative to the CHT ones, but with significantly lower costs [25–28]. So applying a graphite electrode and saturating the surface of a part with carbon, it is possible to carry out the carbonizing process [29], doing the same with the use of aluminum electrode, to perform the aluminizing process [30], etc.

In [31–35], there are described various sulfidizing processes, which are thermal and chemical ones for treating the products made of iron-based alloys to enrich the surface layers with sulfur. The effect of sulfidizing a part consists of creating a sulfide film on the surface of the part. Sulfides increase the surface activity of metals and alloys, as well as provide wetting the same with surface-active substances and improve the resistance to frictional seizure. The sulfide film, which has a lower strength than the base metal, is easily broken down at friction and separated from the base without plastically deforming it, and with preventing the friction surface from frictional seizure thereof. The iron sulfide (FeS) film increases the wear resistance of friction surfaces and improves their workability. The ferrosulfide coating has a rather high porosity and absorbs a large amount of lubricant with imparting self-lubricating properties to the material.

Chrome coatings, due to their high hardness, wear resistance, and corrosion resistance, are widely used in moving joints (chrome–rubber) of piston pumps and hydraulic drives [36]. However, the low coefficient of friction of chrome coatings paired with steel restrains their use in press joints.

To improve hardness, wear resistance and the atmospheric corrosion resistance of steel parts, an aluminizing method [37] is often used. The method includes applying an aluminum layer on a steel surface (usually by spraying), plastering and annealing. Along with positive results, the above-described technology has several disadvantages. Those are: high cost and laboriousness of the process; the need to be monitored at all stages of the technology; heating the entire part, and accordingly, structural changes in the metal; deformation and warpage; process duration being more than 8 h; high power consumption; negative impact on the environment, etc. Ceramic-aluminum single-layer oxide and two-layer coatings are also used [38–41], which provide comprehensive protection of steel against hydrogen sulfide cracking and wear. A layer of aluminum is applied to the surface of the part and subjected to plasma electrolytic oxidation. This provides an increase in the coefficient of friction and the quality of the fixed connection of parts.

In accordance with [42], the coatings containing intermetallic compounds of the Ti–Al system were created by electrospark deposition of titanium on aluminum and aluminum on titanium. Using the methods of electron microscopy, X-ray diffraction and X-ray microanalysis, the structure, and composition of the coatings have been studied. At forming the electrospark coatings in the air, aluminum oxide and titanium nitride are additionally created. This technology is performed in a protective environment, such as argon, and it is used only for titanium parts. The alloying process of the Ti–Al alloy was also studied in [43]. In [44], there was studied the effect of the compositions of the synthesized Ni–Al alloys on the process for forming the coatings obtained by the ESA method on the stainless steel of 30X13 grade.

According to [1], it is often observed that a soft material as in contact with a harder material (for example, an aluminum alloy with nitrated steel) turns out to be damaged by fretting to a lesser degree than the hard material.

As the results obtained by a lot of researchers have shown, the intensity of fretting development depends on many factors. They include parameters of external mechanical action (contact pressure, amplitude, and frequency of vibrations); the physicochemical and mechanical properties of the surface layers and the nature of their material, the composition and properties of the medium (temperature, humidity, composition) [45–47]. In addition, it is also necessary to conduct studies of the electrochemical properties of materials and coatings for the details of the compounds [48, 49].

The analysis of the references has shown that to protect the surfaces of parts of the compression joints of the 'hub-shaft' type from FC, the most promising method might be the ESA method, which allows applying a protective coating in a local place on the surface of one, or if necessary, both of the mating parts. At the same time, there is no need in protecting the remaining parts from the impact of the ESA process. In addition, taking into consideration the positive role of such methods as carbonizing, aluminizing and sulfidizing in protecting the mating surfaces of the parts from the FC, there is occurred a need to solve the problem aimed at creating a new process for forming complex multicomponent C–S–Al coatings for steel parts by the ESA method.

Thus, the aim of the work is to develop a new process for protecting steel parts from the FC with the use of the electrospark alloying method by forming multicomponent

Table 18.1 Dependence of the ESA productivity on the discharge energy

Discharge energy (Wu), J	0.13	0.52	4.6
Productivity, cm^2/min	0.5–0.7	1.0–1.3	2.0–2.5

complex C–S–Al coatings to provide for an increase in the parts wear resistance values, prevent frictional seizure and improve the resistance to atmospheric corrosion.

18.3 Research Methods

To determine the effect of the energy parameters of the ESA equipment on the quality parameters of coatings, there were made specimens of 38Х2МЮА steel grade and with the size of the specimens of 15 × 15 × 8 mm, onto which a consistent matter was applied in the form of a sulfur ointment with 33.3% sulfur content. Before applying, aluminum powder of PAD-0 (ПАД-0 (GOST 5494-95)) grade had been added to the sulfur ointment. The maximum amount of powder was 56%. The further increase in the amount of powder led to decreasing adhesion with the surface to be aluminized. After that, without waiting for the consistent matter to dry out, the process with the use of the EEA method was produced applying the graphite electrode of ЭГ-4 mark at the unit of ≪Элитрох – 52А≫ model using various modes of operation. In that case, the discharge energy values were $W_u = 0.13$; 0.52, and 4.9 j.

Each mode of the ESA method corresponded to its own discharge energy and productivity, that is, the area of the formed coating per unit of time (Table 18.1).

It should be noted that decreasing productivity of the process by the ESA method entails a decrease in the qualitative parameters of the surface layer, that is, the occurrence of burns, and what is most importantly, the destruction of the formed layer, which especially affects when processing on 'rougher' modes with the discharge energy Wu > 1 J. Increasing productivity leads to a decrease in coating continuity.

The metallographic analysis of the coatings was performed using the optical microscope of MIM-7 (МИМ-7) model; the durametric studies were performed on the instrument of PMT-3 (ПМТ-3) model. The surface roughness after processing it by the ESA method was determined using the profilograph-profilometer of 201 model of the "Kalibr" plant production by reading and processing profilograms.

To study the distribution of the elements over the layer depth, a local X-ray microanalysis was performed using the scanning electron microscope of Joel JSM-5400 type equipped with a microanalyzer provided by the ISIS 300 Oxford instruments.

18.4 Research Results

Figure 18.1 shows the microstructure (a) and the distribution of the microhardness for the surface layer (b) made of 38Х2МЮА steel when processing by the ESA

Fig. 18.1 Microstructure **a** and distribution of microhardness **b** for the surface layer of 38X2MЮA steel at the ESA method with Wu = 0.13 J

method with Wu = 0.13 J. The metallographic analysis showed that the obtained layer is not uniform and of low continuity (about 70%). The C–S–Al coating consists of three zones: the white layer of ~10 μm depth, which is not amenable to etching with ordinary reagents, below, there is a transition (diffusion) layer having a depth of ~20 μm, and the base metal, namely, 38X2MЮA steel. At analyzing the microhardness distribution over the depth of the layer, it should be noted that, as a result of electrospark alloying, there is formed a coating characterized by the highest hardness value (up to Hμ = 6100 MPa) for the white layer. As moving away from the surface, the hardness decreases to the hardness of the transition zone (Hμ = 4500–5500 MPa) and the base (Hμ = 3100–3200 MPa).

The results of the micro X-ray spectral analysis (Fig. 18.2) indicate that after processing by the ESA method with Wu = 0.13 J, in the surface layer, there is observed the increased content of carbon, sulfur, and aluminum at a distance of 20, 40 and 35 μm, respectively, and the amount of iron slightly decreases to 25 μm.

At increasing the discharge energy to Wu = 0.52 J, there is the observed formation of a more continuous coating layer (up to 85–90%). Just as with Wu = 0.13 J, the layer consists of 3 zones (Fig. 18.3a). However, with an increase in the discharge energy, the hardness and dimensions of the zones increase (Fig. 18.3). Thus, the hardness of the white layer Hμ ~6600 MPa and its value (h) ~20–30 μm, the transition layer—Hμ = 4500–5000 MPa, h ~20 μm.

Figure 18.4 shows the results of a local micro X-ray spectral analysis of the complex C–S–Al coatings after processing them by the ESA method with Wu = 0.52 J. As it can be seen from the figure, as a result of the ESA, carbon, sulfur, and aluminum diffuse deep into the substrate, namely, 38X2MЮA steel. There is observed an increased content of C, S, and Al, respectively, at a distance of 45, 65 and 57 μm beginning from the surface, with a slight decrease in the iron content (at a distance of ~55 μm).

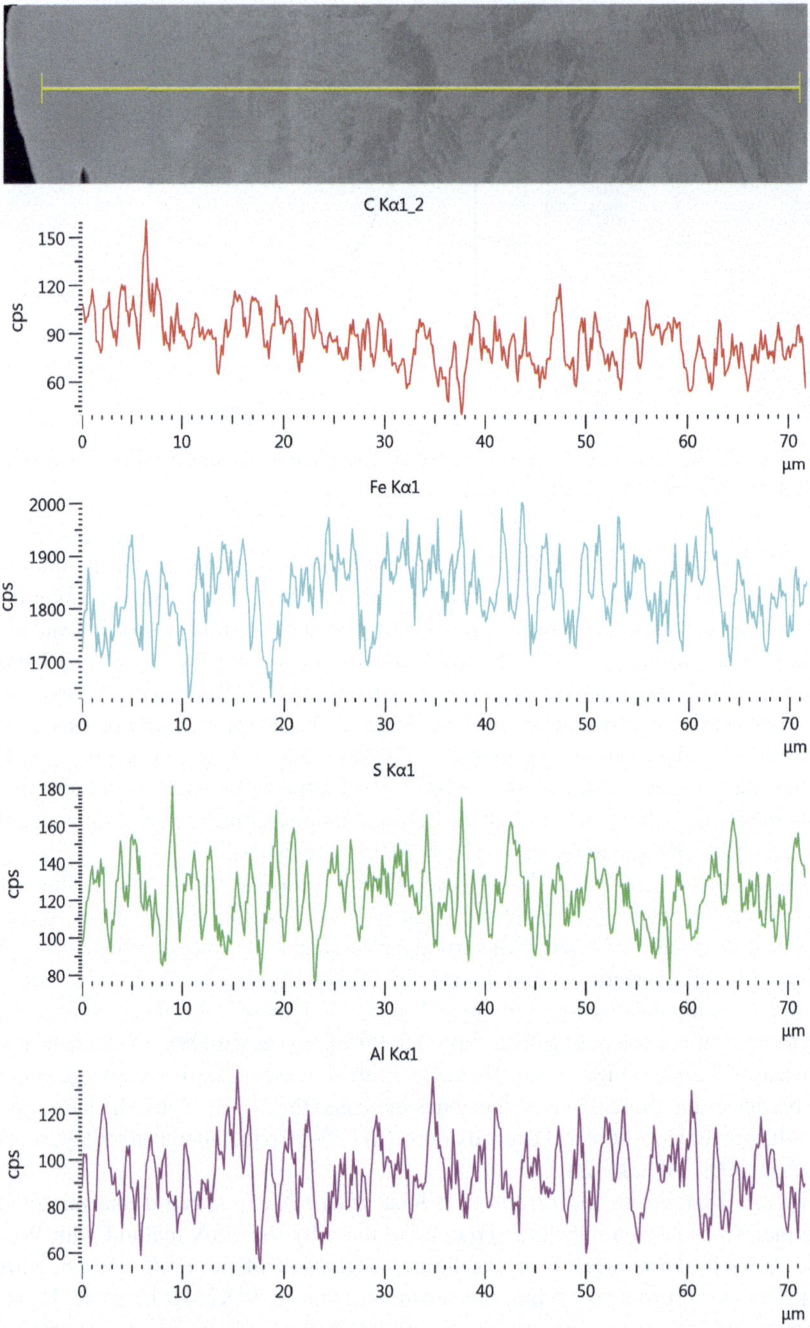

Fig. 18.2 Distribution of elements in the surface layer of 38X2MЮA steel after the ESA method with Wu = 0.13 J

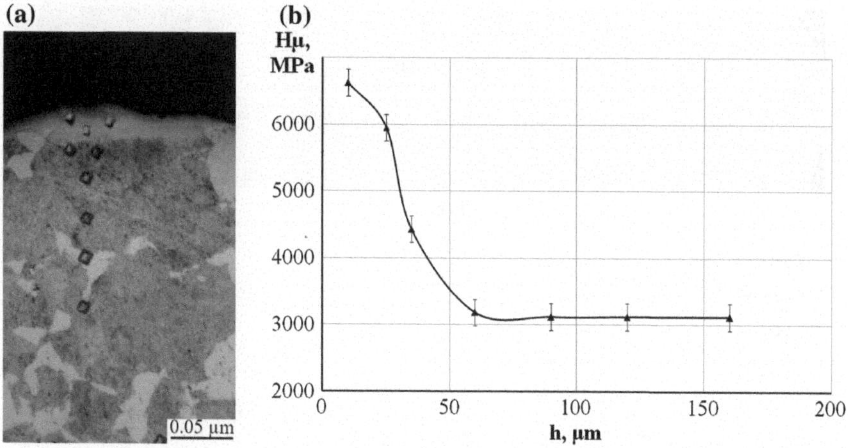

Fig. 18.3 Microstructure **a** and microhardness distribution **b** of 38X2MЮA steel surface layer at processing by the ESA method with Wu = 0.52 J

Figure 18.5 shows the microstructure (a) and the distribution of the microhardness of the surface layer (b) of 38X2MЮA steel at processing the same by the ESA method with Wu = 4.9 J. In this case, the continuity of the layer makes up 100%.

The coating consists of 4 zones. Those are: 1—The sufficiently massive top white layer making up to 50 μm thick and with Hμ ~ 8900 MPa, 2—The light sublayer (20–30 μm) with microhardness of about 6500 MPa, 3—The transition zone of ~20 μm thick, Hμ = 4000–4500 MPa and 4—The base metal of Hμ = 3100–3200 MPa.

Figure 18.6 shows the results of the electron microscopic studies and the local X-ray microanalysis.

On the microstructure obtained with the use of a scanning microscope, there are also observed 4 zones: the surface layer, the sublayer, the transition layer, and the base metal. Studies of the distribution of the elements over the depth of the coating showed that sulfur and carbon are predominantly concentrated on the surface and over the depth up to 100 μm. Aluminum is not evenly distributed in the surface layer, and the greatest amount of its content is observed at a distance of 60–110 μm from the surface, while the iron content herein decreases. It can be assumed that the carbon- and sulfur-containing phases are predominantly formed in the near-surface layers, and the aluminum-containing phases are in the transition zone. Such a distribution of the elements in the coatings obtained by the ESA method seems to be due to the peculiarities of the diffusion processes under the impulse function and the parameters of mutual diffusion of the electrode materials [50].

In Table 18.2, there are shown the qualitative parameters of the complex C–S–Al coatings on the 38X2MЮA steel, obtained by the method of electrospark alloying on various modes.

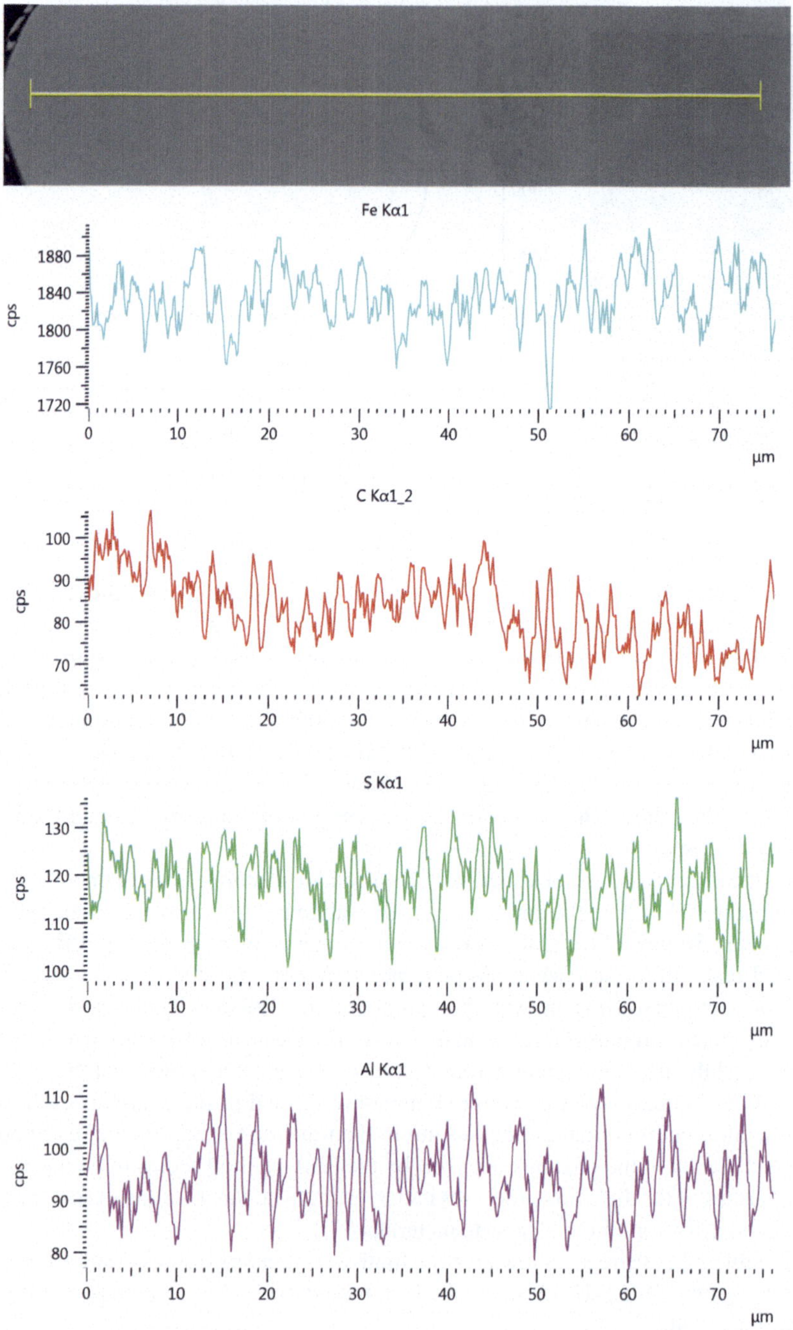

Fig. 18.4 Distribution of elements in the surface layer of 38X2MЮA steel after processing by the ESA method with Wu = 0.52 J

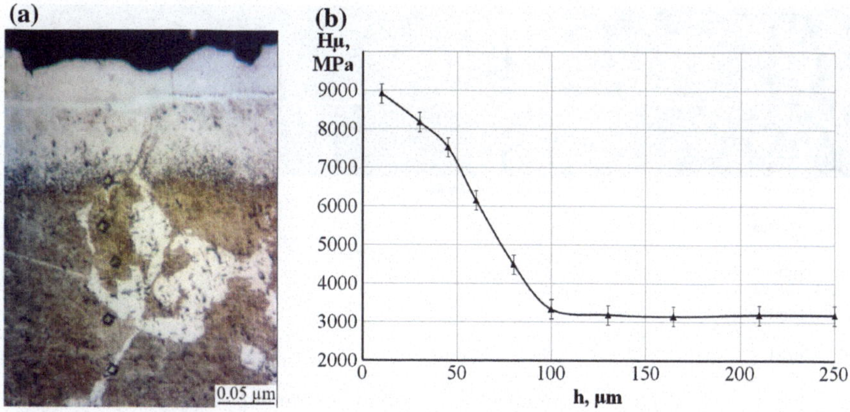

Fig. 18.5 Microstructure (**a**) and microhardness distribution (**b**) of 38X2MЮA steel surface layer at processing the same by the ESA method with Wu = 4.9 J

18.5 Conclusions

A new environmentally-friendly method has been proposed to protect the steel parts mating surfaces of compression joints from fretting corrosion by simultaneously saturating them with carbon, sulfur, and aluminum while processing the same using the method of electrospark alloying.

There have been carried out the metallographic and durametric analyses of the surface layers for the structural steel of 38X2MЮA grade after simultaneously saturating them with carbon, sulfur, and aluminum by the ESA method. It has been shown that the layer structure consists of three zones, namely, the white layer, the diffusion zone, and the base metal. With an increase in the discharge energy from 0.13 to 4.9 J, the thickness, microhardness and continuity of the white layer increase, respectively, from 10 to 50 µm, from 6100 to 8900 MPa and from 70 to 100%, as well as the surface roughness (Ra, Rz and R_{max}) from 1.7; 3.3 and 8.1 to 4.2; 8.7 and 30.2

The presence of sulfur in a consistent matter contributes to the provision of the sulfidizing process. At processing by the ESA method with the discharge energy of 4.9 J, the sulfur content decreases as deepening from the surface, and at the depth of 100 µm, it corresponds to the sulfur amount at the base.

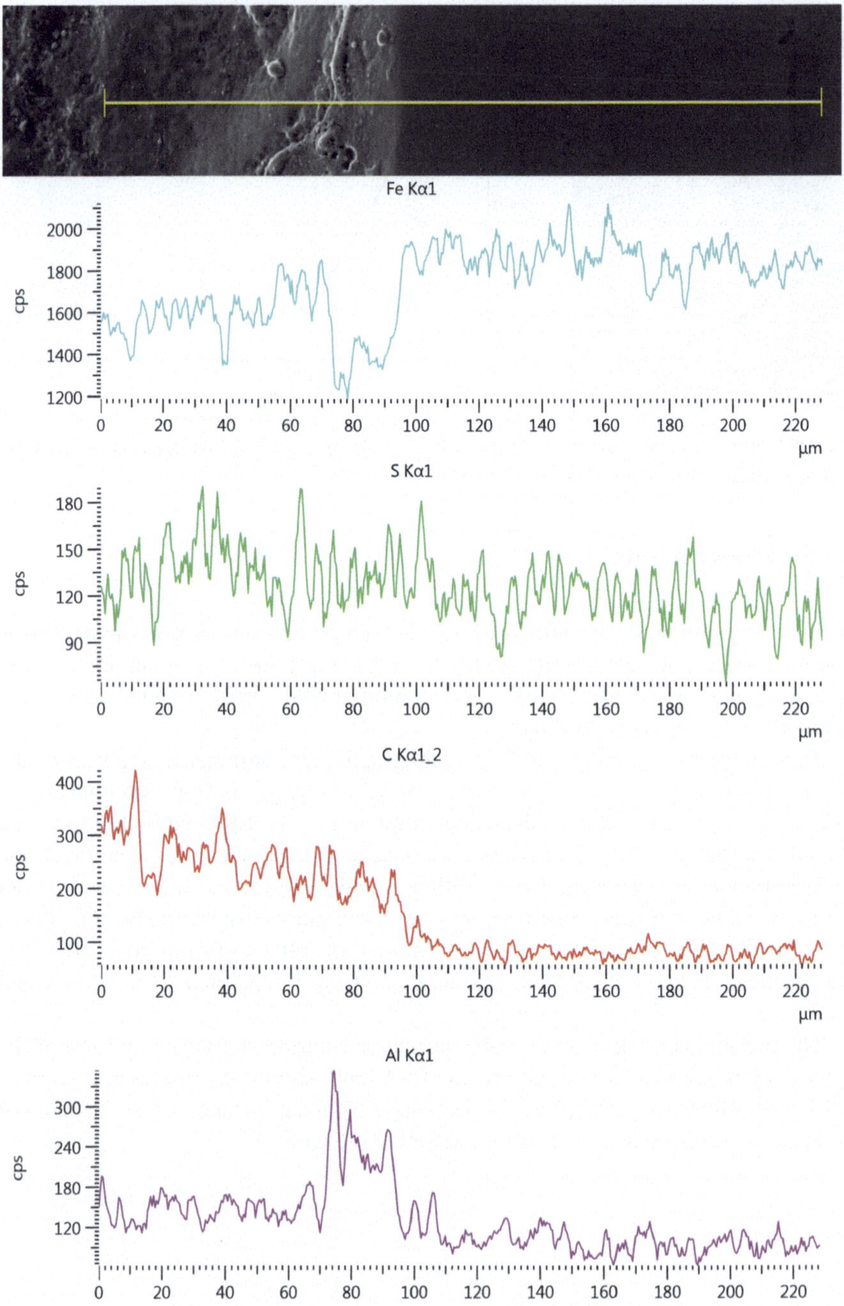

Fig. 18.6 Distribution of elements in the surface layer of 38X2MЮA steel after the ESA method with Wu = 4.9 J

Table 18.2 Qualitative parameters of the complex C–S–Al-coatings on the 38X2MЮA steel after processing by the ESA method

Discharge energy (j)	White layer thickness (μm)	White layer microhardness (MPa)	Roughness (μm)			White layer continuity (%)
			Ra	Rz	R_{max}	
0.13	10	6100 ± 50	1.7	3.3	8.1	70
0.52	30	6600 ± 50	2.3	3.9	9.1	90
4.9	50	8900 ± 50	4.2	8.7	30.2	100

References

1. M.S. Ostrovsky, in *Collection of Scientific Reports* (MSTU, Moscow, 2011), pp. 214–228
2. V.T., Lebed, A.A. Kulachenko, Progressive Technol. Syst. Mech. Eng. **28**, 84 (2004)
3. L. Ropyak, I. Schuliar, O. Bohachenko, Influence of technological parameters of centrifugal reinforcement upon quality indicators of parts. Eastern-Eur. J. Enterp. Technol. **1**(5), 53–62 (2016). https://doi.org/10.15587/1729-4061.2016.59850
4. P.V. Skigoev, A.D. Petrov, A.I. Karmanov, Surveys NIIINFORMTYaZhMASh **1**, 67 (1967)
5. M.V. Gedeon, *Research, Design Improvement, Testing and Implementation of Manufacturing and Restoring Technologies for Large Back-up Rolls* (State Registrations Number 0185.0046681, Kramatorsk, 1988) (in Ukrainian)
6. V.M. Minko et al., in *Proceeding of XIII International Scientific and Engineering Conference "HERVICON+PUMPS"* (Sumu State University, Sumy, 2011)
7. A.N. Petukhov, J. Aerospace Eng. Technol. **7**(15), 128 (2004)
8. A.M. Smyslov, K.S. Selivanov, *Increasing Durability of Machine Parts Under Fretting Conditions* (Gilem, Ufa, 2005)
9. I.M. Panova, A.D. Panov, Naukovedenie **9,** 1 (2017)
10. I.F. Malitsky, E.V. Chernyatina, Mach. Eng. **13**, 149 (2014)
11. A. Panda, K. Dyadyura, J. Valíček, M. Harničárová, J. Zajac, V. Modrák, I. Pandová, P. Vrábel, E. Nováková-Marcinčínová, Z. Pavelek, Manufacturing technology of composite materials—principles of modification of polymer composite materials technology on polytetrafluoroethylene. Materials **10**(377), 1–20 (2017). https://doi.org/10.3390/ma10040377
12. R.B. Waterhouse, *Fretting—Corrosion* (Mechanical Engineering, Leningrad, 1976)
13. O. Maksakova, S. Simoẽs, A. Pogrebnjak, O. Bondar, Y. Kravchenko, V. Beresnev, N. Erdybaeva, The influence of deposition conditions and bilayer thickness on physical-mechanical properties of CA-PVD multilayer ZrN/CrN coatings. Mater. Charact. **140**, 189–196 (2018). https://doi.org/10.1016/j.matchar.2018.03.048
14. A.D. Pogrebnjak, V.M. Beresnev, O.V. Bondar, B.O. Postolnyi, K. Zaleski, E. Coy, S. Jurga, M.O. Lisovenko, P. Konarski, L. Rebouta, J.P. Araujo, Mater. Des. **153**, 47 (2018)
15. V. Tarelnyk, V. Martsynkovskyy, Appl. Mech. Mater. **630**, 397 (2014). http://doi.org/10.4028/www.scientific.net/AMM.630.397
16. V.B. Tarel'nik, V.S. Martsinkovskii, A.N. Zhukov, Chem. Petrol. Eng. **53**, 266 (2017). https://doi.org/10.1007/s10556-017-0333-7
17. V. Martsinkovsky, V. Yurko, V. Tarelnik, Yu. Filonenko, Procedia Eng. **39**, 157 (2012). https://doi.org/10.1016/j.proeng.2012.07.020
18. B. Antoszewski, V. Tarelnyk, Appl. Mech. Mater. **630**, 301 (2014). http://doi.org/10.4028/www.scientific.net/AMM.630.301
19. A.D. Pogrebnjak, V.M. Beresnev, K.V. Smyrnova, Ya.O. Kravchenko, P.V. Zukowski, G.G. Bondarenko, Mater. Lett. **211**, 316 (2018). https://doi.org/10.1016/j.matlet.2017.09.121
20. R.B. Waterhouse, P.A. Brook, G.M. Lee, Wear **5**, 235 (1962)
21. L.T. Balatsky, *Strength of Compression Joints* (Tehnika, Kiev, 1982)

22. B.R. Lazarenko, N.I. Lazarenko, *Electrospark Processing of Conductive Materials* (Izd. Academy of Sciences of the USSR, Moscow, 1968)
23. S.N. Khimukhin, K. Ri, EKh Ri, *Structure and Properties of Metals and Alloys under Electrospark Effect* (Pacific ocean State Univ. Pub, Khabarovsk, 2015)
24. Y.I. Mulin, A.D Verkhoturov, *Electrospark Alloying of Working Surfaces of Tools and Machine Parts with Electrode Materials Obtained from Mineral Raw Materials* (Dal'nauka, Vladivostok, 1999)
25. V.B. Tarelnyk, O.P. Gaponova, I.V. Konoplianchenko, M.Y. Dovzhyk, Metallofiz. Noveishie Tekhnol. **39**, 363 (2017). https://doi.org/10.15407/mfint.39.03.0363
26. V.B. Tarel'nyk, O.P. Gaponova, Y.V. Konoplyanchenko, M.Y. Dovzhyk, Metallofiz. Noveishie Tekhnol. **38**, 1611 (2016). https://doi.org/10.15407/mfint.38.12.1611
27. V. Tarelnyk, I. Konoplianchenko, N. Tarelnyk, A. Kozachenko, Mater. Sci. Forum **968**, 131 (2019). http://doi.org/10.4028/www.scientific.net/MSF.968.131
28. V.B. Tarelnyk, O.P. Gaponova, I.V. Konoplianchenko, V.A. Herasymenko, N.S. Evtushenko, Metallofiz. Noveishie Tekhnol. **40**, 235 (2018). https://doi.org/10.15407/mfint.40.02.0235
29. V.B. Tarel'nik, A.V. Paustovskii, Y.G. Tkachenko, V.S. MartsinkovskiiA. V. Belous, E.V. Konoplyanchenko, O.P. Gaponova, Surf. Eng. Appl. Electrochem. **54**, 147 (2018). https://doi.org/10.3103/s106837551802014x
30. G.V. Kirik, O.P. Gaponova, V.B. Tarelnyk, O.M. Myslyvchenko, B. Antoszewski, Powder Metall. Met. C+ **56**, 688 (2018). https://doi.org/10.1007/s11106-018-9944-6
31. V.D. Zozulya, E.L. Shvedkov, D.I. Rovinsky, E.D. Brown, *Dictionary-Reference to Friction, Wear and Lubrication of Machine Parts*, 2nd edn. (Sci. Dumka, Kiev, 1990)
32. N.E. Denisova, V.A. Shorin, I.N. Gontar, N.I. Volchihina, N.S. Shorina, *Tribotehnicheskoe materialovedenie i tribotehnologiya (Tribotechnical Materials Science and Tribotechnology)* (Penz. State University Publ, Penza, 2006)
33. N.I. Lazarenko, *Electrospark Alloying of Metal Surfaces* (Machine Eng, Moscow, 1976)
34. A.E. Gitlevich, V.V. Mikhailov, N.Ya. Parkanskii, V.M. Revutskii, *Electrospark Alloying of Metal Surfaces* (Shtintsa, Chisinau, 1985)
35. A.G. Shcherbinsky, SU Patent 139336, 30 Dec 1961
36. L. Ropyak, V. Ostapovych, Eastern-Eur. J. Enterp. Technol. **2**(5), 50 (2016). https://doi.org/10.15587/1729-4061.2016.65719
37. M.A. Elizavetin, E.A. Satel, *Technological Ways to Increase Durability of Machines* (Mach. Eng, Moscow, 1969)
38. L.Y. Ropyak, I.P. Shatskyi, M.V. Makoviichuk, Metallofiz. Noveishie Tekhnol. **39**(4), 517–524 (2017). https://doi.org/10.15407/mfint.39.04.0517
39. L.Y. Ropyak, I.P. Shatskyi, M.V. Makoviichuk, Metallofiz. Noveishie Tekhnol. **41**(5), 647 (2019)
40. A.D. Pogrebnjak, V.I. Ivashchenko, P.L. Skrynskyy, O.V. Bondar, P. Konarski, K. Załęski, S. Jurga, E. Coy, Compos. Part B Eng. **142**, 85 (2018)
41. A.D. Pogrebnjak, A.A. Bagdasaryan, A. Pshyk, K. Dyadyura, Usp. Phys. Nauk **60**, 586 (2017)
42. S.A. Pyachin, A.A. Burkov, V.S. Komarova, J. Surf. Invest.: X-Ray, Synchrotron Neutron Tech. **6**, 16–24 (2013). https://doi.org/10.1134/s1027451013030336
43. T.M. Radchenko, V.A. Tatarenko, H. Zapolsky, *Solid State Phenom.* **138**, 283 (2008). http://doi.org/10.4028/www.scientific.net/SSP.138.283
44. S.N. Khimukhin, I.A. Astapov, M.A. Teslina et al., in Formation of heat-resistant coatings by method of electrospark alloying using Ni-Al intermetallic alloys. Paper presented at the 15th International Scientific—Practical Conference Technical Sciences—From Theory to Practice, SibAK, Novosibirsk, 12 Nov 2012
45. N.L. Golego, A.Y. Alyabyev, V.V. Shevel, *Fretting Corrosion of Metals* (Technika, Kiev, 1974)
46. M.S. Ostrovsky, *Tribotechnical Bases for Providing Quality of Functioning of Mining Machines* (MGI, Moscow, 1993)

47. A. Pogrebnjak, K. Smyrnova, O. Bondar, Coatings **9**(3), 155 (2019)
48. L.S. Saakiyan, A.P. Efremov, L.Y. Ropyak, A.V. Gorbatskii, Sov. Mater. Sci. **23**(3), 267 (1987)
49. L.S. Saakiyan, A.P. Efremov, L.Y. Ropyak, Prot. Met **25**(2), 185 (1989)
50. A.D. Verkhoturov, I.S. Anfimov, Phys. Chem. Mater. Process. **3**, 93 (1978)

Chapter 19
Preparation, Properties and Application of Miscanthus Nanocellulose as Coating Layer

V. A. Barbash, O. V. Yashchenko and O. A. Vasylieva

Abstract The data on the preparation of nanocellulose from organosolvent miscanthus pulp and its application as coating layer in the production of bag paper are reported. The influence of technological factors on the properties of nanocellulose films was determined. The obtained nanocellulose films had density up to 1.46 g/cm^3, transparency up to 80%, crystallinity degree up to 77% and tensile strength up to 70 MPa. The properties of nanocellulose films studied by means of electron absorption spectra, XRD and AFM methods. The bag paper with consumption of nanocellulose of 5 g/m^2 has mechanical properties of 12–68% higher and the elongation increases 2.9 times than paper without the use of surface treatment. The using of obtained nanocellulose as a coating layer exhibit great potential its application for improving fiber-based mechanical and barrier properties of paper and cardboard.

19.1 Background

In recent years, many studies have been conducted on the replacement of synthetic materials with natural substances. It is, in particular, concern to the production of nanocellulose and composite materials on its basis. Cellulose, as one of the most common natural polymers on Earth, is widely used for the production of bio-based nanoparticles. Nanocellulose has unique properties such as nanosized, renewable, low toxicity, biocompatibility, biodegradation, availability and low cost of raw material, which allows it to be used in multitude spheres [1, 2]. Nanocellulose is used in optoelectronics, in the production of chemical current of sources, sorbents, for reinforcement and improving the thermal stability of polymeric and paper composites [3, 4]. Nanocellulose has been incorporated into polymer matrices to produce reinforced composites with several tens to hundreds folds higher mechanical strength [5] as well

V. A. Barbash (✉) · O. V. Yashchenko · O. A. Vasylieva
National Technical University of Ukraine "Igor Sikorsky Kyiv Polytechnic Institute", Kiev, Ukraine
e-mail: v.barbash@kpi.ua; vabarbash53@gmail.com

A. D. Pogrebnjak and O. Bondar (eds.), *Microstructure and Properties of Micro- and Nanoscale Materials, Films, and Coatings (NAP 2019)*, Springer Proceedings in Physics 240, https://doi.org/10.1007/978-981-15-1742-6_19

as enhanced optical transparency [6]. These specific characteristics of nanocellulose use to improve the mechanical or barrier properties of nanocomposites [7].

Nanocellulose usually are obtained by acid hydrolysis of pulp from wood and non-wood plants. The main raw material for the production of pulp in the world is wood. For countries that do not have large stocks of free wood, the development of alternative sources of pulp remains an urgent problem. Such sources of raw materials include various types of non-wood plant materials, in particular, Miscanthus x giganteus. The high yield of Miscanthus x giganteus makes it a promising crop for the production of bioenergy and biomaterials, including non-wood pulp [8].

There are few methods to make pulp from wood and non-wood plant materials with different environmental effects [9]. Organosolvent delignification has been suggested as an environmentally friendly process and an alternative method for pulp obtaining. Among organic solvents, acetic acid and hydrogen peroxide should be considered as potential agents for the delignification of plant raw materials. The cooking process, in this case, is carried out at a low temperature and in a relatively short time, which reduces energy costs [10].

Therefore, the purpose of this study is to obtain organosolvent pulp and nanocellulose from Miscanthus x giganteus, investigation their properties and application of nanocellulose as a coating layer for improving mechanical and barrier properties of paper and cardboard.

19.2 Methods

We have used the second year Miscanthus x giganteus biomass as starting. Before research, the miscanthus stalks were ground to 5–7 mm and stored in desiccator for maintenance of constant humidity and chemical composition. The chemical composition of miscanthus chips was determined according to TAPPI standards [11]: cellulose—49.7%; lignin—27.7%, resins, fats, waxes—1.8%; holocellulose—61.9%; ash—1.1% from mass of absolutely dry raw materials (a.d.m.).

The glacial acetic acid, hydrogen peroxide, sulfate acid, NaOH were purchased from Khimlaborreaktiv Ltd. (Ukraine). The reagents were chemically pure.

The preparation of nanocellulose from miscanthus was carried out in 3 stages. At the first stage, cooking was done. To this end, crushed miscanthus chips are loaded into a glass-flask, a cooking solution is added at a liquid to solid ratio of 10:1 containing glacial acetic acid and hydrogen peroxide at a concentration of 35% and a volume ratio of 70:30%, at a temperature of 95 ± 2 °C during 90 min. This cooking regime was determined optimally on the basis of previous studies [12]. In the second stage, the alkaline treatment of miscanthus pulp by the solution of NaOH concentration of 7% during 15–240 min, at the liquid to solid ratio 12:1 at temperature 95 ± 2 °C was carried out. The organosolvent miscanthus pulp (OMP) was washed with hot distilled water to a neutral pH. The quality parameters of the obtained OMP samples are determined according to standard methods [11].

In the third stage, hydrolysis of the obtained OMP was carried out. To obtain nanocellulose, hydrolysis of never dried OMP with a solution of sulfate acid with a concentration of 43 and 50% was carried out, at a liquid to solid ratio of 10:1, at a temperature of 40 and 60 °C for 30–90 min. The calculated amount of sulfuric acid with the corresponding concentration was slowly added into the flask with the cellulose suspension. The hydrolysis was carried out at a temperature of 40, 50 and 60 ± 1 °C. Upon expiration of the reaction time, the hydrolysis was stopped by tenfold dilution with distilled water and cooling of the suspension to the room temperature. The nanocellulose was rinsed with distilled water three times by means of centrifugation at 4000 rev/min and subsequent dialysis until reaching neutral pH. Ultrasonic processing of nanocellulose was carried out using an ultrasonic disintegrator UZDN-A (SELMI, Ukraine) from 22 kHz for 30, 45 and 60 min. The nanocellulose dispersion was placed in an ice bath to prevent overheating during treatment. Eventually, the suspension had the form of a homogenous gel-like dispersion.

The decrease of the size of the cellulose particles and the increase of its dispersity were assessed by measuring the changes in the dimensions of miscanthus pulp. Topographical characterization of nanocellulose samples was investigated using atomic force microscopy (AFM). The measurements were accomplished with Si cantilever, operating in the tapping mode on the device Solver Pro M (NT-MDT, Russia). The scanning speed and area were 0.6 line/s and $2 \times 2\ \mu m^2$, respectively. Before AFM investigation, dilute nanocellulose suspensions with a concentration of 0.01 wt% were ultrasonically treated for 10 min. Subsequently, one drop of nanocellulose dispersion for sample was injected onto a freshly cleaned glass-ceramic and air-dried at room temperature.

Transparency of the nanocellulose films was determined by electron absorption spectra, which were registered in regions from 200 to 1100 nm. Electron absorption spectra of the nanocellulose films in UV, visible and near-infrared regions were registered on two-beam spectrophotometer 4802 (UNICO, USA) with the resolution of 1 nm.

X-ray diffraction patterns of different cellulose samples were obtained by Ultima IV diffractometer (Rigaku, Japan). The method proposed in [13] was used to determine the crystallinity degree (CD) of the samples, in terms of which CD = [(I200 − Iam)/I200] × 100%, where I200 is an intensity of (200) reflex about 22.5°, and Iam is an intensity of amorphous scattering at 18.5°.

The mechanical properties of the nanocellulose films were measured at controlled temperature (23 ± 1 °C) and humidity (50 ± 2%) according to ISO 527-1. Breaking strength tests of nanocellulose films were performed at a crosshead speed of 0.5 mm/min on the TIRAtest-2151 (Germany) instrument equipment with 2-N load stress. For testing, test strips with 10 ± 2 mm width and 25 ± 5 mm long were used. Each composition was tested with a minimum of five specimens to extract an average and standard deviation for each property.

A nanocellulose suspension was applied directly to the paper samples using a coating process. Samples of bag paper were made from sulfate unbleached pulp according to standard procedures. A layer of nanocellulose was applied to received samples of paper with the consumption of nanocellulose from 1 to 5 g/m^2 on each side. Each sample of paper was kept at a temperature of 23 °C and relative humidity of 50% for at least 24 h before determining its properties. All characteristics of the paper have been determined in accordance with the standards of TAPPI [11].

19.3 Results and Discussion

The results of the peracetic cooking of miscanthus pulp showed that an increase in the duration of cooking naturally leads to a decrease in the yield of pulp and the residual content of lignin and mineral substances in it. Therefore, to remove lignin and minerals from organosolvent pulp, it was alkaline treated with NaOH solution. For alkaline treatment, pulp from Miscanthus x giganteus was used after cooking for 90 min. Dependences of properties of organosolvent miscanthus pulp on the duration of alkaline treatment are shown in Table 19.1.

As can be seen from Table 19.1, increasing the duration of alkaline treatment leads to a monotonous decrease in the yield of OMP, the residual content of lignin and the ash content. From the obtained results it can be seen that alkaline treatment of more than 60 min does not significantly reduce the content of lignin and ash. Therefore, to obtain nanocellulose, cellulose from miscanthus was used after organosolvent cooking for 90 min and alkaline treatment with a duration of 60 min with minimum content of lignin (0.08%) and mineral substances (0.07%).

In previous studies, the process of hydrolysis of pulp from non-wood plants has been shown that the action of a solution of sulfate acid at a concentration of 64% leads to almost complete dissolution of the nanocellulose from flax fibers [12], wheat straw [14] and kenaf fibers [15]. When the acid acts at a concentration of 50%, a low yield of nanocellulose (about 10%) is observed. In the case of a 43% solution of $H2SO_4$, the yield of nanocellulose was 25–30%. Therefore, under these conditions, further studies on the production of nanocellulose from OMP have been carried out.

In this case, the effect of the hydrolysis by acid at a concentration of 43 and 50%, temperature is from 40 to 60 °C, the duration from 30 to 90 min and the duration of ultrasonic treatment from 30 to 60 min on the quality indexes nanocellulose films was determined. The influence of the duration of the hydrolysis process by acid H2SO4

Table 19.1 Properties of organosolvent miscanthus pulp after alkaline treatment, % from mass of a.d.m

Duration of treatment (min)	Yield OMP	Residual lignin	Ash
15	72.3	0.19	0.13
30	67.0	0.13	0.12
45	64.2	0.10	0.09
60	55.5	0.08	0.07
120	54.7	0.04	0.06
240	51.0	0.01	0.05

concentration 50% at 60 °C and the duration of ultrasonic treatment on the properties of the nanocellulosic films are shown in Fig. 19.1.

From Fig. 19.1 it can be seen that with increasing prolonged hydrolysis and ultrasonic treatment, all the investigated parameters of nanocellulose films are improved. This is due to the fact that in the process of hydrolysis the amorphous part of the macromolecule of the cellulose begins to rise and thereby increases the percentage of the crystalline part. To confirm this, the crystallinity degree (CD) of the miscanthus stems and various samples of organosolvent miscanthus pulp was determined on the basis of the processing of X-ray diffraction patterns (Fig. 19.2).

The analysis of X-ray investigated diffraction patterns showed increasing of crystallinity degree as a result of thermochemical stages of miscanthus stalks treatment.

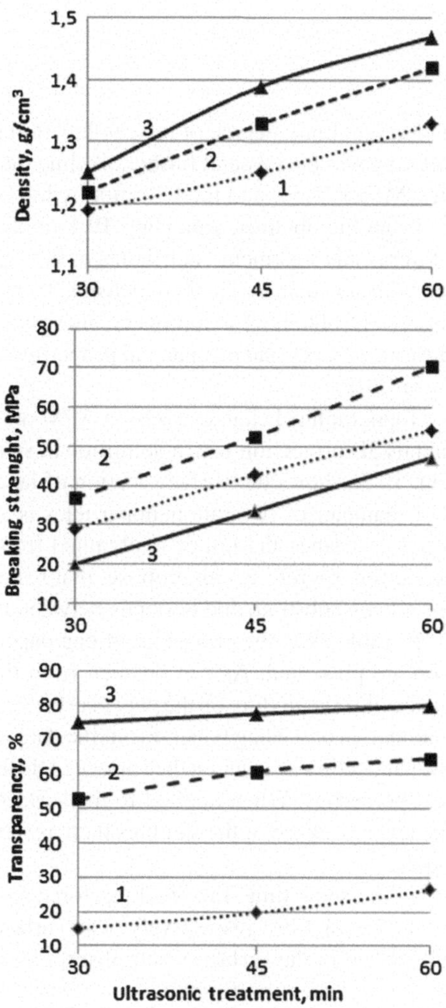

Fig. 19.1 The dependences of density (**a**), transparency (**b**) and tensile strength (**c**) of nanocellulose films on the duration of ultrasonic treatment after hydrolysis of OMP for different duration: 30 min (1), 60 min (2) and 90 min (3)

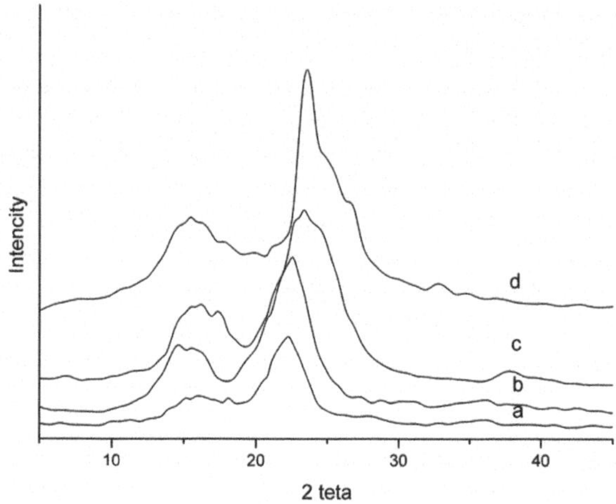

Fig. 19.2 X-ray diffraction patterns of miscanthus stalks (**a**), OMP (**b**), OMP after further alkaline treatment (**c**) and nanocellulose (**d**)

The crystallinity degree of miscanthus stalks was 64.7%, CD of pulp after peracetic cooking was 68.8%, and further alkaline treatment increases the crystallinity degree of OMP to 75.5% and the CD after hydrolysis was up to 77.0%.

From the obtained data (Fig. 19.1) it also follows that an increase in the action of ultrasonic treatment contributes to the grinding of cellulose nanoparticles, which leads to an increase in their density. Ultrasonic treatment also contributes to the process of obtaining a homogeneous suspension of nanocellulose, which increases the transparency and mechanical properties of the obtained particles of nanocellulose from OMP.

Topographical characterization AFM of organosolvent miscanthus nanocellulose and its 3D projection with a definition of sample height are presented in Fig. 19.3. In Fig. 19.3a shows the surface section of layer of forming nanocellulose aggregates. The diameter of separate nanoparticles is within the range to 20 nm and possibly much less since the image is obtained from nanocellulose particles located not in one layer. Therefore, can propose, that nanocellulose forms a film on the surface of the silicon substrate due to bonds between the nanocellulose particles.

In Table 19.2, the properties of bag paper with different nanocellulose consumption are presented. As can be seen from the data in Table 19.2, deposition of the nanocellulose mixture on the paper surface gives an increase in the physic-mechanical parameters of the bag paper. From the data obtained it follows that the application of the nanocellulose layer on the surface of the bag paper naturally increases its mechanical properties. When applied to the surface of the paper 5 g/m^2 of nanocellulose, the tear resistance of the samples increases by 68%, and the elongation increases 2.9 times.

At the same time, the breaking force and burst resistance of bag paper increase by 12% and 14%, respectively. The surface application of nanocellulose leads to a decrease in the surface water absorption (Cobb30) of samples from 136 to 62 g, which indicates the formation of a dense film on the surface of the casting. The

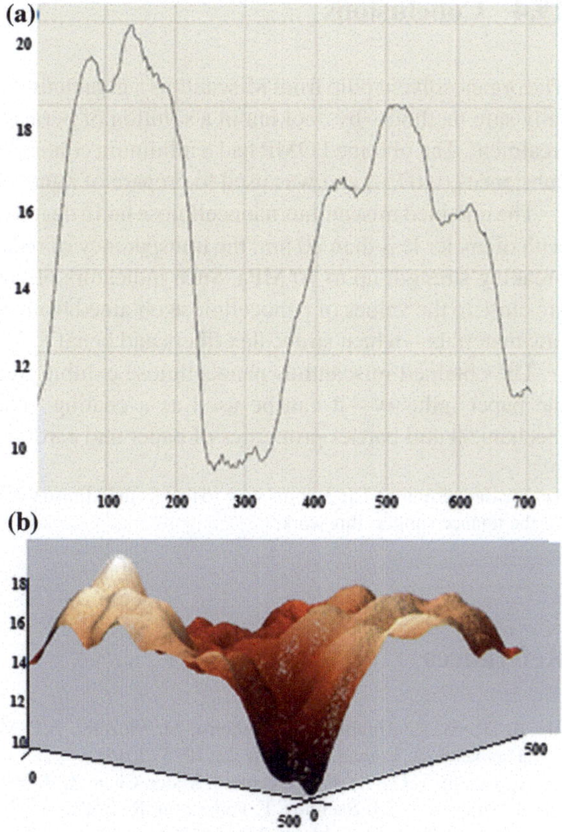

Fig. 19.3 The AFM images of the surface section of a layer of the nanocellulose (**a**) and 3D projection with a definition of sample height (**b**) tapping mode (values are in nm)

Table 19.2 Properties of bag paper with different nanocellulose consumption

Indicators	Nanocellulose Consumption (g/m^2)					
	1	2	3	4	5	6
Tear resistance, MN	890	1050	1200	1300	1400	1500
Burst resistance, MPa	523	533	560	570	580	600
Breaking strength, N	81	82	84	86	88	91
Elongation, %	1.6	2.5	3.2	3.7	4.2	4.6
Water adsorption, Cobb30, g	136	110	97	76	69	62

obtained values of the indicators exceed the requirements of standards for bag paper and confirm the possibility of using nanocellulose to improve the properties of paper and cardboard.

19.4 Conclusions

The organosolvent pulp from Miscanthus x giganteus was obtained in an environmentally safe method—by cooking in a solution of peracetic acid, followed by alkaline treatment. The obtained OMP had a minimum content of lignin (0.08%) and mineral substances (0.07%) and was used to prepare of nanocellulose.

The obtained miscanthus nanocellulose had a degree of crystallinity 77%, particles with diameter less than 20 nm; the transparency of received films is up 80%, and the breaking strength up to 70 MPa. Such indicators of nanocellulose from miscanthus are close to the values of nanocellulose obtained from other kinds of non-wood plant raw materials—wheat straw, flax fibers and kenaf [12, 14, 15].

The obtained miscanthus nanocellulose exhibit great potential its application in the paper industry—it can be used as a coating layer for improving fiber-based mechanical and barrier properties of paper and cardboard.

Acknowledgements The authors wish to thanks the Ministry of Education and Science of Ukraine for the finance support this work.

References

1. B. Deepa, E. Abraham, N. Cordeiro, M. Mozetic, A.P. Mathew, K. Oksman, M. Faria, S. Thomas, L.A. Pothan, Cellulose **22**, 1075 (2015)
2. K. Uetani, T. Okada, H.T. Oyama, J. Mater. Chem. C **4**, 9697 (2016)
3. S. Thiemann, S.J. Sachnov, F. Pettersson, R. Bollström, R. Österbacka, P. Wasserscheid, J. Zaumseil, Adv. Func. Mater. **24**, 625 (2014)
4. A.M. Adel, A.A. El-Gendy, M.A. Diab, R.E. Abou-Zeid, W.K. El-Zawawy, A. Dufresne, Ind. Crops Prod. **93**, 161 (2016)
5. J. Luo, H. Chang, A.A.B. Davijani, H.C. Liu, P.-H. Wang, R.J. Moon, S. Kumar, Cellulose **24**, 1745 (2017)
6. Y. Su, Y. Zhao, H. Zhang, X. Feng, L. Shi, J. Fang, J. Mater. Chem. C **5**, 573 (2017)
7. K.-Y. Lee, Y. Aitomaki, L.A. Berglund, K. Oksman, A. Bismarck, Compos Sci Technol. **105**, 15 (2014)
8. O. Daraban, Ş. Jurcoane, Sci. Pap. Agron. UASVM Iaşi **58**(2), 185 (2015)
9. X.J. Pan, D. Xie, R.W. Yu, D. Lam, J.N. Saddler, Ind. Eng. Chem. Res. **46**(8), 2609 (2007)
10. A. Ferraz, J. Rodriguez, J. Freer, J. Baeza, J. Chem. Technol. Biotechnol. **75**(12), 1190 (2000)
11. *TAPPI Test Methods* (Tappi Press, Atlanta, Georgia, 2004)
12. V.A. Barbash, O.V. Yashchenko, A.N. Kedrovska, J. Sci. Res. Rep. **16**(1), 1 (2017)
13. L.A. Costa, A.F. Fonseca, F.V. Pereira, J.I. Druzian, Cell Chem Technol. **49**, 127 (2015)
14. V.A. Barbash, O.V. Yaschenko, O.M. Shniruk, Dig. Nanoscale Res. Lett. **12**, 241 (2017)
15. V.A. Barbash, O.V. Yashchenko, V.O. Opolsky, Theor. Exp. Chem. **54**(3), 193 (2018)

Chapter 20
Effect of Thermomechanical and Ultrasonic Treatment on the Properties of Amorphous Alloys

Tatiana L. Tsaregradskaya, Alla M. Kuryliuk, Galina V. Saenko, Olexander O. Kalenyk, Vasyl V. Kuryliuk and Mykola V. Okonchuk

Abstract The study of the mechanism of deformation and the formation of a deformation structure in amorphous alloys is one of the most important and interesting in the search for new ways of modifying the properties of such materials. In this regard, the study of the influence of thermomechanical and ultrasonic processing on the thermal stability and properties of amorphous alloys is an urgent task. The intervals of thermal stability for number of multicomponent amorphous alloys based on iron, nickel, and cobalt have been determined. It is shown that thermomechanical treatment of amorphous alloys leads to expansion of the intervals of their thermal stability: for alloys based on nickel and cobalt on 20 K, and for alloy based on iron on 50 K. It is shown that after ultrasound processing of samples of an amorphous alloy on the basis of iron the interval of its thermal stability extends to (30–50) K, depending on the intensity of ultrasonic processing. The microhardness decreases after the thermomechanical treatment carried out by (7–18)% and after ultrasonic treatment by (9–17)%, which indirectly confirms the fact that the percentage of the crystalline phase in the alloy decreases due to the reduction of the size of the frozen centers of crystallization. Expansion of the intervals of thermostability of amorphous alloys occurs at the expense of both mechanical treatment (destruction of frozen-in crystallization centers) and thermal treatment (transition to the temperature range, where, according to the thermodynamic theory of temperature stability, dissolution of the frozen-in crystallization centers in an amorphous matrix occurs).

T. L. Tsaregradskaya (✉) · A. M. Kuryliuk · G. V. Saenko · O. O. Kalenyk · V. V. Kuryliuk · M. V. Okonchuk
Department of Physics, Taras Shevchenko National University of Kyiv, Volodymyrska 64/13, 01601 Kiev, Ukraine
e-mail: tsar_grd@ukr.net

A. D. Pogrebnjak and O. Bondar (eds.), *Microstructure and Properties of Micro- and Nanoscale Materials, Films, and Coatings (NAP 2019)*,
Springer Proceedings in Physics 240, https://doi.org/10.1007/978-981-15-1742-6_20

20.1 Introduction

Amorphous metal alloys are a class of materials that have been intensively studied in recent decades and provide a unique opportunity to study disordered systems in a condensed state. The structural disorder of amorphous metal alloys causes the formation of essentially different combinations of magnetic, mechanical, and electrochemical properties of their crystalline analogs. The magnetic and mechanical properties of amorphous metal alloys have the greatest practical interest. High mechanical strength and high degree of plasticity in combination with impact strength make amorphous metal alloys promising for use as sealing, protective, reinforcing materials. The scope of applied applications of amorphous alloys is constantly expanding, they are also considered to be promising for use as resistive and acoustic precision alloys and materials. Thus, there is every reason to believe that amorphous metal alloys play a significant role in the development and production of a new generation of electronic and microelectronic, radio engineering and instrumentation equipment products and devices [1–5].

Amorphous metallic alloys are in a thermodynamically unstable state, hence external actions such as annealing at temperatures well below the crystallization temperature, intense plastic deformation or irradiation with particles of different nature result in a change in the structure and, consequently, physical properties: electrophysical, magnetic, mechanical characteristics. One of the effective methods of intensive plastic deformation is a method based on the use of ultrasonic vibration energy [6–8]. The effect of ultrasonic vibrations can reduce the size of crystalline grains that are formed and substantially reduce the size distribution of grains. Ultrasonic treatment affects the formation of secondary phases and their distribution by volume of the alloy, that is, it allows for a more homogeneous structure of the material, which significantly improves both the physical and mechanical properties of the alloys.

The question of the mechanism of deformation and the formation of a deformation structure in amorphous metallic alloys is one of the most important and interesting in the search for new ways of modifying the structure and properties of such materials [9–11]. Mechanical or ultrasonic treatment of alloys in an amorphous state can both contribute to extending the intervals of thermal stability of such alloys, and be one of the controlled methods for obtaining nanocrystalline materials with new properties. In this regard, the study of the influence of thermomechanical and ultrasonic processing on the thermal stability and properties of amorphous alloys is an urgent task.

20.2 Experimental Technique

The objects of research were samples of amorphous alloys obtained at the Institute of Metal Physics by the method of spinning of the melt. For alloys, the study of the influence of thermomechanical and ultrasonic processing on their thermal stability and microhardness was conducted. By the parameter, which determines the thermal stability of amorphous alloys, there is the temperature of the intensive crystallization beginning, which was determined by means of high sensitivity dilatometry. It is known that the molar volume of most alloys in amorphous and crystalline states differs by (1–3)%. When heated amorphous alloy, its volume monotonically increases. When the temperature reaches the temperature of the beginning of intensive crystallization, the transition of the amorphous alloy to the crystalline state occurs; this process is accompanied by a sharp decrease in volume. Dilatometric technique allows determining with great accuracy the given parameter of thermal stability of an amorphous alloy. In order to determine the intervals of thermostability of a number of multicomponent amorphous alloys, temperature dependences of the relative volume change were obtained for heating and cooling both for the initial amorphous alloys and for alloys that were thermomechanical or ultrasonic processing. The thermomechanical treatment of amorphous alloys was carried out in two stages. The first stage is a thermal treatment, which consisted of short-term (10 min) isothermal annealing at temperatures below the temperature of the beginning of the intensive crystallization of the alloy. The second stage is the mechanical processing, was 5 times the load pressure of 225 MPa. Ultrasonic treatment of amorphous alloys was carried out in ultrasonic bath Ya Xun YX-3560, which creates the sound pressure and the effect of cavitation under influence. The bath was filled with distilled water, samples of amorphous alloys, which were held under ultrasound load for a certain time, were immersed in it. Ultrasonic mechanoactivation of amorphous alloys was carried out at room temperature, which is below the structural relaxation temperature of the material.

As a parameter for comparing the elastic properties of amorphous alloys in the initial state and after processing, the value of microhardness H was used.

For the row of amorphous alloys in the initial amorphous state and after treatment, measuring of microhardness by Vickers was executed. The method of measuring microhardness is based on measuring of linear size of diagonal of imprint, that appears at pressing of diamond pyramid in the probed material under certain loading. The device allows to measure microhardness at pressing of diamond pyramid with square basis and corner at a top 136° between opposite verges with the appendix of loadings (2–200) g. As a result of measurings length of diagonal of got imprint is determined by eyepiece-micrometer. At measuring microhardness, volume, which becomes deformed by pressure must be less than volume of grain which is measured. The sample was under pressure of about 10 s.

20.3 Experimental Results and Discussion

In accordance with the theory of thermodynamic stability of amorphous alloys [12], external influences can lead to a violation of the thermodynamic equilibrium in the heterogeneous system: amorphous matrix—frozen-in crystallization centers. When amorphous alloys are obtained by the method of spinning the melt, the nuclei of the crystalline phase are formed in the temperature region $T/T_p = 0,7$, with the

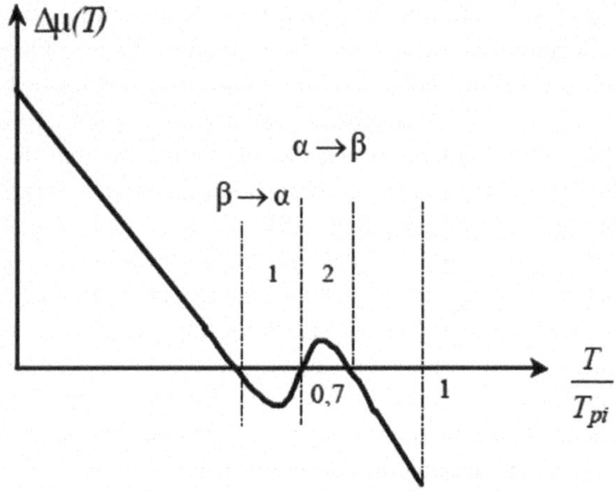

Fig. 20.1 Temperature dependence of the difference of chemical potentials of the i-th component in the amorphous (α) and crystalline (β) phases. (T_{pi}-melting point of the i-th component)

supercooled liquid passing into solid amorphous phase. Thus, the heterogeneous system is formed of amorphous matrix—frozen-in crystallization centers.

The condition of the thermodynamic equilibrium in a heterogeneous system is an amorphous matrix—frozen-in crystallization centers is the equality of the chemical potential of each component in all phases. Thus, a decrease in the chemical potential of the i-th component in the amorphous and crystalline phases contributes to an increase in the thermal stability of the alloys. In the system, an amorphous matrix—frozen-in centers of crystallization, there is elastic stress, therefore, on the surface of the section of the crystalline nuclei—amorphous phase, there is an additional pressure.

The temperature dependence of the difference between the chemical potentials of the i-th component in the amorphous (α) and crystalline (β) phases Δμ, calculated according to the equations of the thermodynamic theory of high-temperature stability of amorphous alloys, is presented in Fig. 20.1.

The temperature range below 0.7 T_{pi} is elastic. In this area, the pressure caused by elastic stresses is taken into account because the volume of the amorphous phase does not equal the volume of the crystalline. As a result, the transition of crystalline nuclei to the amorphous phase (i.e., the dissolution of the nuclei of the crystalline phase in the amorphous matrix, which is physically explained by the ascending diffusion process) is observed in the interval of temperatures where the value is negative (region 1 in Fig. 20.1) [13].

Considering this additional pressure leads to a significant decrease in the difference of the chemical potential of the i-th component in the amorphous and crystalline phases compared to the values of the same magnitude in the case when no elastic stresses are present. Displacement of the equilibrium in the system: an amorphous matrix—frozen-in crystallization centers can occur not only due to thermal effects but also mechanical.

Table 20.1 The temperature of the beginning of intensive crystallization of the original samples and after the thermomechanical processing

Composition of an amorphous alloy	T_k, K	T_k^1, K
$Fe_{76}Ni_4Si_6B_{14}$	720	770
$Co_{55}Fe_5Ni_{14}Si_{16}B_{10}$	790	810
$Ni_{78}Si_4B_{18}$	730	750

Table 20.2 The microhardness of the original samples and after thermomechanical treatment

Composition of an amorphous alloy	H_0 (kgf/mm^2)	H_1 (kgf/mm^2)	$(H_1 - H_0)/H_0$ (%)
$Fe_{76}Ni_4Si_6B_{14}$	790	640	18
$Co_{55}Fe_5Ni_{14}Si_{16}\ B_{10}$	840	765	9
$Ni_{78}B_{18}Si_4$	720	670	7

Table 20.1 is given using dilatometric method of the temperature of the beginning of intensive crystallization for the output amorphous alloys and alloys after the thermomechanical processing.

After thermomechanical treatment, the temperature of the beginning of intensive crystallization for amorphous alloys based on nickel and cobalt increased by 20 K, and for an amorphous alloy based on iron by 50 K. The value of microhardness is reduced on (7–18)% after thermomechanical treatment, which indirectly confirms the fact that the percentage of the crystalline phase in the samples decreases (Table 20.2).

The peculiarities of changing the physical properties of amorphous alloys under the influence of pressure can be explained by the effects of dilatation in strips of shear. The structure of amorphous alloy is characterized by the presence of excess free volume, which reflects the difference between the density of amorphous and crystalline states of 3%. In a crystalline material, under the influence of pressure vacancies drifting in the region where the compressive stresses are applied. By analogy with vacancies, the source of the flow of regions of excess free volume drift is the entire volume of amorphous alloy. A trap for them is the main strip of displacement, which is characterized by the effect of dilatation. There is "pumping" of the displacement band that is spreading, excess free volume, which leads to a decrease in effective viscosity. The effect of reducing viscosity is enhanced by spontaneous heat release and local temperature rise. The temperature in the shift band increases in proportion to the magnitude of the deformation of the sample.

Consequently, the processes of partial destruction and dissolution of the frozen-in crystallization centers are due to the formation in the strips of displacement in the amorphous matrix of mechanical stresses as a result of plastic deformation and increase of the local temperature by almost 400 K. Significant stress leads to mechanical destruction of the frozen-in crystallization centers, and local increase of temperature in the shift bands, they are translated into the region of temperature, where, according to the theory of thermodynamic stability frozen-in crystallization centers dissolution.

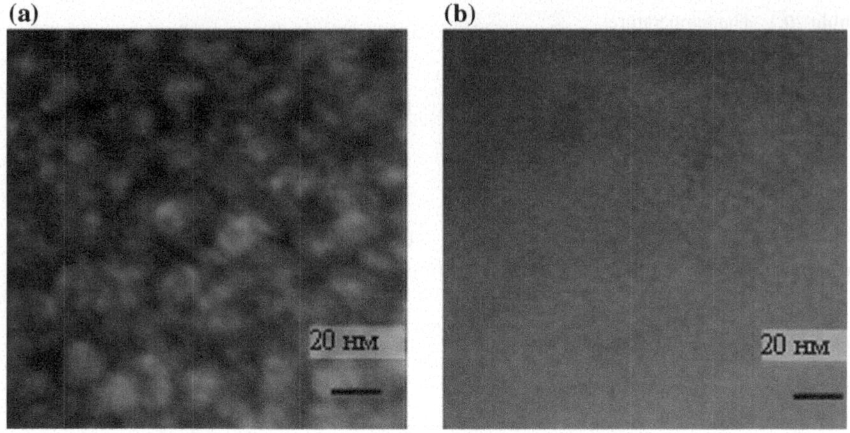

Fig. 20.2 Electronmicroscopic (dark-field) images for $Fe_{76}Ni_4Si_6B_{14}$ alloy in the initial state (**a**) and after thermomechanical treatment (**b**)

To verify this assertion, electron microscopic investigations of the initial slices and those that have undergone a preliminary thermal treatment have been carried out. In Fig. 20.2 shows the results of electronmicroscopic investigations for $Fe_{76}Ni_4Si_6B_{14}$ alloy.

The analysis of the results of electronmicroscopic studies for the $Fe_{76}Ni_4Si_6B_{14}$ alloy (Fig. 20.2) showed that the initial amorphous state is characterized by the presence of frozen-in crystallization centers with an average size (15–20) nm (Fig. 20.2a). After thermomechanical treatment, the crystalline nuclei are dissolved in an amorphous matrix.

One of the types of mechanical treatment is the ultrasonic mechanoactivation of amorphous alloys. For $Fe_{76}Ni_4Si_{14}B_6$ alloy, experimental investigations of the influence of ultrasonic treatment carried out under different modes on the thermal stability of an amorphous alloy have been carried out. The obtained results indicate that after the ultrasound treatment of amorphous alloy samples the temperature range of the amorphous alloy expands.

At an intensity of ultrasonic treatment of 1 W/cm^2, the temperature of the beginning of intensive crystallization increases by 30 K, the microhardness is reduced by 9%. At an intensity of ultrasonic treatment 2 W/cm^2, the temperature of the beginning of intensive crystallization increases by 50 K, the microhardness is reduced by 17%.

Decrease of microhardness after performed ultrasonic treatment indirectly confirms the fact of decreasing the part of the crystalline phase in the alloy, due to the dissolution of frozen-in crystallization centers. Consequently, under the influence of ultrasound processing, the size of frozen-in crystallization centers may also decrease. This leads to an increase in the thermal stability interval and a decrease in the microhardness of the alloy.

20.4 Conclusions

It is shown that the thermomechanical treatment of amorphous alloys on the basis of iron, cobalt, and nickel leads to the expansion of their thermal stability intervals at (20–50) K and reduction of microhardness by (7–18)%.

An increase in the thermal stability of amorphous alloys after thermomechanical treatment is due to a decrease in the size or dissolution of the frozen-in crystallization centers, resulting in the homogenization of the structure of the amorphous alloy.

Expansion of the intervals of thermostability of amorphous alloys occurs at the expense of both treatments by pressure (mechanical destruction of frozen-in crystallization centers) and thermal treatment (transition to the temperature range, where, according to the thermodynamic theory, dissolution of frozen-in crystallization centers in an amorphous matrix occurs).

Due to the effect of ultrasonic treatment, the size of the frozen-in crystallization centers can also decrease, which causes the growth of the thermal stability interval of the alloy under investigation (30–50) K and the decrease of the microhardness of the alloy by (9–11) %.

References

1. D.L. Zhang, Progr. Mater Sci. **49**(3–4), 537 (2004)
2. J. Schroers, Adv. Mater. **22**(14), 1566 (2010)
3. Z. Kovacs, P. Henits, A.P. Zhilyaev, A. Revesz, Scr. Mater. **54**, 1733 (2006)
4. W.H. Jiang, F.E. Pinkerton, M. Atzmon, J. Appl. Phys. **93**, 9287 (2003)
5. R.J. Hebert, N. Boucharat, J.H. Perepezko, H. Rösner, G. Wilde, J. Alloys Compd. **434–435**, 18 (2007)
6. S. Sharma, C. Suryanarayana, J. Appl. Phys. **102**, 083544 (2007)
7. L. Feng, S. Fei, L. Kangsen, G. Feng, L. Xiong, W. Xiaoyu, M. Jiang, Mater. Res. Lett. **6**(10), 545 (2018)
8. W.H. Wang, C. Dong, C.H. Shek, Mat. Sci. Eng. R. **44**(2), 45 (2004)
9. Z.P Lu, C.T. Liu, Acta Mater. **50**(13), 3501 (2002)
10. C.A. Schuh, T.C. Hufnagel, U. Ramamurty, Acta Mater. **55**(12), 4067 (2007)
11. L.I. Jie, W.J. Chen, Univ. Sci. Technol. B **29**, 1246 (2007)
12. V.I. Lysov, T.L. Tsaregradskaya, O.V. Turkov, G.V. Saenko, Russian. J. Phys. Chem. A **87**(10), 1778 (2013)
13. V.I. Lysov, T.L. Tsaregradskaya, O.V. Turkov, G.V. Saenko, Russian. J. Phys. Chem. **88**(12), 1981 (2014)

Chapter 21
Nanostructured Magnesium Oxide Layers Synthesized on Flexible Substrates for Filtering Elements

Z. Tsybrii, F. Sizov, I. Korotash, E. Rudenko, K. Svezhentsova, M. Vuichyk and D. Polotskiy

Abstract The technological modes of MgO films deposition by helicon-arc ion-plasma method for decreasing of thermal stresses on the polymeric substrate was developed and realized. AFM studying of the MgO surface morphology showed the complicated microrelief covered by nanorelief from the ordered crystallites of 50 nm size and 10 nm height. Infrared (IR) transmission and reflection spectra demonstrated a significant signal suppression in the long-wavelength spectral range (due to the location of the Restrahlen band of MgO). Such MgO coatings in combination with other materials on flexible polymeric substrates can be effective filters to reduce unwanted infrared radiation in wide IR spectral range both from the background and from the heated objects.

21.1 Introduction

Thin film oxide materials find a wide application as alternative ones for many purposes in microelectronics. The MgO (magnesium oxide) is of interest oxide that can have a broad range of potentials for applications in medicine, refractory materials, heating apparatus, infrared optics, etc. [1]. It is a stabile wide band-gap insulator ($E_g \approx 7.3$–7.8 eV for bulk MgO [1, 2]) and it has a cubic crystal structure with Fm3m space group. The bulk material MgO has a relatively high dielectric constant ($\varepsilon \sim 9.8$) and higher breakdown field (12 MV·cm^{-1}) compared to commonly used silicon dioxide (SiO_2) [3]. The MgO films are also useful as a protective layer in AC-plasma display panels to improve discharge characteristics and panel lifetime [4]. In addition, they are capable to provide very stable and high secondary electron emission combined with low refractive index (1.735). The natural high transparency

Z. Tsybrii (✉) · F. Sizov · K. Svezhentsova · M. Vuichyk
V. E. Lashkaryov Institute of Semiconductor Physics NASU, Kyiv, Ukraine
e-mail: tsybrii@isp.kiev.ua

I. Korotash · E. Rudenko · D. Polotskiy
Institute for Metal Physics NASU, Kyiv, Ukraine
e-mail: korotash@imp.kiev.ua

A. D. Pogrebnjak and O. Bondar (eds.), *Microstructure and Properties of Micro- and Nanoscale Materials, Films, and Coatings (NAP 2019)*,
Springer Proceedings in Physics 240, https://doi.org/10.1007/978-981-15-1742-6_21

makes this material a suitable one for buffer layer [5] applications in thin-film solar cells (Si and Cu (In, Ga) Se2 (CIGS)). The other areas of emerging applications of MgO involve infrared window material [3] and due to low dielectric losses, it can be applied as a material for producing microwave devices [2, 6].

Recently, many studies have been focused on the synthesis methods of MgO [1, 7]. The methods include ion-plasma deposition, sol–gel process, chemical gas phase deposition, combustion chemical vapor deposition, and aqueous wet chemical methods, etc. However, the morphology and characteristics of the MgO materials grown differ much and depend largely on the method of synthesis, plus processing condition. Here, aiming to grow the MgO thin films as non-transparent in the IR filters, which, because of the Restrahlen band at $\lambda \sim 14$–$26\ \mu$m [8], an attempt is initiated on the application of the low-temperature helicon-arc ion-plasma synthesis to MgO nanostructured thin films on the polymeric substrates. For the characterization of their properties, the optical and morphological techniques were used.

21.2 Magnesium Oxide Thin Films Growth

The helicon-ark ion-plasma method of film synthesis occurs in essentially non-equilibrium conditions under the influence of intensive high-energy plasma flow, condensed on the substrate. This method has been successfully applied to synthesize AlN films on flexible polymeric substrates [8–11]. Many factors, such as deposition temperature, a geometrical location of substrate, a time of substrate cleaning, an arc discharge current, etc., affect the properties and characteristics of the growing films. However, in the case of vacuum-arc deposition of films on polymeric substrates, the important is controlling and limiting the level of thermal heating of the substrate due to plasma surface treatment. As it was established in previous studies [8, 9], significant thermal stresses in thin flexible substrates, both during the processing stage in the plasma of the helicon source and during the vacuum-arc deposition, can lead to destruction of the substrate or its significant damage. Therefore, various technological methods for reducing thermal load were used. Among them are:

1. Breakdown deposition cycles, when the deposition is carried out in a one-time interval and in the next time interval, the cooling of the substrate occurs;
2. Deposition on a substrate rotating during plasma processing;
3. Deposition on a substrate rotating during the plasma processing with the simultaneous scanning of the flux of precipitated particles by controlling the direction of deviation of the plasma flow by changing the configuration of the magnetic field in the drift chamber.

Application of the above-mentioned technological techniques allowed to reduce the level of thermal stresses in thin flexible substrates and to realize the deposition of robust thin films on these substrates.

The feature of MgO film coatings growth, using a plasma arc accelerator and a magnesium cathode, is that the magnesium does not have a liquid phase; it belongs to subliming materials that immediately transform from a solid to a gaseous state. Because one of the properties of the gas is the uniform filling of the entire volume, in which it is located, the ionization of gaseous magnesium burning in a characteristic green color occurs throughout the volume of the discharge chamber (Fig. 21.1). This leads to spraying magnesium not only to the surface of the substrate but also to the walls of the discharge chamber. To avoid the undesired Mg or MgO deposition on the walls of the discharge chamber of the hybrid ion-plasma reactor, the design of the anode of the plasma-arc accelerator (PAA) was changed. The new anode eliminates the effect of combustion of plasma magnesium in the entire volume by "pulling" magnesium ions into a high-energy beam. As the experiments have shown, the use of this design of the anode of the PAA significantly reduces the MgO deposition on the walls of the chamber and increases the mass transfer to the surface of the substrate due to the formation of a high-energy beam, which originates directly on the surface of the cathode. The inflammation of the ion-plasma arc PAA, for obtaining on the substrate surface of the nanostructured coatings, was carried out in the argon atmosphere. Then the arc (for filtration from micro-droplet inclusions) was rotated to the surface of the substrate by a magnetic field. The magnesium oxide films were grown on the surface of polymeric materials (Teflon) with an exposure time of 2 to 10 min.

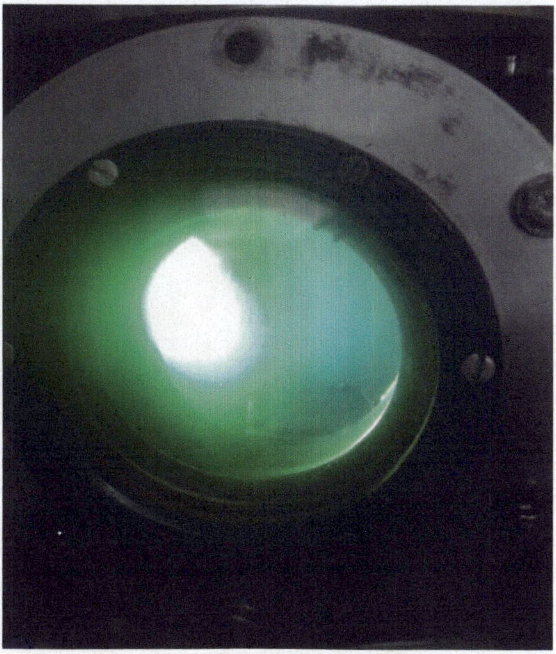

Fig. 21.1 The process of mass transfer of sublimating magnesium material to a polymeric substrate with a modified anode construction

21.3 MgO/Teflon Samples Characterization

Thin film ion-plasma condensates, based on magnesium, were applied to thin flexible Teflon substrates. The initial Teflon surface is characterized by hilly and chaotic morphology (Fig. 21.2).

The sample of MgO/Teflon is characterized by an ordered regular structure and a high degree of homogeneity. The relief is repeated throughout the scan surface. The ordered structure is characterized by crystallites, the lateral dimensions of which range from 170 to 230 nm at the height of 25–40 nm. The main distinguishing feature of this sample is the presence of a special nanorelief on an ordered microrelief, which is clearly visible on scans of the surface 1 × 1 μm (Fig. 21.3). The nanorelief does not cover the entire microrelief but visualizes itself in the form of single nanocrystallites with dimensions of 50 nm and a height of 10 nm, which are formed on the top of the crystallites of the microrelief. This peculiarity of morphology can be caused by the small time of the MgO film growth, during which the nanoparticles do not have time for association.

The sizes of crystallites of micro and nanoreliefs are determined by the technological parameters of growth.

Thin films of ion-plasma condensates (AlN, MgO, Al_2O_3), which were synthesized by the method of hybrid helicon-arc ion-plasma deposition on various substrates, are of interest for the formation on their basis the functional elements—passive infrared filters, which are also transparent in the microwave and THz ranges. Since the main part of the blackbody radiation power density at T~300–400 K falls on a range of wavelengths $\lambda = 5$–25 μm, the spectral characteristics of ion-plasma condensates were studied in this range. The measurements of the unpolarized reflection and transmission spectra of film condensates, to establish areas of their transparency,

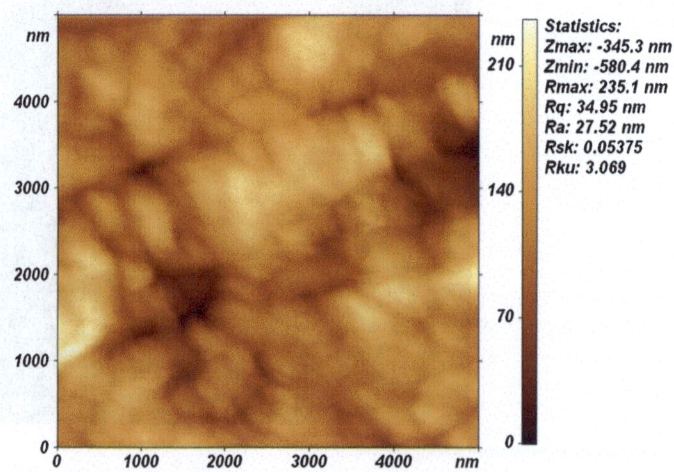

Fig. 21.2 The AFM image of the Teflon film surface with thickness ~220 microns

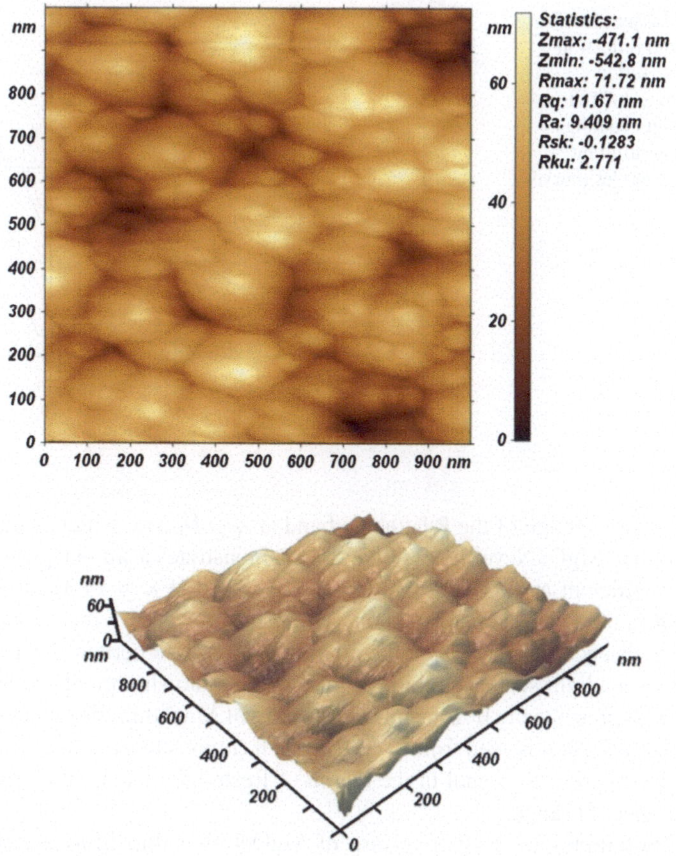

Fig. 21.3 The AFM (2D and 3D) images of MgO/Teflon thin film

were carried out at T = 300 K in the IR spectral range. Spectral characteristics were measured using a Fourier spectrometer (FTIR Spectrum BX-II model) with a resolution in the middle of the IR range of no less than 0.8 cm^{-1}, and a useful signal/noise ratio greater than 15,000/1.

As possible coatings for the formation of functional filters are ion-plasma condensates based on MgO, AlN, Al_2O_3 thin films. That is due to their Restrahlen bands spectral position (the region of the spectrum between the frequencies of longitudinal and transverse optical phonons).

For these materials, they are positioned at $\lambda \approx 14$–$26\,\mu m$ (for MgO), $\lambda \sim 11$–$16\,\mu m$ (for AlN) and $\lambda \approx 13.65$–$18.62\,\mu m$ (for Al_2O_3) [8]. It is important to study their spectral characteristics to find the position of the Restrahlen bands and determine the degree and area of blocking infrared radiation by filters of such type.

The IR reflection spectra of MgO thin films, which were synthesized on Teflon substrates, are shown in Fig. 21.4. The spectral reflection dependencies do not show

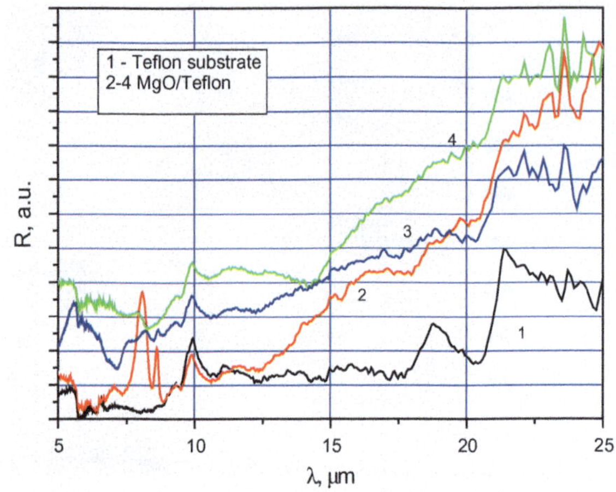

Fig. 21.4 Measured at T = 300 K infrared reflection spectra of the Teflon substrate (curve 1), Teflon films with MgO coatings synthesized at different technological modes (curves 2–4)

a sharp short-wave edge of the Restrahlen band at $\lambda \approx 14\ \mu\text{m}$, which is inherent to monocrystalline MgO. Nevertheless, Fig. 21.4 demonstrates a smooth growth of the reflection coefficient at $\lambda \approx 12.5\ \mu\text{m}$ (curves 2, 3) and at $\lambda \approx 14.3\ \mu\text{m}$ (curve 4) of magnesium oxide coatings formed on Teflon substrate at different technological modes. The absence of clear Restrahlen band position, to a certain degree, maybe connected with a light scattering in these micro- and nano-structured coatings.

Figure 21.5 presents the IR transmission spectra of MgO thin film synthesized on Teflon substrates. As it is seen from Fig. 21.5, such composite structure MgO/Teflon significantly suppress IR signal in the region of Restrahlen band but is transparent in the THz spectral range.

The photoluminescence (PL) spectra of MgO/Teflon thin films are shown in Fig. 21.6. To excite the PL spectra, an argon laser with a wavelength of 488 nm

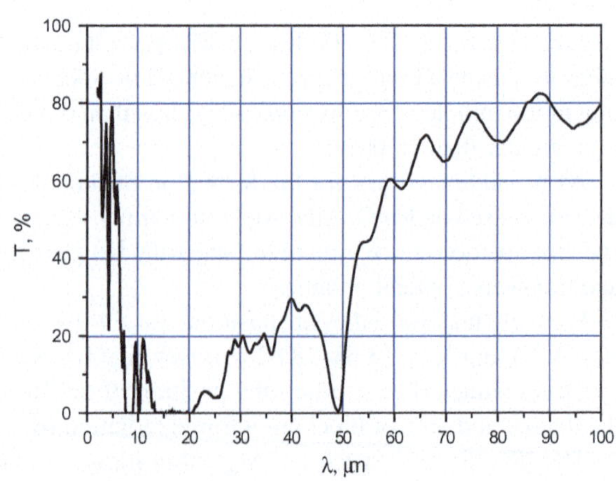

Fig. 21.5 The IR transmission spectra of MgO/Teflon film at T = 300 K

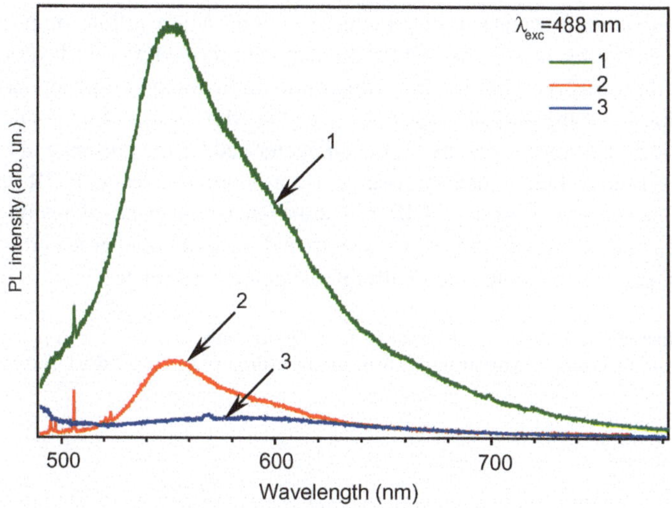

Fig. 21.6 The PL spectra of MgO samples grown on Teflon polymeric substrates under various technological parameters. T = 300 K. $\lambda_{exc} = 488$ nm

was used. Registration of the PL spectra was carried out at room temperature using the Horiba Jobin-Yvon T64000 triple monochromator.

PL spectra of sample 1 (curve 1, Fig. 21.6) is characterized by an intense asymmetric band, with a maximum at 552 nm and a shoulder in the range of 570–600 nm. In the spectrum of sample 2 (curve 2, Fig. 21.6), the band is less intense, but the same is asymmetric. The PL band of sample 3 (curve 3, Fig. 21.6) differs from others by very weak intensity. Since the energy position of these bands is far from the edge of emission, which corresponds to the bandgap of MgO (7.8 eV), they can be interpreted as peaks from the local defect centers inherent this material [2, 12]. Based on the morphological studies, it can be assumed that the reason for the broad PL band with a maximum of 552 nm is the local centers in nanocrystallites and in microcrystallites related to the oxygen ion vacancies.

21.4 Conclusions

The aim of this investigation was to show the possibility of thin layers growth on the flexible polymeric substrates with a high level of homogeneity and adhesion, by hybrid helicon-arc ion-plasma method. Combining the features of different compounds' optical characteristics, it is possible to create effective filtering elements, for example, to block IR radiation and transparent in the THz spectral region. It can be useful for protecting mm-wave and THz detectors with the performance that is close to the upper limit one, from the background parasitic radiation in astronomical and scientific applications.

Here were demonstrated some characteristics of the MgO/Teflon composite structures. The structures were obtained by the deposition of MgO thin layers onto the flexible Teflon substrates by the low-temperature helicon-arc ion-plasma technique. The peculiarity of the surface morphology of the MgO ion-plasma condensates is the complicated surface topology that is characterized by the presence of two types of relief: a microrelief, which is covered by a nanorelief layer. FTIR spectra of MgO/Teflon showed blocking of IR transmission in the band of residual beams (Restrahlen band). Investigation of PL spectra exhibited a broad green emission at $\lambda \approx 552$ nm, presumably connected with the oxygen ion vacancies.

Acknowledgements Authors are thankful to Prof. Viktor Strelchuk for some optical measurements. This work is partly support by NAS of Ukraine Project № IMP-2018/1 and Volkswagen Project#A11597.

References

1. N. Badar, N.F. Chayed, R. Rusdi, N. Kamarudin, N. Kamarulzaman, Adv. Mater. Res. **545**, 157 (2012)
2. O.V. Diachenko, A.S. Opanasuyk, D.I. Kurbatov, S.S. Patel, R.R. Desai, D. Lakshminarayana, C.J. Panchal, P. Suryavanshi, Invertis J. Sci. Technol. **9**(2), 72 (2016)
3. P. Maiti, P. Sekhar Das, M. Bhattacharya, S. Mukherjee, B. Saha, A. Kumar Mullick, A. Kumar Mukhopadhyay, Mater. Res. Express. **4**, 086405 (2017)
4. S.H. Cho, S.M. Lee, W.H. Kim, K.C. Choi, ACS. Appl. Mater. Interfaces. **7**, 7559 (2015)
5. P. Płóciennik, D. Guichaoua, A. Zawadzka, A. Korcala, J. Strzelecki, P. Trzaska, B. Sahraoui, Opt. Quantum Electron. **48**, 277 (2016)
6. Magnesium Oxide (MgO) Data Sheet (Crystran Ltd., 2012), https://www.crystran.co.uk/optical-materials/magnesium-oxide-mgo. Accessed 11 Aug 2019
7. A.A. Pilarska, Ł. Klapiszewski, T. Jesionowski, Powder Technol. **319**, 373 (2017)
8. Z. Tsybrii, F. Sizov, M. Vuichyk, K. Svezhentsova, E. Rudenko, I. Korotash, D. Polotskiy, in *Advances in Thin Films, Nanostructured Materials, and Coatings*, ed. by A. Pogrebnjak, V. Novosad. 8th IEEE International Conference on Nanomaterials: Applications & Properties, Odessa, September 2018. Lecture notes in mechanical engineering (Springer, Singapore, 2019), pp. 235–247
9. E. Rudenko, Z. Tsybrii, F. Sizov, I. Korotash, D. Polotskiy, M. Skoryk, M. Vuichyk, K. Svezhentsova, J. Appl. Phys. **121**, 135304 (2017)
10. V.F. Semeniuk, E.M. Rudenko, I.V. Korotash, L.S. Osipov, DYu. Polotsky, K.P. Shamray, V.V. Odinokov, G.Ya. Pavlov, V.A. Sologub, Metallofiz Noveishie Tekhnol. **33**(2), 223 (2011). (in Russian)
11. V. Korotash, V.V. Odinokov, G. Pavlov, E. Rudenko, D. Polotsky, V. Semenyuk, V. Sologub, Nanoindustriya (Nanoindustry) **4**, 14 (2010). (in Russian)
12. M. Li, X. Wang, H. Li, H. Di, X. Wu, C. Fang, B. Yang, Appl. Surf. Sci. **274**, 188 (2013)

Chapter 22
Surface Phosphorylated Activated Carbons: Preparation and Acidity Studies

V. E. Diyuk, A. V. Yatsymyrskyi, L. M. Grishchenko, D. S. Horodetska, A. V. Vakaliuk, A. V. Mischanchuk, A. N. Zaderko and V. V. Lisnyak

Abstract Here we report on the phosphorylation of activated carbon (AC) at 500–800 °C. From thermal, titrimetric, and catalytic data, high surface acidity of the prepared solids is attributed to the total concentration of the protogenic groups. Transmission electron microscopy and nitrogen physisorption measurements showed a chemisorption-induced contraction of microporosity. The most active catalysts prepared at 700 °C contain 1.04 mmol g^{-1} of the phosphonic groups and supports dehydration of isopropanol with about 100% conversion and 100% selectivity to propylene at 170–180 °C.

22.1 Introduction

Ethylene and propylene produced by high-temperature oil cracking are important unsaturated hydrocarbons for the chemical industry. In the modern paradigm, it has become clear that catalytic dehydration of alcohols is a promising alternative to oil refining [1–3]. Compared to thermal cracking, this reaction runs at lower temperatures with high yields of target products. Using alcohol dehydration as an alternative to oil refining, we contribute to the significant development of biochemical and renewable technologies for energy and fuels [4, 5]. By fermentation of natural carbohydrates, such as plant biomass (lignocellulose wastes) and by recycling of sugar processing residuals, one can prepare the required amounts of alcohols from renewable sources [5, 6]. Convenient alcohol dehydration catalysts are carbon-based solid acids. Their acidity depends on the surface concentration of different protogenic groups [7–11]. Among them, one should pay special attention to phosphonic groups

---

V. E. Diyuk · A. V. Yatsymyrskyi · L. M. Grishchenko · D. S. Horodetska · A. V. Vakaliuk · A. N. Zaderko · V. V. Lisnyak (✉)
Taras Shevchenko National University of Kyiv, Kyiv 01601, Ukraine
e-mail: lisnyak@univ.kiev.ua

A. V. Mischanchuk
Chuiko Institute of Surface Chemistry of the National Academy of Science of Ukraine, Kyiv 03164, Ukraine

A. D. Pogrebnjak and O. Bondar (eds.), *Microstructure and Properties of Micro- and Nanoscale Materials, Films, and Coatings (NAP 2019)*, Springer Proceedings in Physics 240, https://doi.org/10.1007/978-981-15-1742-6_22

that have the highest thermal stability [12–14]. By treatment of different raw materials with H_3PO_4, phosphonic groups have introduced at different stages of production of Activated Carbon (AC) or even by phosphorylation of the already prepared carbon materials.

Puziy et al. [15] and Rosas et al. [16] showed that the main feature of chemical activation is the correlation between the parameters of the porous structure and the phosphorus content. In the past decade, carbons activated by phosphoric acid have found application for adsorption of contaminants, e.g., heavy metal ions, dyes, and medicals [17–20]. We should notice that phosphorus can incorporate into the carbon matrix, during chemical activation, in addition to acid phosphonic groups grafting. The passage of reduction processes on the carbon surface can be a reason for the forming of phosphonates and other related products. On this background, an alternative approach could be the phosphorylation of ready-made ACs. By phosphorylation carried out in such a way, one can graft phosphonic groups with high selectivity while maintaining structural porosity. To the best of our knowledge, materials scientists and process engineers frequently use phosphoric acid at the stage of chemical activation and pretreatment of biomass [21–26], but the aforementioned approach is rare in use for the surface functionalization of nanoporous carbon materials.

In the present chapter, we show the results of high-temperature phosphorylation of the carbon surface. Here we also report about the surface acidity of the prepared samples tested as catalysts for the dehydration of isopropanol.

22.2 Experimental

22.2.1 *Materials*

For preparing catalysts, commercial nanoporous AC, which has the Brunauer-Emmett-Teller specific surface area of ~1,300 $m^2\ g^{-1}$ and the sorption volume V_S of 1.09 $cm^3\ g^{-1}$, was used without further purification. Concentrated phosphoric acid (H_3PO_4, 85%), sodium hydroxide (NaOH, min 99.5%), sodium nitrite ($NaNO_2$, min 99.9%), and isopropyl alcohol (i-C_3H_7OH, 99.7%) were purchased at Khimlaborreaktiv Ltd. (Brovary, Ukraine). Aqueous solutions for washing and chemical analysis were prepared from Merck's ammonium nitrate (NH_4NO_3, p.a.) and ammonium heptamolybdate ($(NH_4)_6Mo_7O_{24} \cdot 4H_2O$, min 99.9%). High purity argon (Ar, Linde Gas 99.998%) was used as an inert medium during phosphorylation.

Phosphorylation was conducted by the following procedure. AC (0.5 g) powder loaded in a glass beaker was impregnated with 85% H_3PO_4 (0.8 ml) for a certain time. After impregnation, the resulting AC was loaded in an alumina boat and placed in an electrically heated tubular reactor made of mullite. The reactor was closed with Teflon® end caps sealed with Silicon-gaskets and stainless steel clamps. The caps at both ends have provisions for charging/discharging the reactor with gases. The closed reactor was connected to a system for gas feeding from a steel bottle containing argon gas at STP. An electric regime of the resistance furnace used for reactor heating was operated by a digital programmable Omega PID temperature controller of the CNi8 series. The reactor temperature was controlled by thermopiles inserted into the reactor close to the boat with the sample. Their measurements were monitored by a personal computer that connected to them with a data acquisition interface and the software automatically converting the resistance measurement to a temperature. Before heating, the air was extruded from the reactor with argon gas flow. Under argon flow of 20 ml min^{-1}, the temperature of the reactor was raised at a rate of 10 °C min^{-1}. When the temperature reached 500, 600, 700, or 800 °C, the sample was heat-treated at the constant temperature for 60 or 120 min. The impregnation time of 4, 24, and 24 h was used to prepare the samples of the first, second, and third series for which the heat-treatment time was 60, 60, and 120 min, correspondingly. The samples of AC(I), AC(II), and AC(III) series were labeled as AC(#)-T, where T is the heat-treatment temperature.

22.2.2 *Methods*

22.2.2.1 Analysis

The total phosphorus content (C_P) in the samples was determined by the gravimetric method. Phosphorylated AC (0.3 g) was subjected to pyrolytic decomposition in molten NaOH–$NaNO_2$ mixture. When AC decomposed in the melt, the melt was cooled to room temperature and dissolved in water. After what HNO_3 was added to set pH of this solution to 2 that followed by adding aqueous solutions of 25% (w/v) NH_4NO_3 (10 ml) and 10% (w/v) $(NH_4)_6Mo_7O_{24}$ (10 ml). To precipitated ammonium phosphomolybdate hydrate, the resulting solution was stirred at 50 °C for 1 h. The yellow precipitate was filtered off, washed with NH_4NO_3 solution and double-distilled water, and then dried to a constant mass at 100–120 °C and weighed.

22.2.2.2 Characterization

Thermal decomposition of the surface oxygen-containing groups was analyzed by ThermoProgrammed Desorption (TPD) method with IR registration of desorption products. Concentrations of carbon dioxide and carbon monoxide ($C(CO_2)$ and $C(CO)$) in the argon-diluted evolved gas mixture were found, as reported in [27]. Measurements were conducted when samples were heated in a dynamic argon atmosphere at a heating rate of 10 °C min^{-1}. TPD mass-spectrometry (TPD MS) method was used to register water (H_2O), CO_2, and CO that are products of thermal decomposition of the carbon surface layer in a vacuum [28]. The total titration with NaOH was used to measure the concentration of the surface acidic groups (C_A). Nitrogen adsorption-desorption isotherms were measured by a TriStar Micromeritics C10900A porosimeter, up to $p/p_0 = 1$ atm at -196 °C. Before measurements, the sample was degassed at 120 °C for 6 h under vacuum. All texture parameters were determined by using the software supplied to the instrument.

The surface morphology was studied using scanning electron microscopy (SEM). The samples were imaged with a Zeiss EVO 60 scanning electron microscope, and the images were captured at 20.00 kV. Transmission electron microscopy (TEM) images were acquired using a JEOL JEM-2100F instrument at an acceleration voltage of 200 kV.

22.2.2.3 Catalysis

The gas-phase dehydration of i-C_3H_7OH was studied in a flow catalytic reactor with an infrared (IR) spectrometric monitoring of the outlet gas mixture. For catalytic tests, the argon gas-carrier flow of 45 $cm^3\ min^{-1}$ was saturated with i-C_3H_7OH vapor. The alcohol concentration in the argon gas flow that fed to the heated reactor packed with 0.1 g of catalyst was 1.22×10^{-3} mol L^{-1}.

22.3 Results and Discussion

Figure 22.1a shows the microspheres of the pristine AC. These microspheres are a compact form convenient for loading into and unloading from industrial catalytic reactors.

Figures 22.1b and 22.1c reveal that the smooth surface maintains after phosphorylation. From the SEM image of the AC phosphorylated at 700 °C, it is clear that the sponged areas with visible macropores are resulted by evolved organic volatiles (cf. Figures 22.1b and 22.1c).

Figure 22.2a shows selected TPD MS profiles for the pristine AC. According to the TPD MS data, the thermal decomposition products are H_2O, CO, and CO_2 released from the surface layer of AC. Dehydration of AC takes place between 30 and 800 °C, with the maximum weight loss rate observed at 100 °C. Respective TPD MS profile sums contributions from desorbed H_2O (Fig. 22.2, profile (1)). These contributions are from physisorbed H_2O and chemisorbed H_2O forms, which are the products of thermal decomposition of oxygen-containing surface groups. We observed a slight gas emission of CO_2 and CO between 200 and 600 °C. A few carboxylic, anhydride, and lactone groups present on the surface of the pristine AC. This corresponds to conclusions reached before in [29, 30]. Between 600 and 800 °C, we observed the most intense CO release. The thermal decomposition of phenolic surface groups [28] can explain this observation. TPD IR data (CO^+ and CO_2^+ profiles) confirmed the prevalence of phenolic groups on the carbon surface.

TPD MS data showed that decomposition products for the phosphorylated AC are H_2O and CO for the most (Fig. 22.3). A significant release of H_2O is found between 150 and 400 °C. This release can be explained by the cleavage of water from phosphonic groups caused by the formation of polyphosphoric acids. And, so,

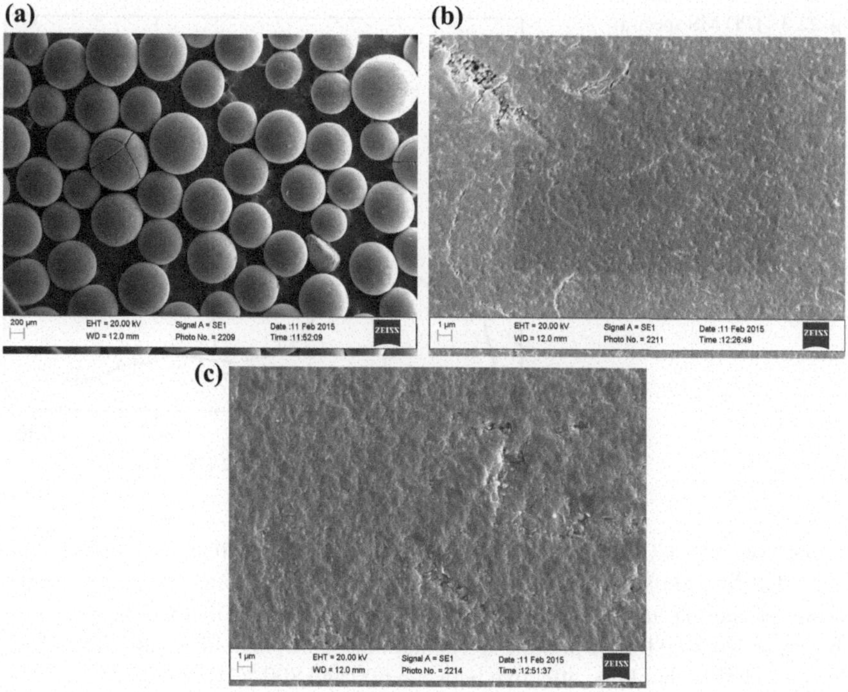

Fig. 22.1 SEM images of the AC microspheres (**a**) and the surface of **b** AC and **c** AC(I)-700

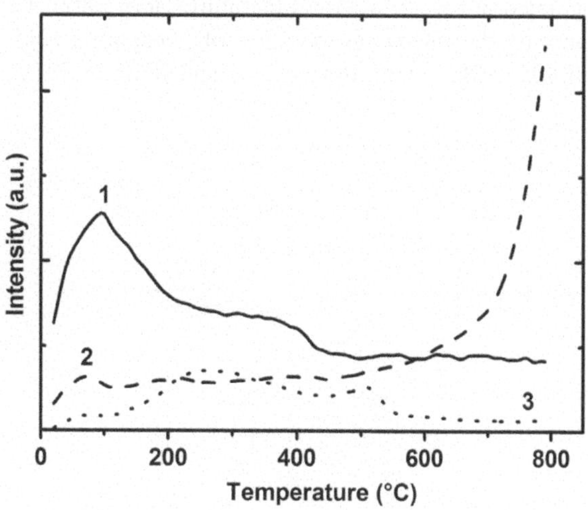

Fig. 22.2 TPD MS spectra for the AC: (1) H_2O^+, (2) CO^+, and (3) CO_2^+

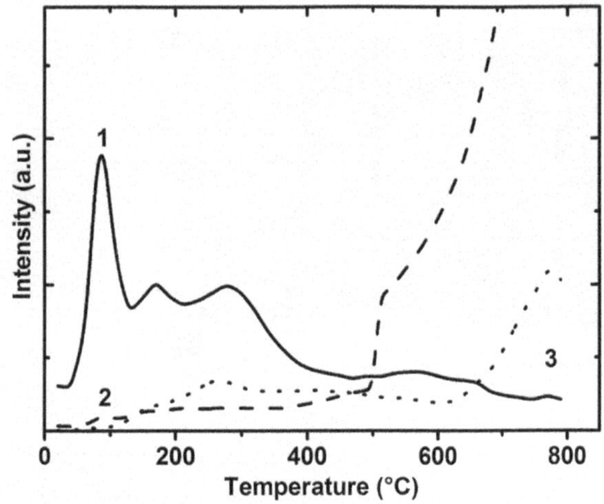

Fig. 22.3 TPD MS spectra for the AC(I)-700: (1) H_2O^+, (2) CO^+, and (3) CO_2^+

we observed only CO_2 traces released from the phosphorylated carbon surface. This means that the carboxylic, anhydride, and lactone groups are completely decomposed during the thermal treatment. From the TPD data, it is clear that CO-bearing sites have prevailed even for the pristine AC containing some phenolic groups. Therefore, phosphorylation should be effective between 600 and 800 °C (Fig. 22.3).

Under the stated conditions, the phenolic groups will decompose and free additional active sites. They will be involved in the surface phosphorylation reactions, as depicted in Fig. 22.4. At temperatures above 800 °C, phosphorylation will accompany by the intense evaporation of phosphoric acids and by the release of phosphorus oxides. According to Wu and Radovic [31] and Rosas et al. [32, 33], the

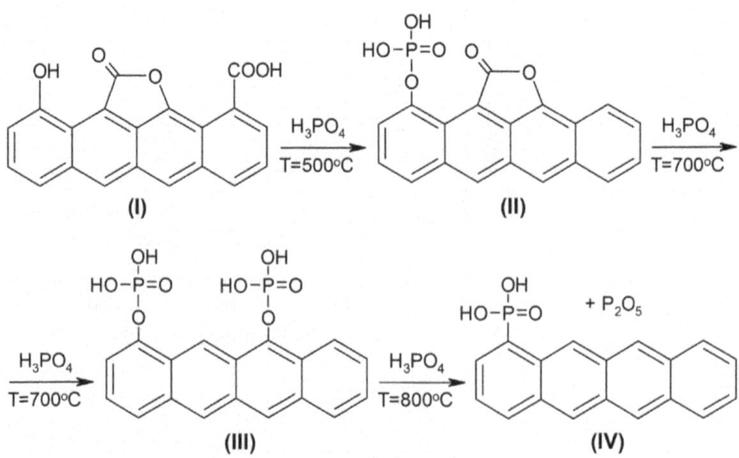

Fig. 22.4 Carbon phosphorylation schema and possible surface groups formed during thermal treatment with H_3PO_4, as C–O–P and C–P–O groups

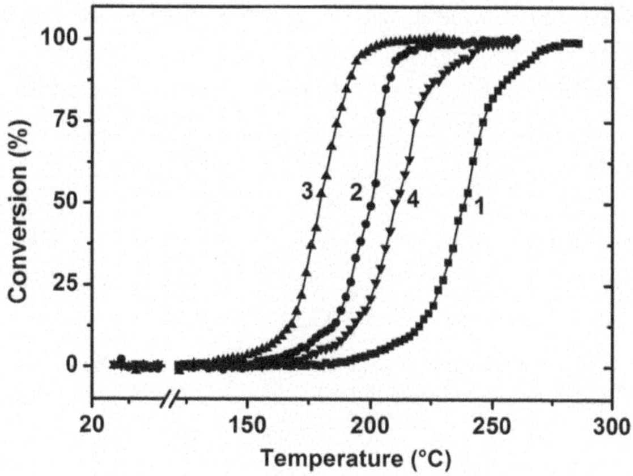

Fig. 22.5 Conversion of isopropanol into propylene against reactor temperature for (1) AC(I)-500, (2) AC(I)-600, (3) AC(I)-700, and (4) AC(I)-800

phosphorus-containing surface groups remain very stable at relatively high temperatures. Phosphorylation confers to different carbon materials a high oxidation resistance caused by the inhibition effect [33, 34] of the phosphorus-oxygen surface coverage. Improve oxidation resistance opens new possibilities for the use of nanoporous ACs as catalysts for reactions that take place at moderate temperatures. However, high-temperature treatment will initiate red-ox reactions that produce phosphonates and elemental phosphorus. The latter is an undesirable oxidizing agent.

All prepared phosphorylated ACs show significant catalytic activity and provide the total conversion of i-C_3H_7OH to propylene. Figure 22.5 demonstrates the typical results of temperature dependence measurements.

The characteristic behavior of the temperature dependence is a rapid conversion increase with temperature rise. An increase from 10 to 90% is observed with a 30 °C temperature increase from 150 °C. A rather steep progression of dependencies at high conversions indicates on availability of most active, strong acid sites. The activity of samples of the AC(I) series varies greatly depending on the phosphorylation temperature. In the tables below, to analyze all obtained data, we reported the temperature at 50% ($t_{50\%}$) and 100% ($t_{100\%}$) conversion of i-C_3H_7OH to propylene as a measure of the catalytic activity [8]. For comparison reasons, we calculated the effective activation energies (E_a) of catalytic i-C_3H_7OH dehydration determined from the Arrhenius equation applied to the initial ranges of the kinetic curves. Table 22.1 sums the data collected by miscellaneous test methods.

As seen in Table 22.1, the TPD MS results, in the form of C(CO), are in a positive correlation with the titration data (C_A). High-temperature treatment of AC at 700 °C with H_3PO_4 for 60 min of reaction time resulted in the highest phosphorylation efficiency, cf. tabulated values of C_P. Since we observed a significant CO emission at above 600 °C, only phenolic groups can remain under the stated conditions. Despite this observation, the tabulated data show that $C_A > C_P$. This means that the remained oxygen-containing groups have a prominent acidity. The value of $t_{50\%}$ is the highest for AC(I)-500 catalyst, which is because of the insufficient reactivity of the carbon matrix regarding the addition of H_3PO_4 at this temperature.

Table 22.1 Chemical analysis, TPD MS, titration, and catalytic parameters for the AC and the samples of AC(I) series

Sample	C_P (mmol g^{-1})	C_{CO} (°C)	C_A (mmol g^{-1})	$t_{50\%}$ (°C)	$t_{100\%}$ (°C)	E_a (kJ mol^{-1})
AC	0	0.85[a]	0.61	365[b]	–	–
AC(I)-500	0.21	0.39	0.31	237	280	130
AC(I)-600	0.62	1.11	1.12	200	226	117
AC(I)-700	0.79	1.32	1.37	181	207	103
AC(I)-800	0.56	0.32	0.74	209	248	136

[a] About 0.13 mmol g^{-1} of CO_2 was evolved from the surface
[b] The conversion degree is 19%

Because of heating to 500 °C, this catalyst has a few oxygen-containing groups that may act as reactive sites for the interaction with H_3PO_4. The highest activity has the AC(I)-700 catalyst having the lowest $t_{50\%}$. We associated this observation with the highest content of phosphorus in it. Since $t_{50\%}$ for the AC(I)-700 catalyst is 181 °C, we reached a better catalytic activity in comparison with the results reported earlier in [14]. We suggest that the optimum temperature for phosphorylation should be 700 °C. This temperature provides the highest rate of interaction of H_3PO_4 with the carbon matrix and intense thermal decomposition of phenolic groups. With further temperature increase, the phosphorus oxides desorption becomes noticeable, and so the activity of the AC(I)-800 catalyst decreases as compared with that of the AC(I)-700 catalyst. When one considers the samples of the AC(II) series (Table 22.2), it will be clear that C_p in the AC(II)-700 reaches the highest value of 1.04 mmol g^{-1}.

The samples of AC(II) series prepared at higher impregnation time of 24 h contain a larger number of newly formed oxygen-containing groups and show high catalytic activity. Among the samples of AC(I) and AC(II) series, the highest catalytic activity has the AC phosphorylated at 700 °C (Table 22.2). For this catalyst, the values of $t_{50\%}$ and $t_{100\%}$ are lower, by 9 and 13 °C, than that of the AC(I)-700 catalyst (Table 22.1). A prolonged impregnation time (24 h) has a positive effect on phosphorylation efficiency and improves the activity of the prepared catalysts. To prepare the samples of the AC(III) series, we increased the phosphorylation time up to 2 h.

Table 22.2 Chemical analysis, TPD MS, titration, and catalytic parameters for the samples of AC(II) series

Sample	C_P (mmol g^{-1})	C_{CO} (°C)	C_A (mmol g^{-1})	$t_{50\%}$ (°C)	$t_{100\%}$ (°C)	E_a (kJ mol^{-1})
AC(II)-500	0.23	0.36	0.42	232	273	190
AC(II)-600	0.66	1.36	1.29	194	216	122
AC(II)-700	1.04	1.50	1.71	170	194	93
AC(II)-800	0.68	0.40	0.87	211	247	155

Table 22.3 Chemical analysis, TPD MS, titration, and catalytic parameters for the samples of AC(III) series

Sample	C_P (mmol g^{-1})	C_{CO} (°C)	C_A (mmol g^{-1})	$t_{50\%}$ (°C)	$t_{100\%}$ (°C)	E_a (kJ mol^{-1})
AC(III)-500	0.32	0.50	0.56	222	261	258
AC(III)-600	0.52	0.96	0.97	206	237	187
AC(III)-700	0.59	1.43	1.22	191	217	112
AC(III)-800	0.34	0.14	0.40	236	–	186

As seen in Table 22.3, an increase in the phosphorylation time reduces the values of C_P and C_{CO}. This has a negative impact on catalytic activity. Within a much longer exposure period, a lot of H_3PO_4 can desorb or react with the carbon surface forming elemental phosphorus that can embed in the carbon matrix. From data comparison in Tables 22.2 and 22.3, it is clear that the values of $t_{50\%}$ and $t_{100\%}$ for the samples of the AC(III) series are higher by 20–25 °C. An exception is the AC(III)-500 catalyst. This catalyst showed an activity higher than the AC(II)-500 catalyst. One can explain this fact by the positive effect of higher processing time at the lowest high temperature. According to the data listed in Tables 22.1, 22.2, and 22.3, there is a fairly clear correlation between E_a and the catalytic activity ($t_{50\%}$ and $t_{100\%}$) for all the samples obtained. For the least active catalysts having low C_P, the value of E_a is 150–200 kJ mol^{-1}. The most active catalysts showed lower E_a values of 93–112 kJ mol^{-1}. To explain these observations, we proposed the reaction mechanism similar to that reported in [35]. This mechanism involves i-C_3H_7OH interaction with several acidic surface groups, and, the multi-point adsorption greatly enhances the turnover frequency. A semi-linear correlation between $t_{50\%}$ and C_A is a clear proof of that proposition (Fig. 22.6).

Fig. 22.6 Correlation between $t_{50\%}$ and C_A

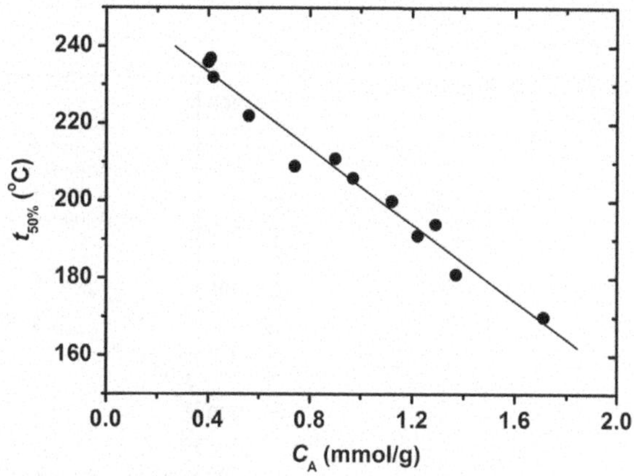

To a large extent, catalytic activity depends on the content of phosphorus-containing groups, but also the impact has how well acidic oxygen-containing groups speed up the dehydration reaction when act as adsorption sites for alcohol.

Below we will consider nanoporosity and related textural issues on an example of the samples of the AC(II) series. The pore diameter found from the pore size distribution is a useful parameter to differentiate nanoporous carbon solids. The International Union of Pure and Applied Chemistry (IUPAC) proposed the following classification by the pore diameter: micropores with diameter below 2 nm, mesopores whose diameter ranges between 2 and 50 nm, and macropores having a diameter of above 50 nm [36]. Mays proposed alternative classification and differentiated them as nanopores, micropores, and millipores, which diameters are between 0.1–100 nm, 0.1–100 um, and 0.1–100 mm, correspondingly [37]. Both classifications have advantages and weaknesses, but, according to Mays's scheme, one should consider any pore which size is below 10 nm as nanopores that cover the IUPAC micropore range and a part of the IUPAC mesopore range.

The nitrogen adsorption isotherms for the pristine AC and the phosphorylated ACs have characteristic features of types I and IV isotherms classified by the IUPAC (Fig. 22.7). They have branches that correspond to the filling of nanopores at low pressures. Hysteresis loops registered at high pressures are from nitrogen adsorption/desorption in mesopores. The hysteretic branching occurs when the mechanism of mesopore filling by capillary condensation varies from mesopore emptying.

From the nitrogen adsorption isotherms (Fig. 22.7), the Brunauer–Emmett–Teller surface area (S_{BET}) and micropores surface area (S_{micro}) were calculated through the Brunauer–Emmett–Teller and t-plot models. The total pore volume (V_{tot}) was estimated at a relative pressure of $p/p_0 = 0.98$. The micropore volume (V_{micro}) was found by the t-plot method. Both the mesopores surface area (S_{meso}) and mesopores volume (V_{meso}) were estimated from a difference: $S_{\mathrm{meso}} = S_{\mathrm{BET}} - S_{\mathrm{micro}}$ and $V_{\mathrm{meso}} = V_{\mathrm{tot}} - V_{\mathrm{micro}}$. The Dubinin-Radushkevich limiting micropore volume (V_{DR}) was found as in [27].

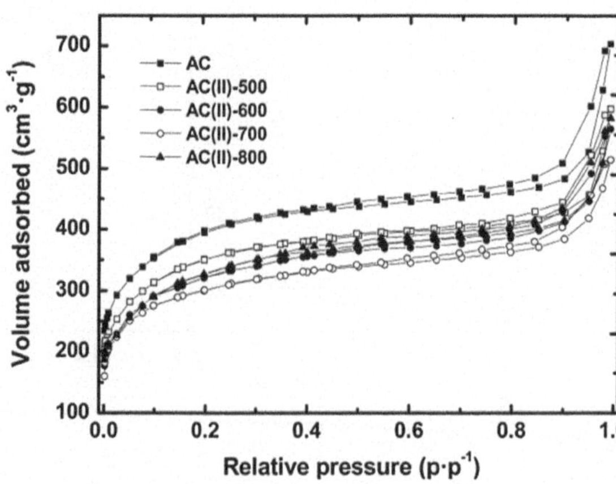

Fig. 22.7 N_2 adsorption isotherms

Table 22.4 Textural parameters for the AC and the samples of AC(II) series

Sample	Specific surface area (m^2 g^{-1})			Specific volumes (cm^3 g^{-1})			
	S_{BET}	S_{micro}	S_{ext}	V_{tot}	V_{micro}	V_{meso}	V_{DR}
AC	1304	1164	140	1.09	0.59	0.50	0.45
AC(II)-500	1098	979	119	0.91	0.53	0.38	0.44
AC(II)-600	887	652	235	0.86	0.44	0.42	0.40
AC(II)-700	813	527	286	0.79	0.38	0.41	0.37
AC(II)-800	991	851	140	0.89	0.48	0.41	0.44

All obtained textural parameters are shown in Table 22.4. As seen from the tabulated S_{micro} and S_{ext} data, phosphorylation suppresses microporosity while increases the mesoporosity of the AC(II)-600 and AC(II)-700 samples. Nevertheless, the contribution of mesopores to the total porosity is relatively low. Depending on the phosphorylation temperature, the textural parameters showed nonlinear changes. An increase in the values of S_{BET} and S_{micro} for the AC(II)-800 sample can be in part because of emptying nanopores under high-temperature conditions.

For pore size distribution modeling, we used the Tikhonov regularization method with the regularization parameter chosen by the L-curve criterion implemented in the SAIEUS program (http://www.nldft.com). The SAIEUS program evaluates the fit of the density functional theory (DFT) model to the experimental isotherm by calculating the root-mean-square deviation [38]. The SAIEUS program calculates the cumulative volume distribution and the pore size distribution for each Non-Local Density Functional Theory (NLDFT) model. This cumulative volume distribution shows how the pore volume is distributed across the pore sizes in the carbon solid. However, for correct modeling, the cumulative volume graphs should reflect the

Fig. 22.8 Pore size distributions

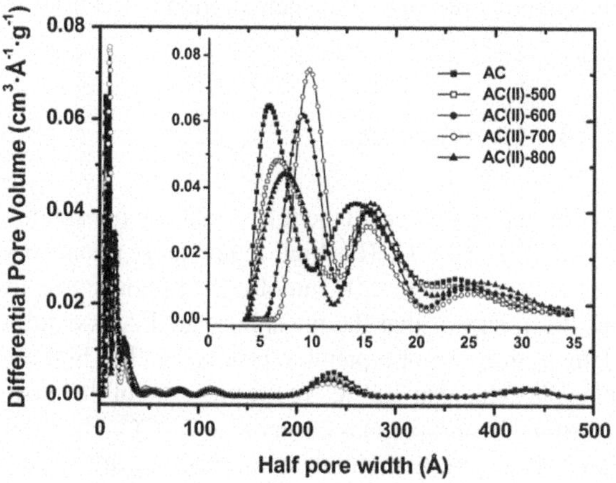

trends in the isotherms [39]. Figure 22.8 sums and presents the pore size distributions obtained by using the NLDFT model for slit pores [38, 40].

As can be seen from Fig. 22.8, all the samples examined have several types of pores, and the phosphorylation leads to some redistribution between them. Large macropores of size 100 Å and larger pores (half pore width peaked at 55, 81, 113, 238, and 430 Å) contribute to a small amount to the total porosity. However, these pores are important when transporting reagents to small nanopores and for removal of the catalytic reaction products. They do not undergo noticeable changes because of phosphorylation. In contrast, we registered the transformation of smaller mesopores of 20–70 Å in size. The same is true for micropores of up to 20 Å in size. Such transformations showed that this part of the porous structure is subjected to modification at grafting of phosphonic groups. As seen from the above Fig. 22.8 insert, the number of micropores with a half-width of 5.8 Å and mesopores with a half-width of 14 Å significantly reduces because of phosphorylation. Instead, there arose larger pores of 7.3–9.7 Å and 15.1–15.7 Å in width, correspondingly. The largest pore size variations are found for the AC(II)-600 and AC(II)-700 samples, which is associated with the highest content of the introduced phosphonic groups. A probable explanation for the changes in pore size is the partial or complete blocking of the smallest slit pores by the grafted phosphonic groups.

According to Table 22.4, the specific volumes show a decrease because of phosphorylation. Nevertheless, the residual free inner volume is sufficient for effective transport in catalysis and adsorption processes. The smoother surface of the phosphorylated carbon particles on the TEM micrographs confirms the reduction of porosity (cf. Figs. 22.9a and 22.9c). From the TEM studies, the nanoporous structure formed by the pore walls composed of graphene layers (Figs. 22.9b and 22.9d). These layers have a different thickness that varies from 1 to 5 graphene monolayers.

Regardless of porosity contraction, the protogenic groups confined in the nanopores are accessible for isopropyl alcohol. We found that the nanoporous structure showed reasonable dehydration activity after phosphorylation and can support the effective passage of the dehydration reactions.

22.4 Conclusions

In this paper, we treated the pristine AC by phosphoric acid in the temperature range from 500 to 800 °C. This high-temperature phosphorylation method is effective and can introduce up to 1.04 mmol g^{-1} of phosphorus into the carbon surface layer. It has been shown that the phosphorylated ACs exhibit high catalytic activity in the dehydration of isopropanol to propylene. The highest catalytic activity has the ACs phosphorylated at 700 °C. In the presence of the most active catalysts, the total conversion of isopropanol is observed at 200 °C. The catalytic activity of the phosphorylated ACs increases with the total surface acidity. The results proved the remarkable catalytic performance of the prepared catalysts, which is attributed to the synergetic effect of the acidic phosphorus- and oxygen-containing groups.

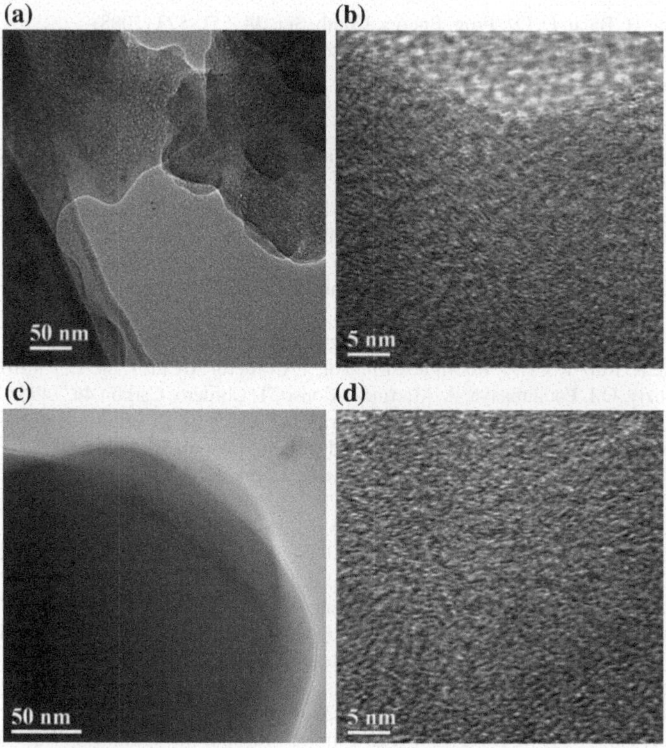

Fig. 22.9 TEM micrographs of **a**, **b** AC and **c**, **d** AC(II)-700

Acknowledgements V.V. Lisnyak thanks the Slovak Academic Information Agency (SAIA) for grant ID [20917] and acknowledges the International Visegrad Fund for scholarship ID number [51810574] in 2018.

References

1. V.E. Diyuk, V.V. Lisnyak, R. Mariychuk, in *Renewable Energy Sources: Engineering, Technology, Innovation, Springer Proceedings in Energy*, ed. by M. Wróbel, A. Szlęk, M. Jewiarz et al. (Springer Nature, Cham, 2020), pp. 59–67
2. L.M. Grishchenko, V.E. Diyuk, R.T. Mariychuk, A.V. Vakaliuk, V.Z. Radkevich, S.G. Khaminets, O.V. Mischanchuk, V.V. Lisnyak, Appl. Nanosci. 1–17 (2019). https://doi.org/10.1007/s13204-019-01069-3
3. L.M. Grishchenko, A.V. Vakaliuk, V.E. Diyuk, O.V. Mischanchuk, O.Yu. Boldyrieva, T.M. Bezugla, V.V. Lisnyak, Mol. Cryst. Liq. Cryst. **673**, 1 (2018)
4. A. Singh, P.S. Nigam, in *Microbial Biotechnology: Progress and Trends*, ed. by F.D. Harzevili, H. Chen (CRC Press, Boca-Raton, 2017), pp. 155–169
5. K. Bulkowska, Z. Gusiatin, E. Klimiuk, A. Pawlowski, T. Pokoj, *Biomass for Biofuels*, 1st edn. (CRC Press, London, 2016), p. 188

6. M. Balat, H. Balat, C. Öz, Prog. Energy. Comb. Sci. **38**, 551–573 (2008)
7. M. Toda, A. Takagaki, M. Okamura, J.N. Kondo, S. Hayashi, K. Domen, M. Hara, Nature **438**, 178 (2005)
8. V.E. Diyuk, A.N. Zaderko, L.M. Grishchenko, A.V. Yatsymyrskiy, V.V. Lisnyak, Catal. Commun. **27**, 33 (2012)
9. T.M. Bezugla, L.M. Grishchenko, A.V. Vakaliuk, V.E. Diyuk, O.V. Mischanchuk, V.V. Lisnyak, Mol. Cryst. Liq. Cryst. **661**, 58 (2018)
10. J. Bedia, J.M. Rosas, D. Vera, J. Rodríguez-Mirasol, T. Cordero, Catal. Today **158**, 89 (2010)
11. J. Bedia, R. Barrionuev, J. Rodríguez-Mirasol, T. Cordero, Appl. Catal. B Environ. **103**, 302 (2011)
12. A. Villa, M. Schiavoni, P.F. Fulvio, S.M. Mahurin, S. Dai, R.T. Mayes, G.M. Veith, L. Prati, J. Energy Chem. **22**, 305 (2013)
13. L. Wang, X. Dong, H. Jiang, G. Li, M. Zhang, Catal. Commun. **56**, 164 (2014)
14. J. Bedia, R. Ruiz-Rosas, J. Rodriguez-Mirasol, T. Cordero, AIChE J. **56**, 1557 (2010)
15. A.M. Puziy, O.I. Poddubnaya, A. Martinez-Alonso, T. Cordero, Carbon **40**, 1493 (2002)
16. J.M. Rosas, R. Ruiz-Rosas, J. Rodriguez-Mirasol, T. Cordero, Kinetic study of the oxidation resistance of phosphorus-containing activated carbons. Carbon **50**, 1523 (2012)
17. M.M. Rahman, S.H. Samsuddin, M.F. Miskon, K. Yunus, A.M. Yusof, Green Chem. Lett. Rev. **8**, 9 (2015)
18. J. Bedia, M. Peñas-Garzón, A. Gómez-Avilés, J.J. Rodriguez, C. Belver, Carbon **4**, 63 (2018)
19. R.G. Pereira, C.M. Veloso, N.M. da Silva, L.F. Sousa, R.C.F. Bonomo, A.O. Souza, M.O. da Guarda Souza, R. da Costa Ilhéu Fontan, Fuel Proc. Technol. **126**, 476 (2014)
20. N. Suganthi, K. Srinivasan, J. Environ. Sci. Eng. **53**, 163 (2011)
21. R.H. Hesas, A. Arami-Niya, W.M.A.W. Daud, J.N. Sahu, BioResources **8**, 2950 (2013)
22. C.-X. Chen, B. Huang, T. Li, G.-F. Wu, BioResources **75**, 5109 (2012)
23. J. Xu, L. Chen, H. Qu, Y. Jiao, J. Xie, G. Xing, Appl. Surf. Sci. **320**, 674 (2014)
24. M.A. Nahil, P. Williams, Biomass Bioenerg. **37**, 142 (2012)
25. C.L. Lee, P.S. H'ng, M.T. Paridah, K.L. Chin, U. Rashid, M. Maminski, W.Z. Go, R.A.R. Nazrin, S.N.A. Rosli, P.S. Khoo, R. Soc. Open Sci. **5**, 180775 (2018)
26. S. Zuo, Chem. Indus. Forest. Prod. **37**, 1 (2017)
27. V.E. Diyuk, A.N. Zaderko, K. Veselovska, V.V. Lisnyak, J. Thermal. Anal. Calorim. **120**, 1665 (2016)
28. V.E. Diyuk, R.T. Mariychuk, V.V. Lisnyak, J. Thermal. Anal. Calorim. **124**, 1119 (2016)
29. L.M. Grishchenko, V.E. Diyuk, O.P. Konoplitska, V.V. Lisnyak, R.T. Maryichuk, Adsorption Sci. Technol. **35**, 884 (2017)
30. J.L. Figueiredo, M.F.R. Pereira, M.M.A. Freitas, J.J.M. Órfão, Carbon **38**, 1379 (1999)
31. X. Wu, L.R. Radovic, Carbon **44**, 141 (2006)
32. J.M. Rosas, J. Bedia, J. Rodríguez-Mirasol, T. Cordero, Fuel **88**, 19 (2009)
33. J.M. Rosas, R. Ruiz-Rosas, J. Rodríguez-Mirasol, T. Cordero, Carbon **50**, 1523 (2012)
34. J.M. Rosas, R. Berenguer, M.J. Valero-Romero, J. Rodríguez-Mirasol, T. Cordero, Front. Mater. **1**, 29 (2014)
35. V. Ye Diyuk, R.T. Mariychuk, V.V. Lisnyak, Mater. Chem. Phys. **184**, 138 (2016)
36. T.J. Mays, Stud. Surf. Sci. Catal. **160**, 57 (2007)
37. M. Thommes, K. Kaneko, A.V. Neimark, J.P. Olivier, F. Rodriguez-Reinoso, J. Rouquerol, K.S.W. Sing, Pure Appl. Chem. **87**, 1051 (2015)
38. J. Jagiello, J.P. Olivier, Carbon **55**, 70 (2013)
39. A.M. Puziy, O.I. Poddubnaya, B. Gawdzik, M. Sobiesiak, Adsorption **22**, 459 (2016)
40. J. Jagiello, J.P. Olivier, Adsorption **19**, 777 (2013)

Chapter 23
Flexible Thermostable Metal Spin-Valves Based on Co, Cu, Fe, Au, Ru Thin Films

A. M. Lohvynov, I. V. Cheshko, O. P. Tkach, Yu. M. Shabelnyk and S. I. Protsenko

Abstract According to study results of the structural-phase state, electrophysical and magnetoresistive properties of multilayer film systems based on Co, Cu, Fe, Au and Ru the possibility of their application as functional layers of metal spin-valve structures with stable operating characteristics in the temperature range 300–900 K is shown. Clarified principles and concept of spin-valve nanostructures formation technique with thermostable properties for the creation of functional elements of flexible sensor electronics in the form of metallic thin-film systems of spin-valve type using multilayers $[Cu(Au)/Co(Fe)]_n$ or synthetic antiferromagnetic layer based on Co and Ru.

23.1 Introduction

The development of flexible electronics involves the transition to new electronic devices and gadgets. For example, small and lightweight electronic storage and display devices [1], photovoltaic panels and reconfigured antennas, flexible biological electronic implants [2], printed batteries and batteries are created through the use of advanced technologies and materials of flexible functional electronics [3, 4]. The crucial point is that the use of flexible electronics technologies means cheaper production, unlike the technologies used to produce traditional solid-state semiconductor devices by sophisticated manufacturing processes. Flexible items can be printed on cheap plastic film using conventional printing techniques on paper or plastic. Distinctive features of devices of flexible electronics can be called transparency, stability, low coefficient of thermal expansion, elasticity, etc. This trend has been developing rapidly in recent years and is finding more and more new applications [5, 6]. Based on flexible electronics, many applications can be developed. Simplicity and compatibility with traditional solid-state electronics technologies the flexible electronics method have become widespread in the manufacture of flexible and translucent

---

A. M. Lohvynov · I. V. Cheshko (✉) · O. P. Tkach · Yu. M. Shabelnyk · S. I. Protsenko
Sumy State University, 2, Rymsky-Korsakov Str, Sumy 40007, Ukraine
e-mail: i.cheshko@aph.sumdu.edu.ua

A. D. Pogrebnjak and O. Bondar (eds.), *Microstructure and Properties of Micro- and Nanoscale Materials, Films, and Coatings (NAP 2019)*,
Springer Proceedings in Physics 240, https://doi.org/10.1007/978-981-15-1742-6_23

information display devices, thin-film transistors, memory cells, solar cells, various sensors, including radio or radio equipment. tags and more. Flexible batteries based on organic and inorganic materials are becoming widespread. Although the efficiency of such elements does not exceed the values of traditional solid state elements ~11–40%, such as modern silicon solar cells, they have other advantages: low cost, flexibility, elasticity, lightness, increased resistance to environmental factors. Further development of flexible electronics technologies is aimed at the search for new substrate materials with a wide range of operating temperatures, acceptable optical and mechanical parameters and the development of new technologies for the fabrication of metal and non-metal functional structures.

The main directions of the development of flexible electronics can be called a search for new materials and flexible substrates known adaptation and development of unique methods of creating functional elements of flexible electronics based on thin films of metals, dielectrics, and semiconductors. The most well-known technologies are used for the creation of metal films and printing technologies for metal nanoparticles. Spin-dependent electron-scattering materials based on Co, Cu, Fe, Au or Ru in the form of multilayer spin-valve type film systems can be used to create effective functional elements of flexible sensor electronics in the form of spin-valve structures with thermally stable working characteristics. For this purpose, it is necessary to carry out a detailed study of the dependence of the crystalline structure and phase composition of such film metallic nanosystems on the conditions of formation and conditions and modes of temperature treatment. The purpose of the work is to create thermostable functional structures in the form of Co, Cu, Fe, Au or Ru-based spin valves for the implementation of communication channels in sensory flexible electronics.

23.2 Experiment Details

During the experiment, a series of single-, two-, three- and multilayer film systems based on Co, Cu, Fe, Au, and Ru were obtained. Samples were condensed by electron beam method in vacuum (10^{-5} Pa) at temperature $T_s = 300$ K with the condensation rate $\omega = 0.2$–0.4 nm/s. The thickness of the individual layers was controlled by a quartz resonator during the deposition process. To investigate the thermal stability of the physical properties, the samples were annealed in a vacuum chamber to temperatures $T_a = 600$ and 900 K.

As the substrate used in various substrates depending on the hearth-lshyh methods of research samples. To control the phase composition and structure of the flexible substrate samples were used as a carbon film thickness of 20 nm. To study the electrophysical properties, polished glass ceramic plates 2 × 1 cm were used.

The study of the electrical properties (resistivity and thermal coefficient of resistance) samples was carried out in a vacuum. In obtaining the dependence of the specific resistance of the samples on the temperature $R(T)$, two stabilization cycles of “heating and cooling” were used at a constant speed of 8–10 K/min. Then, using the ratio:

$$\beta = (1/R_{in})(\Delta R/\Delta T), \tag{23.1}$$

were R_{in}—initial resistance value, the value of the temperature coefficient of resistance (TCR or β) was calculated and dependencies $\beta(T)$ were obtained.

Magnetic resistance (*MR*) studies of the proposed elements were carried out at room temperature using a four-point scheme in the range of 0 to 500 mT [7]. During the measurement, the electric current was directed parallel to the sample plane, and the measurement of the magnetic resistance was carried out in three geometries—longitudinal (magnetic induction lines B directed along the direction of the current flow), transverse (line *B* perpendicular to the direction of flow) and perpendicular (line *B* perpendicular to the sample plane). The value of *MR* was determined by the ratio:

$$MR = \Delta R/R_0 = (R_B - R_0)/R_0, \tag{23.2}$$

were R_B and R_0—the resistance of the sample in the given field and in its absence.

The sensitivity of the samples to the magnetic field S_B was calculated for the maximum value of *MR* obtained by the formula:

$$S_B = |MR\text{max}/\Delta B|, \tag{23.3}$$

were MR_{max}—the maximum value of the magnetoresistance; ΔB—the change in magnetic induction from the saturation field B_S to the demagnetization field.

23.3 Experiment Results

Particular attention in the study of the electrical properties of the proposed spin-valve structures is given to systems based on Co and Cu, as this classic combination of metals. A number of dependences of resistivity and temperature coefficient of resistance of single and multilayer films based on Co and Cu were obtained (Fig. 23.1). It was found that the resistivity dependence on temperature in all considered cases of the layer thicknesses had a metallic nature with a pronounced change in the resistivity (50–1.2) × 10^8 Om·m during heating. Measurements were made at 2 annealing cycles. TCR dependence on temperature is nearly linear, while dependent decreases with increasing temperature from (4–3.5) × 10^4 K^{-1}. The slightest decrease in the resistivity at the first heating cycle was observed for the system made in the form of a multilayer $[Cu(3)/Co(3)]_7/S$ (S—substrate, in brackets the thickness in nm). This can be explained by the large number of interfaces compared to two- and three-layer specimens, which leads to activation of healing by heating of completely different types of defects with less healing energy.

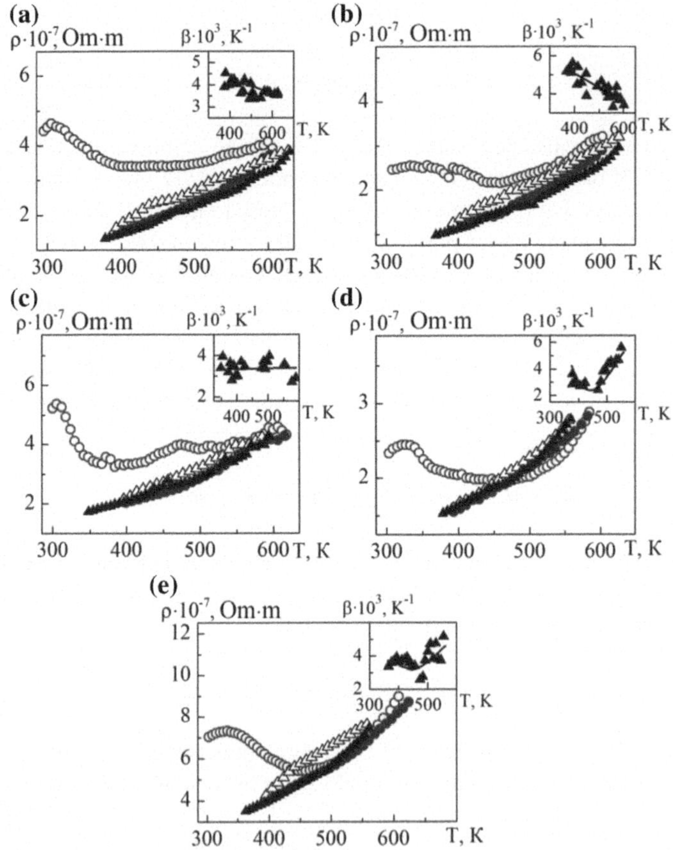

Fig. 23.1 Temperature dependences of resistivity and thermal resistance (on inserts) of film systems: **a** Co(20)/Cu(5)/Co(20)/S; **b** Co(20)/Cu(10)/Co(20)/S; **c** Co(20)/Cu(20)/Co(20)/S; **d** Cu(10)/Co(20)/S; **e** $[Cu(3)/Co(3)]_7$/S. The first cycle of heating (○) and cooling (●); the second cycle of heating (Δ) and cooling (▲)

In the general case, the tendency to form solid solutions (s.s.) in this film system can be traced. This Cu layer leads to a significant change in the type of dependency as a result of the formation s.s. already at the stage of sample formation. The results of the study of electrophysical properties completely correspond to the results of the study of the phase composition and structural characteristics of multilayer film systems based on Co and Cu, which indicate the formation Co atoms in them in the hcc-Cu matrix. As shown in our work [8–10] the formation process s. s. Cu (Co) at the interface between the layers occurs in the process of forming samples at room temperature as a result of condensation-stimulated diffusion and subsequent heat treatment to as a result of condensation-stimulated diffusion and subsequent heat treatment to $T_a = 400–700$ K. At $T_a \geq 900$ K there is a partial decay of the solid solution to form Co granular state. Electrograms of film systems based on Co and

Cu in the form of spin-valve structures Au(10)/Co(5)/Cu(10)/Co(20)/S at $T_a = 300$ K there are two groups of lines corresponding to the phases of hcp-Co + hcc-Co (packing defects) and hcc-s.s. Cu (Co) (parameter $a_{s.s} = 0.3588$ nm). After annealing the sample to $T_a = 700$ K parameter $a_{s.s}$ decreases to 0.3575 nm as a result of further mixing of the layers Co and Cu.

Investigation of the phase composition of multilayer films in the form of a pseudo spin valve Au(10)/Co(3)/Cu(6)/Co(20)/Cr(3)/S complicated. Electrograms from such structures are complex. However, according to our data, the tendency of the formation of solid solutions persists in spin-valve structures. The unburned system consists of hcc-Au, hcp-Co + hcc-Co phases (packing defects) and s.s. Cu (Co). The annealed film additionally forms s.s. Au (Co). Before and after annealing multilayer films have dispersion structure.

For temperature dependences of film systems: Fe(10)/Au(20)/S and Au(10)/Fe(20)/Au(10)/S, as for all metals, characterized decrease resistance with increasing temperature during the heating cycle in the first study of the temperature dependence of the resistivity. This can be explained by the treatment of defects of the vacancy-impurity atom type of the crystalline structure in the freshly condensed samples. When heated during the first cycle in the temperature range of 300–600 K there is a slight increase in resistivity with a decrease in the interval of 600–700 K, which may be related to the healing of the defective structure and the recrystallization processes, which are accompanied by a decrease in the contribution of the grain boundary scattering of the current carriers to the overall resistance of the system. During cooling, the resistivity decreases monotonically, which is explained by the completion of the relaxation processes already in the first cycle of heat treatment. Such specific resistivity dependences correlate with the data obtained from the results of the study of the phase composition and crystalline structure of Fe and Au based film systems. As shown in [10], after condensation of such samples, the layers retain their identity and only after annealing does it gradually begin to s. s. Au (Fe) form and at high temperatures partial oxidation of Fe occurs. Electronograms from unburned specimens have two ring systems corresponding to the hcc-Au and bcc-Fe phases with the corresponding lattice parameters of 0.4074 and 0.2875 nm. These values are very close to the table values for the bulk samples Au $a_0 = 0.4078$ nm and Fe $a_0 = 0.2870$ nm [11]. After annealing up to 700 K Au (Fe) formation occurs.

Another important combination is systems based on Co and Ru as systems in which layers retain their identity over the entire range of temperatures considered. And this guarantees high thermal stability of the functional characteristics of the functional structures made on their basis. The temperature dependences of the resistivity of single-layer specimens with a thickness of 20 nm and 40 nm were obtained, which were condensed on an unheated substrate. These dependencies are nonmetallic character. This is due to the fact that the substrate temperatures and the thickness of the films are insufficient for the mechanisms inherent in the metal films to start working. When the substrate is heated, it receives additional energy that contributes to the formation of a homogeneous structure and uniform deposition of the film, and during annealing, thermostabilization processes occur in grains and domains, healing structural defects of the crystalline structure (vacancy, packing defects). Most likely, in this case, we are dealing with carbide and ruthenium, which is inherently a semiconductor. It has been found that the production of oxide-free single-layer Ru films is only possible by heating the substrate to $T_s = 400–700$ K at thicknesses d_{Ru} > 40 nm.

For single-layer films based on Ru over the entire thickness range of the resistivity by an order of magnitude greater than the table value for mass samples ($\rho_0 = 7.15 \times 10^{-7}$ Om•m). The closest value was observed for Ru (40)/S, that in the condensed at room temperature condition of the sample was $\rho = 1.79 \times 10^{-5}$ Om•m, and after annealing to 1058 K—$\rho = 7.28 \times 10^{-6}$ Om•m. The average value of the temperature coefficient of resistance for this sample is $\beta = 4 \times 10^{-3}$ K^{-1}.

Co- and Ru-based two- and multilayer film samples are characterized by a metallic temperature dependence, which has a typical Vendian character. Such a result is confirmed by the conclusions drawn from the results of the investigation of the crystalline structure of single and multilayer film systems based on Co and Ru. In all the cases considered, Co and Ru based systems maintain the individual duality of the layers in the temperature range of 300–900 K at all thicknesses considered.

The analysis of the experimental results obtained in all the considered cases was performed on the basis of the comparative characteristic of the formation of solid solutions in different film systems. Such an analysis makes it possible to establish the fact of solid solution formation in different materials and at different temperatures. Namely, for the film systems, the analysis was performed in a fresh condensed state and during annealing.

From the analysis, we can conclude that the effect of temperature on the fact of the formation of a solid solution in film systems. In Au and Fe films, the possibility of solid solution formation begins after annealing, which is not typical of Co and Cu based film systems, in which the solid solution can be seen in the early stages of film condensation. This can be explained by the fact that the lattices Fe and Au are less favorable for the formation of a solid solution, they are denser than the lattices Co and Cu. The above features of the order of formation of solid solutions completely determine the thermal stability of the considered nanostructures, including structures of spin-valve type with a different combination of working layers. Thermal stabilization of such structures can be ensured using Ru layers or multilayers.

23.4 Application

The analysis of the experimental results made it possible to offer some practical applications for the needs of modern flexible electronics. An embodiment of an effective and simple pseudo-spin-valve structure is a sandwich multilayer structure, which is performed according to the scheme "magnetic layer 1/non-magnetic layer/magnetic layer 2", a schematic of which is shown in Fig. 23.2. The magnetic layer 1 and 2 are made in the form of thin films of Co, Fe, Ni or their alloys. The non-magnetic layer

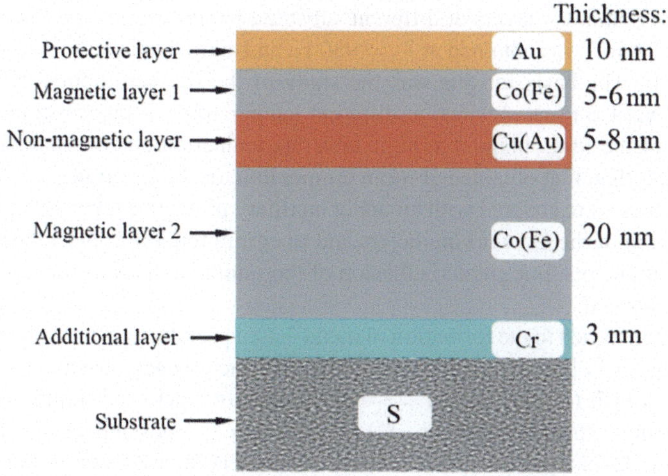

Fig. 23.2 The forming scheme of pseudo spin-valve based on Co or Fe magnetic layers

is made of films of metal with high electro-liquidity, for example, Cu, Au, Ag, etc. The peculiarity of such a functional pseudo-spin-valve structure is the use of one material for the magnetic layer 1 and 2 (see, for example, [12–14]). The difference of coercivity between them is provided by the dimensional effect since the ability to magnetize thin films depends essentially on the thickness of the sample and the conditions of its production.

As shown in [15, 16] the parameters of the functional elements of spin-valve structures based on working magnetic layers Co, separated by a non-magnetic layer with Cu thickness of 5–6 nm, are as follows: the thickness of the upper "soft" layer should be approximately 15–20% of the thickness of the lower "hard" layer (3–6 and 20 nm, respectively). This combination ensures the operation of a magnetic field all valves. The upper thin Co layer is more sensitive to the variation of the magnetic field inductance B and it begins the process of remapping its domain structure with smaller values, while the lower thick layer has a larger remapping field. The spin-valve effect is manifested in the interval ΔB between the onset of magnetization of the upper and lower layers. Moreover, the larger this gap, the greater the value of the magnetoresistance. Thus, when changing the external magnetic field, this system is capable of producing a high-amplitude bipolar electrical signal.

The choice of materials of magnetic layers and non-magnetic layer is determined by the possible degree of blurring of the interfaces between them. In our research [17, 18] magnetoresistive properties of systems based on Co and Cu and Fe and Au films were studied. In the considered systems, the scheme of which is shown in Fig. 23.2, the maximum values of magnetoresistance are possible [19] among the variants of pseudo spin valves.

Further enhancing the difference between the characteristics of the magnetic layer 1 and 2 is possible through different heat treatment. We are in the works [20] it was proposed to obtain Co layers at different substrate temperatures. The lower thicker layer ($d = 20$ nm) was obtained at $T_s = 950$ K, and a top thin layer ($d = 5$–6 nm)—$T_s = 300$ K. The basis for this was the study of the magneto-optical Kerr effect in single-layer Co films obtained at different temperatures of the substrates. So the coercivity of the film obtained at high substrate temperature is approximately 5–6 times greater than that obtained at room temperature of the substrate.

All samples were covered with an upper auxiliary protective layer Au (10) to prevent oxidation of the Co working layers and to ensure reliable contact during resistance measurements. For greater adhesion of the sample to the substrate, a sublayer Cr (3) was formed.

Another approach to the formation of metal-based pseudo-spin-valve systems was proposed by us in [21] by analogy with [22, 23], the essence of which is to a [Co (3)/Cu (3)]$_n$ or [Fe (3)/Au (3)]$_n$, ($n = 3$–18—number of repetitions) multilayer using. When forming instrument systems of spin-valve type it is possible instead of one of the magnetic layers, provided that the other is preserved, as shown in Fig. 23.3. A solid Co or Fe layer with a high value of coercivity compared to the multilayer will play the role of a fixed magnet-rigid layer to provide a spin-valve effect.

To study the effectiveness of such systems, the magnetoresistive properties of a series of multilayers Au(10)/[Cu(3)/Co(3)]$_n$/S and a series of pseudo spin-valve type samples using multilayers Au(10)/[Cu(3)/Co(3)]$_n$/Cu(5)/Co(20)/S and Au(10)/Co(6)/Cu(5)/[Cu(3)/Co(3)]$_n$/S were investigated. In the general case, the result of this change was an increase in the value of *MR* to 0.3–0.5% but a decrease in sensitivity S_B to values (0.1–0.2) × 10^{-2} %/mT. However, in such nanosystems, the annealing temperature significantly affects the nature of the dependencies *MR*. However, when using a multilayer, an anisotropic character MR is observed in the unburned specimens instead of the upper magnetic layer, which after annealing to $T_a = 700$ K disappears. Figure 23.4 shows the dependence of the *MR* value on

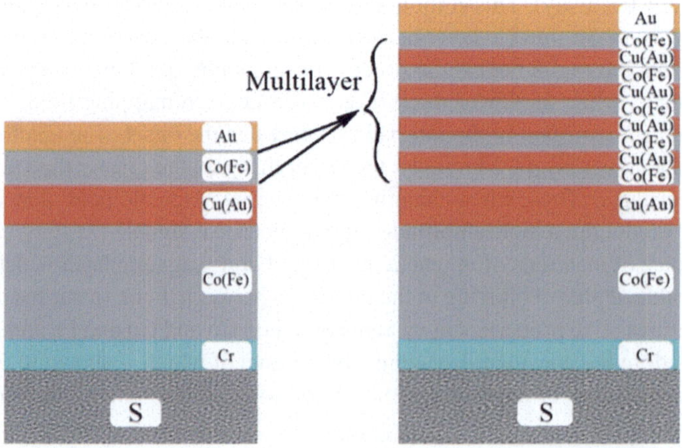

Fig. 23.3 The formation scheme of a pseudo spin-valve based on multilayer

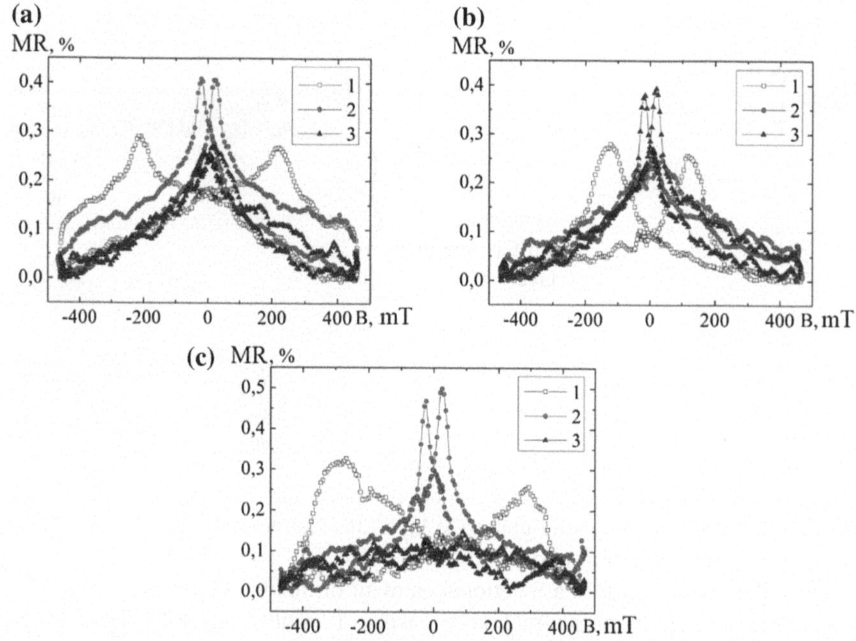

Fig. 23.4 *MR* dependences on the magnetic field induction of a multilayer-based pseudo-spin valve Au(10)/Co(5)/Cu(6)/[Co(3)/Cu(3)]$_8$/П at $T_a = 300$ (**a**), 700 (**b**) and 900 K (**c**) in perpendicular (1), transverse (2) and longitudinal (3) measurement geometries

the induction of an external magnetic field for film structures of spin-valve type Au(10)/Co(5)/Cu(6)/[Co(3)/Cu(3)]$_8$/S at $T_a = 300$ K (Fig. 23.4 a), 700 K (Fig. 23.4 b) and 900 K (Fig. 23.4 c) in perpendicular (1), transverse (2) and longitudinal (3) measurement geometries.

Such a system operates in the range of 0–0.5 T magnetic fields and has the following operating parameters: maximum *MR* values for perpendicular (0.50%) and transverse (0.3%) geometries are observed at $T_a = 900$ K, and for longitudinal geometry (0.26%)—at $T_a = 700$ K for film structure Au(10)/Co(5)/Cu(6)/[Co(3)/Cu(3)]$_8$/S. For structure Au(10)/[Co(3)/Cu(3)]$_4$/Cu(5)/Co(20)/S the *MR* values are slightly smaller and are 0.3%, 0.27% and 0.12%, respectively. The same situation is observed for the magnitude of sensitivity (S_B), maximum values of which are 1.4–2.4 × 10^{-2} %/mT for the first type of structure. Moving maxima on the given dependencies when T_a changing. This is especially evident in the perpendicular geometry of the measurement associated with the change in coercivity B_c of the upper magnetic layer Co and its magnetic anisotropy due to the change of the effective thickness of the layer in the process of thermodiffusion and recrystallization and change of the domain structure. In the transverse and longitudinal geometries of the measurement, the magnetoresistive properties up to $T_a = 700$ K stable. When T_a raising up to 900 K The substantially decreasing *MR* values can be attributed to the destruction of the

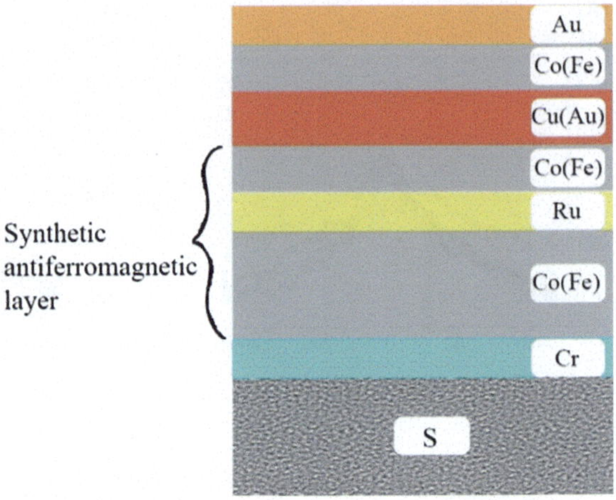

Fig. 23.5 The forming scheme of a pseudo spin-valve with synthetic antiferromagnetic layer based on Co and Ru

interface between the magnetic upper Co layer and the nonmagnetic Cu layer as a result of the formation of a solid solution.

Thus, the manufacture of a functional element of flexible electronics spin-valve type, in which one of the structural layers consists of a multilayer by the above design and manufacturing technology allows it to be used in a variety of sensor devices in a wide temperature range. Small values of the sensitivity of spin-valve nanosystems using Co and Cu based multilayers at higher *MR* values are associated with a high rate of switching from one magnetic state to another.

Figure 23.5 shows a scheme of manufacturing a spin valve using a layer Ru. Optimal conditions in terms of the temperature stability of the crystalline structure and, accordingly, the operating characteristics of the working element of the spin valve can be realized using film functional antiferromagnetic synthetic layers based on Ru and Co. They consist of a layer of Ru 5 $_{\mathrm{Ta}}$ 20 nm and two layers of Co 20 nm, obtained with the pre-heating of the substrate to 500 K and the subsequent thermal annealing to 900 K.

23.5 Conclusions

The results of the study of the structural-phase state, magnetoresistive properties of pseudo spin-valve type nanosystems and their fragments based on Co, Cu, Fe, Au, and Ru are presented. It is established that in the process of obtaining and thermal annealing to temperatures $T_a = 700$ and 900 K in these systems, solid solutions of Co or Fe atoms in the Cu or Au matrix are formed.

It has been shown that a pseudo-spin-valve type film system is reasonably modifiable by using instead of one of the multilayer magnetic layers $[Co/Cu]_n$ or $[Fe/Au]_n$. This modification, in the case of Co and Cu based systems, increases the magnetoresistance to 0.3–0.5%, the switching speed from one magnetic state to another, and the temperature stability of the entire nanosystem to a temperature of 700 K, although it reduces the magnetic sensitivity to values $S_B = (0.1\text{–}0.2) \times 10^{-2}\,\%$/mT. It can be said that the changes in the dependence of the MO on the total concentration of the components in the systems under consideration make it possible to trace the formation of solid solutions. Substantial changes after thermal annealing up to 800–900 K are observed for systems based on Fe and Au, since in this system processes are formed of solid solutions when in the case of systems based on Co and Cu they have occurred at the stage of condensation of components.

Based on the results obtained, we can conclude that it is advisable to use the proposed method of forming pseudo spin-valve structures with different thicknesses of magnetic working layers in a particular field of science or technology. Those are experimental specimens that have a magnetic sensitivity greater 8.0%/mT, it is advisable to use magnetic fields in the range of 200–600 mT. They have a low magnetic sensitivity, which will provide a more accurate measurement process and less error, increased resolution. At the same time, samples that showed the highest values of the magnetic sensitivity can be used in both analog circuits and digital in origin to respond quickly to the corresponding change in magnetic flux. This can be used in non-volatile memory, control, automation and control systems and more.

Acknowledgements The work has been performed under the financial support of the Ministry of Education and Science of Ukraine (state registration number 0117U003925).

References

1. V. Luchinin, Nanoindustry **8**, 46 (2013)
2. V. Zardetto, T.M. Brown, A. Reale, J. Polym. Sci. B Polym. Phys. **49**, 638 (2011)
3. A.V. Lukashyn, A.A. Eliseev, *Physical Methods for the Synthesis of Nanomaterials* (MSU, Moscow, 2007)
4. Siya Huang, Yuan Liu, Yue Zhao, Zhifeng Ren, C.F. Guo, Adv. Funct. Mater. **29**(6), 1805924 (2019)
5. D. Karnaushenko, D. Makarov, M. Stöber, Adv. Mater. **27**(5), 880 (2015)
6. S. Khan, L. Lorenzelli, R.S. Dahiya, I.E.E.E.T. Semicond, Manufacturer **28**(4), 486 (2015)
7. IYu. Protsenko, I.V. Cheshko, L.V. Odnodvorets, D.M. Kondrakhova, O.V. Pylypenko, YuM Shabelnyk, O.V. Vlasenko, Usp. Fiz. Met. **14**, 229 (2013)
8. I. Protsenko, I. Cheshko, Ja Javorsky, Funct. Mater. **13**(2), 219 (2006)
9. I. Cheshko, S. Protsenko, L. Odnodvorets, Tech. Phys. Lett. **35**(10), 903 (2009)
10. I.V. Cheshko, I.M. Pazukha, S.I. Protsenko, D.V. Shapko, Yu.M. Shabelnyk, *in NAP-2013: Magnetoresistive and Magnetooptical Properties Nanosize Film Systems Based on Fe and Au*, ed. A. Pogrebnjak. Proceedings of the International Conference Nanomaterials: Applications & Properties 2013, Alushta, September, vol. 2(**1**) (SSU, Sumy, 2013), p. 01NTF40
11. S.S. Gorelik, L.N. Rastorguev, YuA Skakov, *X-ray and Electron-optical Analysis* (MISIS, Moscow, 1994)
12. P.V. Leksin, N.N. Garif'yanov, A.A. Kamashev, Ya.V. Fominov, J. Schumann, C. Hess, V. Kataev, B. Büchner, I.A. Garifullin, Phys. Rev. B **91**, 214508 (2015)
13. A.V. Svalov, A.H. Sorokin, P.A. Savin, A. García-Arribas, A. Fernández, V.O. Vas'kovskiy, G.V. Kurlyandskaya, Key Eng. Mater. **644**, 211 (2015)

14. A.V. Svalov, G.V. Kurlyandskaya, V.O. Vas'kovskiy, Appl. Phys. Lett. **108**, 063504 (2016)
15. W.H. Butler, X.-G. Zhang, D.M.C. Nicholson, J.M. MacLaren, Phys. Rev. B **52**, 13399 (1995)
16. S. Luby, E. Majkova, Appl. Surf. Sci. **248**, 316 (2005)
17. I.V. Cheshko, S.I. Protsenko, P. Shifalovich, Visnyk SumDU. Ser. Phys. Math. Mech. **2**, 106 (2008)
18. I.V. Cheshko, I.Y. Protsenko, Metallofiz. Nov. Tekhnol. **31**(7), 963 (2009)
19. S. Zsurzsa, L. Péter, L.F. Kiss, I. Bakonyi, J. Magn. Magn. Mater. **421**, 194 (2017)
20. P. Matthes, M. Albrecht, Sensor. Actuat. A **233**, 275 (2015)
21. I.V. Cheshko, M.V. Kostenko, V.I. Grebynakha, A.M. Lohvynov, S.I. Protsenko, J. Nano Electron. Phys. **8**(3), 03041 (2016)
22. P. Perna, C. Rodrigo, M. Muñoz, J.L. Prieto, A. Bollero, D. Maccariello, J.L.F. Cuñado, M. Romera, J. Akerman, E. Jiménez, N. Mikuszeit, V. Cros, J. Camarero, R. Miranda, Phys. Rev. B **86**, 024421 (2012)
23. F. Zhang, Z. Liu, F. Wen, J. Superconduc. Nov. Magn. **30**(2), 533 (2017)

Chapter 24
Formation of Composite Reinforced Coating by Chemical Deposition and Chemical-Thermal Treatment of Boron and Carbon

A. E. Stetsko and Y. T. Stetsko

Abstract Complex method of chemical treatment and diffusive boride-carbon-plating is offered for the strengthening details of machines, diffusive layers restored method the given are investigated.

24.1 Introduction

The reason for breakdowns of 80% of machines and mechanisms—breakage of tribological parts. Therefore, parts with special tribo-surfaces with a significant resource must be manufactured. Particularly good for the growth of the resource is the use of boron.

In this work [1] a diffusional boron heat treatment was carried out at 950 °C by 6 h on the AISI M2 steel sample. The boride diffusional layer achieved 50 μm with an increment of 7 times in the microhardness regarding untreated sample and 4 times regarding the adjacent zone to the boride layer. In addition, a reduction of the friction coefficient and better wear response was achieved as a result of boride heat treatment.

The study compares the formation and characterization of various boron layers or boride coatings [2]. The technological aspects of boring processes are analyzed, advantages and disadvantages of many methods of boring, their influence on the microstructure of boron materials are presented. The mechanism of formation of active atoms or boron ions and phenomena during the melting of the doping material together with the substrate is described. Particular attention is paid to powder boring, electrochemical boring in a storm, gas treatment, plasma or paste boring, as well as laser or plasma surface boron doping, which the author recognized as the most important modern methods of struggle.

A. E. Stetsko (✉) · Y. T. Stetsko
Faculty of Computer Engineering Printing, Ukrainian Academy of Printing, UAP, Lviv, Ukraine
e-mail: andrew73@ukr.net

Y. T. Stetsko
e-mail: yaryna75@ukr.net

A. D. Pogrebnjak and O. Bondar (eds.), *Microstructure and Properties of Micro- and Nanoscale Materials, Films, and Coatings (NAP 2019)*,
Springer Proceedings in Physics 240, https://doi.org/10.1007/978-981-15-1742-6_24

Research [3] of the Cr-Rare earth-boronized layers were fabricated on the steel 45 at 650 °C for 6 h., showed the Cr-Rare earth-boronized layer was composed of single Fe_2B phase. A sawtooth morphology was obtained in the Cr-Rare earth-boronized layer and the microstructure of the Cr-Rare earth-boronized layer was compact and dense. The thickness of the boride layer is about 23 μm. The boride tooth was thin and straight. The microhardness of the Cr-Rare earth-boronized layer was 1200 ~ 1700 HV.

The influence of chemical elements on the formation of borides has been investigated [4]. γ'-Strengthened Co-based single crystal superalloy was jointed with B-Ni76CrWB alloy powder by transient liquid phase bonding. The growth mechanism of the borides of preferred orientation formed in the diffusion effect zone of the joint was investigated. Results indicated that the prepared borides were $(Co, W, Ta)_3B_2$ and were long plate-shaped in three dimensions.

The author [5] gives an account of the central aspects of diffusion in solids. Diffusion-controlled phenomena, including ionic conduction, grain-boundary, and dislocation pipe diffusion are considered.

24.2 Materials and Methods

Diffusion carbide coating is an extremely effective method of protecting metals and alloys. Promising is the use of Complex diffusive coatings when the positive properties are combined in one-component coatings.

A problem of worked surfaces repair of machine parts, manufactured of carbon steels by methods that are technologically simple, using universal equipment and existing qualifications of employers, studying the types of recovered layers to establish their application specifically to the working conditions of machine parts.

The purpose of this work is the formation of surface diffusion layers on the parts of machines made by a Complex method of chemical processing (sedimentation) and diffusive boron and carbon.

A complex method developed for repairing machine parts, an offer of chemical processing and diffusion chemical elements is proposed [6–8]. It consists in depositing on the prepared surface of a Ni–Co–P chemical coating in the aqueous solution of certain formulations and diffusion boron. As a result of strengthening, a diffusion hardened layer on the surface of the parts is formed. Its structure, depending on the applied modes of the method, consists of several zones, which working zones are outer composite zones, which reaches 150–500 μ. The large thickness of the strengthened layer allows it to be updated several times, which is very cost-effective.

24.3 Discussion

The diffusion of elements in the surface layers of machine parts forms solidified layers of small thickness and high hardness (Figs. 24.1, 24.2). Also, these diffusion layers are very fragile. This prevents the strengthening of parts that work under dynamic conditions, reduces their resources. The use of steel with a small amount of carbon (0.2% C) does not guarantee the solidification of the hardened layer during loading due to the low hardness of the metal base.

Application for the measurement of machine parts The complex method forms on the surface the reinforced composite layers of a large thickness up to 500 μ and integral microhardness 12000 MPa, which withstand significant dynamic loads (Fig. 24.3).

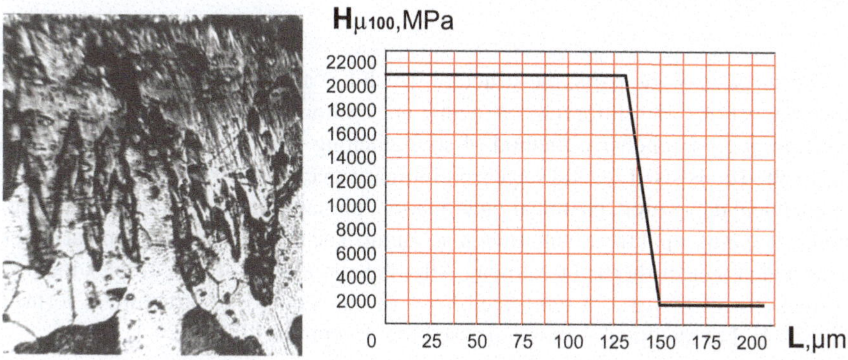

Fig. 24.1 Microstructure and microhardness of the hardening layer of steel (0.2% C) after boriding. Magnification ×600

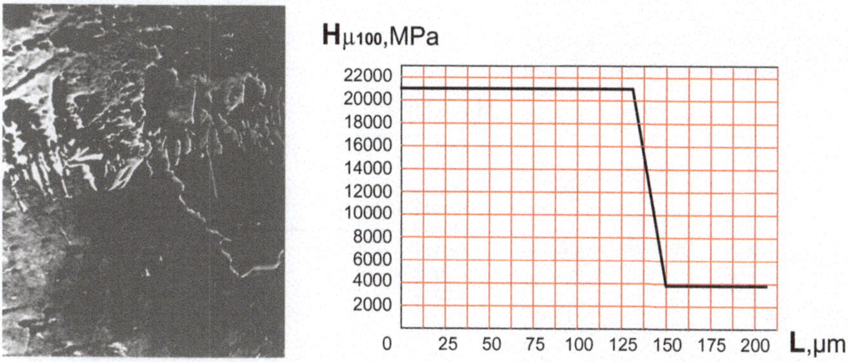

Fig. 24.2 Microstructure and microhardness of the hardening layer after Boriding of steel AISI 52100. Magnification ×600

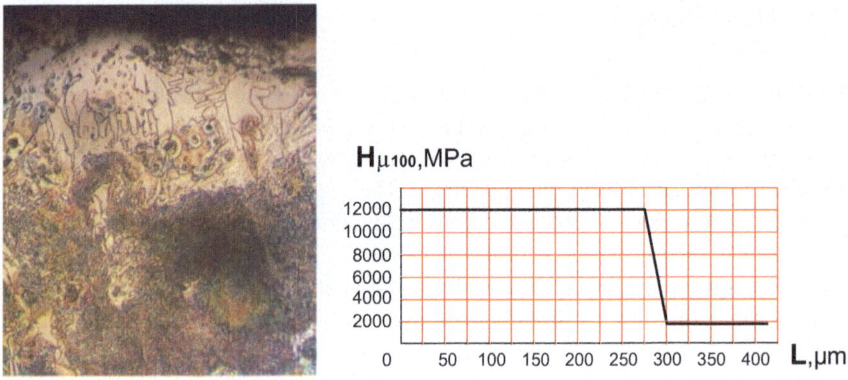

Fig. 24.3 Microstructure and microhardness of the hardening layer of the Complex method by boriding. Magnification ×450

The surfaces of the parts, the structure which has a composite structure with the new connections of reinforcing elements, are optimal. For the formation of such structure, a comprehensive method of strengthening the surfaces of parts is used. The method consists of several stages: preliminary mechanical treatment, preparation for chemical deposition, chemical precipitation, preparation for chemical treatment, chemical and thermal treatment, final machining. The main ones are chemical deposition and chemical-thermal treatment. After the application of the complex method on the surfaces of parts, a reinforced layer of nanocomposite structure is formed (Fig. 24.4). For chemical depositions used few receptions (Tables 24.1, 24.2).

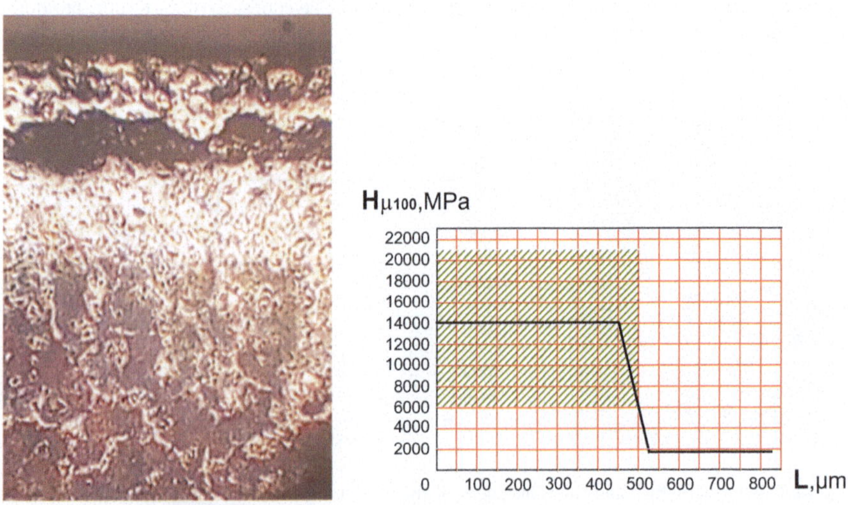

Fig. 24.4 Microstructure and microhardness of the hardening layer of the Complex method by chemical sedimentation diffusion boron and carbon. Magnification ×450

Table 24.1 Receptions of chemical deposition

#	Recipes					
	Chemical substance	#1	#2	#3	#4	#5
1	$CoCl_2$, (g/l)	–	15	15	30	–
2	$NiCl_2$, (g/l)	–	30	30	30	–
3	$Na_3C_6H_5O_7$, (g/l)	84	100	100	80	–
4	NaH_2PO_2, (g/l)	30	20	60	10	25
5	NH_4Cl, (g/l)	–	50	50	50	50
6	$CoCO_3$, (g/l)	7	–	–	–	–
7	$CoSO_4$, (g/l)	–	–	–	–	30
8	$NiSO_4$, (g/l)	15	–	–	–	30
9	CH_3COONa, (g/l)	–	–	–	–	100
10	H_2SO_4, (g/l)	15	–	–	–	–
11	NH_4OH, (ml)	90	60	60	60	50

Table 24.2 Receptions of chemical deposition

#	Recipes					
	Chemical substance (g/l)	#6	#7	#8	#9	#10
1	$CoCl_2$	15	30	–	–	–
2	$NiCl_2$	30	30	–	–	–
3	$Na_3C_6H_5O_7$	90	–	–	–	–
4	NaH_2PO_2	25	25	25	30	–
5	$K_2H_2PO_2$	–	–	–	–	30
6	NH_4Cl	–	50	50	45	–
7	$CoCO_3$	–	–	–	15	–
8	$NiCO_3$	–	–	–	30	–
9	$CoSO_4$	–	–	20	–	–
10	$NiSO_4$	–	–	30	–	–
11	$Co(NO_3)_2$	–	–	–	–	18
12	$Ni(NO_3)_2$	–	–	–	–	30
13	CH_3COONa	–	90	–	90	–
14	$(CHCOONa)_2$	–	–	90	–	–
15	$C_4H_6O_6$	15	–	–	–	–
16	CH_2NH_2COOH	25	–	–	–	–
17	$Na_2C_4H_4O_4$	–	–	–	–	100

Chemical-thermal treatment with boron and carbon in the complex method provides strengthened layers (integral microhardness of 14000 MPa) of a composite nanostructure and a large thickness (up to 500 μ). Due to the composite structure, the resource life of the components strengthened by this method significantly increases. The considerable thickness of the strengthened layer allows repeated use of parts after wear due to their mechanical processing to repair sizes.

Strengthened diffusion layer is divided into 6 zones (Fig. 24.5): zone 1 solid boride layer, thickness up to 10 μ; zone 2 composite zone of predominantly boride grains FeB in solid B solution in α-Fe; zone 3 intermediate; zone 4 composite zone of predominantly boron beans Fe_2B in solid solution B in α-Fe; zone 5 transitions

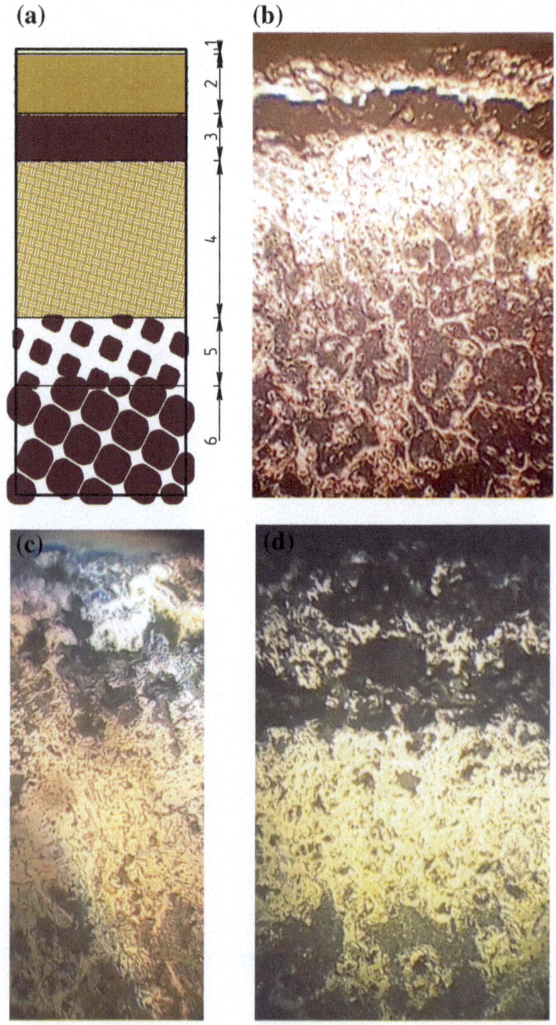

Fig. 24.5 The microstructure of the hardening layers of Complex method by chemical sedimentation diffusion boron and carbon: **a** Schematic structure of the strengthened layer; **b** Recipe chemical sedimentation #8; **c** Recipe chemical sedimentation #9; **d** Recipe chemical sedimentation #10. Magnification ×450

with boride grains in solid B solution in α-Fe and ferrite and perlite grains; zone 6 is the base metal.

Various formulations of chemical deposition form reinforced layers of different structures (Fig. 24.5). The morphology of zones 2 and 3 are mainly changed.

The morphology of zones 4 and 5, which have the largest thickness, has interesting boride grains in solid B solution in α-Fe. They have different orientations and different geometric shapes and sizes: globular type, elongated, fused together (Fig. 24.6).

The use of recipe #6 chemical sedimentation for the Complex method allows obtaining a strengthened layer, which does not have a zone 2 (Fig. 24.7). This structure is very good for the finishing machining and spinning of the conjugated parts.

Applying the recipe #7 of the chemical precipitation of the Complex method, a reinforced layer with small intermittent zones 1, 2 and 3 was obtained (Fig. 24.8). The

Fig. 24.6 The microstructure of the hardening layer of the Complex method by chemical sedimentation diffusion boron and carbon: the boundary of zone 4 and zone 5, recipe #3. Magnification ×600

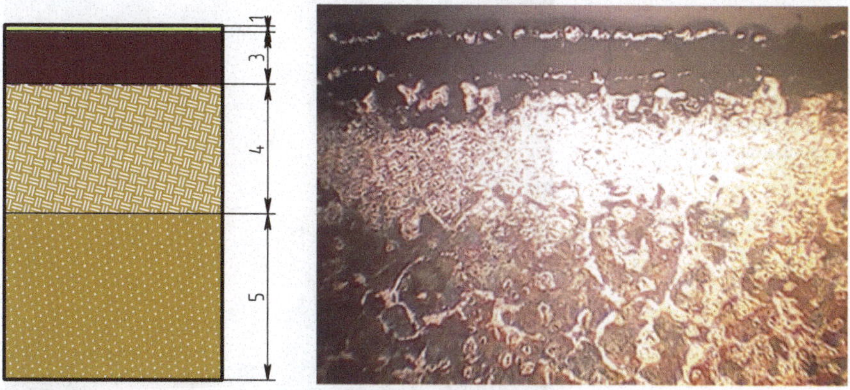

Fig. 24.7 Schematic structure of the strengthened layer microstructure of the hardening layer of the Complex method by chemical sedimentation diffusion boron and carbon: recipe chemical sedimentation #6. Magnification ×450

Fig. 24.8 Schematic structure of the strengthened layer microstructure of the hardening layer of the Complex method by chemical sedimentation diffusion boron and carbon: recipe chemical sedimentation #7. Magnification ×450

thickness of the zone 4 is very large (at least 200 μ) and has an integral microhardness of 14000 MPa. Such a strengthened layer is also beneficial for finishing machining. The large thickness of the strengthened layer can provide multiple repairs by method of Repair Size.

Applying the recipe #3 of the chemical precipitation of the Complex method is formed composite zone 2 of predominantly boron beans FeB in solid solution B in α-Fe pass to composite zone 4 of predominantly boron beans Fe_2B in solid solution B in α-Fe (Fig. 24.9). The intermediate zone 3 is non-uniform and of a small thickness (up to 30 μ). Such a structure of the strengthened layer is good for the operation of machine parts.

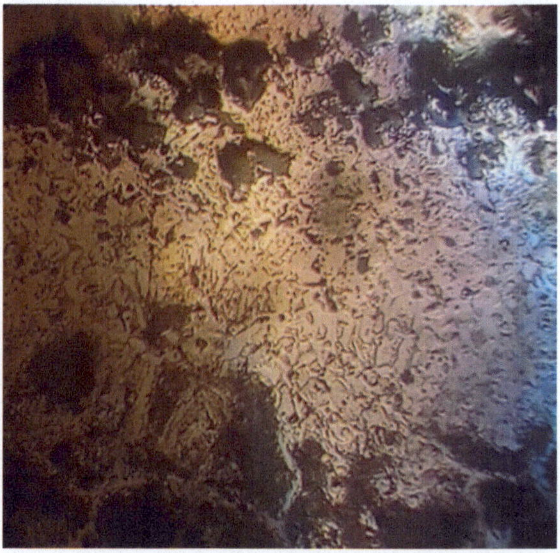

Fig. 24.9 The microstructure of the hardening layer of Complex method by chemical sedimentation diffusion boron and carbon: zone 4, recipe chemical sedimentation #3. Magnification ×450

During the carbon-boronation complex and intensive processes of diffusion, saturation occurs. It shows the structure of boride grains of complex configuration (Fig. 24.10). These grains, which were formed in a different direction, their sizes and thickness, testify to complex diffusion processes of borage. It is precisely the preliminary chemical deposition that allows for such a composite structure of the strengthened layer and its significant thickness of at least 500 μ.

The application of chemical deposition leads to the intensification of diffusion processes. Borides in the composite layer are both globular and elongate, multi-directional (Fig. 24.11). Also formed are packages of acute borides in the middle of the composite zone, although such a structure is characteristic for the surface of

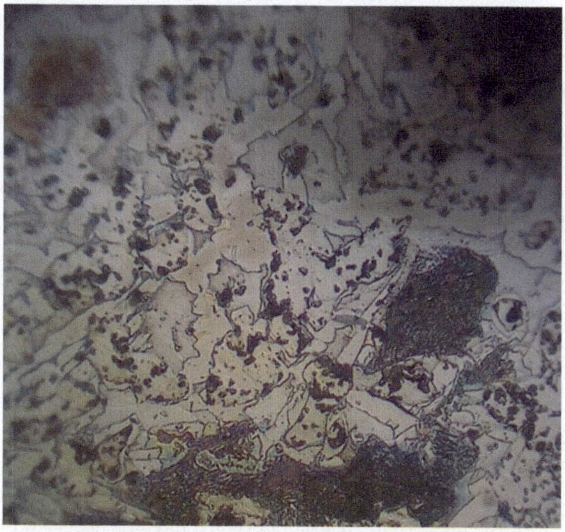

Fig. 24.10 The microstructure of the hardening layer: a boundary of zone 4 and zone 5, recipe chemical sedimentation #4. Magnification ×600

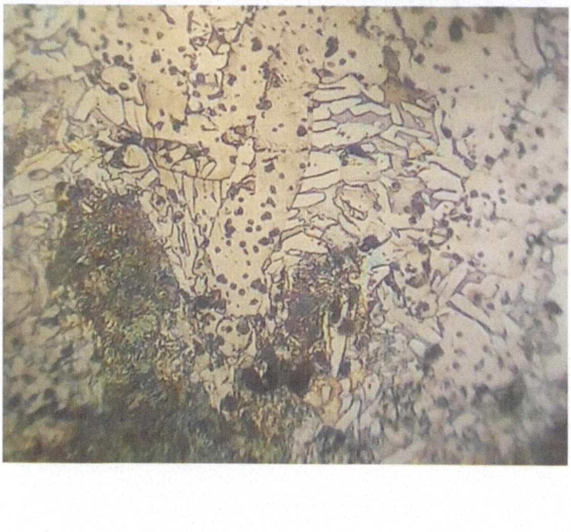

Fig. 24.11 The microstructure of the hardening layer of Complex method by chemical sedimentation diffusion boron and carbon: the boundary of zone 4 and zone 5, recipe chemical sedimentation #7. Magnification ×600

reinforcing parts in the usual boring. Spliced grains of borides are also available, which increases the integral microhardness of the layer and allows you to strengthen the parts that work by a heavy load.

24.4 Conclusions

The Complex method chemical sedimentation and diffusive boron and carbon for manufacturing or repair parts of machine forms on their surfaces reinforced layers of a large thickness up to 500 μ.

Surface hardened layers are divided into several zones, the outer ones of which have a composite structure (solid micro and nanoparticles of borides in a soft matrix) with an integral microhardness of 14000 MPa. Borides in the composite layer are both globular and elongate, multi-directional.

Adjusting the modes of the Complex method the morphology of the composite layer with borides of different structures, sizes, and density of placement is obtained. The specific structure and characteristics of the reinforced layer are used for specific parts of machines and operating conditions.

The large thickness of the strengthened composite layer allows them to be restored several times to repair sizes, which is very cost-effective.

References

1. L. Gutierrez-Noda, C.A. Cuao-Moreu, O. Perez-Acosta, E. Lorenzo-Bonet, P. Zambrano-Robledo, M.A.L. Hernandez-Rodriguez, Wear **426–427**(Part B), 1667 (2019)
2. M. Kulka, *Current Trends of Boriding, Engineering Materials* (Springer, Switzerland, 2019)
3. X. Yuan, B. Xu, Y. Caib, Phys. Procedia **50**, 82 (2013)
4. S.Y. Wang, Y. Sun, X.Y. Hou, C.Y. Cui, X.F. Sun, Y.Z. Zhou, Mater. Lett. **246**, 190 (2019)
5. H. Mahrer, in *Springer Series in Solid-State Sciences*, vol. 155 (Springer-Verlag, Berlin, Heidelberg, 2007), p. 654
6. A.E. Stetsko, in *Nanophysics, Nanomaterials, Interface Studies, and Applications*, ed. by O. Fesenko, L. Yatsenko. 4th International Conference Nanotechnology and Nanomaterials, Lviv, August 2016. Springer Proceedings in Physics, vol. 195 (Springer, Cham, Switzerland, 2017)
7. Stetsko, in *Nanochemistry, Biotechnology, Nanomaterials, and Their Applications*, ed. by O. Fesenko, L. Yatsenko. 5th International Conference Nanotechnology and Nanomaterials, Chernivtsi, August 2017. Springer Proceedings in Physics, vol. 214 (Springer, Cham, 2018), pp. 345–356
8. A.E. Stetsko, Y.T. Stetsko, in *Advances in Thin Films, Nanostructured Materials, and Coatings*, ed. by A. Pogrebnjak, V. Novosad. 8th IEEE International Conference on "Nanomaterials: Applications & Properties", Odessa, September 2018. Lecture Notes in Mechanical Engineering (Springer, Singapore, 2019), pp. 179–188

Chapter 25
The Effect of the Transition from Layered to Granular Structures on the Magnetoresistive Properties in the Systems Based on Fe_xNi_{100-x} and Cu

I. M. Pazukha, Yu. O. Shkurdoda, V. V. Shchotkin, S. R. Dolgov-Gordiichuk, A. M. Chornous and S. I. Protsenko

Abstract The conducted electron diffraction and electron microscopy studies have shown that the phase state of as-deposited and annealed at 700 K three-layer film systems based on ferromagnetic alloy Fe_xNi_{100-x} (F-layer) and Cu (N-layer) ($x > 60$ wt%, $d_F = 10$–40 nm, $d_N = 2$–20 nm) corresponds to BCC-(Fe–Ni) + FCC-Cu. The study of diffusion processes by the SIMS method showed that the individuality of the layers after annealing at a temperature of 700 K is retained only for systems with relatively thick layers ($d_F = 30$–40 nm, $d_N = 5$–20 nm). After thermostabilization of film systems with $d_F = 10$–20 nm, $d_N = 5$–20 nm at 700 K, a granular state is observed. It is shown that spin-dependent scattering of electrons is carried out in as-deposited samples. Annealing at temperatures of 400–500 K leads to a significant decrease in the isotropic magnetoresistance magnitude. Further increase the annealing temperature up to 700 K causes the growth of the isotropic magnetoresistance due to the formation of a granular state.

25.1 Introduction

The discovery of the giant magnetoresistance (GMR) effect gave an extremely powerful impulse to the study of layered and granular magnetic film systems [1–3]. Physical properties of nanostructures are investigated especially active on the basis of ferromagnetic Fe_xNi_{100-x} and copper alloy [4–8]. The interest in such film systems is primarily due to their widespread use as sensitive elements of magnetic field sensors, digital devices for magnetoresistive memory, automotive electronics, biomedical technologies, and others. By this time, a large number of studies have been conducted on obtaining and studying the magnetoresistive properties of multilayer

I. M. Pazukha (✉) · Yu. O. Shkurdoda · V. V. Shchotkin · S. R. Dolgov-Gordiichuk · A. M. Chornous · S. I. Protsenko
Department of Physics, Sumy State University, Rimsky-Korsakov 2, Sumy 40007, Ukraine
e-mail: iryna.pazukha@gmail.com

A. D. Pogrebnjak and O. Bondar (eds.), *Microstructure and Properties of Micro- and Nanoscale Materials, Films, and Coatings (NAP 2019)*, Springer Proceedings in Physics 240, https://doi.org/10.1007/978-981-15-1742-6_25

films based on Fe_xNi_{100-x} [9–11]. But in the vast majority of works the structures are studied on the basis of ferromagnetic alloys $Fe_{20}Ni_{80}$ and $Fe_{50}Ni_{50}$ (permalloy). Practically there are no works where physical properties of multilayered structures based on Fe_xNi_{100-x} in a wide range of concentrations, obtained under the same conditions, would be conducted.

The purpose of this work was to study the influence of heat treatment conditions on the structure-phase state, diffusion processes and magnetoresistive properties of film systems based on Fe_xNi_{100-x} ($x > 60$ wt%) and Cu in a wide range of layer thicknesses.

25.2 Method and Technique of Experiment

Multilayer film systems with layer thicknesses 1–50 nm were obtained in a vacuum chamber at pressures of gases in the residual atmosphere of 10^{-4} Pa. Alternate condensation of films was carried out as a result of evaporation of metals from independent sources (Cu—from a tungsten ribbon, Fe_xNi_{100-x}—an electron beam gun). The starting materials for the Fe_xNi_{100-x} layers were massive alloys of the respective composition.

The film condensation was conducted at room temperature of the substrate at a rate $\omega = 0.5$–1 nm/s, depending on the mode of work of the evaporators. To study the magnetoresistive properties, glass plates with pre-applied contact surfaces were used as a substrate. The geometric dimensions of the films for measuring their electrical resistance were given by windows in mechanical masks with nichrome foil.

The results of the study of the chemical composition of the original alloy and the films obtained by X-ray spectral microanalysis indicate that they coincide within the measurement error, which does not exceed 2%. Layered analysis of the three-layer films was conducted by the SIMS method.

Measurement of the longitudinal and transverse magnetic resistivity (magnetic field in the plane of the film) and the thermomagnetic processing of the films were conducted in a special installation, described in [7], in conditions of ultrahigh oil-free vacuum 10^{-6}–10^{-7} Pa in a magnetic field with induction to $B = 200$ mT. The films were annealed according to the scheme "Heating—Exposure at temperatures 400, 500, 600 and 700 K for 15 min—Cooling".

The magnitude of the longitudinal and transverse magnetic resistances (MR) of the film samples was calculated by the formula $(R(B) - R(B_c))/R(B_c)$, where $R(B)$ is the resistance of the sample in a magnetic field with induction B; $R(B_c)$ is the resistance of the sample in the field of the coercive force B_c.

25.3 Results and Discussion

It is well known [12, 13] that the conditions of heat treatment of film systems affect their phase composition, crystalline structure, which in turn determine the magnetic and magnetoresistive properties. Consider the features of the structural and phase state of as-deposited and thermostabilized 700 K three-layer systems based on the Fe_xNi_{100-x} and Cu alloy.

According to the results of electron diffraction and electron microscopic studies of three-layer films Fe_xNi_{100-x}/Cu/Fe_xNi_{100-x}, all as-deposited films were polycrystalline with grain sizes less than 10 nm (Fig. 25.1a). According to results of electron diffraction research at $c_{Ni} < 40$ wt%, phase composition of as-deposited three-layer samples corresponds to BCC-(Fe–Ni) + FCC-Cu (Fig. 25.1b, Table 25.1). The parameter of the bcc lattice α-phase (Ni–Fe) is $a = 0.285$–0.287 nm. Annealing at a temperature of 700 K does not lead to a change in the phase composition. The lattice parameter also does not significantly change (Fig. 25.1c, d).

Necessary condition for the effect of the GMR in three-layer magnetic films is the presence of a structurally-continuous non-magnetic layer. Therefore, a layer analysis of three-layer films Fe_xNi_{100-x}/Cu/Fe_xNi_{100-x} with different thicknesses of layers and concentration of components in magnetic layers by secondary-ion mass spectrometry was conducted.

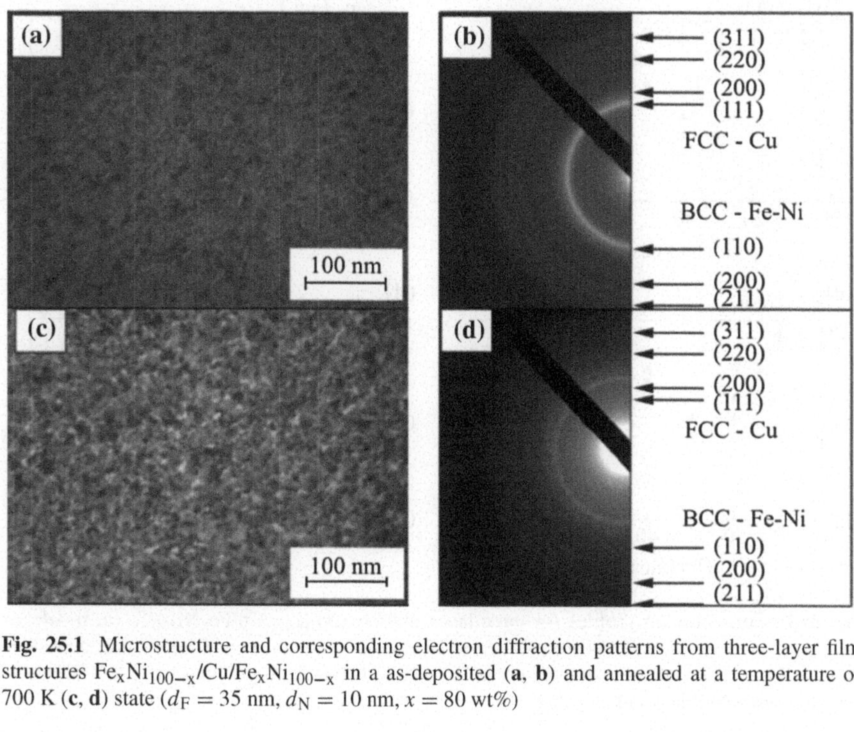

Fig. 25.1 Microstructure and corresponding electron diffraction patterns from three-layer film structures Fe_xNi_{100-x}/Cu/Fe_xNi_{100-x} in a as-deposited (**a**, **b**) and annealed at a temperature of 700 K (**c**, **d**) state ($d_F = 35$ nm, $d_N = 10$ nm, $x = 80$ wt%)

Table 25.1 Analysis of the electron diffraction patterns from as-deposited and annealed in an ultrahigh vacuum (~10^{-6} Pa) at a temperature of 700 K film Fe_xNi_{100-x}/Cu/Fe_xNi_{100-x} (x ≅ 80, d_F = 35 nm, d_N = 10 nm)

As-deposited				Annealed to 700 K			
I, a.u.	d, Å	hkl	Phase	I, a.u.	d, Å	hkl	Phase
Very strong	2.070	110, 111	FCC-Cu, BCC-Fe–Ni	Strong	2.080	110 111	FCC-Cu, BCC-Fe–Ni
Weak	1.780	200	FCC-Cu	Medium	1.781	200	FCC-Cu
Medium	1.440	200	BCC-Fe–Ni	Weak	1.150	200	BCC-Fe–Ni
Medium	1.270	220	FCC-Cu	Weak	1.280	220	FCC-Cu
Medium	1.170	211	BCC-Fe–Ni	Weak	1.170	211	BCC-Fe–Ni
Medium	1.070	311	FCC-Cu	Weak	1.070	311	FCC-Cu

The results of the study showed that in the as-deposited samples, which were obtained at room temperature, a relatively wide area of mutual diffusion was observed (Fig. 25.2a, c). Although it should be noted that at such concentrations the component in the magnetic layers the individuality of the as-deposited layers is preserved in the thicknesses greater than 4 nm due to the low solubility of the components. In general,

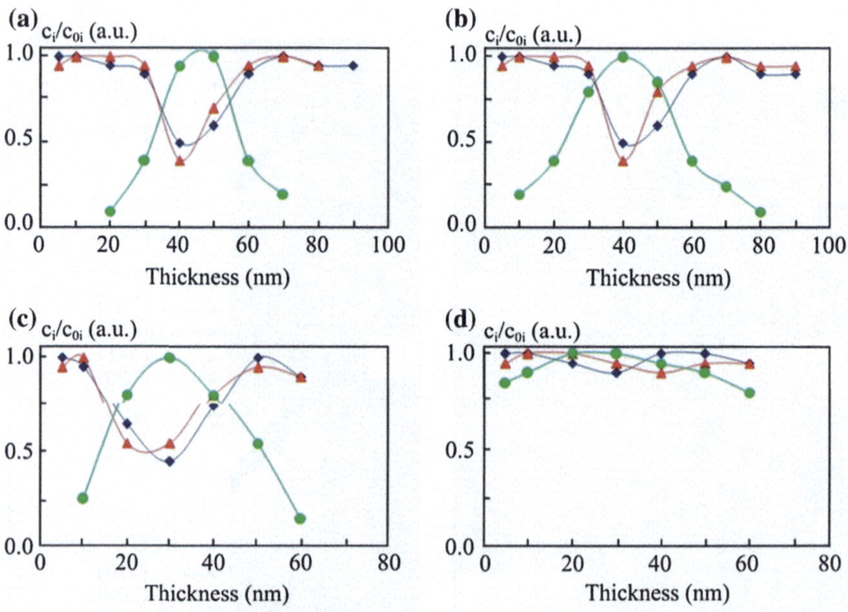

Fig. 25.2 Concentration profiles for three-layer films Fe_xNi_{100-x}/Cu/Fe_xNi_{100-x} (**a**, **b**—d_F = 35 nm, d_N = 20 nm, x = 80 wt%, **c**, **d**—d_F = 20 nm, d_N = 20 nm, x = 80 wt%) in as-deposited (**a**, **c**) and annealed state at a temperature of 700 K (**b**, **d**) (black up-pointing triangle—Fe, black diamond suit—Ni, black circle—Cu)

it is possible to specify the following reasons for the existence of a diffusion zone in as-deposited films. Firstly, it is the presence of gradient concentrations of components close to the boundaries of the layers. The confirmation of the implementation of such a mechanism is the diffusion of atoms in the film system both in the direction of the lower and upper layers. Secondly, as-deposited films have a higher surface roughness compared to thermostabilized ones. In third place, the as-deposited samples are finely divided and have a higher concentration of defects in the crystalline structure. The fourth reason is caused by condensation-stimulated diffusion.

Annealing samples at a temperature of 700 K causes recrystallization and mutual grain-bound diffusion, which is the main mechanism of mass transfer. Although it can be argued that in this system, with sufficiently thick layers ($d_N > 10$ nm and $d_F > 30$ nm), the individuality of the layers is maintained even after heat treatment at a temperature of 700 K (Fig. 25.2b). The boundaries of the crystallites become diffusely saturated already when the upper layer is condensed, so the effect of annealing on diffusion processes in such film systems is insignificant.

Consider the dependence of a negative isotropic magnetoresistance on the annealing temperature for three-layered films $Fe_xNi_{100-x}/Cu/Fe_xNi_{100-x}/S$ (Fig. 25.3). At concentration of Ni in magnetic layers up to 40 wt% and with relatively large thicknesses of the magnetic layers ($d_F = 30$–35 nm), dependencies are similar and are not monotonous in nature, and the maximum value of the GMR is observed after annealing at a temperature of 500 K. The dependence $\Delta R_{max}/R_0 = f(T_{ann})$ for films with relatively thin magnetic ($d_F = 15$–20 nm) and nonmagnetic layers ($d_N = 10$–15 nm) at $c_{Ni} = 10$–20 wt% in magnetic layers (Fig. 25.3b) are also not monotonic in nature—at low temperature annealing, the isotropic magnetoresistance decreases, and high temperature annealing leads to its higher increase. This dependence behaviour $(\Delta R/R_0)_{max} = f(T_{ann})$ can be explained as follows. Low temperature annealing ($T_{ann} = 400$–500 K) is accompanied by a mutual diffusion of atoms and a violation of the structural continuity of both magnetic layers and a nonmagnetic layer (simple diffusion mixing of

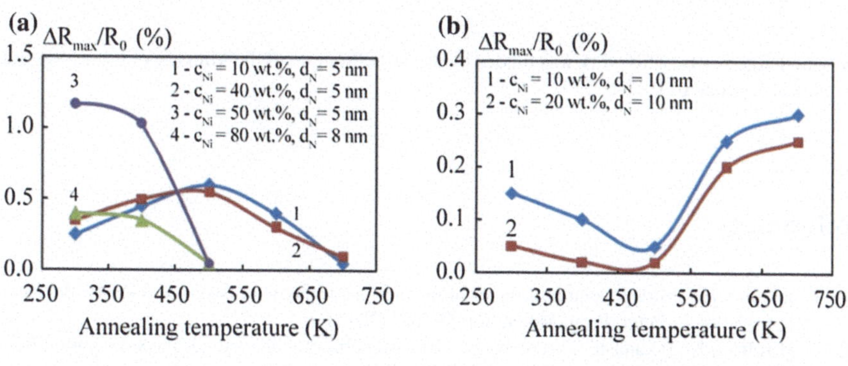

Fig. 25.3 The dependence of the isotropic magnetoresistance on the annealing temperature for three-layer films $Fe_xNi_{100-x}/Cu/Fe_xNi_{100-x}/S$ with different Ni concentrations in magnetic layers and different thicknesses of the layers (**a**—$d_F = 35$ nm; **b**—$d_N = 10$ nm)

magnetic and nonmagnetic components). A further increase in the annealing temperature (600, 700 K) can lead to the formation of a granular state and, as a consequence, the strengthening of the role of spin-dependent scattering of electrons [14, 15]. It should be noted that the results of electron microscopic and electron diffraction studies confirm the formation of granular state.

25.4 Conclusions

1. For as-deposited and annealed at a temperature of 700 K films based on the Fe_xNi_{100-x} ($x > 60$ wt%) phase composition correspond to BCC-(Fe–Ni) + FCC-Cu. The parameter BCC of the lattice α-phase (Ni–Fe) is $a = 0.285$–0.287 nm. At $d_N > 10$ nm and $d_F > 30$ nm ($x > 60$ wt%), the individuality of the layers after heat treatment at a temperature of 700 K is preserved. In the case of relatively thin magnetic ($d_F = 10$–20 nm) and non-magnetic layers ($d_N = 10$–15 nm) in the process of heat treatment there is a violation of their structural continuity and the formation of granular alloy.
2. For as-deposited and thermostabilized samples at 700 K, isotropic field dependencies of MR are observed. The maximum value of the isotropic magnetic resistivity at room temperature is 0.6% for 550 K films $Fe_{80}Ni_{20}$/Cu/$Fe_{80}Ni_{20}$/S with $d_F = 30$ nm and $d_N = 6$ nm; for samples with $d_F = 15$ nm and $d_N = 10$ nm maximum value of MR is 0.5% after heat treatment at 700 K. The relatively small values of isotropic MR are due to the low concentration of copper in the formed granular alloys.
3. The obtained experimental results indicate the possibility of using annealing for the thermostabilization of the characteristics of the magnetoresistive elements on the basis of three-layer films Fe_xNi_{100-x}/Cu/Fe_xNi_{100-x}/S ($c_{Ni} < 40$ wt%) in vacuum at a temperature of 500 K directly after their condensation.

Acknowledgements This work was funded by the State Program of the Ministry of Education and Science of Ukraine 0119U100777.

References

1. I. Bakonyi, Electrodeposited multilayer films with giant magnetoresistance (GMR): Progress and problems, L. Peter, Prog. Mater. Sci. **55**, 107 (2010)
2. A. Hirohata, K. Takanashi, Future perspectives for spintronic devices, J. Phys. D Appl. Phys. **47**, 193001 (2014)
3. K.G. Patrin, V.Y. Yakovchuk, G.S. Patrin, S.A. Yarikov, Magnetism and Magnetoresistance in NiFe/Bi/NiFe Films , Solid State Phenom. **190**, 439 (2012)
4. C. Wang, X. Xiao, H. Hu, Y. Rong, T.Y. Hsu, Nanoparticle morphology in FeNi-Cu granular films with giant magnetoresistance, Phys. B **392**, 72 (2007)

5. Y.-T. Chen, J.-Y. Tseng, S.H. Lin, T.S. Sheu, Effect of grain size on optical and electrical properties of Ni80Fe20thin films, J. Magn. Magn. Mater. **360**, 87 (2014)
6. M.A. Corrêa, R. Dutra, T.L. Marcondes, T.J.A. Mori, F. Bohn, R.L. Sommer, Exploring the magnetization dynamics of NiFe/Pt multilayers in flexible substrates, Mater. Sci. Eng., B **211**, 115 (2016)
7. Y.O. Shkurdoda, I.M. Pazukha, A.M. Chornous, Effect of thermomagnetic treatmenton magnetoresistive properties of trilayer thin films based on FexNi100-x and Cu, Intermetallics **93**, 1 (2018)
8. A.M. Chornous, Y.O. Shkurdoda, V.B. Loboda, Y.M. Shabelnyk, V.O. Kravchenko, Influence of the surface morphology on the magnetoresistance of ultrathin films of ferromagnetic metals and their alloys, Eur. Phys. J. Plus. **132**, 58 (2017)
9. Z. Zhou, Y. Zhou, Y. Cao, The investigation of giant magnetoimpedance effectin meander NiFe/Cu/NiFe film, J. Magn. Magn. Mater. **320**, e967 (2008)
10. Y.O. Shkurdoda, I.M. Pazukha, A.M. Chornous, Peculiarity of magnetoresistance of discontinuous ferromagnetic thin films, Int. J. Miner. Metall. Mater. **24**, 1459 (2017)
11. T. Shinjo, *Nanomagnetism and Spintronics*, 2nd edn. (Elsevier Science and Technology, Japan, 2013)
12. L.A. Chebotkevich, Y.D. Vorob'ev, I.N. Burkova, A.V. Kornilov, Structure and magnetic properties of annealed Co/Cu/Co films, Phys. Metal. Metallogr. **89**, 263 (2000)
13. A.G. Basov, S.I. Vorobiov, Y.O Shkurdoda, L.V. Dekhtyaruk, Structure and Effect of Giant Magnetoresistance in Three-Layered Films FeNi/Cu/Co, J. Nano- Electron. Phys. **2**, 69 (2010)
14. V.O. Vas'kovskii, A.A. Yuvchenko, V.N. Lepalovskii, N.N. Shchegoleva, A.V. Svalov, Elements of the Granular State in Multilayered Co/Cu Films, Phys. Metal. Metallogr. **93**, 232 (2002)
15. I.M. Pazukha, O.V. Pylypenko, L.V. Odnodvorets, A comprehensive investigation of electrophysical and magnetoresistive properties of thin films based on permalloy and silver, Mater. Res. Express. **5**, 106409 (2018)

Chapter 26
Enhancement of the Wear Resistance of Tungsten Cobalt Carbide Plates Using Ion Implantation and Al–Si–N Coatings

B. P. Gritsenko, D. S. Rechenko, E. A. Rogachev, K. V. Smyrnova, A. A. Bagdasaryan, V. P. Sergeev, A. Yu. Popov, G. Zh. Nogaibekova, M. V. Fedorischeva and Alexander D. Pogrebnjak

Abstract Producing of cutting tools with high sharpness of the cutting edge (characterized by a small conditional inscribed roundness radius) improves the quality of edge cutting processing of hard-to-cut parts. However, existing carbides do not withstand increased contact loads and quickly destroy. This leads to a deterioration of the quality of the processed surface and to failure of the tool. The paper investigates the influence of different types of processing on the wear of carbide plates. It is found that an increase in the grinding speed leads to a decrease in roughness and an increase in tool life. The efficient ways of the improvement of plate wear resistance are the processing of plates with rhenium and argon ions and the application of thin Al–Si–N coatings.

26.1 Introduction

Constantly increasing demands on the quality of processed details and hard-to-process materials lead to a need for cutting tools with the principally new characteristics. One of the ways to increase the processing quality is production of a sharp instrument with a small radius of the edge roundness (ρ smaller than 5 μm). The

B. P. Gritsenko · V. P. Sergeev · G. Zh. Nogaibekova · M. V. Fedorischeva
Institute of Strength Physics and Materials Science, Siberian Branch of the Russian Academy of Sciences, Tomsk, Russia
e-mail: gritsenko@ispms.tsc.ru

D. S. Rechenko · E. A. Rogachev · A. Yu. Popov
Omsk State Technical University, Omsk, Russia

B. P. Gritsenko · V. P. Sergeev · G. Zh. Nogaibekova
Tomsk Polytechnic University, Tomsk, Russia

K. V. Smyrnova · A. A. Bagdasaryan · A. D. Pogrebnjak (✉)
Sumy State University, Sumy, Ukraine
e-mail: alexp@i.ua

A. D. Pogrebnjak and O. Bondar (eds.), *Microstructure and Properties of Micro- and Nanoscale Materials, Films, and Coatings (NAP 2019)*,
Springer Proceedings in Physics 240, https://doi.org/10.1007/978-981-15-1742-6_26

problem of the sharp instrument is their quick failure at contact loads and temperature [1, 2].

A possible solution to this problem is an application of hard coatings [3–7]. However, traditional coatings, which are quite numerous now, appear to be unsuitable for this purpose. The application of coatings increases the edge radius, that leads to decrease in the accuracy and quality of the edge cutting. In addition, these coatings are quite fragile. At the edge radius (near 1 μm), the coating thickness should be comparable (traditional coatings have optimal thickness of about 5 ÷ 10 μm). Consequently, new coatings should withstand high contact loads and temperature at the very small thickness for providing the long service life of cutters.

It is well-known that ion implantation can modify the surface of carbide cutters to increase the wear resistance, hardness, improve tribological characteristics [3, 8–10]. The wear resistance improves at ion implantation due to alloying with implanted elements and changes in the properties of implanted materials owing to structure changes [11]. At ion implantation, radiation defects, whose number is two to three orders of magnitude larger than that of implanted elements, are formed, along with new structures, nanocrystal and amorphous states of the near-surface layers [12]. The decrease in the crystallite size increases the strength, corrosion resistance, and wears resistance [13]. It seems interesting to separate the effects associated with alloying and defect formation. This is especially important when two technologies are used, namely, coating and ion implantation into the coatings [14–17]. The latter can impart extra properties to the coatings. The effect of implanted ions depends not only on the ion type but also on the material, they are implanted in [18–20].

We have studied the influence of the atomic weight of implanted elements in series C, Si, Ar, Ti, Mo, and Re on the wear resistance of U8A steel [9]. The AISI W108 steel was chosen as the substrate. The hardening effects in it are mostly connected with formation of the defect structure and grain refining in near-surface layers of specimens. The study has shown that the wear resistance increases nearly linearly with an increase in the atomic weight.

The roughness of surfaces of a cutting wedge also significantly affects the quality of material processing. The influence of the mentioned factors on the service life of carbide tools in processing of hard-to-treat materials was studied in this paper. The objective of this study is to determine how the wear resistance of cutting tools depends on different kinds of processing.

26.2 Experimental Technique

The widely used fine-grain GC1105 carbide material with tungsten carbide size of 0.1 ÷ 1 μm was chosen as an object of study. The composition of the material is given in Table 26.1.

Table 26.1 Composition of *GC*1105 carbide material

Chemical element	Weight (%)
Carbon	5…6
Cobalt	10…11
Tungsten	83…85

The chemical composition of the coatings and the near-surface layers of carbide plates was determined by the method of X-ray fluorescence analysis on the spectrometer «BRUKER S4 Pioneer». The state of the cutting edge of carbide plates was examined with a ZEISS LEO EVO-50XVP scanning microscope.

Carbide plates were sharpened at the modernized VZ-326F4 tool and cutter grinding machine with the specially designed and produced St-140-40/7.5 s high-speed spindle, GMT-1 oil mist generator, SO-0.5 cooling system, and PFVMK-16.2 system for compressed air cleaning. The sharpening was conducted with a cutting speed of 350÷400 m/s with using of the designed grinding wheels having the following characteristic: synthetic diamond micron powder (SDMP) 28/20 (25%), glass microspheres (50%), phenol-formaldehyde resin (SFP) (25). The height profile of the surface pattern of processed replaceable carbide plates was measured with the MarSurf PS1 profilometer. The edge sharpness was analyzed with MikroCad Premium (GF Messtechnik GmbH).

Coatings were deposited with the UVN-05MI «KVANT» vacuum setup (Techimplant ltd) in the argon-nitrogen gas mixture at a total pressure of 0.3 Pa. The partial pressure ratio was Ar: N = 3: 1. The bipolar pulsed magnetron sputtering with a composite target based on aluminum with the addition of 20 at.% Silicon was used. The magnetron discharge power was 1.2 kW, the current pulse repetition frequency was 50 kHz. Substrates were heated and the constant temperature of substrates in the coating process was maintained with a 10-kW resistive molybdenum heater.

Implantation was carried out at the same setup with the DIANA-3 vacuum-arc ion source (Techimplant ltd). The fluence for rhenium ions was $\Phi = 3 \times 10^{17}$ ion/cm^2, the accelerating voltage was U = 80 kV. The ion energy was determined by the accelerating voltage and the mean charge state of ions in the arc of the ion source [21]. The processing by argon ions was conducted with the "DIONIS" gas ion source (Techimplant ltd). The fluence of argon ions was $\Phi = 1 \times 10^{18}$ ion/cm^2, the accelerating voltage was 10 kV.

The carbide plates were tested for wear resistance at the turning of VT3-1 titanium alloy at the Okuma ES-L8II turning center, the BlasoCut 2000 lubricating-cooling fluid was applied. The elemental composition of the alloy is given in Table 26.2.

Table 26.2 Elemental composition of VT3-1 alloy

Material grade	Chemical element, weight %				
	Al	Mo	Cr	Fe	Si
VT3-1	6.2	2.7	2.1	0.7	0.3

The processing modes at the turning were the following: cutting speed V = 150 m/min, longitudinal feed F = 0.1 mm/rev and cutting depth t = 0.25 mm. These cutting modes correspond to finish machining. The tests were conducted until the wear at the rear surface reached 0.1 mm.

The structure and the phase composition of a coating were investigated using the DRON-7 X-ray diffractometer (Burevestnik, Russia) and the JEOL-2100 transmission electron microscope (Jeol Ltd., Japan). X-ray investigation was carried out under continuous 2θ—scanning with the Bragg-Brentano focusing at Co Kα radiation. The database JCPDS PDF-2 was used for interpretation of the diffractograms. The foils for TEM studies were prepared by the "cross-section" of a coating using the Ion slicer EM-09100IS (Jeol Ltd., Japan).

26.3 Results and Discussion

Table 26.3 demonstrates the influence of the grinding speed, ion implantation, and coating on the surface roughness and tool life at the turning.

Table 26.3 Parameters of surface roughness of front faces and life of studied carbide plates

	Processing	Mean roughness (Ra), μm	Root-mean-square roughness (Rq)*, μm	Tool life (t), min
1	After grinding at V = 30 m/s	0.44	0.58	18
2	After grinding at V = 120 m/s	0.40	0.52	20
3	After grinding at V = 300 m/s	0.38	0.50	22
4	After grinding at V = 300 m/s and processing Re	0.48	0.65	30
5	After grinding at V = 300 m/s and processing Ar	0.69	0.92	37
6	After grinding at V = 300 m/s and with Si–Al–N coating	0.54	0.72	209
7	After grinding at V = 300 m/s and with Si–Al–N + Mo coating	0.54	0.80	104
8	After grinding at V = 30 m/s and with TiN coating	0.65	0.99	23

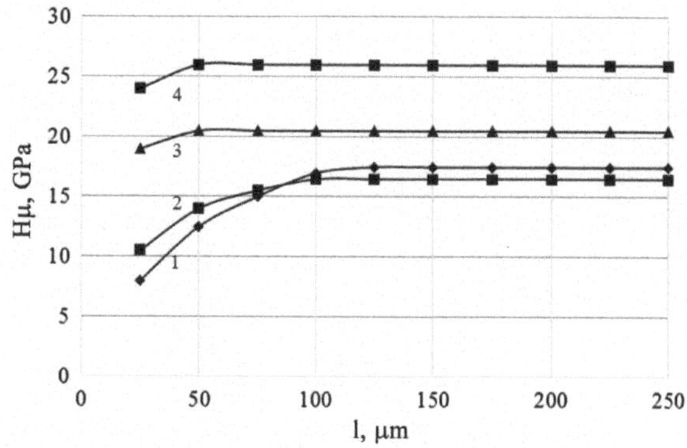

Fig. 26.1 Microhardness of the cutting wedge of carbide plate specimens: (1) unprocessed plates, (2) after processing by argon; (3) after processing by rhenium, (4) with Si–Al–N coating

26.3.1 Effect of the Ion Implantation on the Wear Resistance of Carbide Plates

The front faces of R300-0828E-PM 1030 plates were processed by argon and rhenium ions for testing the influence of ion implantation on the wear resistance of carbide plates 9. The measurement of microhardness from the plate edge to the center along the front face has shown that the processing by argon ions did not change considerably the microhardness in comparison with the unprocessed plates (Fig. 26.1). The microhardness increased gradually from 10 to 17 GPa within the point of the cutting wedge from 25 to 100 μm. The microhardness remained unchanged up to 250 μm. After the processing by rhenium, the microhardness changed only slightly, about 20.5 GPa.

As was mentioned above, the other way of increasing the wear resistance is the deposition of thin coatings on the cutting edge. Among the great variety of coatings, for example, TiC, TiN, Ti–Al–Zr–N, Ti–Al–N, Ti–Al–Y–Cr–N, Si–Ti–N, Al–Si–N, Si–B–C–N, and others, only Si–Ti–N, Al–Si–N, and Si–B–C–N are suitable for studied modes because of the local temperature of cutting reaching about 1300 K and even higher, the rest coatings fail to withstand such a high temperature [22–24]. In this connection, we have studied the influence of thin Al–Si–N coatings on the wear resistance of carbide plates.

The elemental composition of the Al–Si–N coating was the following: Al—34, Si—13, and N—53 at.%. The coating thickness determined by the Calotest CAT-S-0000 device was 2.3 μm at the front face and 0.26 μm at the rear edge (due to the extra spraying).

The X-ray structural analysis has shown that the Al–Si–N coatings consist of three phases: α- and β-phases of Si3N4 with P31c and P63/m HCP lattices, respectively, with volume fractions of ~30% and AlN phase with wurtzite-type P63mc HCP lattice with the fraction ~40 volume % (Fig. 26.2). We have obtained the close phase composition of Al–Si–N coatings at the similar modes of magnetron deposition on silica substrates [25].

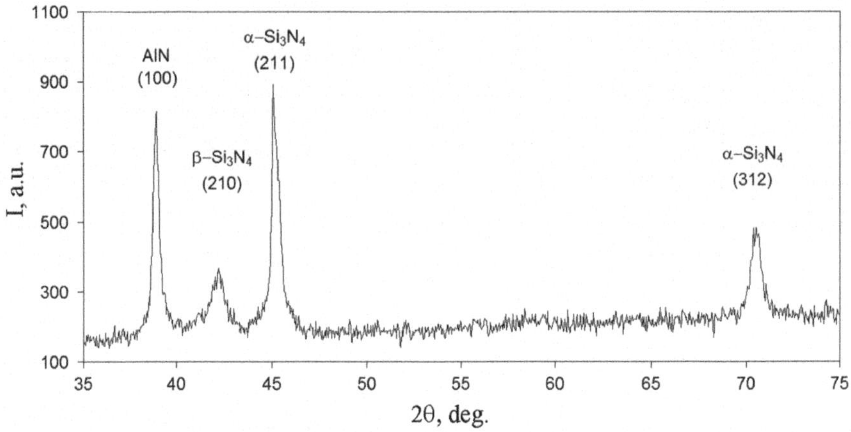

Fig. 26.2 X-ray pattern of Al–Si–N coating characterizing the phase composition

It was found that depending on the application modes, grains in Al–Si–N coatings have the average size from 10 to 30 nm, that is, they have the nanocrystal structure. The analysis of the microhardness Hμ of the 6-μm thick coatings and the adhesion Fa of the coatings to the substrate has shown high values: Hμ ~ 31.2 GPa and Fa ~ 20.4 N. We have calculated the coefficient of restitution k_r of the coatings with the use of the "load Fn—indenter penetration depth Pd" curves obtained at nanoindentation of the coatings under the load on the indenter 20 mN at the NanoTest 600 device (CSM Instruments, Switzerland). It was found that the Al–Si–N coatings have the high value of the restitution coefficient $k_e = 82\%$. Figure 26.1 demonstrates the measured microhardness of thin 2.6-μm coatings. It is shown to achieve 26 GPa. The difference in microhardness of thick and thin coatings is caused by the strong effect of the smooth substrate (carbide). The adhesion value of the coatings was determined by scratching the surface of the coating with a diamond indenter with a continuously increasing load with aid of the Revetest-RST macro scratch tester (CSM Instruments, Switzerland).

Some coatings were additionally implanted by molybdenum ions. The fluence of molybdenum was 1×10^{17} ion/cm^2, and the average energy was Eav = 120 keV. The results of surface processing and tests are summarized in Table 26.3. The analysis of the results (Table 26.2) shows that the plate wear is a multifactor process, which is influenced by the grinding speed and the type of processing. The influence of the grinding speed is demonstrated by positions 1–3 in Table 26.3. The higher the grinding speed, the longer the plate life. The analysis of the surfaces of grinded plates demonstrates that at ultra-high-speed grinding the mechanism of material removal from the plate surface alternates. At the low grinding speed, microchipping of tungsten carbide crystals is present.

At the ultra-high-speed of grinding (higher than 300 m/s), tungsten carbide crystals destroy. This favors the creation of a sharper edge of the cutting wedge and formation of small tungsten carbide crystals just on the surface. At the cutting by plates grinded in this way, the microchipping of tungsten carbide crystals is lower.

The processing of carbide plates by rhenium and argon ions elongates the tool life (positions 4, 5). It is interesting to note that in the case of irradiation of carbide plates by argon ions, the effect of increase in service life was more pronounced than after processing by heavier rhenium ions.

The X-ray spectral analysis of the plates implanted with rhenium has shown its presence up to 0.9 wt%. This relatively low concentration is caused by the fact that the probing X-ray penetrates much deeper than the layer of implanted rhenium. In addition, the analysis has shown a decrease in the cobalt concentration by approximately 2 wt% in comparison with the case of processing by argon. The analysis of TEM images has demonstrated that tungsten carbide crystals are clearly seen in the surface layer of the cutting wedge after processing by rhenium ions. At the same time, this pattern is not observed after processing by argon. We believe that this is caused by the large difference in atomic weights of tungsten, cobalt, argon, and rhenium. As a result, the intense sputtering of cobalt occurs in the case of processing by rhenium, since rhenium is threefold heavier than cobalt. Tungsten carbide crystals lose their binding, and microchipping is observed during cutting.

The deposition of thin Si–Al–N coatings increases significantly the wear resistance of carbide plates (positions 6, 7) in Table 26.3. It seems interesting that extra implantation of the coating by molybdenum ions failed to increase the wear resistance. This is likely connected with the fact that after molybdenum implantation the adhesion to the processed VT3-1 alloy is increased. The widely used TiN coating insignificantly increased the service life of carbide plates.

26.4 Conclusions

The studies have shown that to increase the service life of carbide plates and the quality of processing of VT3-1 alloy, the following actions are needed:

- An increase in the quality of grinding of carbide plates can be achieved through the use of ultra-high-speed grinding, at which the grinding mechanism changes. No microchipping of tungsten carbide crystals occured.
- An increase in the wear resistance by the method of ion implantation. However, additional studies are needed to find the optimal regimes of processing, because implantation changes many properties of the processed material.
- Deposition of thin coatings like Al–Si–N efficiently increases the service life of carbide plates.

Acknowledgements The work was supported by the Ministry of Education and Science of Ukraine within the framework of the state budget program No.0119U100787 and by the National Research Tomsk Polytechnic University development program within the scope of the basic scientific research of the State Academies of Sciences for 2013–2020.

References

1. A.Y. Popov, D.S. Rechenko, *Technology of Ultra-High-Speed Grinding of Carbide Tools* (Tonkie Naukoemkie Tehnologii, Staryi Oskol, 2015), p. 160
2. A.S. Yanyushkin, *Technology of Electric Powered Diamond Grinding of Cutting Tools and Methods of Its Implementation* (Tonkie Naukoemkie Tehnologii, Staryi Oskol, 2013), p. 336
3. V.A. Gribkov, F.I. Grigor'ev, B.A. Kalin, V.L. Yakushin, *Promising Radiation-Beam Materials Processing Technologies* (Kruglyi God Publishing House, Moscow, 2001), p. 528
4. S. Pal Dey, S.C. Deevi, Mat. Sci. Eng. A **342**, 58 (2003)
5. P.H. Mayrhofer, C. Mitterer, L. Hultman, Prog. Mater Sci. **51**, 1032 (2006)
6. A.D. Pogrebnjak, A.P. Shpak, N.A. Azarenkov, V.M. Beresnev, Phys.-Usp. **52**, 29 (2009)
7. T.N. Oskolkova, A.M. Glezer, Steel Transl. **47**(12), 788 (2017)
8. K.K. Kadyrzhanov, F.F. Komarov, A.D. Pogrebnyak, V.S. Rusakov, T.E. Turkebaev, *Ion-Beam and Ion-Plasma Modification of Materials* (Moscow State University, Moscow, 2005), p. 640
9. B.P. Gritsenko, A.Yu. Popov, D.S. Rechenko, in Ion Implantation: Synthesis, Applications and Technology, ed. by A.D. Pogrebnyak (Nova Science Publishers, New York, 2018) p. 93
10. Z. Qing, Z. Feng, T. Hong-Hui, G. Man, Z. Yong-Zhen, Plasma Sci. Technol **3**(4), 897 (2001)
11. S.-X. Wang, B.-J. Xiong, C. Li, S.-Q. Yang, J. Harbin Inst. Technol. **43**(7), 102 (2011)
12. ISh Abdulin, V.S. Zheltuhin, V.V. Kudinov, I.R. Sagbiev, R.F. Sharafeev, Perspektivnii materiali **6**, 88 (2008)
13. S. Wang, C. Li, B. Xiong, X. Tian, S. Yang, Appl. Surf. Sci. **257**, 5826 (2011). https://doi.org/10.1016/j.apsusc.2011.01.113
14. A. Panckow, D. Sladkov, P.K. Singh, C. Genzel, Surf. Coat. Tech. **188**(1), 214 (2004)
15. P.W. Shum, K.Y. Li, Y.G. Shen, Surf. Coat. Tech. **198**, 414 (2005)
16. J.H. Yang, M.F. Cheng, X.D. Luo, T.H. Zhang, Mat. Sci. Eng. A **445–446**, 558 (2007)
17. O.V. Sergeev, M.P. Kalashnikov, M.V. Fedorischeva, V.P. Sergeev, V.E. Panin, in *AIP Conference Proceedings*, vol. **1783**, p. 020203 (2016). https://doi.org/10.1063/1.4966497
18. V. Uglov, G. Remnev, A. Kuleshov, M. Saltymakov, Inorg. Mater. Appl. Res. **2**, 242 (2011)
19. L. Ward, P.P. Kavuri, R. Manory, Nucl. Instr. Meth. Phys. Res. B **368**, 37 (2016)
20. A.D. Pogrebnjak, S.O. Bor'ba, Ya.O. Kravchenko, E.O. Tleukenov, C.V. Plotnikov, V.M. Beresnev, Y. Takeda, K. Oyoshi, A.I. Kupchishin, J. Superhard Mater. **38**(6), 393 (2016)
21. I. Brown (ed.), *The Physics and Technology of Ion Sources* (Wiley, 2004), p. 380
22. J. Musil, M. Šašek, P. Zeman, R. Čerstvý, D. Heřman, J.G. Han, V. Šatava, Surf. Coat. Tech. **202**, 3485 (2008)
23. J. Musil, Surf. Coat. Tech. **207**, 50 (2012)
24. J. Musil, G. Remnev, V. Legostaev, V. Uglov, A. Lebedynskiy, A. Lauk, J. Procházka, S. Haviar, E. Smolyanskiy, Surf. Coat. Tech. **307**, 1112 (2016)
25. I.A. Bozhko, E.V. Rybalko, M.V. Fedorischeva, V.P. Sergeev, A.I.P. Conf, Proc. **1909**, 020020 (2017)

Chapter 27
Contact Melting in Ag/Ge Layered Nanofilms: Critical Thickness and Onset Temperature

Alexey Minenkov, Aleksandr Kryshtal and Sergey Bogatyrenko

Abstract We present the results of a rigorous investigation of contact melting in thin Ag/Ge bilayers fulfilled via ex-situ TEM techniques based on tracing the abrupt changes of sample morphology and structure at melting. It has been revealed that the liquid-phase formation in the system at the eutectic temperature takes place only if the silver film mass thickness value is greater than the critical one (≈1.2 nm), which corresponds to ≈8 nm-sized Ag nanoparticles. The onset temperature of liquid nuclei formation at the metal-semiconductor interface was estimated to be 200 °C.

27.1 Introduction

In a frame of continuous miniaturization of semiconductor device fabrication, the quality of metal-semiconductor contacts plays an ever-growing role [1–4]. For instance, good ohmic contacts are vital for achieving cutting edge electrical characteristics of high-speed Photonic-Electronic integrated circuits [5–7]. The demand for stable data transfer networks motivates scientists to seek progressive solutions for developing nanoscale devices with low power consumption and high bandwidth. According to the requirements of modern nanoelectronics, it is necessary to search the metal-semiconductor systems for contacts with minimal and thermally stable resistance. However, the temperature treatment, which is common during the contact functioning, also leads to an increase in the contact resistance due to interface degradations. Hitherto, Au/Ge based n-type ohmic contacts are of widespread use in electronic devices [8–11] since they have low contact resistances after annealing [11]. Another perspective system that is under consideration nowadays is Ag/Ge [5, 9]. It has been revealed in high electron mobility transistors (HEMTs) [9] that Ag/Ge based ohmic contacts can provide contact resistances as low as of Au/Ge based ones.

A. Minenkov (✉) · S. Bogatyrenko
V. N. Karazin Kharkiv National University, Kharkiv, Ukraine
e-mail: alexey.a.minenkov@univer.kharkov.ua

A. Kryshtal
AGH University of Science and Technology, Kraków, Poland

A. D. Pogrebnjak and O. Bondar (eds.), *Microstructure and Properties of Micro- and Nanoscale Materials, Films, and Coatings (NAP 2019)*,
Springer Proceedings in Physics 240, https://doi.org/10.1007/978-981-15-1742-6_27

Moreover, their thermal stability is also improved compared to Au/Ge systems due to a much higher eutectic temperature of the Ag/Ge alloy [9, 12]. Evidently, successive development and application of such contacts cannot be even imagined without understanding the nature of components interaction at the metal-semiconductor interface while heat-treating [3, 13–17].

It is well-known [18–20] that heating two contacting solid entities, which form a eutectic couple, may cause melting at the interphase boundary at a temperature well below the melting points of pure materials. This effect is named contact melting. However, the mechanism of the effect is still an issue. In the case of nanosystems, a limited volume of components' material may constraint the nucleation and growth of new phases [21, 22]. Theoretical calculations along with experimental data indicate the considerable changes of binary phase diagram at the nanoscale [23–27]. For instance, in [28] we have revealed for Au/Ge films the existence of a thickness, below which the liquid phase of eutectic composition was not formed in the contact zone. To be more accurate, the liquid phase was not formed in systems with gold films mass thickness lower than 0.2–0.3 nm. This thickness was named a critical thickness of the contact melting. Unfortunately, we were unable to determine the size of the Au nanoparticles corresponding to the 0.3 nm thick Au film.

Current research is a logical successor of a rigorous investigation of the eutectic melting mechanism [13, 22, 28, 29] and is in line with recent in situ TEM studies of the eutectic melting in Ag–Ge bi- and tri-layer nanofilms [30]. In the long run of that investigation, we have faced the methodological problem: it was not possible to detect the eutectic temperature for films with germanium layer thinner than 5 nm since self-supported sample collapsed during in situ TEM heating experiments. Thus, in current work we aimed to fill this gap and have focused on determining both: the so-called critical thickness of contact melting and the minimum temperature which is required for liquid phase formation at the Ag–Ge interface. We have used a well-proven morphological criterion [22, 28] as the basis of an experimental routine.

27.2 Objects and Experimental Techniques

Nanosized Ag–Ge films have been chosen as an object under study. Ag and Ge form a phase diagram of a simple eutectic type and have a rather restricted mutual solubility in a solid state. The eutectic composition in a bulk is formed at 24.5 at.% Ge and at a temperature of 651 °C [12]. Layered Ag–Ge films are convenient model objects for studying the contact melting phenomena. First, a few monolayers thick germanium films being grown on an amorphous substrate at room temperature are continuous. This ensures a true atomic contact between pure Ag and Ge that is crucial for the valid investigation of components interaction at metal-semiconductor interface. Secondly, we can vary the mass ratio of the components in a wide range by changing the thickness of silver and germanium layers.

Ag–Ge layered films were formed at room temperature by sequential thermal evaporation of components from independent sources at 5×10^{-8} Torr vacuum.

Fresh KCl monocrystal cleavages served as a substrate for the films. The substrate was degassed at 200–250 °C for 2 h prior to the film deposition. Mobile shutters, on which initial portions of the metal for formation of evaporation stationary mode were condensed, were put between the evaporators and the substrate holder. The film mass thickness was measured by means of a pre-calibrated piezo-quartz sensor.

The experimental approach was a well-proven morphological criterion for measuring the eutectic temperature [28], i.e. variation of film morphology at phase transformation. The methodology of these studies involves: (i) heating films with a thickness varying continuously along a substrate or (ii) heating a film of constant thickness on a substrate with a temperature gradient. These methods allow us to determine both the thickness of a film which melts at a selected temperature (i) and melting temperature for a given film thickness (ii). Consequently, two sets of samples were formed. In such experiments, KCl monocrystals were additionally covered with 2 nm thick amorphous carbon film. The latter served to prevent the interaction between the Ag–Ge film and the salt crystal during a further heating-cooling cycle. We have used an extended copper block as a substrate holder. The block included the built-in heater and was fitted with thermocouples and a system of mobile shatters. The accuracy of the temperature measurement was estimated to be 2% across the whole studied temperature range. After cooling down to room temperature the Ag/Ge/C samples of both sets were separated from the substrate by dissolving the salt in distilled water and were picked up on Cu electron-microscopic grids for immediate TEM inspection.

27.3 Experimental Results and Discussion

27.3.1 Critical Thickness of Contact Melting in Ag/Ge Films

In this part the investigation, the set of Ag/Ge/C samples consisted of 5 nm thick Ge film and silver film of variable thickness (0.5–5 nm) were annealed at the eutectic temperature for 30 min without breaking the vacuum. Figure 27.2 shows the typical TEM images of Ag/Ge/C films with different Ag film mass thickness after annealing. One can clearly see that the samples with relatively thick silver films exhibit spherical particles over the whole area of the film system (see Fig. 27.1a–c). These particles, judging from their form, are the result of the liquid phase formation followed by its crystallization while cooling. Such behavior is expected since the formation of a liquid phase always affects morphology of films drastically [28, 31–34]. In case the film has a sufficient amount of Ag, the eutectic strips the carbon sublayer (see selected area I in Fig. 27.1). This is because the eutectic particles do not wet the amorphous carbon sublayer but wet the Ge film. As a result, these particles move toward the germanium film and gather in large islands. The traces of this motion are clearly seen in Fig. 27.1a–c (region II).

As it is seen in the microphotographs, the size of these spherical particles decreases gradually with the thickness of the silver film reduction and, below a certain threshold

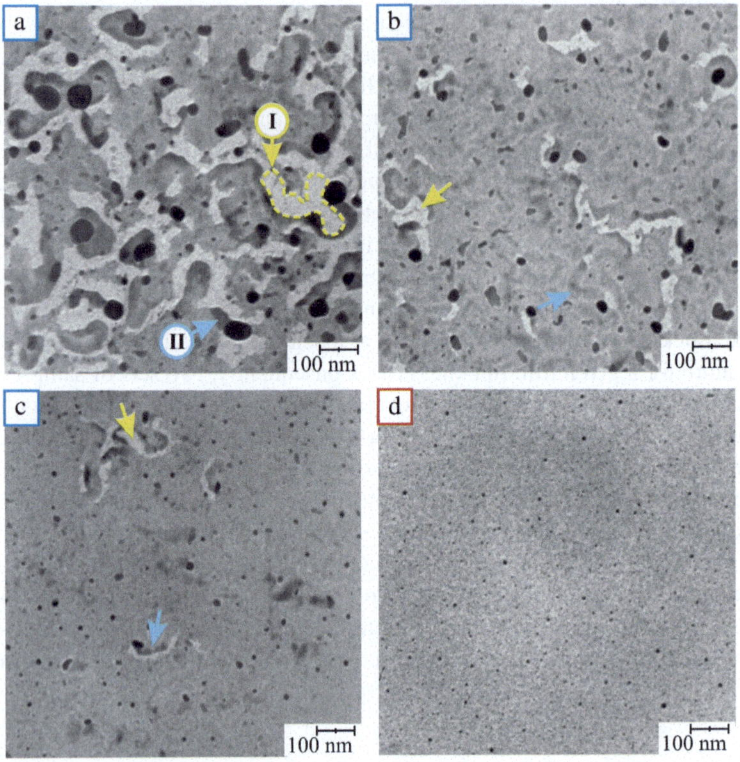

Fig. 27.1 Bright-field TEM images of the Ag/Ge/C films formed on the substrate at room temperature and annealed at the eutectic temperature. The Ge film thickness is 5 nm; the Ag film mass thicknesses are **a**—3.5 nm, **b**—2 nm (**b**), **c**—1.5 nm, and **d**—1.2 nm. I—striped carbon sublayer area, II—eutectic particle trace

thickness, which is $\approx$1.2 nm, the eutectic movement traces, as well as the huge eutectic particles, have not been observed any more (Fig. 27.1d). At the same time, the film morphology undergoes abrupt changes. This fact is clearly seen on the histogram of the particles' footprint $N \cdot S(d)$ (Fig. 27.2). The 1.5 nm thick silver film on amorphous Ge is the disordered system with a large spread of particles' sizes. One can see two maxima on the histogram. The former corresponds to eutectic particles with a size of 11 nm, the later represents the ones of 24 nm. While the histogram obtained for the film with silver film thickness of 1.2 nm has only one pronounced maximum at the particles' size of $\approx$7.5 nm. Kinetics of the film morphology alteration in the binary system when heated and the formation of a liquid phase will be discussed in detail in the forthcoming article. In the present paper we highlight the fact of abrupt morphology change and use it as a criterion for the determination of phase transition.

Thus, it can be argued that the liquid phase formation at the eutectic temperature does not occur for Ag/Ge films with the silver film thinner than 1.2 nm. It is necessary to note the following fact. The notion of the mass thickness of a film implies that

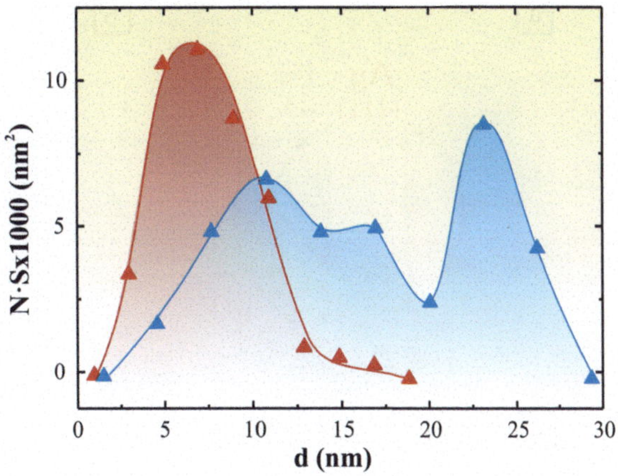

Fig. 27.2 Distribution of the particles' footprint in the Ag/Ge films. The silver film mass thickness is 1.5 nm (▲)(see Fig. 27.1c) and 1.2 nm (▲)(see Fig. 27.1d). d—mean diameter of a particle, N—number of particles of a given size, S—a particle projection area

the film material is distributed evenly over the substrate surface, in our case, over the amorphous germanium. It is observed in the as-deposited films. Despite the fact that condensation of silver on amorphous germanium film occurs via Volmer-Weber growth mode, a silver film is becoming continuous at early stages of growth, due to a relatively high surface energy of germanium. At the same time, the solid-state de-wetting process takes place while heating the Ag/Ge film. This process leads to a spontaneous transformation of continuous Ag film into an array of isolated crystal islands and is driven by the film–substrate surface energy reduction [33–35]. Naturally, we have implied the negligible solubility of Ag in the Ge film [12]. Hence, upon the eutectic temperature achievement, the Ag/Ge film system consists of silver-based solid solution islands on the continuous and amorphous germanium film. If the size of these islands is larger than the nucleus size, the liquid phase of eutectic composition will be formed. This size can be evaluated by electron microscopic images of the Ag/Ge film in which melting at the eutectic temperature was not observed (Fig. 27.1d). The estimation gives the size of 5–8 nm, which is reasonable for a film with mass thickness of 1.2 nm [33, 34].

It should be also noted that along with the abrupt change in morphology (Fig. 27.1c, d) the change of crystal structure also occurs under melting in the Ag/Ge film. Thus, all diffraction patterns of the binary system with the Ag film thicker than 1.2 nm exhibit not only the amorphous germanium haloes but also the diffraction rings corresponding to crystalline germanium (Fig. 27.3a). While the diffraction patterns taken from the samples with the silver film mass thickness below the critical value show only diffuse haloes corresponding to amorphous germanium at all temperatures studied (Fig. 27.3b). However, elemental analysis (see Fig. 17.3c), unambiguously indicates the presence of silver in these samples.

This clearly shows that the amorphous germanium film crystallization (metal-induced crystallization) occurs simultaneously with the formation of a liquid phase in the binary film. Since the formation of chemical compounds has not been detected throughout the whole investigated temperature range, the converse is probably also

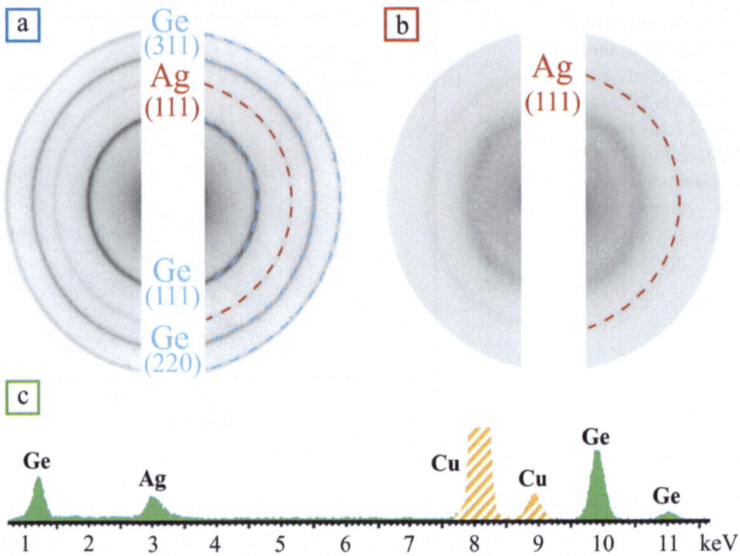

Fig. 27.3 SAED patterns of the Ag/Ge films formed on the substrate at room temperature and annealed at the eutectic temperature. The Ge film thickness is 5 nm; the Ag film mass thicknesses are 1.5 nm (**a**) and 1.2 nm (**b**). **c**—EDX spectrum of the Ag/Ge film system presented in Figs. 27.1d and 27.3b (Cu peaks correspond to TEM grid)

true for the Ag/Ge system. Viz. the crystalline germanium emergence in the system indicates the formation of a liquid phase at the metal-semiconductor interface, i.e. it testifies the manifestation of the contact melting effect. This observation is of fundamental importance since it evidences in favor of contact melting as the most probable mechanism of metal-induced crystallization of a semiconductor film in eutectic couples [3, 13–17, 36, 37].

27.3.2 Onset Temperature of Contact Melting in Ag/Ge Films

To determine the onset temperature of germanium crystallization, the Ag/Ge system with germanium and silver film thicknesses of $\approx$3 nm and $\approx$1.5 nm respectively has been annealed at different temperatures for half an hour. The formation of crystalline germanium was registered at temperatures of approximately 200 °C. I.e. we believe that the liquid phase in the Ag/Ge system with 1.5 nm thick silver film was formed at this temperature. In the same way, when the thickness of the silver film was lower than the critical value, a liquid phase was not formed at the eutectic temperature too. This point is shown in Fig. 27.4 (red square), which, together with in situ TEM research data [30], describes the whole eutectic temperature size dependence of the Ag/Ge binary film at the nanoscale.

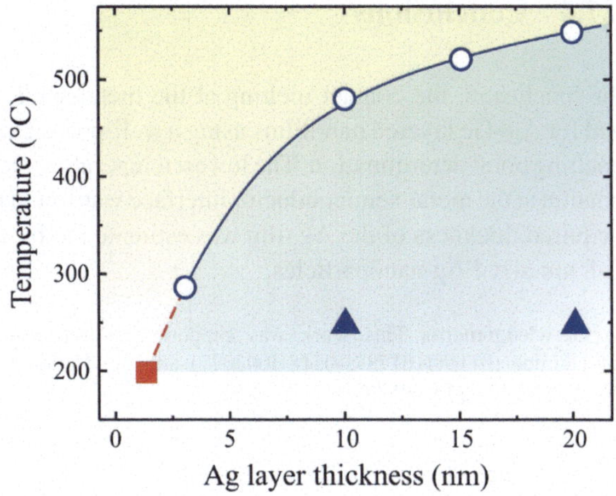

Fig. 27.4 Size dependence of the eutectic temperature T_E for Ag–Ge film system. ■—the point determined by means of morphological criterion. ○, ▲—the previous in situ TEM studies results [30] (○—the eutectic temperature, ▲—the silver-induced a-Ge crystallization temperature)

27.3.3 General Discussion

Considering the above-mentioned data and the results of the previous in situ TEM studies [30] the melting process in the Ag/Ge binary film is as follows. During the process of film heating, at a temperature of about 200 °C the liquid phase nucleation, i.e. contact melting, occurs at the metal-semiconductor interface. The liquid phase nuclei are unstable due to the eutectic temperature size dependence. However, at the moment we cannot definitely say whether this phase remains in a metastable state or its crystallization, accompanied by the decomposition into silver and germanium based solid solutions, occurs. As the temperature rises, the amount of a liquid phase in the system increases, and the complete transition to equilibrium liquid state occurs at the eutectic temperature of the particles of a given size. It is evident that if the film thickness is below the critical value a liquid phase at the interface will not be formed. Consequently, the amorphous germanium crystallization will not occur, as it has been observed experimentally. The existence of a critical thickness of contact melting was observed by us previously in Au-Ge, Bi-Ge and Bi-Sn systems [22, 28, 29]. This result indicates that such effect is probably featured all eutectic systems.

The difference between the amorphous germanium film crystallization temperatures for the samples annealed at the eutectic temperature (current investigation) and rapidly heated to the eutectic temperature (our previous research [30]) should be additionally discussed. In the first case, the crystallization temperature was estimated to be 200 °C, while in the second one it was measured as 250 °C (see Fig. 27.4). It should be also mentioned that crystalline germanium lines appeared at about 250 °C in all the samples studied by means of in situ TEM heating. We believe the essential differences between these series of samples can be the annealing kinetics and the thickness of Ge film. I.e. the amorphous germanium film crystallization is not only a thermally activated process but it is also dependent on the interfacial interactions kinetics. Evidently, the phenomenon requires further investigation.

27.4 Conclusions

In conclusion, the contact melting at the metal-semiconductor interface was studied for Ag–Ge layered nanofilms using a well-proven morphological criterion for the melting point determination. The lowest temperature required for the liquid phase formation at the metal-semiconductor interface was found to be 200 °C, while the lowest required thickness of the Ag film was estimated to be 1.2 nm, which corresponds to $\approx$8 nm-sized Ag nanoparticles.

Acknowledgements This work was supported by the Ministry of Education and Science of Ukraine (0119U101787, 0118U002027) and the National Science Centre, Poland (project No. 2016/23/B/ST8/00537).

References

1. S. Li Sheng (ed.), *Semiconductor Physical Electronics* (Springer, New York, 2006)
2. M. Grundmann (ed.), *The Physics of Semiconductors* (Springer International Publishing, Switzerland, 2016)
3. A.F. Marshall, P.C. McIntyre, Appl. Phys. Lett. **97**(8), 082104 (2010)
4. M.V. Fischetti, W.G. Vandenberghe, in *Advanced Physics of Electron Transport in Semiconductors and Nanostructures* (Springer, 2016)
5. L. Shen, V. Dolores-Calzadilla, C.W.H.A. Wullems, Y. Jiao, A. Millan-Mejia, A. Higuera-Rodriguez, D. Heiss, J.J.G.M. van der Tol, H.P.M.M. Ambrosius, G. Roelkens, M.K. Smit, Opt. Mater. Express **5**(2), 393 (2015)
6. J.J.G.M. Tol, van der, R. Zhang, J. Pello, F. Bordas, G. Roelkens, H.P.M.M. Ambrosius, P.J.A. Thijs, F. Karouta, M.K. Smit, IET Optoelectron. **5**(5), 218 (2011)
7. M. Smit, J. Van der Tol, M. Hill, Laser Photon. Rev. **6**(1), 1 (2012)
8. P. Fay, in *Encyclopedia of Materials: Science and Technology (Second Edition)* (Pergamon Press, 2001)
9. W. Zhao, S. Kim, J. Zhang, I. Adesida, IEEE Electron Device Lett. **27**(1), 4 (2006)
10. D. G. Ivey, D. Wang, D. Yang, R. Bruce, G. Knight, J. Electron. Mater. **23**, 441 (1994)
11. A.G. Baca, F. Ren, J.C. Zolper, R.D. Briggs, S.J. Pearton, Thin Solid Films **308–309**, 599 (1997)
12. H. Okamoto (ed.), *Phase Diagrams for Binary Alloys* (ASM International, United States, 2000)
13. A.P. Kryshtal, A.A. Minenkov, P.J. Ferreira, Appl. Surf. Sci. **409**, 343 (2017)
14. V.B. Neimash, A.O. Goushcha, P.Y. Shepeliavyi, V.O. Yukhymchuk, V.A. Danko, V.V. Melnyk, A.G. Kuzmich, J. Mater. Res. **30**(20), 3116 (2015)
15. E. Sutter, P. Sutter, Nano Lett. **8**(2), 411 (2008)
16. Z. Wang, L.P.H. Jeurgens, E.J. Mittemeijer, in *Metal-Induced Crystallization: Fundamentals and Applications* (Jenny Stanford Publishing, 2015)
17. Z.M. Wang, J.Y. Wang, L.P.H. Jeurgens, E.J. Mittemeijer, Phys. Rev. B **77**(4), 45424 (2008)
18. L.K. Savitskaya, P.A. Savintsev, in *Surface Phenomena in Metallurgical Processes*, ed. by A.I. Belyaev (Springer, Boston, 1965)
19. V.M. Zalkin, *Nature of Eutectic Alloys and Effect of Contact Melting* (Metallurgiya, Moscow, 1987)
20. O.V. Bystrenko, V.V. Kartuzov, J. Alloys Compd. **617**, 124 (2014)
21. M. Cui, H. Jiang, Z. Cao, X. Meng, Sci. Rep. **7**, 41990 (2017)
22. N.T. Gladkikh, A.P. Kryshtal', R.V. Sukhov, Phys. Solid State **52**(3), 633 (2010)

23. S. Bajaj, M.G. Haverty, R. Arróyave, S. Shankar, Nanoscale **7**(21), 9868 (2015)
24. J. Sopoušek, A. Kryštofová, M. Premović, O. Zobač, S. Polsterová, P. Brož, J. Buršík, Calphad **58**, 25 (2017)
25. S. Bogatyrenko, A. Kryshtal, A. Minenkov, A. Kruk, Scr. Mater. **170**, 57 (2019)
26. W.A. Jesser, G.J. Shiflet, G.L. Allen, J.L. Crawford, Mater. Res. Innov. **2**(4), 211 (1999)
27. L.S. Palatnik, B.T. Boiko, Phys. Met. Met. **11**, 119 (1961)
28. P. Kryshtal, R.V. Sukhov, A.A. Minenkov, J. Alloys Compd. **512**(1), 311 (2012)
29. A.A. Minenkov, A.P. Kryshtal, S.I. Bogatyrenko, J. Alloys Compd. **756**, 50 (2018)
30. A. Kryshtal, S. Minenkov, S. Bogatyrenko, A.Gruszczyński, J. Alloys Compd. **786**, 817 (2019)
31. S.I. Petrushenko, S.V. Dukarov, Z.V. Bloshenko, I.G. Churilov, V.N. Sukhov, in *Advances in Thin Films, Nanostructured Materials, and Coatings* (Springer, 2019), pp. 159–167
32. S.I. Petrushenko, S.V. Dukarov, V.N. Sukhov, Vacuum **142**, 29 (2017)
33. P. Kryshtal, A.A. Minenkov, S.S. Dzhus, J. Nano-Electron, Phys. **7**(1), 1024 (2015)
34. A.P. Kryshtal, Appl. Surf. Sci. **321**, 548 (2014)
35. S.I. Petrushenko, S.V. Dukarov, V.N. Sukhov, I.G. Churilov, J. Nano- Electron. Phys. **7**(2), 2033 (2015)
36. R. Sinclair, T.J. Konno, Ultramicroscopy **56**(1–3), 225 (1994)
37. N. Wilson, A.K. Petford-Long, R.C. Doole, J. Appl. Phys. **84**(9), 5283 (1998)

Chapter 28
Characteristics of Nanocomposites Formed on the Steel Surface Contacting with Precious Metal Solutions

Olena M. Lavrynenko, Olesia Yu. Pavlenko, Yuriy S. Shukin, Nataliia O. Dudchenko, Aleksandr B. Brik and Tatiana S. Antonenko

Abstract The properties of the core&shell type nanocomposites formed on the steel surface via the rotation-corrosion dispergation route in the presence of precious metal aquahydroxoforms have been studied using physical-chemical multi-methods of the research. Depending on the kind of precious metal only magnetite or magnetite and lepidocrocite are formed in the system. The size of magnetite cores does not exceed 25 nm. The precious metal fraction on the magnetite surface varies from 0.04 to 0.5 wt% and it depends on the initial concentration of the solutions. All samples belong to superparamagnetics and do not show the remanence magnetization and coercitivity. The last parameter equals for Ag-containing system 24, for Au-bearing—48, for Pd-bearing—62, and for Pt-bearing—49 to 60 A m^2 kg^{-1}.

28.1 Introduction

Recently biology and biomedicine belong to the nanoresearch priority due to the requirement for the creation of new functional materials for diagnostic and therapeutic aims [1]. The nanosized core&shell type composites that include magnetite or maghemite core and aurum, argentum, platinum, or palladium shell are adopted for medical-biological application due to their unique optical properties, bioavailability, and selectivity in relation to the range of biological objects [2, 3]. Such kinds of nanocomposites showed suitable properties for catalysis, analytical chemistry, separation technology, environmental protection, etc. [4–6].

The most typical methods to form core&shell composite particles are co-precipitation ferric and ferrous salts in low-alkaline media in the presence of precious

O. M. Lavrynenko · O. Yu. Pavlenko (✉) · Y. S. Shukin
Frantsevich Institute for Problems of Material Science, National Academy of Science of Ukraine, Kyiv, Ukraine
e-mail: scorpioless@gmail.com

N. O. Dudchenko · A. B. Brik · T. S. Antonenko
Semenenko Institute of Geochemistry, Mineralogy and Ore Formation, National Academy of Science of Ukraine, Kyiv, Ukraine

A. D. Pogrebnjak and O. Bondar (eds.), *Microstructure and Properties of Micro- and Nanoscale Materials, Films, and Coatings (NAP 2019)*,
Springer Proceedings in Physics 240, https://doi.org/10.1007/978-981-15-1742-6_28

metal aquahydroxoforms with their following reduction [7]; the method of reversive micelles [8]; the separate precipitation of iron oxide particles and primary particles of precious metal and making-up of the composites in organic liquids [9]. The application of the alternative rotation-corrosion dispergation method under the oxidative conditions shows the possibility of the one-stage synthesis procedure [9, 10].

Origin and growth of the Fe(II)-Fe(III) LDH (Green Rust) structures take place on the steel—water solution interface in the presence of air oxygen and carbon dioxide due to undergoing the reductive-oxidative reaction [11]. Due to presence of ferrous iron in the Green Rust crystal lattice it can reduce the electropositive complexes from the dispersion medium to metal state and simultaneous oxidize GR phase to the iron oxides and/or hydroxides. When the phase formation process carries out via co-precipitation two separate phases (iron oxide and precious metal particles) are usually obtained in the dispersion solution [12], but under specific physicochemical RCD conditions the precious metal shell may be formed on the iron oxide surface [13]. In the last case we obtained the nanocomposite particles that composed an iron oxide (magnetite) core and precious metal shell.

The core&shell type composite nanoparticles differ by quantity of core particles (one or more), the phase composition of core (magnetite or maghemite) and shell (aurum, argentum, platinum, palladium), presence or absence of the additional SiO_2 layer between core and shell phases, the degree of the filling of core surface with precious metal (full or partial), the general quantity of composite layers (two or more) and other. The complicated (multiple) composites which, for example, contain two precious metals in their shell or a few alternate layers containing organic and precious metal particles are considered as a separate group.

The nanocomposite particles obtained via the rotation-corrosion dispergation route are characterized by partial shell filling and they correspond to the simple core&shell structures. The particles of such nature not only preserve the ferrimagnetic properties of iron oxide core and optical properties of the precious shell but will acquire the new unique colloidal-chemical and physicochemical behavior and they are perspective for their practical usage as a functional material for varies medical-biological purposes [14].

The aim of the present work is the study of the physical-chemical properties of the nanosized core&shell type composites based on ferrimagnetic cores and precious metal shells and formed on the steel surface contacting with a water solution containing aquahydroxoforms of precious metals under open-air conditions.

28.2 Objects and Methods of the Research

To obtain core&shell type nanocomposite particles on the iron-carbon alloy we used the rotation-corrosion dispergation method performed under the free entrance of air oxygen and carbon dioxide onto the electrode surface. The rotating disk electrode was made of iron-carbon alloy—steel 3 (St3). Before every experiment the surface of St3 was exposed to mechanical treatment and chemical activation procedure. The water solution of precious metals: argentums, aurum, platinum, palladium, in the range of initial concentrations from 0.5 to 20 mg dm^{-3} has been chosen as dispersion medium. The process of phase formation was carried out until the system reached the stationary state when the chemical composition of dispersion medium and value of pH in the system was stabilized. The chemical analysis was performed according to the standard methodic.

To analyze phase and chemical compositions of the surface structures we used X-ray diffraction (XRD) and X-ray fluorescence spectroscopy (XRFS). The morphology of the nanoparticles was described according to scanning electron microscopy (SEM) images. The magnetic properties of the iron oxides and oxyhydroxides formed in the presence of Au^{3+}, Pt^{4+}, Pd^{2+}, and Ag^{+} cations were estimated via magnetometry that was performed using a magnetometer equipped by Hall sensor.

The phase composition of the surface structures was determined using a computer-aided X-ray diffractometer (DRON—UM1) equipped with two Soller's slits and Ni filtered radiation of cobalt anode. The rate of recording was 1°/min, and the interfacial Woolf-Bragg's angle made up 80°. The coherent scattering region (CSR) was calculated according to the standard Debye-Scherrer's formula. A scanning electron microscopy with the aid of JOEL-6700 microscope was chosen as the main visual method of the research. The iron-to-precious metal mass ratio in the samples was analyzed by means of XRFS and carried out by the automatic spectrometer «ElvaX».

28.3 Results and Discussion

28.3.1 A Study of the Phase Composition of Nanocomposite Particles Formed on the St3 Surface Contacting with Precious Metal Solutions

The main mineral phases formed on the steel surface contacting with water medium containing precious metal aquaforms are lepidocrocite (JCPDS file No 08-0098) and magnetite (JCPDS file No 19-0629) with an admixture of goethite (JCPDS file No 17-536). As show the obtained data (Fig. 28.1), the main phases formed in the presence of argentum and palladium species are lepidocrocite and magnetite, whereas the single magnetite phase corresponds to the aurum and platinum-bearing systems. In additions the small reflexes of precious metals are clearly seen in the XRD-patterns.

The average size of the primary magnetite particles (or their coherent scattering region) formed in the aurum-bearing system equals 21 nm. But the minimal size of magnetite cores formed under the reductive conditions is less and it reaches 9 nm into the near-surface film and 16 nm on the steel surface. Also, in FeF_2O_4&Au^0 systems when cAu^{3+} was increased from 1 to 20 mg/dm^3 the magnetite size is gradually increased from 21 to 24 nm. The size of magnetite particles obtained in the presence of platinum is the same (21 nm), but in the presence of palladium the size of magnetite

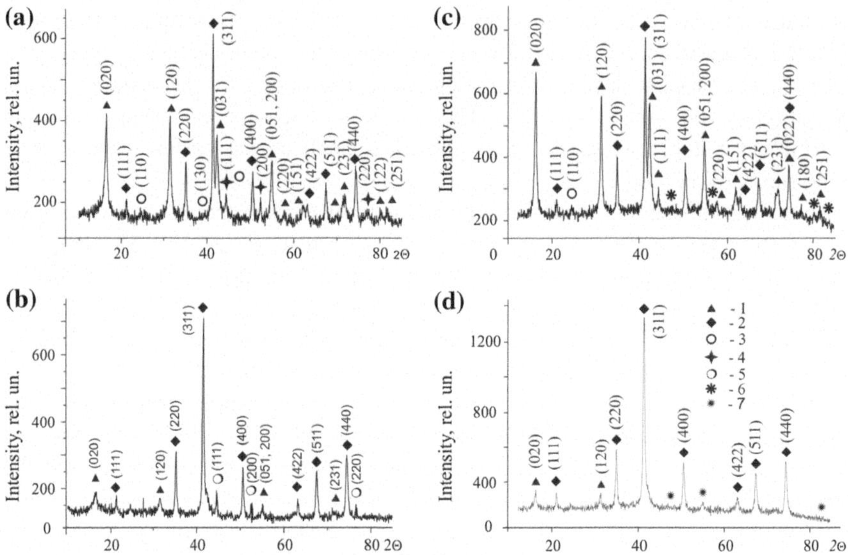

Fig. 28.1 The X-ray diffraction patterns of the nanosized structures formed on the steel 3 surface contacting with precious metal water solutions: **a** argentum; **b** aurum; **c** palladium; **d** platinum. Numbers correspond to: 1—lepidocrocite (gamma-FeOOH); 2—magnetite Fe_3O_4; 3—goethite (alpha-FeOOH); 4—silver; 5—gold; 6—palladium; 7—platinum

particles decreases to 18 nm. The oxidative conditions lead to increase in the particle size up to 32 nm. For argentum-containing systems are typical the increasing the magnetite size from 19 to 25 nm when the initial concentration of the solution is increased from 1 to 10 mg dm^{-3}.

The X-ray fluorescence spectroscopy data helps us to estimate the weight ratio of iron to precious metal parts. So, the content of platinum on the surface of composite particles reaches 0.14 wt% when the initial concentration of $c(Pt^{4+}) = 10$ mg/dm^3. The gold fraction in the corresponding composite particles increases from 0.05 to 0.9 wt% in the range of aurum concentrations from 1 to 20 mg dm^{-3}. Whereas in both presented systems the precious metal parts are associated with the surface structures only, palladium is detected in the composition of surface structures and phases of the near-surface film. Its weight percent equals, corresponding, 0.05 and 0.04 wt%. The study of the chemical composition of silver-bearing samples was performed depending on the initial argentum concentrations and pH values of the solutions. Also, in the range of $c(Ag^+)$ from 0.5 to 10 mg dm^{-3} the silver fraction in the composition of nanoparticles is increasing from 0.03 to 0.46 wt%. Simultaneously, silver-bearing particles formed in dispersion medium include 40 wt% at $c(Ag^+) =$ 0.5 mg dm^{-3} to 64 wt% at $c(Ag^+) = 5$ mg dm^{-3}. The silver fraction in the magnetite particles formed at the pH in the range 5.6–9.5 equals 0.04–0.06 wt%. On the contrary, at pH values <5.6 and >9.5 the ferric oxyhydroxides are the main phases that do not contain significant concentration of silver.

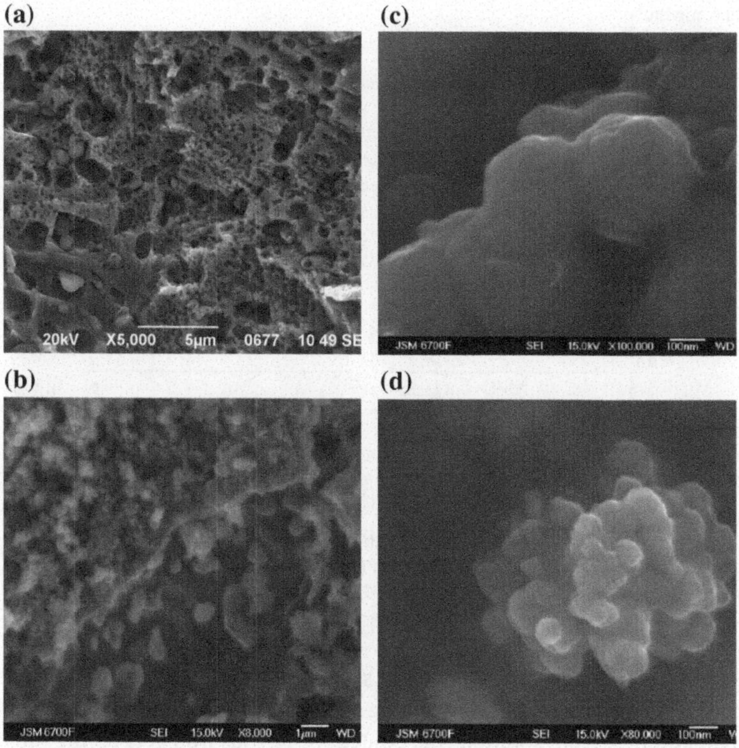

Fig. 28.2 SEM images of the surface structures formed in the presence of $AgNO_3$ solutions: **a** the steel 3 surface; **b** the interface St3-water solutions; **c** nanocomposites formed at c(Ag+) = 5 mg dm^{-3}; **d** nanocomposites formed at c(Ag+) = 15 mg dm^{-3}

SEM images of the products of the phase formation processes are shown in Fig. 28.2. The surface of the steel 3 contacting with argentum nitrate solution within 1 h is seen in Fig. 28.2a. The mineral phases formed at the steel—$AgNO_3$ water solution interface when the system went to the stationary state (Fig. 28.2b). Silver-bearing nanocomposites formed on the St3 surface at c(Ag^+) 5 and 15 mg dm^{-3} are shown in Fig. 28.2c, d, respectively.

28.3.2 Kinetic Regularities of the Phase Formation Processes on the Steel Surface in the Presence of Precious Metals

Kinetic regularities determined for ferric and ferrous cations in the water solutions obtained when the St3 surface was contacted with precious metal solutions are present in Fig. 28.3. Generally, in the presence of precious metals, the corrosion process is

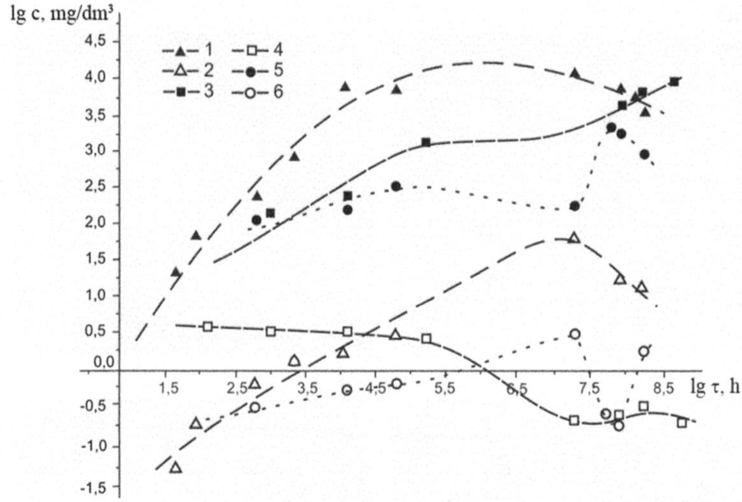

Fig. 28.3 Kinetic regularities of iron concentration when the steel surface was contacting with precious metal water solutions: platinum–Fe sum (1), ferric iron (2); palladium–Fe sum (3), ferric iron (4); argentum–Fe sum (5), ferric iron (6)

increased. So, in the platinum-containing solution the concentration of ferrous ions is 40 mg dm^{-3}, in the argentum-containing solution its concentration is 28 mg dm^{-3}. In the palladium-bearing solutions the content of ferrous iron gets 55 mg dm^{-3}, but in aurum-bearing solution the concentration of ferrous ions increased to 70 mg dm^{-3}. The kinetic regularities obtained for precious metals are present in Fig. 28.4. Whereas the c(Pd^{2+}) sharply decreased from 10 to 2 mg dm^{-3} in 15 min and then it gradually decreased to 0.08 mg dm^{-3}. The concentration of platinum does not change within 24 h and it led 0.06 mg dm^{-3} after 70 h only. The Au^{3+} concentration decreased to 0.4 mg dm^{-3} within 3 h and then it decreased to 0.07 mg dm^{-3}. To compare, c(Ag^{+}) decreased from 10 to 0.01 mg dm^{-3} in 15 min. The pH value 2.5 in the solutions containing Au^{3+}, Pt^{4+}, and Pd^{2+} increased to 5.3, but in argentum-bearing solutions it increased to 5–9.4 depending on the initial pH.

28.3.3 Magnetic Properties of Composite Particles Formed on the Steel Surface When It Was Contacting with Precious Metal Water Solutions

The practical application of core&shell type nanocomposites that include magnetite cores and precious metal shells is closely connected with their unique optical and magnetic properties [15]. The magnetic force microscopy data (MFM) show the magnetic response from magnetite cores, but when lepidocrocite is present in the samples the magnetic properties of nanocomposites are decreased. The typical 2D

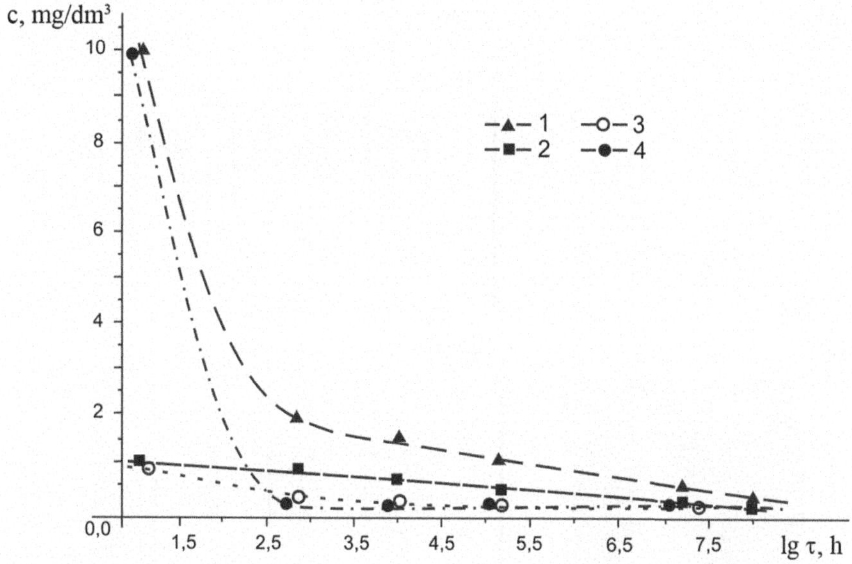

Fig. 28.4 Kinetic regularities of the precious metal concentrations: 1—Pt, 2—Pd, 3—Au, 4—Ag

and 3D images are present in Fig. 28.5. We can see the results obtained when the phase formation process was carried out at argentum concentration 0.5 and 5 mg dm^{-3} (a–d). Other images are characterized by Pt, Pd, and Au-bearing nanocomposites.

The following magnetic study point to the absence of remained magnetization and coercitivity. Hence, all samples belong to superparamagnetics. The kind of precious metal and its concentration determine the single parameter – saturation magnetization. So, the best results for aurum-bearing composites showed $M_s = 48$ A m^2 kg^{-1} at $c(Au^{3+}) = 10$ mg dm^{-3} (Fig. 28.6a), and for argentum-bearing particles for same concentration $M_s = 24$ A m^2 kg^{-1} (Fig. 28.6b). But the saturation magnetization of composite particles formed in the presence of platinum ions does not influence the initial concentration of the solutions. So, in all cases it increased to 62 A m^2 kg^{-1}. The value of M_s for palladium depending on the initial concentration varies from 49 to 60 A m^2 kg^{-1}. Hence, we obtain the intensity row of saturation magnetization for nanocomposite particles formed under the rotation-corrosion dispergation conditions on the steel surface contacted with precious metals: $Ag^+ < Au^{3+} < Pd^{2+} < Pt^{4+}$.

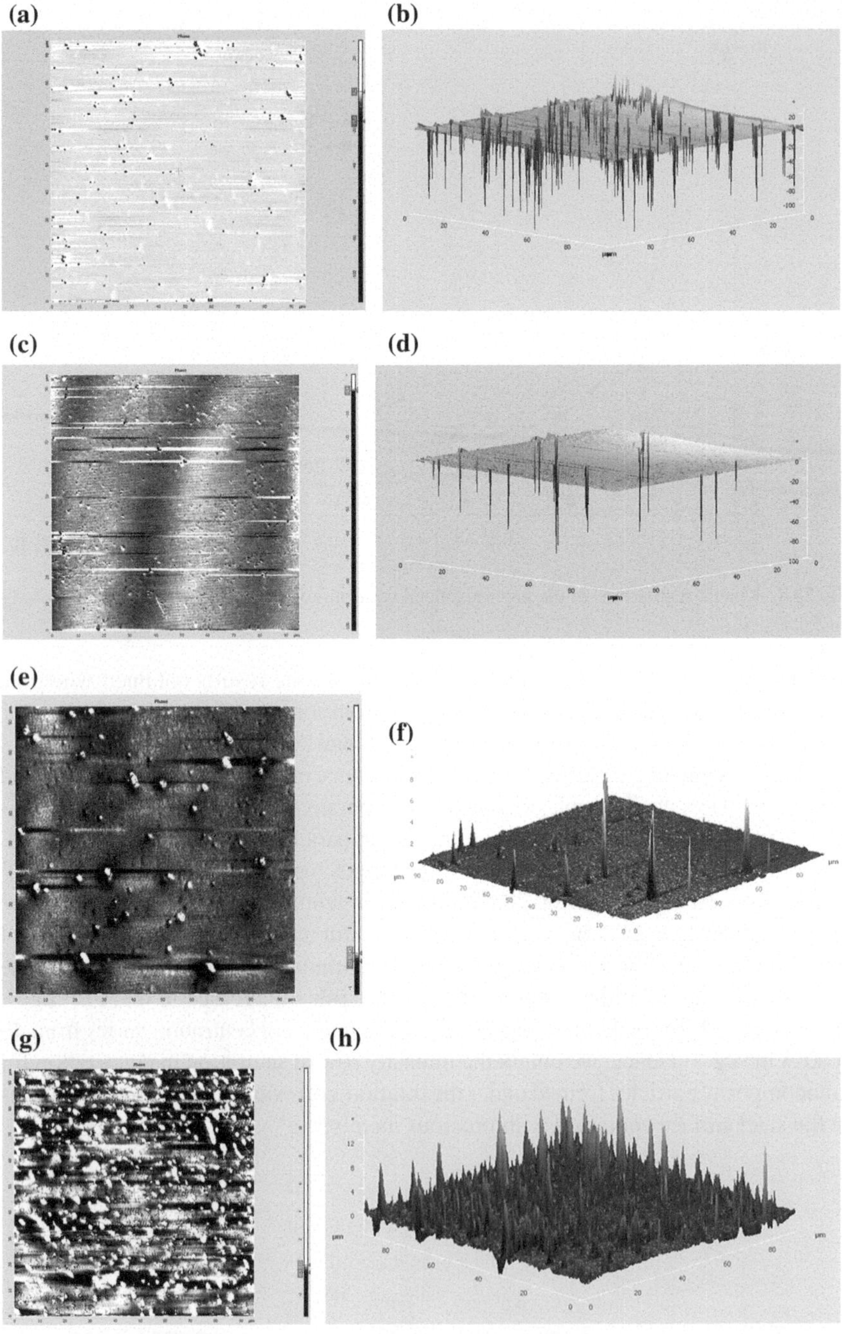

Fig. 28.5 Magnetic force microscopy data (2D and 3D images) of the composite nanoparticles formed when the steel surface was contacted with water solutions in the presence: argentum, where **a**, **b** c(Ag) = 0.5 mg/dm^3; **c**, **d** c(Ag) = 5 mg/dm^3; **e**, **f** platinum; **g**, **h** palladium; **j**, **k** aurum

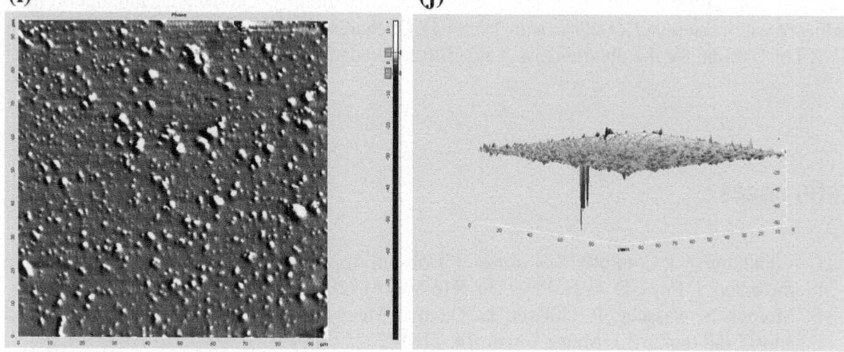

Fig. 28.5 (continued)

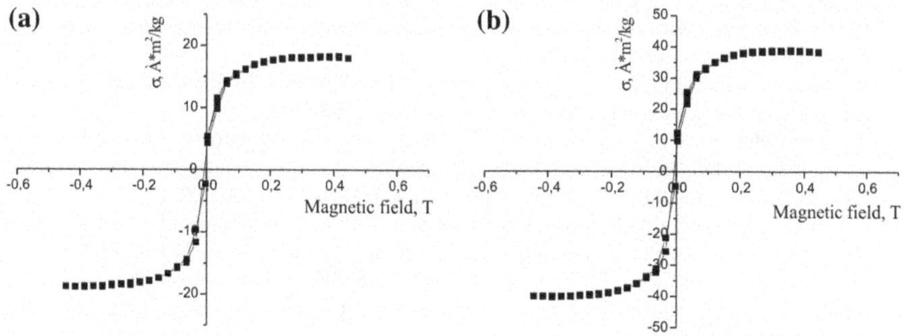

Fig. 28.6 The magnetization curves of the nanosized surface structures formed on the steel surface contacting with water solution where the metal concentration was 10 mg/dm^3: **a** argentum; **b** aurum

28.4 Conclusions

The precious metal solutions are used to perform the phase formation process on the steel surface under the rotation-corrosion dispergation conditions. The composition of the obtained particles includes magnetite cores and precious metal shells that supply unique optical and magnetic properties of nanoparticles. The composite structures formed on the steel surface are characterized as superparamagnetics. The maximal saturation magnetization for argentum-containing system is 24 A m^2 kg^{-1}, for the aurum-bearing system such parameter is higher—48 A m^2 kg^{-1}, for platinum-bearing system it reaches 62 A m^2 kg^{-1}, and for palladium-bearing system it varies from 49 to 60 A m^2 kg^{-1}. So, we think that nanocomposite particles formed under the RCD conditions may be perspective magnetic material for biomedical applications.

Acknowledgements We acknowledge leading research scientist, candidate of geological and mineralogical sciences *O.A. Vyshnevskyi* of Mass-spectrometric centre of solid phase, gas isotopic trace element analysis (M.P. Semenenko Institute of Geochemistry, Mineralogy and Ore Formation of

NAS of Ukraine) for the obtaining of the SEM images. We gratefully acknowledge the support from leading research scientist *P.O. Kosorukov* from F.D. Ovcharenko Institute of Bio-Colloid Chemistry of NAS of Ukraine for the obtaining of X-ray diffraction data.

References

1. Q.A. Pankhurst, J. Connolly, S.K. Jones, J. Dobson, Applications of magnetic nanoparticles in biomedicine. J. Phys. D Appl. Phys. **36**, R167–R181 (2003)
2. S. Mornet, S. Vasseur, F. Grasset, E. Duguet, Magnetic nanoparticle design for medical diagnosis and therapy. J. Mater. Chem. **14**, 2161–2175 (2004)
3. D. Zhang, C. Zhou, Z. Sun, L.Z. Wu, C.H. Tung, T. Zhang, Magnetically recyclable nanocatalysts (MRNCs): a versatile integration of high catalytic activity and facile recovery. Nanoscale **4**(20), 6244–6255 (2012)
4. M. Zhu, Ch. Wang, D. Meng, G. Diao, In situ synthesis of silver nanostructures on magnetic Fe3O4-C core–shell nanocomposites and their application in catalytic reduction reactions. J. Mater. Chem. A **1**, 2118–2125 (2013)
5. W. Bao, T. Chen, Y. Liu, P. Zhu, K. Jin, L. Wang, J. Liu, Y. Wei, Y. Li, Bifunctional Au-Fe3O4 nanoparticles for protein separation. ACS Nano **1**, 293–298 (2007)
6. J.T. Kinoshita, S. Seino, Y. Mizukoshi, T. Nakagawa, T.A. Yamamoto, Functionalization of magnetic gold/iron-oxide composite nanoparticles with oligonucleotides and magnetic separation of specific target. J. Magn. Magn. Mater. **311**(1), 255–258 (2007)
7. L. Lyon, D.A. Fleming, M.B. Stone, P. Schiffer, M.E. Williams, Synthesis of Fe oxide Core/Au shell nanoparticles by iterative hydroxylamine seeding. Nano Lett. **4**(4), 719–723 (2004)
8. D.K. Kim, M. Mikhailova, M. Toprak, Y. Zhang, B. Bjelke, J. Kehr, In-situ gold coating of superparamagnetic nanoparticles by microemulsion method. Mater. Res. Soc. Symp. Proc. **704**, W6.28.1–6.28.5 (2002)
9. O.M. Lavrynenko, S.V. Netreba, The formation of nanoscale structures in the system Steel3-H2O-O2-AgNO3. Nanostrukturnoe materialovedenie **2**, 9–24 (2009). (in Russian)
10. O.M. Lavrynenko, The mechanism of the formation of the ultra disperse iron oxide minerals on the steel surface in presence of Cu(II), Ag(I), Au(III), Pt (IV), Pd(II). Nanostrukturnoe materialovedenie **3**, 3–13 (2010). (in Russian)
11. O.M. Lavrynenko, Ya.D. Korol, S.V. Netreba, V.A. Prokopenko, Kinetic regularity of the formation of Fe (II)–Fe (III) LDH structures (Green Rust) on the steel surface in presence of the FeSO4 and Fe2(SO4)3 water solutions. Him. Fiz. Tehnol. Poverhni **1**(3), 338–342 (2010)
12. E.O. Loughlin, S. Kelly, K. Kemner, R. Csencsits, R.E. Cook, Reduction of AgI, AuIII, CuII, and HgII by FeII/FeIII hydroxysulfate green rust. Chemosphere **53**(5), 437–446 (2003)
13. J.K. Lim, S.A. Majetich, Composite magnetic-plasmonic nanoparticles for biomedicine: Manipulation and imaging. Nano Today **8**, 98–113 (2013)
14. O.M. Lavrynenko, D.A. Dolynskyi, Nanosized core & shell composites (iron oxide-noble metal): production, structure, properties, prospects of biomedical application. Nanostrukturnoe materialovedenie **3**, 3–34 (2011). (in Russian)
15. O.M. Lavrynenko, Yu.S. Shchukin, N.O. Dudchenko, O.B. Brik, in *Proceedings of the IEEE 8th International Conference Nanomaterials: Applications & Properties*, Zatoka, Ukraine, 9–14 September 2018

Chapter 29
Increase of Mechanical Properties of Weld Metal of High-Strength Low-Alloy Steels

Valery Kostin, Olena Berdnikova, Victor Zukov and Georgy Grigorenko

Abstract The article presents test results concerning the structure of welds made of high strength low-alloy steel 14HGNDC which, in the molten state, was provided with nanoparticles of various refractory compounds including oxides, carbides, and nitrides (TiC, TiN, SiC, VC, NbC, TiO_2, Al_2O_3, MgO and ZrO_2). The performed tests revealed the effective use of the nanooxides of titanium (TiO_2) and zirconium (ZrO_2) enabling the obtainment of high mechanical properties of the weld metal (R_m—708 MPa and 621 MPa, KCV_{-20}—60 J/cm^2 and 72.9 J/cm^2, a—21 and 19%). The use of a Gleeble 3800 welding cycle simulator made it possible to determine the dependence between temperature ranges of transformations, amount of structural constituents and types of modifying nanoparticles.

29.1 Introduction

The issue of increasing the reliability and load-carrying capacity of various building structures is primarily related to the reliability of their welded structures. The popularity of high-strength low-alloy steels in pipeline/gas pipeline-based transport, building engineering and mechanical engineering results from relatively low production-related investments and costs without compromising high mechanical properties of end-products. An increase in the strength of the above-named steels to 700…800…900 MPa (steel X80, X90, and X100) could significantly widen their application potential in welded structures. The problem of reducing the properties of welded joints made of high-strength low-alloy steels (HSLA) can be solved by the controlled effect on the crystallization of liquid metal, phase transformations as well as on the parameters of grain structure, non-metallic inclusions, and phases.

V. Kostin (✉) · O. Berdnikova · V. Zukov · G. Grigorenko
E. O. Paton Electric Welding Institute of the National Academy of Sciences of Ukraine, Kiev, Ukraine
e-mail: valerykkos@gmail.com

A. D. Pogrebnjak and O. Bondar (eds.), *Microstructure and Properties of Micro- and Nanoscale Materials, Films, and Coatings (NAP 2019)*,
Springer Proceedings in Physics 240, https://doi.org/10.1007/978-981-15-1742-6_29

One of the promising methods enabling the control of liquid metal crystallization processes during welding and casting is the use of refractory particles from the nano dimensional range (≤100 nm) [1–3].

The use of large particles (≥10 μm) in the metallurgy of the production of steels and alloys is known adequately well [4], whereas the use of nanodispersive particles aimed to control the structure of metal during arc welding poses numerous difficulties including the necessity to ensure the uniform arrangement of particles in the volume of liquid metal, to prevent the removal of particles from the bath through flowing out or interacting with fluxes, to ensure the appropriate wet ability of molten metal as well as to prevent against coagulation, dissolution and oxidation during welding.

Previous research [5] revealed that automatic arc welding can be applied as a physical model enabling the examination of crystallization processes and the development of phase transformations in metals subjected to welding. The results of such research could be used in conventional metallurgy during the production of steel.

29.2 General Information

29.2.1 Objectives

The research work aimed to investigate the effect of nanodispersive particles of refractory metals on the crystallization of the weld pool liquid metal and the kinetics of austenite transformation in the weld metal of high strength steels.

29.2.2 Materials and Conditions for Obtaining of Welded Joints

In order to study the effect of nanoparticles on the formation of the structure and mechanical properties of the weld metal, butt-welded joints of sheets of high-strength 14HGNDC steel (0.18% C, 0.33% Si, 0.98% Mn, 1.20% Cr, 2.07% Ni, 0.22% Mo, 0.08% V, <0.018% P, <0.005% S) with a thickness of 20 mm were made. In accordance with the requirements of ISO 14171 at a direct current of reverse polarity 240–250 A at arc voltage 31–32 V. Welding speed was 10–12 m/h, the heat input of welding at the same time was 26–28 J/cm. Arc welding was performed in $Ar + CO_2$ shielding gases using flux-cored wire with a diameter of 1.6 mm.

Figure 29.1 shows the filling scheme for cutting the edges of butt joints, in accordance with which passes from 1 to 9 were made with cored wire with a basic doping system (C–Mn–Cr–Ni–Mo–Si–Cu), and passes from 10 to 18 according to the "cold input" technology, flux-cored wires into the core of which different nanopowders were inputting.

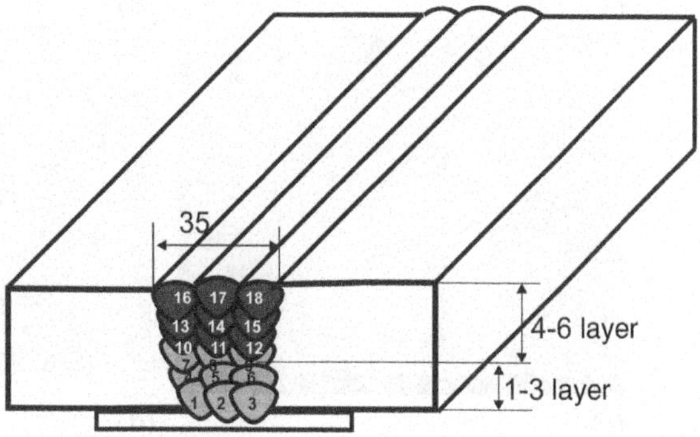

Fig. 29.1 Filling scheme for cutting the edges of butt joints (1–3 without nanopowder, 4–6 layers with nanopowder)

To modify the structure of the weld metal were used various refractory compounds including oxides, carbides, and nitrides of various metals (TiC, TiN, SiC, VC, NbC, TiO_2, Al_2O_3, MgO and ZrO_2). The modifier particles were provided to the liquid metal using arc welding and flux-cored wires containing the flux mixture including particles of required composition. The use of welding technology for the modification of the molten metal composition required the protection of fluxes against the direct effect of welding arc which could trigger their entire melting or evaporation. For this reason, the flux-cored wires were entered into the relatively "cool" part of the weld pool having a temperature restricted within the range of 1600–1800 °C, i.e. lower than the melting points of most compounds.

The test nanopowders were obtained using various methods (mechanical refinement, reduction synthesis and self-developing high-temperature synthesis (SHS)). As a result, the particles obtained were characterized by an appropriate composition, size, and uniform distribution. The main view of modifier particles is presented in Fig. 29.2.

29.2.3 Techniques and Equipment for Research

Microstructure of weld metals modified by nanopowders were studied by optical metallographic microscope Neophot-32 (Japan), scanning JSM 840 (JEOL, Japan) and transmission JEOL 200CX (JEOL, Japan) electron microscope, and thermo-mechanical simulator Gleeble 3800 (DSI, USA) provided with a quick dilatometer [6].

In accordance with a methodology developed at E.O. Paton Electric Welding Institute, specimens were heated in a vacuum chamber to a temperature of 1170 °C and cooled in accordance with a thermal cycle characteristic of automatic welding under flux performed at a welding rate of 5; 10; 17; 30 and 45 °C/s within the temperature range of 800–500 °C. When testing the kinetics of austenite decomposition, the temperatures of transformation start and end, as well as the amounts of transformation products, were identified using the methodology described in publication [7].

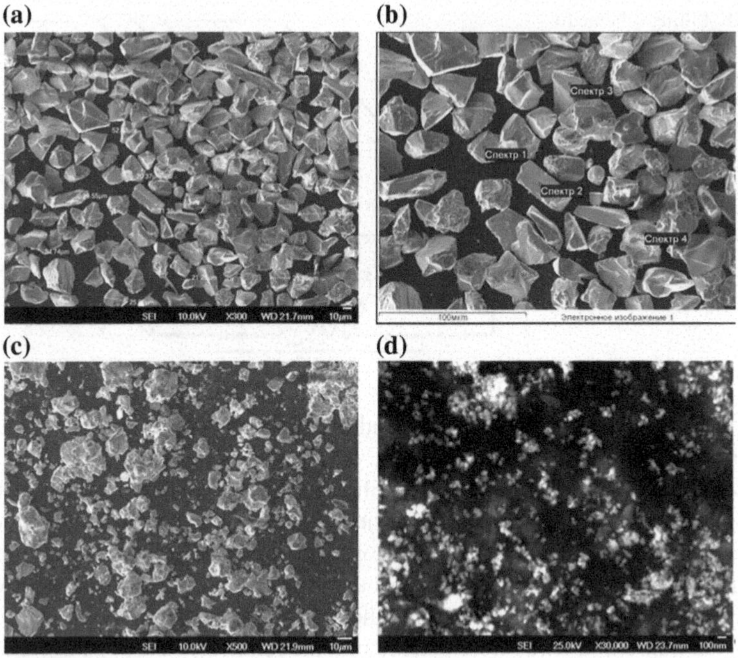

Fig. 29.2 Used nanoparticles: **a** TiC, **b** TiN, **c** TiO_2, **d** ZrO_2; ×30000

The microstructure of the weld metal containing carbide, nitride and oxide modifier particles are presented in Fig. 29.3.

29.3 Results and Discussion

The analysis of the weld metal structure revealed that the use of modifiers containing VC, NbC and SiC led to the formation of primarily upper bainite (Fig. 29.3a) (between 40 and 70%), which significantly reduced the toughness of the welds at sub-zero temperatures (Table 29.1).

The use of titanium nitride (TiN) nanoparticles led to an increase in the content of intra-grain and polygonal ferrite (Fig. 29.3b) to 50%, which also extremely negatively affected the toughness of the weld metal at the relatively high temperature of the tests, i.e. $KCV_{-20} = 40$ J/cm^2.

The use of TiO_2, ZrO_2 and MgO nanoparticles (Fig. 29.3c, d) favored the formation of dispersive acicular ferrite (between 30 and 90%) enabling the obtainment of the more favorable combination of high strength and toughness, particularly at very low temperatures (−40 and −60 °C).

Welded metal markers as "–"was chosen as the base of comparison, in which flux-cored wires with nanoparticles were not used.

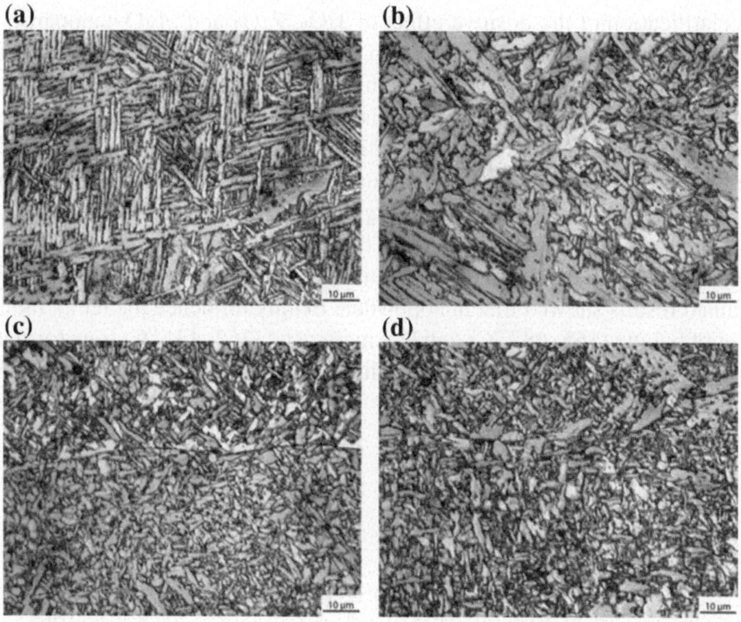

Fig. 29.3 Microstructure of the weld metal with nanoparticles: **a** TiC; **b** NbC; **c** TiO_2; **d** ZrO_2; × 500

Table 29.1 Mechanical properties of welded joints

Type nano	R_m (MPa)	$R_{0.2}$ (MPa)	a (%)	Z (%)	KCV (J/cm^2) at T (°C)			
					+20	0	−20	−40
–	693	605	14.5	48.4	97	87	75	53
TiC	716	644	19.2	63.0	–	–	85	73
TiN	712	580	5.3	14.7	55	47	40	–
SiC	726	650	21.4	62.3	85	72	65	61
VC	780	706	14.2	56.8	57	55	52	–
NbC	544	594	3.0	5.75	44	35	24	–
TiO_2	708	636	19.3	56.7	85	72	60	50
Al_2O_3	728	621	17.5	54.4	82	58	50	36
MgO	644	586	18.6	59.9	103	–	69	60
ZrO_2	621	532	19.5	65.2	120	–	73	65

The clarification of the positive effect of TiO_2, ZrO_2 and MgO nanoparticles on the structure and mechanical properties of the weld metal (in comparison with the effect of carbide and nitride particles) required the performance of tests concerning the kinetics of austenite decomposition. The test results enabled the development of diagrams concerning austenite decomposition in the weld metal (Fig. 29.4).

The analysis of results revealed that austenite decomposition in the weld metal modified using TiO_2, ZrO_2, Al_2O_3, and MgO nanoparticles occurred at significantly higher temperatures (by 100–150 °C) than those at which austenite decomposition in the weld metal modified using carbide or nitride nanoparticles took place (Table 29.2).

Obtained results showed that nanopowders mainly influence the temperatures of the onset of bainite (565–687 °C) and less martensite (363–421 °C) transformations. In this case, the temperature range of bainite transformations varies from 155 °C

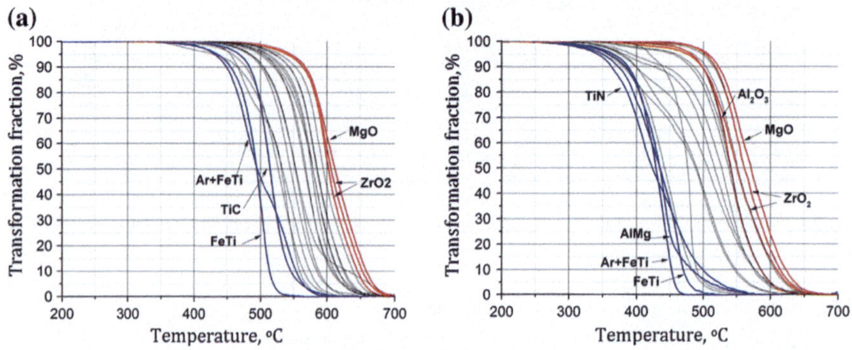

Fig. 29.4 Kinetic diagrams of the decomposition of austenite in the metal of the welds subjected to tests performed using various types of modifiers and the following welding rates: **a** 5 °C/s, **b** 45 °C/s

Table 29.2 Temperatures phase transformation

Type nano	Temperature transformation (°C)						
	Ac_3	Bs	Bf	ΔB	Ms	Mf	ΔM
–	870	673.07	443.07	230.00	363.6	249.5	114.1
TiC	838	583.00	402.10	180.90	399.2	288.2	111.0
TiN	853	589.88	403.04	186.84	403.0	289.6	113.4
SiC	851	579.27	435.06	209.11	420.6	311.4	109.2
VC	840	565.00	410.00	155.00	407.0	296.6	110.4
NbC	893	612.01	431.07	180.94	406.3	295.8	110.5
TiO_2	885	666.42	431.22	235.20	407.7	297.4	110.3
Al_2O_3	890	656.52	437.26	219.26	396.0	284.7	111.3
MgO	873	680.36	457.18	223.18	420.5	311.2	109.3
ZrO_2	867	687.36	458.07	229.29	421.2	311.9	109.3

(VC) to 230 °C (ZrO_2) and 235 °C (TiO_2), but martensite transformation remains almost constant 110–113 °C.

During the modification of weld metal performed using the particles of nanooxides, the initiation of bainite transformation occurs at 670 ÷ 700 °C and at a cooling rate of 5 °C/s, which corresponds to the technological conditions of arc welding performed at 630 ÷ 670 °C and at a high cooling rate of 45 °C/s, which, in turn, corresponds to the technological conditions of laser or hybrid welding.

During the modification of weld metal using carbide and nitride nanoparticles, the transformation of austenite in welds begins at significantly lower temperatures (at a cooling rate of 5 °C/s—550 ÷ 600 °C, whereas at a cooling rate of 45 °C/s—450 ÷ 470 °C). The temperature accompanying the termination of bainite transformation and the initiation of martensite transformation nearly never depends on the types of provided nanoparticles and amounts to 420 ÷ 450 °C.

The results obtained in the tests revealed that, in the weld metal of low-alloy high strength steels, the decomposition of austenite occurred in the high-temperature area of bainite transformation. In contrast to weld metal structures obtained using conventional methods of providing alloying elements leading to the formation of Widmanstätten polygonal ferrite and upper bainite, the modification of weld metal using oxide nanoparticles leads to the formation of acicular ferrite. Such an effect of nanoparticles results from the fact that nanooxides in the weld metal of high strength low-alloy steels constitute additional crystallization nuclei during the formation of acicular ferrite.

The increased resistance of supercooled austenite to transformation when modifying the weld metal using the nanoparticles of carbides and nitrides was probably connected with the difference of surface energy on the phase-particle boundary (γ-phase/carbide and γ-phase/oxide). For this reason, the surface energy on the γ-phase/oxide boundary favors the early initiation of transformation development.

The second probable reason for the higher stability of austenite was the size of micro-stresses generated on the γ-phase/nanoparticle boundary. At low temperatures, the formation of ferritic phases around nanoparticles (through diffusion) is significantly impeded. The hypothesis was verified using micro-diffraction images of areas located on the inclusion/nanoparticle boundary (Fig. 29.5).

The reduction of transformation temperature generates additional micro-stresses around nanoparticles favoring the development of the $\gamma \rightarrow \alpha$ transformation in accordance with mobility kinetics. Micro-stresses generated around globular VC, NbC and SiC nanoparticles are probably significantly higher than those generated around regularly shaped nanoparticles (affecting the character of transformation).

The tests revealed that the angle of mutual orientation on the particle/ferrite boundary amounted to 15° in cases of nanooxides and to 5 ÷ 10° in cases of carbide nanoparticles.

The results correspond well with data presented in the publication [8] stating that the difference between the size of crystal lattice and the phase crystallization lattice should not exceed 10 ÷ 15%.

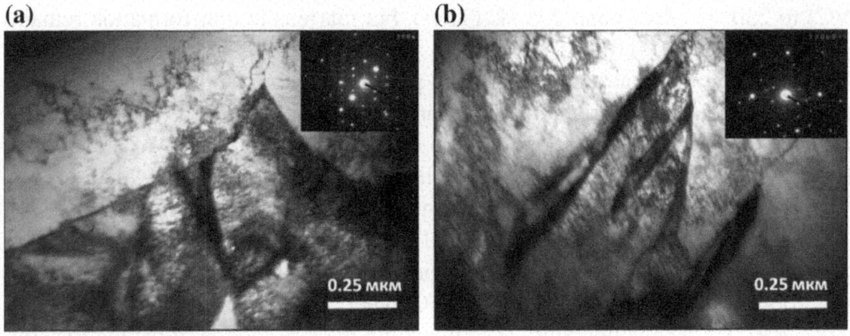

Fig. 29.5 Microstructure and diffraction patten of weld metal with various types of modifiers: **a** TiC, **b** TiO_2

29.4 Conclusions

The effectiveness of modifier particles provided to the metal of the welds in high strength steel 14HGNDC proved the higher, the lower their solubility and the higher their thermodynamic stability, as well as the greater the difference between the melting points of the particles and that of the liquid metal bath, were [9]. The nanoparticles changed the kinetics of austenite transformation favoring the formation of the appropriate weld metal structure (ferrite, bainite or martensite) providing required mechanical properties.

The modification of the weld metal using the nanoparticles of titanium and zirconium oxides (TiO_2 and ZrO_2 respectively) led to the development of transformation in the high-temperature area of the formation of bainite and acicular ferrite characterized by the favorable complex of mechanical and plastic properties. TiC and SiC carbides as well as TiN and NbN nitrides, reducing the initiation temperature of a bainite transformation, led to the formation of bainite-martensite structures (lower bainite, micro-phases, MAC-phase) characterized by lower plastic properties. The VN, VC and ZrC compounds are readily soluble in the liquid metal of the weld pool and, because of that, of little use as effective modifiers of weld metal structures.

References

1. V.P. Saburov, A.N. Cherepanov, *Plasmochemical Synthesis of Ultrafine Powders and their Application for the Modification of Metals and Alloys* (Nauka, Novosibirsk, 1995). (in Russian)
2. V.A. Kostin, V.V. Golovko, G.M. Grigorenko, *Influence of Nanoinclusions on the Formation of the Structure of the Joint Metal of Ferrite-Bainite Steels (Review)*. Collection of Scientific Publications of NUS **4**, 433 (2011). (in Russian)
3. V.P. Saburov, E.N. Eremin, A.N. Cherepanov, *Modification of Steels and Alloys Dispersed Inoculators* (OmGTU, Omsk, 2002), p. 257 (in Russian)

4. I.K. Pokhodnya, V.F. Al'ter, V.S. Shumikhin, I.G. Razdobarin, *Technology of Modifying and Refining Pig Iron using Cored Wire*. Casting Process. **1**, 3 (1993). (in Russian)
5. G.M. Grigorenko, V.A. Kostin, V.V. Golovko, V.V. Zhukov, T.A. Zuber, *Influence of nanopowder inoculators on the structure and properties of cast metal of high-strength low-alloy steels*. Adv. Electrometall. **2**, 32 (2015). (in Russian)
6. G.M. Grigorenko, V.A. Kostin, V.Yu. Orlovsky, *Modern possibilities of modeling austenite transformations in welds of low-alloyed steels*. Autom. Weld. **3**, 31 (2008). (in Russian)
7. S. Van der Zwaag, *Kinetics of phase transformations in steels*. In *Phase Transformations in Steels*, ed. by E. Pereloma, D.V. Edmonds (Woodhead Publishing Limited, 2012), p. 656
8. A.N. Zadiranov, A.M. Kats, *Theoretical Foundations of Crystallization of Metals and Alloys*. (MGIU, Moskva, 2008), p. 194 (in Russian)
9. V.T. Kalinin, V.Ye. Khrychikov, V.A. Krivosheyev, *On the criterial evaluation of the effectiveness of modifiers when processing cast irons*. Theor. Practi. Metall. **2**, 25 (2004) (in Russian)

Chapter 30
Effect of Growth Conditions on Structural and Optical Properties of SnS_2 Thin Film

M. M. Kolesnyk, A. A. Voznyi, Yu. Skyriya, A. Čerškus, Yu. S. Yeromenko, A. P. Voznyi and A. S. Opanasyuk

Abstract Effect of growth conditions on structural and optical properties of SnS_2 thin film was investigated. It was shown, that samples deposited at substrate temperature of 573 K have SnS_2 single phase structure. It was also determined that optical band gap of the SnS_2 decreases with the film thickness.

30.1 Introduction

SnS_2 thin films are of considerable scientific interest since its unique properties allow to produce the heterojunction solar cells (SCs) [1–3].

Due to the large value of the wide of the band gap (BG) $E_g = 2.2–2.8$ eV and n-type conductivity, the SnS_2 compound is considered to be a promising material for the creation of SCs window layers [3–8]. At the same time, SnS_2 films do not contain rare (In, Ga) and environmentally hazardous (Cd) elements that are part of the traditional buffer and absorption layers of the SCs—CdTe, $CuInSe_2$ (CIS), Cu(In, Ga)Se_2 (CIGS), CdS. SnS_2 films often have low structural quality, which may significantly affect the optical and electrical properties of the material. This can be explained by a large number of recombination centers and presence of secondary phases ($SnS_2 \geq Sn_2S_3$) [9–12].

Among a large number of methods for producing SnS_2 thin films, the method of thermal evaporation in close-spaced vacuum sublimation (CSS) deserves special attention. This method allows obtaining high crystalline quality and purity semiconductor layers [13–15]. Also, this method is technologically simple and cheap enough.

M. M. Kolesnyk · A. A. Voznyi · Yu. Skyriya · Yu. S. Yeromenko (✉) · A. P. Voznyi · A. S. Opanasyuk
Sumy State University, Sumy, Ukraine
e-mail: yuri.yeromenko@gmail.com

A. Čerškus
Department of Electronics State Research Center, Center for Physical Sciences and Technologies, Vilnius, Lithuania
e-mail: a_cerskus@yahoo.com

A. D. Pogrebnjak and O. Bondar (eds.), *Microstructure and Properties of Micro- and Nanoscale Materials, Films, and Coatings (NAP 2019)*, Springer Proceedings in Physics 240, https://doi.org/10.1007/978-981-15-1742-6_30

It allows receiving semiconductor films in conditions close to thermodynamically equilibrium. It should be noted that, despite the advantages of this method, for the application of SnS_2 compounds, the CSS method was used only in several works [16, 17].

Thus, the control of phase composition, structural and optical properties of SnS_2 films and the study of the effects of growth conditions on these properties for the development of a new generation of SCs are relevant.

30.2 Experimental Details

The SnS_2 thin films were obtained in a vacuum system VUP-5 M. The pressure of residual gas in the chamber was not more than 5×10^{-4} Pa. The temperature of the evaporator during the precipitation of tin disulfide films was the same and amounted to 948 K. This temperature was chosen experimentally, but is slightly lower than the melting point of the SnS_2 compound ($T_s = 1143 \pm 15$ K) [18]. The substrate temperature during the production of thin films varied in the range $T_s = 423$–623 K. Under substrate temperatures above $T_s > 723$ K, the deposition of the SnS_2 compound on the substrate did not occur. This can be explained by the complete evaporation of the material. The time of film deposition was varied in the range t = 0.2–7.0 min and was taken as the main parameter for controlling the thickness of the layer. The films were deposited on glass substrates. The substrates were pre-purified in an ultrasonic bath in a solution of ethyl and isopropyl alcohol and deionized water for 10 min in each.

Determination of the structural properties and phase composition of tin disulfide films was performed using the X-ray diffractometer DRON-4-07. Sampling was carried out in the mode of continuous registration in the range of Bregian angles 2θ from 10° to 80°. The identification of the crystalline phases was carried out using the Crystallography Open Database (COD) database.

Raman spectra were measured at the Renishaw InVia90V727 Raman spectrometer at room temperature.

30.3 Results and Discussions

Figure 30.1 shows typical SEM images of surface and cross-section of the samples, obtained at a substrate temperature of $T_s = 573$ K. The analysis of the surface images of the films obtained in the range of the substrate temperature 423–623 K, showed that the films were polycrystalline, homogeneous with the area and had good adhesion with the substrate. The grains were shaped like plates. A similar form of crystallites was also observed in the works [19–21]. The average size of film crystallites, when the substrate temperature changed, was varied in the range $D = 0.5$–1.4 μm.

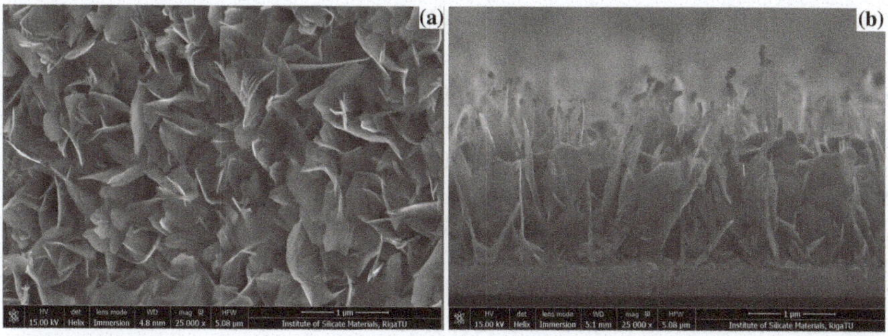

Fig. 30.1 SEM images of SnS_2, obtained with substrate temperature: $T_s = 573$ K (**a**), and typical cross-section (**b**)

The typical cross-section of SnS_2 thin films is shown in Fig. 30.1b. The figure shows the lamellar form of crystallites. It was established that the grain in the layers is oriented along the plane, perpendicular to the surface of the substrate. It was also found that substrate temperature T_s does not significantly affect the thickness of the films. The average thickness of the SnS_2 thin films was 2 μm.

The XRD patterns of SnS_2 films deposited at different substrate temperatures are shown in Fig. 30.2. It was found that the detected peaks corresponded to reflections from the (002) crystallographic planes of hexagonal phases of the SnS_2, (002), (102), (103), (104), (013), (210), (204), (105), (205) crystallographic planes of the orthorhombic phase of Sn_2S_3 and (110) crystallographic planes of the orthorhombic phase of SnS. The positions of these peaks are in good agreement with the Crystallography Open Database (COD) for SnS_2, Sn_2S_3 Ta SnS (card No. 96-900-0614, 96-901-1237, 96-900-8296 respectively).

Fig. 30.2 XRD patterns of SnS_2 obtained with different substrate temperature

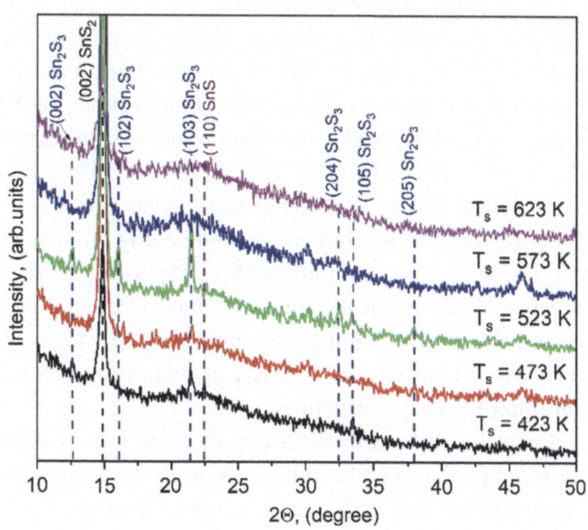

It is established that the reflection from the crystallographic plane (002) of the hexagonal phase of the SnS_2 compound is the most intense. The intensity of peaks corresponding to the Sn_2S_3 and SnS compounds was very weak. It was found that the film obtained at $T_s = 573$ K is the most textured, single-phase, as evidenced by the highest peak intensity from the plane (002) of the hexagonal phase of the SnS_2 compound. Calculation by the inverse pole figures method reveals the [002] growth texture in the SnS_2 films. The quality of this texture depended on the film growth conditions. This direction of growth texture coincides with the growth direction of the most crystallites in the material. A similar growth texture for SnS_2 films was also observed by researchers in the works [22, 23].

The lattice constants of the material were calculated by the NelsoneRiley method. It was found that the lattice parameters of the SnS_2 layers increase monotonically with changes of T_s from 473 to 623 K. The value of lattice parameters a changes in the range of $a = 0.3640$–0.3647 nm; while lattice parameters c varies from 0.5703 to 0.5738 nm.

Experimentally calculated values of the lattice constants a, c, for tin disulfide films, correlate with the values presented in the database (COD) for a single crystal SnS_2 ($a = 0.36380$ nm, $c = 0.58800$ nm). However, the value of the parameter c is less than the reference data. This may be due to the presence of point defects in the crystalline structure in films, which arose due to the deviation of their chemical composition from the stoichiometric (films depleted by sulfur). Also, to change the values of the lattice parameters of tin disulfide films, the macro-stress may be affected. They may arise as a result of the difference in coefficients of thermal linear expansion of the film and the substrate.

One of the most effective methods for phase analysis of tin chalcogenides is Raman spectroscopy [10, 24]. This method is often used in conjunction with the XRD and allows identification of the presence of SnS_2 polytypes in thin films [25, 26].

Typical results of measurements of Raman spectra obtained at $T_s = 423$ K and $T_s = 573$ K are shown in Fig. 30.3.

It is established that Raman spectroscopy results are well correlated with the results of X-ray analysis. On all spectra, there is a clear intense peak at 315 cm^{-1}. It corresponds to the A1g optical phonon mode of SnS_2 polytype with hexagonal symmetry [27]. Were also recorded peaks of significantly less intensity at frequencies 62, 73, 155, 184, 210, 237, 252, 320 cm^{-1}. They correspond to the orthorhombic phase of the Sn_2S_3 compound [28]. Also was found the weak peaks at 97, 225, 288 cm^{-1} of a stable SnS compound having an orthorhombic structure [25].

From the analysis of the results of the X-ray diffraction and Raman spectroscopy, we can conclude that the films obtained in the temperature range $T_s = 423$–623 K contain the predominant hexagonal phase of SnS_2. The optimum substrate temperature was found ($T_s = 573$ K), at which the films become single-phase, with a high-quality crystal structure.

Since the SnS_2 thin films can be used as a window layer of thin-film SCs, we studied the effect of film thickness on its optical properties. To do this, we obtained a series of samples at the same substrate temperature $T_s = 573$ K, but with different thicknesses. The thickness of the layers was determined by SEM directly from the

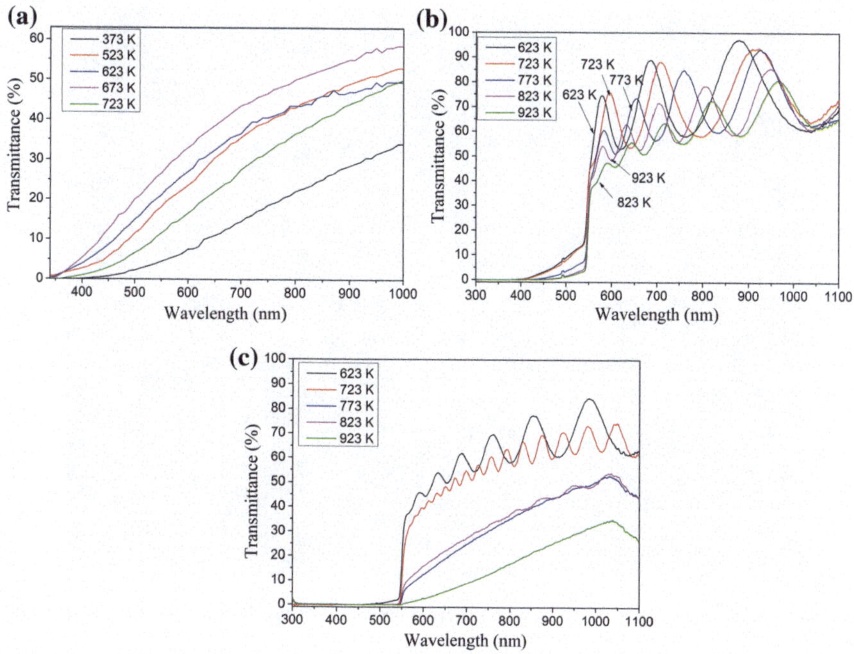

Fig. 30.3 Raman spectra of the SnS_2 thin films obtained at different substrate temperatures

cross-section. Figure 30.4 shows typical cross-sectional SEM images of the samples SnS_2, with different thickness.

As a result, it was found that the thickness of the plate grains decreases from the bottom upwards in the direction of the perpendicular surface of the samples. This indicates a cone-shape of the crystallites and columnar structure of growth, which is required to create a SCs based on polycrystalline films.

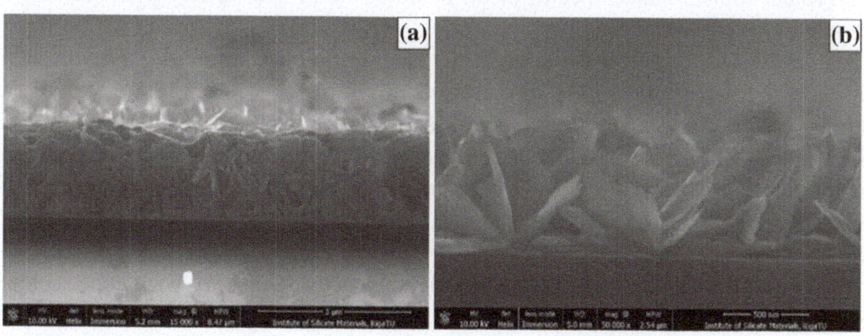

Fig. 30.4 Typical cross-section SEM images of SnS_2 films with different thickness

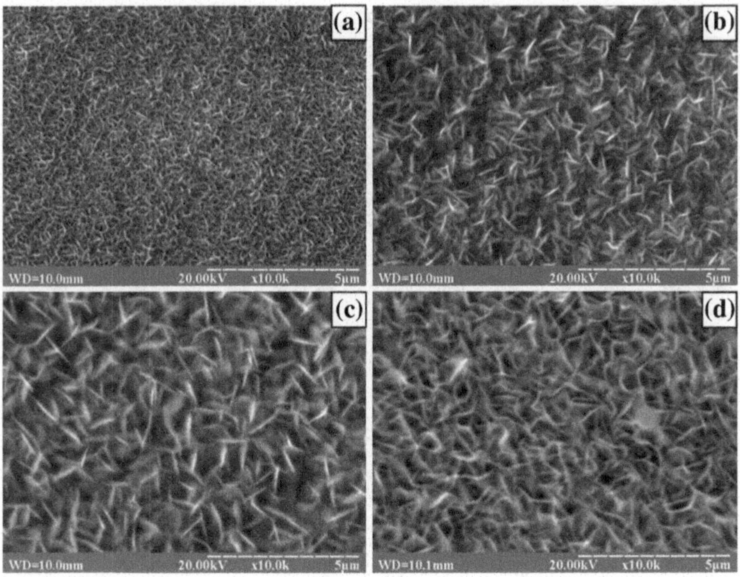

Fig. 30.5 SEM images of SnS_2 films with different thickness, l, nm: 200 (**a**), 500 (**b**), 800 (**c**), 800 (**d**)

The analysis of the surface images of films with different thickness (Fig. 30.5) showed that increasing the thickness of the film increases the size of the grains in their plane and improves their crystalline quality.

Increasing the size of grains more intensively occurs at small film thicknesses. With a further increase in the thickness of the layers, the growth of the size of the crystallites was slowed down due to blockage of closely spaced grains. It was established that the average plate size (d) varies in the range of 0.2–3.0 μm when the thickness of the film increases.

Typical transmittance spectra of tin disulfide films with different thickness are shown in Fig. 30.6a.

It was found, that near the edge of the absorption band, films had a high transmission coefficient of 60–90%. For all studied samples, the spectra have regions of strong ($h\nu < E_g$) and weak ($h\nu > E_g$) passes. It was established that an increase in the thickness of films from 200 to 1700 nm leads to a significant change in the position of the material absorption edge toward larger wavelengths. This is due to the fact that the optical band gap (BG) of the material decreases its value by increasing the thickness of the SnS_2 films.

To determine the optical BG of SnS_2, dependences $(\alpha h\nu)^2$–$h\nu$ were constructed. Figure 30.6b shows the dependence of the calculated optical BG on the thickness of the film. It is established that the width of the BG decreases with an increase in the thickness of the films from the values from $E_g = 2.98$ eV (for 200 nm thick film) to $E_g = 2.44$ eV (for 1700 nm thick film). The change in the form of the spectra

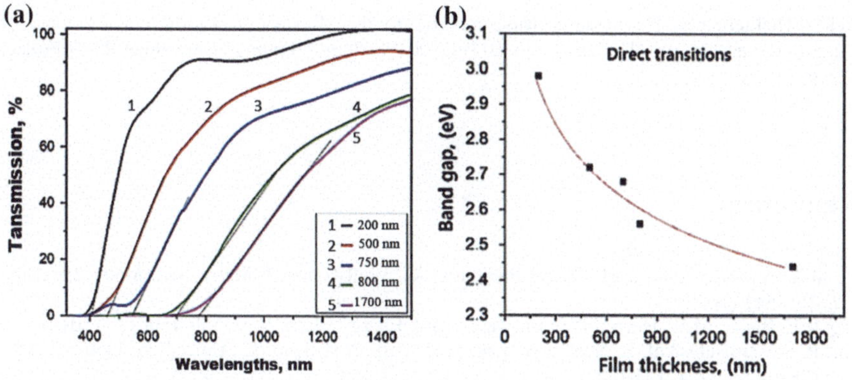

Fig. 30.6 Transmittance spectra SnS_2 thin films (**a**) and dependence of the band gap on the thickness of the film (**b**)

can be explained by the effect of varying the degree of film crystallinity and the orientation of the crystallites in the material. This causes a strong anisotropy of its optical properties.

30.4 Conclusions

In this work, SnS_2 thin films were obtained by the CSS technique. It is shown that the obtained tin disulfide films consist of plate-like grains. Increasing substrate temperature leads to increasing size of crystallites in the range of $D = 0.5–1.4$ μm. The average thickness of the films does not change with increasing substrate temperature and was 2 μm.

X-ray analysis showed that the films obtained in the temperature range of $423 < T_s < 623$ K mainly contain the hexagonal phase of SnS_2. At the same time, a small amount of Sn_2S_3 compound with an orthorhombic structure was detected in the layers.

The optimal substrate temperature was found to be $T_s = 573$ K, at which the films become single-phase, with high crystal quality.

It is established that with increasing thickness of condensates, their grain size increases and their crystalline quality improves.

Investigation of the optical characteristics of condensates has shown that an increase in the thickness of films from 200 to 1700 nm leads to a significant change in the position of the material absorption edge towards the larger wavelengths. The increase in the thickness of the film also reduces the optical width of the BG of the material from 2.98 to 2.44 eV.

Acknowledgements This research was supported by the Ministry of Education and Science of Ukraine (Grant No. 0116U002619, No. 0119U100398, Joint Ukrainian-Lithuanian R&D Project No. 36-2019).

References

1. J.A. Andrade-Arvizu, M. Courel-Piedrahita, O. Vigil-Galán, J. Mater. Sci.: Mater. Electron. **26**, 4541 (2015)
2. R.E. Banai, M.W. Horn, J.R.S. Brownson, Sol. Energy Mater. Sol. Cells **150**, 112 (2016)
3. P. Sinsermsuksakul, L. Sun, S.W. Lee, H.H. Park, S.B. Kim, C. Yang, R.G. Gordon, Adv. Energy Mater. **4**, 1 (2014)
4. K. Abass, H.A. Jassim, K.J. Majeid, R.H. Misho, Phys. Status Solidi **91**, 129 (1985)
5. L.A. Burton, D. Colombara, R.D. Abellon, F.C. Grozema, L.M. Peter, T.J. Savenije, G. Dennler, A. Walsh, Chem. Mater. **25**, 4908 (2013)
6. E. Lee, Y.S. Yoon, D.-J. Kim, ACS Sensors **3**, 2045 (2018)
7. C. Shi, Z. Chen, G. Shi, R. Sun, X. Zhan, X. Shen, Thin Solid Films **520**, 4898 (2012)
8. H. Zhong, G. Yang, H. Song, Q. Liao, H. Cui, P. Shen, C.-X. Wang, J. Phys. Chem. C **116**, 9319 (2012)
9. C. Manoharan, K.S. Kumar, S. Dhanapandian, G. Kiruthigaa, K.R. Murali, in *Proceedings of the International Conference on Nanoscience, Engineering and Technology, ICONSET 2011,* Chennai, Tamilnadu, India, 28–30 November 2011
10. M.G. Sousa, A.F. da Cunha, P.A. Fernandes, J. Alloys Compd. **592**, 80 (2014)
11. A. Voznyi, V. Kosyak, A. Opanasyuk, N. Tirkusova, L. Grase, A. Medvids, G. Mezinskis, Mater. Chem. Phys. **173**, 52 (2016)
12. A. Voznyi, V. Kosyak, P. Onufrijevs, L. Grase, J. Vecstaudža, A. Opanasyuk, A. Medvids, G. Mezinskis, J. Alloys Compd. **688**, 130 (2016)
13. D. Kurbatov, A. Opanasyuk, H. Khlyap, Phys. Status Solidi A **206**, 1549 (2009)
14. M.M. Ivashchenko, I.P. Buryk, A.S. Opanasyuk, D. Nam, H. Cheong, Ja.G. Vaziev, V.V. Bibyk, Mat. Sci. Semicon. Proc. **36**, 13 (2015)
15. V. Kosyak, Y. Znamenshchykov, A. Čerškus, L. Grase, YuP Gnatenko, A. Medvids, A. Opanasyuk, G. Mezinskis, J. Lumin. **171**, 176 (2016)
16. N.R. Paudel, C. Xiao, Y. Yan, in *Proceedings of the 2015 IEEE 42nd Photovoltaic Specialist Conference (PVSC)*, New Orleans, Louisiana, USA, 14–19 June 2015
17. C. Shi, P. Yang, M. Yao, X. Dai, Z. Chen, Thin Solid Films **534**, 28 (2013)
18. H. Gamsjäger, *Chemical Thermodynamics of Tin—Volume 12* (OECD Publishing, 2012)
19. O.V. Bilousov, Y. Ren, T. Törndahl, O. Donzel-Gargand, T. Ericson, C. Platzer-Björkman, M. Edoff, C. Hägglund, Chem. Mater. **29**, 2969 (2017)
20. L.S. Price, I.P. Parkin, A.M.E. Hardy, R.J.H. Clark, T.G. Hibbert, K.C. Molloy, Chem. Mater. **11**, 1792 (1999)
21. S. Wang, S. Wang, J. Chen, P. Liu, M. Chen, H. Xiong, F. Guo, M. Liu, J. Nanoparticle Res. **16**, 2610 (2014)
22. K.S. Kumar, C. Manoharan, L. Amalraj, S. Dhanapandian, G. Kiruthigaa, K. Vijayakumar, Cryst. Res. Technol. **47**, 771 (2012)
23. K. Vijayakumar, C. Sanjeeviraja, M. Jayachandran, L. Amalraj, J. Mater. Sci.: Mater. Electron. **22**, 929 (2011)
24. V.G. Hadjiev, D. De, H.B. Peng, J. Manongdo, A.M. Guloy, Phys. Rev. B **87**, 104302 (2013)
25. H.R. Chandrasekhar, D.G. Mead, Phys. Rev. B **19**(2), 932 (1979)
26. I.P. Parkin, L.S. Price, T.G. Hibbert, K.C. Molloy, J. Mater. Chem. **11**(5), 1486 (2001)
27. A.J. Smith, P.E. Meek, W.Y. Liang, J. Phys. C: Solid State Phys. **10**, 1321 (1977)
28. H.R. Chandrasekhar, R.G. Humphreys, U. Zwick, M. Cardona, Phys. Rev. B **15**, 2177 (1977)

Chapter 31
Effect of Mn Doping on the Optical Properties of Chalcogenide Compounds ZnS and ZnTe

O. V. Klymov and D. I. Kurbatov

Abstract In this paper, the influence of the addition of Mn on the optical properties of the binary compounds ZnS and ZnTe deposited by a close-spaced vacuum sublimation method onto nonoriented glass at different substrate temperatures was considered. The spectral dependences of the transmittance T (λ), reflectance R (λ) and the absorption α (λ), as well as the bandgap for each material (E_g) were measured and calculated.

31.1 Introduction

With the rapid development of technology, more and more types of equipment and devices are being invented that are more material-intensive. Therefore, there is an active search for materials that could replace the pioneers of semiconductors materials such as silicon and germanium, which have great efficiency. Some of the promising materials are semiconductor compounds II–VI based on elements of group II (Cd, Zn, Hg) and group VI (O, S, Se, Te). A variety of compounds makes it possible to obtain materials with a band gap from a few hundredths of eV (for mercury chalcogenides) up to 3.72 eV (ZnS). This feature of group II–VI elements allows, in turn, to cover a very large spectral range and use these materials as photoconductors, field-effect transistors, optical coatings, neutron detectors, light-emitting diodes, electroluminescent materials, optical sensors, and solid-state solar windows layers [1–3]. Careful examination of properties of the above materials will allow us to use these compounds in the electronics, light industry, military, and medical fields. Therefore, it is very important to find optimal conditions in order to avoid structural defects and to obtain the desired electrical and optical properties.

O. V. Klymov (✉)
Department of Applied Fisica, University of Valencia, Valencia, Spain
e-mail: Oleksii.klymov@uv.es

D. I. Kurbatov
Sumy State University, Sumy, Ukraine
e-mail: dkurbatov@sci.sumdu.edu.ua

A. D. Pogrebnjak and O. Bondar (eds.), *Microstructure and Properties of Micro- and Nanoscale Materials, Films, and Coatings (NAP 2019)*,
Springer Proceedings in Physics 240, https://doi.org/10.1007/978-981-15-1742-6_31

Adding additional elements to binary compounds ZnS or ZnTe allows to create new material and change the band gap, which makes it possible to significantly expand the scope. One of such alloying element is manganese (Mn), which allows to vary the width of band gap from $E_g = 3.1$ eV (MnS) to $E_g = 3.68$ eV (ZnS) for ZnMnS, or from $E_g = 2.26$ eV (ZnTe) to $E_g = 2.9$ eV (MnTe) for ZnMnTe.

At present, solid solutions of ZnS and ZnTe alloyed by Mn can be made by using variety of physical and chemical deposition methods, such as hot-wall epitaxy [4], spray pyrolysis [5], magnetron sputtering [6], MOVPE [7], sol-gel method [8], Molecular beam epitaxy [9], chemical co-precipitation [10, 11], and others. In this paper, to apply films we used the close-spaced vacuum sublimation method (CSVS). This method, due to the design features of the evaporator, allows obtaining the stoichiometric films of multicomponent semiconductors high-purity under a well-controlled process [12, 13].

31.2 Experiment Details

Thin $Zn_{1-x}Mn_xS$ and $Zn_{1-x}Mn_xTe$ films were deposited on an ultrasonically cleaned glass substrate by a close-spaced vacuum sublimation method. Samples were obtained in the vacuum equipment of VUP-5M [14]. Residual gas pressure was no more than 5×10^{-3} Pa. For the ZnMnS and ZnMnTe films, the charge of semiconducting purity containing manganese content was used to be about 7% and 5%, respectively. The evaporation time was usually 15 min (hereinafter—series 1) for ZnMnS, and 4 min (hereafter—series 2) and 10 min (hereinafter—series 3) for ZnMnTe films. The evaporator temperature was $T_e = 1473$ K (ZnMnS) and $T_e = 1073$ K (ZnMnTe). The substrate temperature has varied in the interval $T_s = (373–723)$ K and $T_s = (623–923)$ K for ZnMnTe and ZnMnS, respectively.

Measurement of the structural characteristics of condensates was carried out by a *CreCYCSem* group of the University of Valencia (Valencia, Spain) using the Hitachi S-4800 scanning electron microscope and the Philips X'Pert PRO MRD PW3050/65 X-ray diffractometer. When calculating diffractograms were used Ni-filtered K_α radiation of the copper anode in the angle range 2θ from 10° to 160°, where 2θ is the Bragg angle [15]. The average grain size (D) in condensates was found on the Jeffris method [16].

Investigation of the spectral transmittance coefficient was carried out at room temperature by using the spectrophotometer SF-2000. The diameter of the optical cuvette was 1 cm, the size of the light slit for ultraviolet light was 0.5 nm, the size of the light split for visible light was 1.0 nm, the spectral range was 190–1100 nm.

The spectral range for reflectance measurement was 400–800 nm, SFO-2000 console was used to measure reflection. The size of the light spot on the specimen was 2.5 mm.

The geometry of the reflectance measurement of the studied samples was as follows:

Series 1 and 2:

- Diffuse reflection:
  - the angle between the normal to the surface of the sample and the direction of illumination was 8°;
  - the angle between normal to the surface of the sample and the direction of reflection was 45°.
- Mirror reflection:
  - the angle between the normal to the surface of the sample and the direction of illumination was 45°;
  - the angle between normal to the surface of the sample and the direction of reflection was 45°.

Series 2:

- Diffuse reflection:
  - the angle between the normal to the surface of the sample and the direction of illumination was 8°;
  - the angle between normal to the surface of the sample and the direction of reflection was 45°.
- Mirror reflection:
  - the angle between the normal to the surface of the sample and the direction of illumination was 8°;
  - the angle between the normal to the surface of the sample and the direction of reflection was 8°.

The absorption spectra of chalcogenide films what are necessary for the calculation of E_g were founded by using the spectra of transmission (T) and reflection (R) of light and equation:

$$\alpha_i = -\frac{1}{d}\ln\left(\frac{1}{R^2}\left(-\frac{(1-R)^2}{2T} + \frac{\left[(1-R)^4 + 4R^2T^2\right]^{\frac{1}{2}}}{2T}\right)\right), \tag{31.1}$$

where d is the thickness of the sample, R and T are the reflection and transmission coefficients, respectively.

To determine the optical band gap (Eg) of materials from their absorption spectra, we used the following relation, which is valid for direct band semiconductors:

$$\alpha_i h\nu = A_0\left(h\nu - E_g\right)^{1/2} \tag{31.2}$$

where A_0 is a constant, which depends on the effective mass of charge carriers; $h\nu$ is the energy of optical quanta.

From this, it follows that the extrapolation of the linear part of the graph $(\alpha h\nu)^2$-$h\nu$ on the axis of energy allows us to determine the band gap of the semiconductor.

31.3 Results and Discussions

Measuring the surface morphology and sample cross-section, it was established that the growth of Mn-alloyed $Zn_{1-x}Mn_xS$ and $Zn_{1-x}Mn_xTe$ films occurred similarly to the growth of condensates of other chalcogenides [15, 17]. Electron microscopic images of the surface of typical $Zn_{1-x}Mn_xS$ and $Zn_{1-x}Mn_xTe$ films are shown in Fig. 31.1. In the obtained $Zn_{1-x}Mn_xS$ compounds, the grains had a longitudinal shape and relatively small thickness of the films (d_1 ~ 2.25–3.58 μm). Their size was D_1 ~ 0.08–2.5 μm (Series 1), while it increased with increasing substrate temperature.

On the other hand, the $Zn_{1-x}Mn_xTe$ grain had a more rounded shape and a smaller size D_2~ 0.5–0.9 μm. Depending on the series, the thickness of the films was d_2 ~ 0.85–1.69 μm (series 2) or d_3 ~ 1.72–2.84 μm (series 3), however, it increased with increasing substrate temperature (not given in this paper). These values were somewhat lower compared to the results obtained for non-alloyed ZnS and ZnTe films deposited under similar conditions [18].

X-ray studies of $Zn_{1-x}Mn_xTe$ samples showed that the films were single-phase, polycrystalline, with a cubic modification of the sphalerite type (Fig. 31.2). Typically, reflections from planes (111), (222), (333) and (444) of the cubic phase were recorded in the diffractograms. At the same time, in the majority of cases, the dominant intensity was peak (111). On the other hand, $Zn_{1-x}Mn_xS$ films have some differences. At temperatures T_s> 623 K, diffractograms exhibit peaks corresponding to the hexagonal phase, so high-temperature condensates $Zn_{1-x}Mn_xS$ are two-phase. More detailed information is provided in the paper [19].

Transmittance spectra of all 3 series of samples of solid solutions obtained on glass substrates at different substrate temperatures are presented in Fig. 31.3. As shown in Fig. 31.3a, the increase in the transmittance of $Zn_{1-x}Mn_xS$ films with a wavelength of the used laser is smooth, and the transmittance was in the range of T_1~ (32–58)% at a wavelength of 1000 nm. In this case, the growth of the substrate temperature leads to a higher value of the transmittance. However, this value is slightly lower for

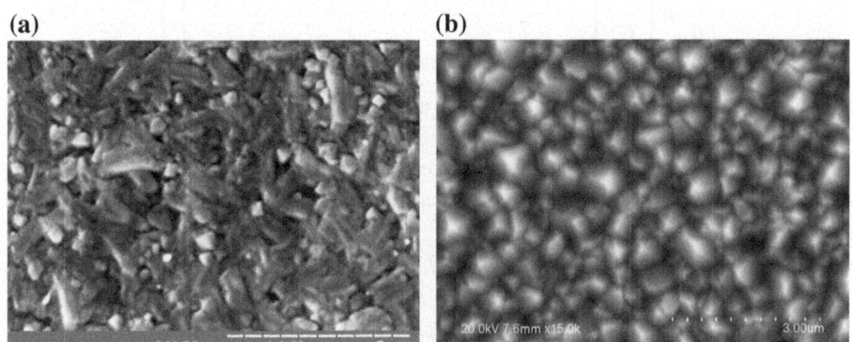

Fig. 31.1 Microphotographs of the surface of the **a** $Zn_{1-x}Mn_xS$ and **b** $Zn_{x-1}Mn_xTe$ films deposited at a substrate temperature Ts: 723 K and 773 K, respectively

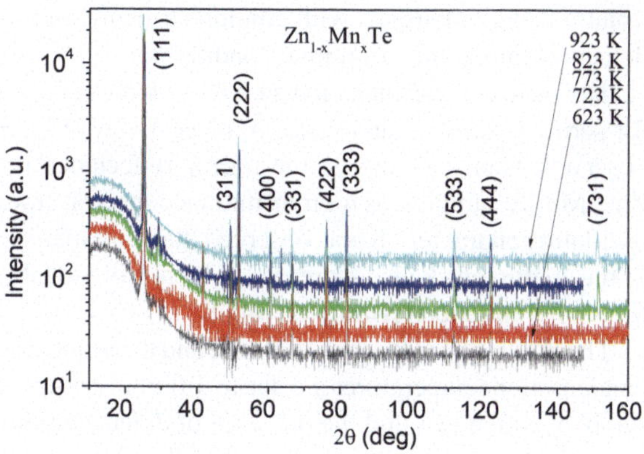

Fig. 31.2 Diffractograms of $Zn_{x-1}Mn_xTe$ films deposited under various substrate temperature, T_s

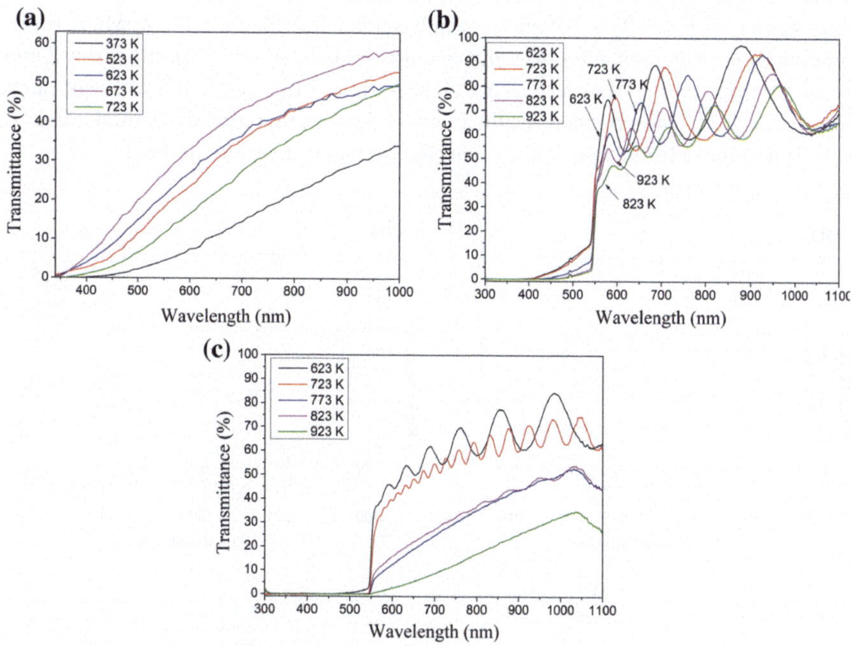

Fig. 31.3 Spectra of transmittance of **a** ZnMnS films (series 1) and **b**, **c** ZnMnTe films (series 2 and series 3, respectively)

two-phase films (T_s> 623 K). On the other hand, for $Zn_{1-x}Mn_xTe$ films can detect a sharp increase in light transmittance at wavelengths $\lambda > 540$ nm. It was found that the transmittance was in the range T_2~ (55–95)% and T_3~ (40–70)% for the series 2 and series 3, respectively. Moreover, if the thickness of the film is commensurate with the wavelength of the incident radiation, then the transmission curves acquire an oscillating character as shown in Fig. 31.3b, c. The best optical characteristics usually had condensates with column structure and low concentration of structural defects obtained under optimal conditions.

The diffuse reflectance reached R_1~ (1–9)%, R_2 ~ (1–15)% and R_3 ~ (15–30)% for series 1, series 2 and series 3, respectively (Fig. 31.4). A similar situation was observed when calculating the mirror reflectance of light (Fig. 31.5), where the $Zn_{1-x}Mn_xTe$ samples, as for transmittance spectra, at wavelengths $\lambda > 540$ nm have an oscillating character . It was observed that with increasing temperature, the substrate of the coefficient R also increased, which is obviously due to the emergence of such films of pronounced pyramidal relief.

The difference in the transmittance and the reflectance of films obtained at different condensation temperatures are due to different crystalline and phase structure of these samples, roughness and the presence of defects on their surface.

Typical absorption coefficients for all series are shown in Fig. 31.6. In the region of radiation energies of larger band gap, the absorption coefficient of ZnMnS thin films was $\alpha_1 = (1.8\text{–}3.2) \times 10^4$ cm^{-1}, which is much less than for thin ZnMnTe films, showing $\alpha_2 = (2\text{–}3.5) \times 10^6$ cm^{-1} and $\alpha_3 = (4\text{–}7.1) \times 10^6$ cm^{-1} for the series 2 and series 3, respectively. These values are close to those observed in the works of other authors [20, 21]. The obtained results confirm the fact that already in thicknesses d ~ (1–3) μm these films absorb almost all the incident radiation on them.

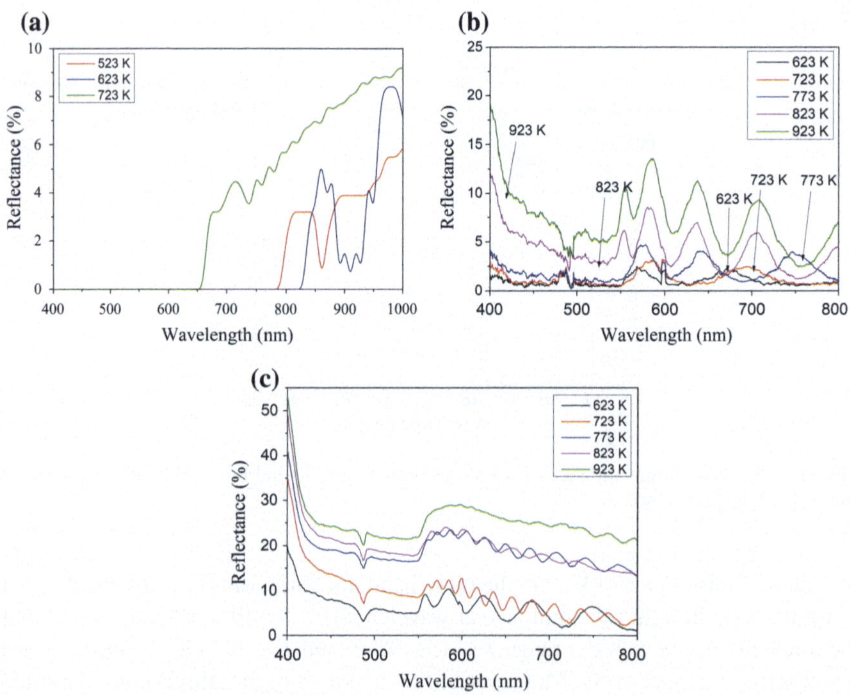

Fig. 31.4 Spectra of diffuse reflectance of **a** ZnMnS films (series 1) and **b**, **c** ZnMnTe films (series 2 and series 3, respectively)

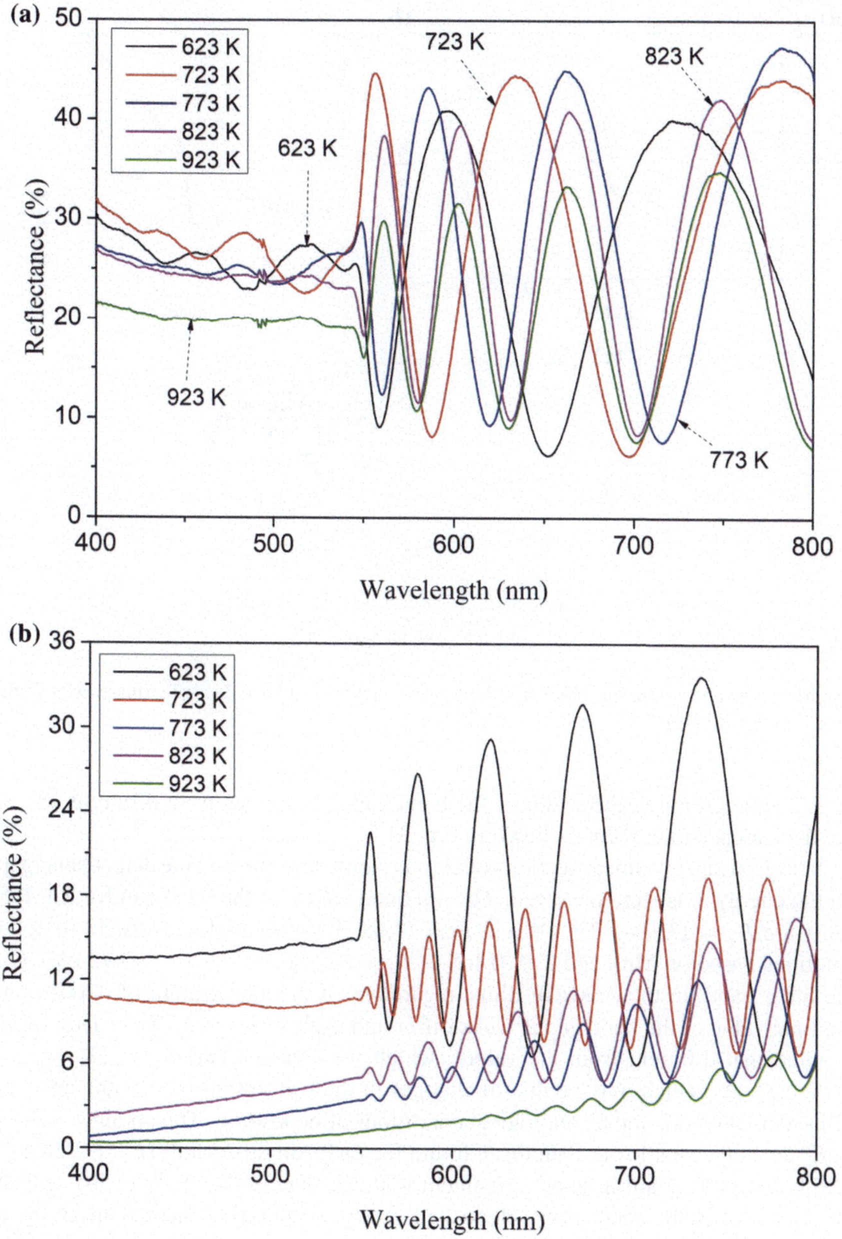

Fig. 31.5 Spectra of mirror reflectance of light of **a**, **b** ZnMnTe films (series 2 and series 3, respectively)

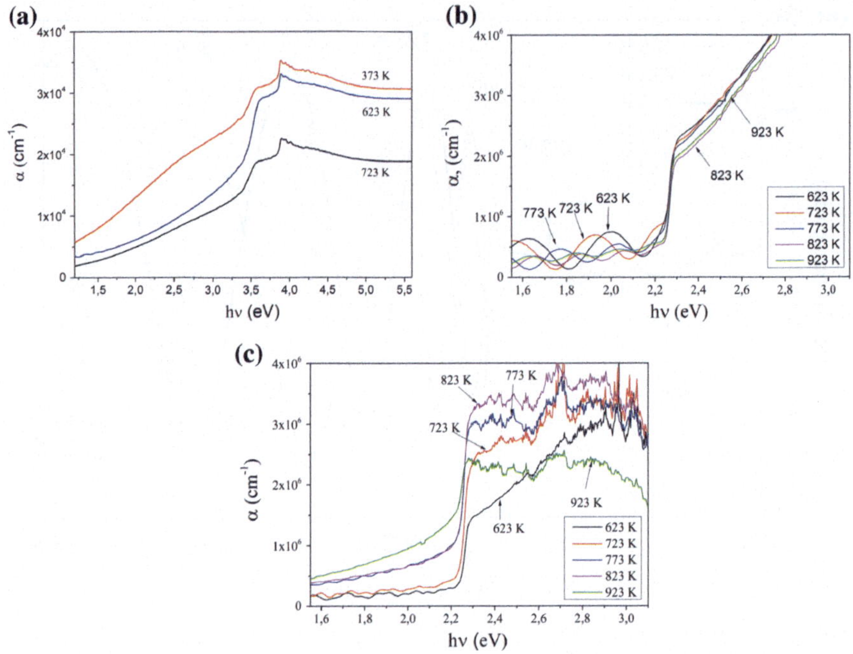

Fig. 31.6 The absorption spectra of **a** ZnMnS films (series 1) and **b**, **c** ZnMnTe films (series 2 and series 3, respectively)

When determining the width of the band gap for samples from different series, the dependence $(\alpha hv)^2$-hv is shown in Fig. 31.7.

In Fig. 31.6a, two linear sections of extrapolation that allowed the detect band gap on the energy axis were observed. The obtained values of the band gap for ZnMnS films are E_g = (3.10–3.23) eV and E_g= (3.68–3.72) eV and correspond to solid solutions based on MnS and ZnS (MnS – E_g = 3.1 eV, ZnS – E_g = 3.72 eV). On the other hand, in the ZnMnTe films, regardless of the temperature of deposition, the final value of the band gap is weak differ and is about Eg = 2.25 eV. This value is somewhat different from the values, which we obtained earlier in [22]—E_g = 2.15–2.23 eV, which may be due to changes in the concentration of manganese in films, depending on the technological conditions of deposition. The obtained values of E_g are somewhat lower than those found for ZnTe single crystals (E_g = 2.28 eV) [23] however they are in good agreement with the data of the study of the optical properties of ZnTe condensates deposited by the evaporation method in an open vacuum (E_g = (2.0–2.2) eV) [24] and CSE (E_g = (2.0–2.2) eV) [25].

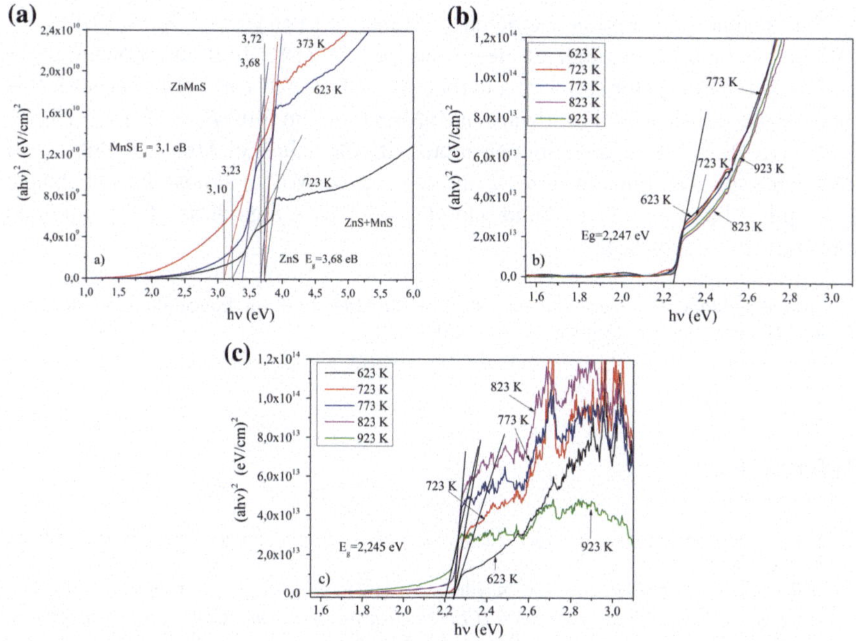

Fig. 31.7 $(\alpha h\nu)^2$-$h\nu$ dependence of **a** ZnMnS films (series 1) and **b**, **c** ZnMnTe films (series 2 and series 3, respectively)

31.4 Conclusions

In this paper, ZnS and ZnTe thin films with a low Mn concentration were obtained by using a close-spaced vacuum sublimation method onto the ultrasonically cleaned glass. Structural studies have shown that the addition of manganese slightly changed the morphology of the films, and the increase in temperature deposition contributed to an increase in the grain size in studied films. Diffraction patterns showed that the films were a single phase with a cubic structure (111). Although at high temperatures ($T_s > 623$ K), a hexagonal phase appeared in the ZnMnS films.

Measurement of the transmittance showed that for ZnMnS films it increased smoothly with increasing incident wavelength and was T_1~ (32–58)%. On the other hand, the thin film of ZnMnTe in the case of irradiation of incident light $\lambda > 540$ nm showed a sharp increase in the transmittance, while it decreased from T_2~ (55–95)% to T_3~ (40–70)% with an increase in deposition time from 4 to 10 min (series 2 and series 3, respectively). The diffuse reflectance of light reached R_1~ (1–9)%, R_2~ (1–15)% and R_3~ (15–30)% for the series from 1 to 3, respectively. At the same time, with increasing substrate temperature the coefficient R also increased. The difference in the transmittance and the reflection of films obtained at different condensation temperatures is due to different crystalline and phase structure of these samples, roughness, and defects on their surface.

The calculated absorption coefficient of the series 1 (ZnMnS) is $\alpha_1 = (1.8\text{–}3.2) \times 10^4$ cm^{-1}, which is quantitatively less than for thin ZnMnTe films, showing $\alpha_2 = (2\text{–}3.5) \times 10^6$ cm^{-1} (for series 2) and $\alpha_3 = (4\text{–}7.1) \times 10^6$ cm^{-1} (for series 3). It is obtained the values optical band gaps of ZnMnS thin film films $E_g = (3.10\text{–}3.23)$ eV та E_g= (3.68–3.72) eV are commensurate with the values of MnS and ZnS based solid solutions. The films in series 2 and series 3 (ZnMnTe) almost did not change their optical band gap of the band gap (E_g ~ 2.5 eV), regardless of the time and temperature of deposition.

Acknowledgements This work was supported by the Ministry of the Education and Science of Ukraine (Grant numbers 0118U003576, 0117U003929).

References

1. S. Kasap, P. Capper (eds.), *Springer Handbook of Electronic and Photonic Materials* (Springer, US, Berlin, 2007)
2. B.V. Grinyov, V.D. Ryzhikov, S.V. Naydenov, A.D. Opolonin, E.K. Lisetskaya, S.N. Galkin, P. Lecoq, in *Proceedings of the 2006 IEEE Nuclear Science Symposium Conference Record*, San Diego, California, October 29–November 4 2006
3. I.I. Gnezdilov, G.L. Dedenko, R.F. Ibragimov, V.A. Idalov, V.V. Kadilin, A.A. Kaplun, A.V. Klemetiev, V.I. Mukhin, A.A. Taraskin, E.M. Turin, R.N. Zaripov, Optimization of the neutron detector design based on the ^{6}LiF/ZnS(Ag) scintillation screens for the GAMMA-400 space observatory. Phys. Procedia **74**, 199 (2015)
4. H. Kuwabara, R. Sakamoto, H. Tatsuoka, Y. Nakanishi, Growth of $Zn_{1-x}Mn_xTe$ films on GaAs(100) by hot-wall epitaxy. Appl. Surf. Sci. **212–213**, 267 (2003)
5. O. Diachenko, A. Opanasuyk, D. Kurbatov, N. Opanasuyk, O. Kononov, D. Nam, H. Cheong, Surface morphology, structural and optical properties of MgO films obtained by spray pyrolysis technique. Acta Phys. Pol. A **130**, 805 (2016)
6. A.N. Georgobiani, A.N. Gruzintsev, Effect of surface morphology on manganese ion luminescence kinetics in ZnS:Mn^{2+} films. Inorg. Mater. **47**, 107 (2011)
7. A. Zozime, M. Seibt, J. Ertel, A. Tromson-Carli, R. Druilhe, C. Grattepain, R. Triboulet, Influence of a ZnMnTe buffer layer on the growth of ZnTe on (001)GaAs by MOVPE. J. Cryst. Growth **249**, 15 (2003)
8. B. Bhattacharjee, D. Ganguli, K. IakoubovskII, A. Stesmans, S. Chaudhuri, Synthesis and characterization of sol-gel derived ZnS:Mn^{2+} nanocrystallites embedded in a silica matrix. Bull. Mater. Sci. **25**, 175 (2002)
9. Y.T. Shih, W.C. Chiang, C.S. Yang, M.C. Kuo, W.C. Chou, Optical properties of $Zn_{1-x}Mn_xTe$ epilayers grown by molecular beam epitaxy. J. Appl. Phys. **92**, 2446 (2002).
10. S. Kumar, H.C. Jeon, T.W. Kang, R. Singh, J.K. Sharma, R.K. Choubey, Structural and optical properties of silica capped ZnS:Mn quantum dots. J. Mater. Sci.: Mater. Electron. **26**, 3939 (2015)
11. K.R. Bindu, E.I. Anila, Structural and optical properties of white light emitting ZnS:Mn^{2+} nanoparticles at different synthesis temperatures. J. Fluoresc. **25**, 795 (2015)
12. Y.S. Yeromenko, Y.P Gnatenko, A.S. Opanasyuk, D.I. Kurbatov, P.M. Bukivskij, M.S. Furier, V. Kuznetsov, A.P. Bukivskii, Photoluminescence of high optical quality CdS:Dy thin films deposited by close-spaced vacuum sublimation. J. Lumin. **197**, 343 (2018)
13. A. Lopez-Otero, Hot wall epitaxy. Thin Solid Films **49**, 3 (1978)

14. V. Kosyak, A. Opanasyuk, P.M. Bukivskij, Y.P Gnatenko, Study of the structural and photoluminescence properties of CdTe polycrystalline films deposited by close-spaced vacuum sublimation. J. Cryst. Growth **312**, 1726 (2010)
15. International Centre for Diffraction Data, American Ceramic Society, *Selected Powder Diffraction Data for Education & Training: Search Manual and Data Cards* (USA, 1988)
16. A.D. Pogrebnjak, A.P. Shypylenko, H. Amekura, Y. Takeda, A.S. Opanasyuk, D.I. Kurbatov, I.A. Kolotova, O.V. Klymova, C. Kozak, Effect of Cu negative ion implantation on physical properties of $Zn_{1-x}Mn_xTe$ films. Acta Phys. Pol. A **123**, 939 (2013)
17. C.J. Panchal, A.S. Opanasyuk, V.V. Kosyak, M.S. Desai, I.Yu. Protsenko, Structural and substructural properties of the zinc and cadmium chalcogenides thin films (a review). J. Nano-Electron. Phys. **3**, 274 (2011)
18. D. Kurbatov, A. Opanasyuk, S.M. Duvanov, A.G. Balogh, H. Khlyap, Growth kinetics and stoichiometry of ZnS films obtained by close-spaced vacuum sublimation technique. Solid State Sci. **13**, 1068 (2011)
19. O. Klymov, D. Kurbatov, A. Frolov, in *Proceedings of the International Conference on Information and Telecommunication Technologies and Radio Electronics* (UkrMiCo'2018), Odessa, Ukraine, 10–14 Sept 2018
20. A.K.S. Aqili, Z. Ali, A. Maqsood, Characterization of zinc telluride thin films deposited by two-source technique and post-annealed in nitrogen ambient. J. Cryst. Growth **317**, 47 (2011)
21. M. Gaceur, M. Giraud, M. Hemadi, S. Nowak, N. Menguy, J.P. Quisefit, K. David, T. Jahanbin, S. Benderbous, M. Boissière, S. Ammar, Polyol-synthesized $Zn_{0.9}Mn_{0.1}S$ nanoparticles as potential luminescent and magnetic bimodal imaging probes: synthesis, characterization, and toxicity study. J. Nanoparticle Res. **14**, 932 (2012)
22. O. Klymov, D. Kurbatov, O. Levchenko, in *International Conference on Advanced Optoelectronics & Lasers*, Sudak, Ukraine, 9–13 Sept 2013
23. M. Imamura, A. Okada, T. Yamaguchi, Magneto-optical properties of wider gap II-VI ZnMnTe and ZnMnCoTe films. J. Appl. Phys. **99**, 08M706 (2006)
24. R. Amutha, A. Subbarayan, R. Sathyamoorthy, K. Natarajan, V. Subramaniam, Conduction Studies on ZnTe thin films. J. New Mat. Electr. Syst. **10**, 27 (2007)
25. G.I. Rusu, P. Prepelita, R.S. Rusu, N. Apetroaie, G. Oniciuc, A. Amariei, On the structural and optical characteristics of zinc telluride thin films. J. Opt. Electr. Adv. Mat. **8**(3), 922 (2006)

Chapter 32
Magneto-Optical and Magnetic Properties of Three-Layer Films Based on Permalloy and Copper

V. B. Loboda, Yu. O. Shkurdoda, Yu. M. Shabelnyk, A. M. Chornous, S. M. Khursenko, L. V. Dekhtyaruk, K. Merkotan and O. O. Drozdenko

Abstract Three-layer nanostructured films $Fe_xNi_{100-x}/Cu/Fe_xNi_{100-x}$ ($x \cong 50$, $d_{FeNi} = 40$ нм; $d_{Cu} = 5$ нм) consisting of $Fe_{0,5}Ni_{0,5}$ alloy layers separated by Cu layer were obtained by layers-by-layers vacuum condensation at the pressure of 10^{-4} Pa. Magnetic hysteresis loops of three-layer films were obtained using a vibration magnetometer. Magnetic hysteresis loops obtained in transverse geometry have an almost rectangular shape with sharp magnetization reversal. The magnetization reversal of the films completely changes in the magnetic field range with induction of −25 to 25 mT for freshly condensed films and −30 to 30 mT for thermostabilized films. The rectangular hysteresis contour is typical for single-layer films of ferromagnetic metals and alloys based on them and indicates the presence of an axis of easy magnetization in the sample plane. Magnetization in the direction perpendicular to the film plane shows the effect of a strong demagnetizing field. After thermal stabilization of the films, the dependence does not change significantly, except for an increase in the residual magnetization and coercive force (more pronounced hysteresis).

V. B. Loboda · S. M. Khursenko
Sumy National Agrarian University, 160, Herasym Kondratiev Str, Sumy 40021, Ukraine

Yu. O. Shkurdoda (✉) · Yu. M. Shabelnyk · A. M. Chornous · O. O. Drozdenko
Sumy State University, Rimsky-Korsakov Str., 2, Sumy 40007, Ukraine
e-mail: yu.shkurdoda@gmail.com

L. V. Dekhtyaruk
Kharkiv National University of Construction and Architecture, 40, Sumska Str, Kharkiv 61002, Ukraine

K. Merkotan
Odesa National Polytechnic University, 1, Shevchenko Ave, Odesa 65044, Ukraine

A. D. Pogrebnjak and O. Bondar (eds.), *Microstructure and Properties of Micro- and Nanoscale Materials, Films, and Coatings (NAP 2019)*, Springer Proceedings in Physics 240, https://doi.org/10.1007/978-981-15-1742-6_32

32.1 Introduction

Currently, an intensive study of the physical properties of new film materials is being carried out in connection with their wide practical use as functional elements of instrument structures [1–8] and protective coatings [9–18]. The study of magnetically inhomogeneous film materials (multilayer structures, granular films, composites, etc.) made it possible at the end of the twentieth century to discover a number of important magnetoresistive effects—giant magnetoresistance (GMS), tunneling magnetoresistance (TMS), colossal magnetoresistance, etc. [19–23].

Multilayer film systems based on permalloy (Py) and copper, in which spin-dependent electron scattering is realized, are widely used as effective magnetic field sensors, digital magnetoresistive memory devices, automotive electronics, biomedical technologies, etc. But despite this, there is a need for further search and experimental study of film structures that would satisfy additional requirements (minimum dimensions, high sensitivity, high reproduction samples, etc.). Also important is the task of predicting the behavior of the electrical and magnetoresistive properties of multilayer systems with a change in the thickness of the metal layers, structure, temperature, and external magnetic field. The solution to such problems is possible only under the condition of using an integrated approach to the study of the physical properties of film systems.

The purpose of this work is a comprehensive study of the magnetic structure and magnetic parameters of magnetic three-layer films based on permalloy and copper.

32.2 Experiment Details

Investigated samples with a thickness of 5–100 nm were obtained by evaporation and condensation in the vacuum chamber of VUP-5 M (at the pressure of 10^{-4} Pa). The films were condensed onto a glass substrate with pre-applied copper contacts with a substrate of chromium at room temperature. The rate of precipitation of the alloy was $\omega = 0.5–1$ nm/s. Three-layer films consisting of $Fe_{0,5}Ni_{0,5}$ alloy layers separated by Cu layer were obtained by layers-by-layers condensation. The construction of the lining holder allowed to receive two film samples in one cycle with a different thickness of the copper layer and with similar thicknesses of the alloy layers. For measuring their electrical resistance, the geometric dimensions of the films were set by windows in the nichrome foil mechanical masks, which made with high accuracy.

The percentage composition of the film was determined by x-ray microanalysis with an energy dispersive spectrometer (EDS).

32.3 Results and Discussion

To establish the relationship between the magnetoresistance of three-layer films with a change in their magnetic structure, a study was made of the domain structure of the films $Fe_{0,5}Ni_{0,5}/Cu/Fe_{0,5}Ni_{0,5}/S$ with simultaneous measurement of the magnetoresistance of the films. To observe the domain structure of the films, the magneto-optical Kerr effect was used, which consists of rotating the plane of polarized light reflected from the surface of a magnetized ferromagnet by a certain angle φ of the plane of polarization. Due to the differences in the angles of rotation of the plane of polarization of light by neighboring domains (depending on the orientation of their magnetic fields), after passing the reflected light through the analyzer, a light contrast is observed between areas with different magnetization. This light contrast is insignificant, since the light reflected from the film surface is not plane-polarized, but polarized elliptically. In Fig. 32.1 shows the loop of magnetoresistive hysteresis and its connection with the domain structure of the upper layer of a freshly condensed three-layer $Fe_xNi_{100\text{-}x}/Cu/Fe_xNi_{100\text{-}x}/S$ ($x \cong 50$, $d_F = 40$ nm; $d_N = 5$ nm) with giant magnetoresistance. The magnitude of the isotropic magnetoresistance for freshly condensed films is 0.5–1%, depending on the thickness of the layers. Small values of isotropic magnetoresistance are caused by small sizes of crystallites, high defectiveness of layers, and shunting of the effect with a relatively thick nonmagnetic interlayer [24, 25].

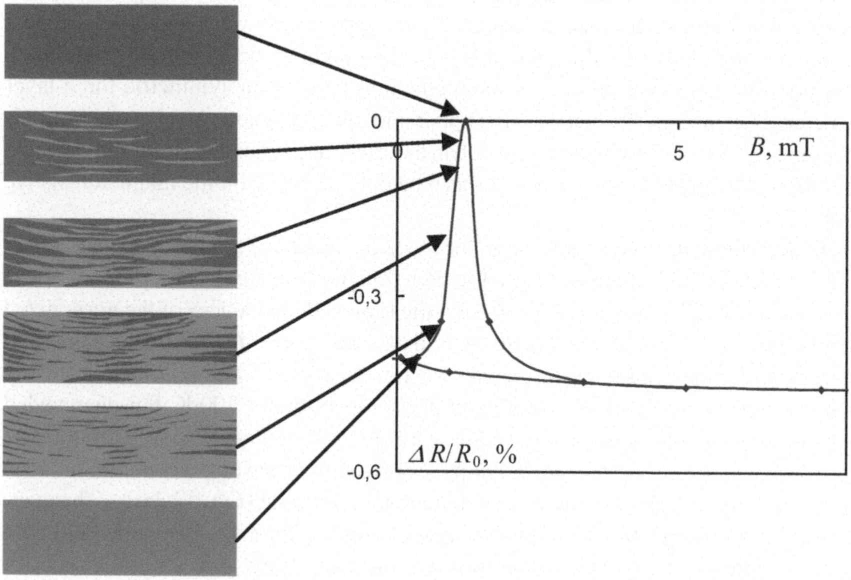

Fig. 32.1 Domain structure and magnetic resistance loop of a freshly condensed three-layer film $Fe_{0,5}Ni_{0,5}/Cu/Fe_{0,5}Ni_{0,5}/S$ ($d_{Py} = 40$ nm; $d_{Cu} = 5$ nm)

In this case, the film was first magnetized to saturation in a magnetic field directed along the axis of easy magnetization. When the magnetic field decreases to zero, the ferromagnetic layers of the film remain single-domain and their magnetization does not change. The magnetoresistance is also practically unchanged. At the first moment of growth of the field directed opposite to the field in which the film was magnetized, the ferromagnetic layers remained single-domain and the magnetoresistance did not change significantly. Then, when the magnetic field becomes sufficiently strong (about 0.8 mT), the magnetization vectors rotate at the edges and throughout the upper layer. As a result, small domains are formed in the form of strips. The domains that were formed are favorably oriented relative to the direction of the external magnetic field. With increasing magnetic field induction, they increase their size. Then, even with a small increase in the magnetic field, a sharp increase in the advantageously oriented domains continues at the expense of the neighboring ones. With a further increase in the magnetic field, a "breakdown" occurs—a rapid magnetization reversal of the central regions. The electrical resistance at the same time reaches a maximum value. It is obvious that the magnetic moments in the lower and upper layers of permalloy are oriented antiparallel (antiparallel configuration). One of the reasons for the delay in the magnetization reversal of the lower layer may be internal stress due to the adhesion of the layer to the substrate, which leads to an increase in the coercive force. With a further increase in the magnetic field induction (over 1 mT), a sharp decrease in the electrical resistance of a three-layer film is observed due to the magnetization reversal of the lower layer and orientation of the magnetizations in the Fe_xNi_{100-x} layers in parallel (parallel configuration). We believe that in these three-layer structures, the upper and lower magnetic layers are re-magnetized at slightly different values of the magnetic field induction. Although it is possible when the process of magnetization reversal in symmetric three-layer structures occurs synchronously—the upper and lower magnetic layers switch simultaneously at the same induction value of an external magnetic field. It should be noted that these values are close because a sharp peak is recorded on the magnetoresistive loop.

To determine the magnetic state of the systems, magnetic studies were performed using an LDJ-9500 vibration magnetometer (magnetic hysteresis loops were determined). Figure 32.2 shows for illustration the typical dependences of the normalized magnetization on the induction of an applied external magnetic field in transverse and perpendicular measurement geometries for $Fe_{80}Ni_{20}/Cu/Fe_{80}Ni_{20}$ films in a freshly condensed state and after heat treatment at a temperature of 700 K. For unannealed and annealed at 700 K three-layer films with $d_F = 25$–40 nm, $d_N = 5$–15 nm, and $c_{Ni} < 70$ at.% magnetic hysteresis loops, obtained in transverse geometry, have an almost rectangular shape with sharp magnetization reversal (Fig. 32.2a, c). The magnetization reversal of films completely changes in the range of a magnetic field with an induction of -25 to 25 mT for freshly condensed films and -30 to 30 mT for thermostabilized films. The rectangular hysteresis contour is typical for single-layer films of ferromagnetic metals and alloys based on them and indicates the presence of an axis of easy magnetization in the sample plane. The magnetization in the direction perpendicular to the film plane shows the influence of a strong demagnetizing field

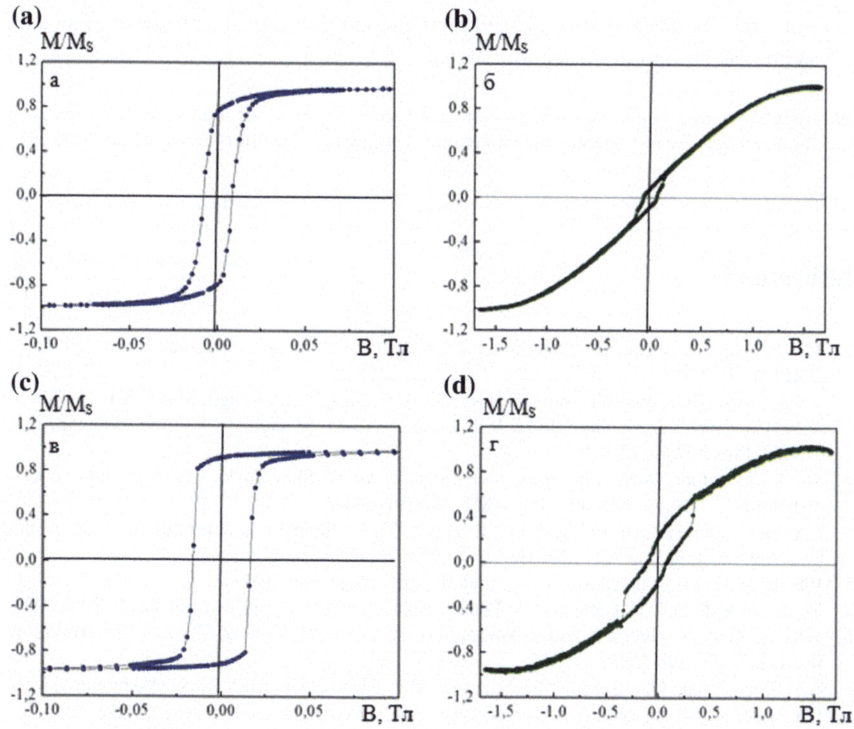

Fig. 32.2 Dependence of the normalized magnetization on the applied external magnetic field in transverse (**a**, **c**) and perpendicular (**b**, **d**) measurement geometries for films $Fe_{0,5}Ni_{0,5}/Cu/Fe_{0,5}Ni_{0,5}/S$ (d_{Py} = 40 nm, d_{Cu} = 10 nm) in a freshly condensed state (**a**, **b**) and after heat treatment at a temperature of 700 K (**c**, **d**)

(Fig. 32.2d). After thermal stabilization of the films, the dependence does not change significantly, except for an increase in the residual magnetization and coercive force (more pronounced hysteresis).

32.4 Conclusion

The maximum value of isotropic magnetoresistance at the room temperature is 1% for as-deposited $Fe_{0.5}Ni_{0.5}/Cu/Fe_{0.5}Ni_{0.5}/S$ thin films with $d_F = 25$ nm and $d_N = 5$ nm. We consider that in these three-layer structures, the upper and lower magnetic layers are re-magnetized at slightly different values of the magnetic field induction. Small values of isotropic magnetoresistance are caused by small sizes of crystallites, high defectiveness of layers, and shunting of the effect with a relatively thick nonmagnetic interlayer. The rectangular hysteresis contour is typical for investigated thin films of ferromagnetic metals and alloys based on them and indicates the presence of an axis of easy magnetization in the sample plane.

Acknowledgements The work has been performed under the financial support of the Ministry of Education and Science of Ukraine (state registration numbers 0119U100777 and 0117U003925).

References

1. M.A. Corrêa, R. Dutra, T.L. Marcondes, T.J.A. Mori, F. Bohn, R.L. Sommer, Mater. Sci. Eng., B **211**, 115 (2016)
2. Yuan-Tsung Chen, Jiun-Yi Tseng, S.H. Lin, T.S. Sheu, J. Magn. Magn. Mater. **360**, 87 (2014)
3. A.M. Chornous, Yu.O. Shkurdoda, V.B. Loboda, Yu.M. Shabelnyk, V.O. Kravchenko, Eur. Phys. J. Plus **132**, 58 (2017)
4. Yu.O. Shkurdoda, A.M. Chornous, V.B. Loboda, Yu.M. Shabelnyk, V.O. Kravchenko, L.V. Dekhtyaruk, J. Nano- Electron. Phys. **8**(2), 02056 (2016)
5. L.V. Odnodvorets, I.Yu. Protsenko, O.P. Tkach, Yu.M. Shabelnyk, N.I. Shumakova, J. Nano-Electron. Phys. **9**(2), 02021 (2017)
6. V.B. Loboda, I.E. Protsenko, Kristall und Technik **16**(4), 489 (1981)
7. Yu.O. Shkurdoda, V.B. Loboda, L.V. Dekhtyaruk, Metallofiz. Nov. Tekhnol. **30**(3), 295 (2008)
8. J. Li, Q. Zhan, S. Zhang, J. Wei, J. Wang, M. Pan, Y. Xie, H. Yang, Z. Zhou, S. Xie, B. Wang, R.-W. Li, Sci. Rep. **7**, 2837 (2017)
9. A.D. Pogrebnjak, O.V. Sobol, V.M. Beresnev, P.V. Turbin, G.V. Kirik, N.A. Makhmudov, M.V. Il'yashenko, A.P. Shypylenko, M.V. Kaverin, MYu. Tashmetov, A.V. Pshyk, Ceram. Eng. Sci. Proc. **31**(7), 127 (2010)
10. V.I. Lavrentiev, A.D. Pogrebnjak, Surf. Coat. Tech. **99**(1–2), 24 (1998)
11. A.D. Pogrebnjak, D.I. Proskurovski, Phys. Status Solidi **145**(1), 9 (1994)
12. A.D. Pogrebnjak, ShM Ruzimov, Phys. Lett. **120**(5), 259 (1987)
13. A.D. Pogrebnjak, A.A. Bagdasaryan, A. Pshyk, K. Dyadyura, Phys. Usp. **60**(6), 586 (2017)
14. A.D. Pogrebnjak, Ivashchenko, P.L. Skrynskyy, O.V. Bondar, P. Konarski, K. Załęski, S. Jurga, E. Coy, Composite. Part B. Eng. **142**, 85 (2018)
15. A.A. Goncharov, V.V. Petuhov, D.N. Terpij, P.I. Ignatenko, V.A. Stupak, Neorg. Mater. **41**(7), 799 (2005)
16. A.A. Goncharov, FMM **3**, 325 (2011)
17. A.I. Bazhin, A.A. Goncharov, A.D. Pogrebnyak, V.A. Stupak, S.A. Goncharova, Phys. Metal. Metallography **117**(6), 594 (2016)
18. A.D. Pogrebnjak, A.A. Goncharov, Metallofiz. Nov. Tekhnol. **38**(9), 1145 (2016)
19. P. Fert, A. Grüunberg, F. Barthelemy, W.Zinn Petro, J. Magn. Magn. Mater. **140**, 1 (1995)
20. M.N. Baibich, J.M. Broto, A. Fert, F. Nguyen Van Dau, F. Petro, P. Etienne, G. Creuzet, A. Friederich, J. Chazelas, Phys. Rev. Lett. **61**, 2472 (1988)
21. E.Y. Tsymbal, D.G. Pettifor, in *Solid State Physics*, vol. 56, ed by H. Ehrenreich, F. Spaepen (Academic, San Diego, CA, 2001), p. 113
22. O.I. Tovstolytkin, M.O. Borovyi, V.V. Kurylyuk, Yu.A. Kunytskyi, *Fizychni Osnovy Spintroniky* (TOV Nilan LTD, 2014)
23. Ia.M. Lytvynenko, I.M. Pazukha, V.V. Bibyk, Vacuum **116**, 31 (2015)
24. A.G. Basov, S.I. Vorobiov, Yu.O. Shkurdoda, L.V. Dekhtyaruk, J. Nano- Electron. Phys. **2**, 69 (2010)
25. S.S.P. Parkin, K.P. Roche, T. Suzuki, Jpn. J. Appl. Phys. **31**(9A), L1246 (1992)

Chapter 33
Size-Dependent Melting Behavior of Silver Nanoparticles: A Molecular Dynamics Study

Bohdan Natalich, Yaroslav Kravchenko, Olga Maksakova and Vadym Borysiuk

Abstract Thermal behavior of ten silver nanoparticles (NP) with spherical shape and diameter from 4.8 to 24.5 nm has been investigated by the molecular dynamics (MD) simulations. The structural changes in nanoparticles have been studied within the temperatures from 300 to 2500 K. The melting point has been detected from the temperature dependencies of Lindemann index and potential energy, which were calculated during the simulation process in the chosen temperature range. Obtained data show that melting of the Ag nanoparticles has occurred at temperatures of about 1000 K for the smallest NP shifting to higher values with the growth of NP size. The investigations reveal that the thermal degradation of the crystal structure of the spherical nanoparticles begins with the surface atoms and propagates to the center.

33.1 Introduction

It is known that nanoparticles (NPs) are a particle with a size ranged from 1 to 100 nm, at least for one dimension [1]. NPs are already shown perspective biological and physicochemical properties [2–4]. Nowadays, they start to be widely used in nanoelectronics as a part of sensors and optoelectronic devices [5, 6]. Initially, the main focus of scientists was directed on the synthesis and investigation of NPs that contained single structures, termed simple NPs. Silver and gold are the most representative among simple NPs due to their unique features and utilization [7–9].

The progress in materials science has helped scientists to design a new class of NPs known as hybrid NPs, which can be defined as well-organized nanomaterials consisting of two, three or more types of single nanocomponents [10]. NPs with a core-shell structure is a kind of hybrid NPs that are composed of two or more nanomaterials, where an inner core surrounded by one shell of a different material. Among hybrid NPs, the investigation of bimetallic NPs composed of two different metal elements increasingly gathers steam [11]. Recent research had shown that

B. Natalich · Y. Kravchenko · O. Maksakova (✉) · V. Borysiuk
Sumy State University, Sumy, Ukraine
e-mail: maksakova.tereshenko@gmail.com

A. D. Pogrebnjak and O. Bondar (eds.), *Microstructure and Properties of Micro- and Nanoscale Materials, Films, and Coatings (NAP 2019)*,
Springer Proceedings in Physics 240, https://doi.org/10.1007/978-981-15-1742-6_33

bimetallic nanoparticles sometimes can be a preferable choice over a monometallic, due to their superior properties [12–14]. Electronic properties of the metallic nanoparticles are strongly depending on particle size [15], structure and compound. Therefore, the controlling of the fabrication process, atomic compound and temperature stability of the nanoparticles are important problems of modern nanotechnology. Numerous methods of synthesis of nanoparticles with different structures, size and shape were suggested in many works [5, 12]. However, some experimental techniques that are widely used in materials science, not always can be used for investigation of the structure and behavior of the nanoscale objects [4]. Hence, various theoretical [16–18] and computational [19–22] investigations can be an additional tool in studying nanostructures.

The researchers have noted that for fcc metals among all crystallographic planes that terminate the nanocrystal surface, high index facets demonstrate excellent activity and selectivity for chemical reactions compared with low index facets [23–27]. Therefore, silver nanoparticle structures with different sizes are of great scientific and research interest.

33.2 Model

In this paper, we present the investigation of the melting behavior of silver nanoparticles with spherical shape and different diameters by classical molecular dynamics simulation. The snapshots that represent the size variation of Ag NPs were prepared using Visual Molecular Dynamics software [28] and shown in Fig. 33.1.

Silver atoms were placed at initial positions according to the Ag face-centered cubic lattice. Ideal vacuum conditions were maintained during the simulation. Free boundary conditions were applied in all three directions. The initial diameter of the nanoparticle varies from 4.8 to 24.5 nm with the number of particles (N) involved in simulation from 3589 to 452,221 atoms. The embedded atom method (EAM)

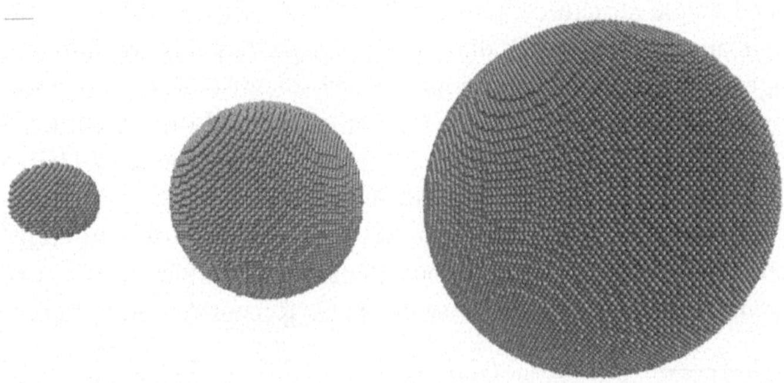

Fig. 33.1 Example of the studied silver nanoparticles with diameter of the NP equal to 4.8, 11.4, and 24.5 nm from left to right

was chosen to describe the interaction between atoms [29]. The generalized EAM potential is widely used in MD simulation of metal alloys and well fitted to reproduce the basic material properties. Within the EAM approach, the potential energy of the metal crystal is expressed as follows:

$$E = \frac{1}{2}\sum_{i,j,i\neq j}\varphi\left(r_{ij}\right) + \sum_{i} F(\rho_i), \tag{33.1}$$

where $\varphi\left(r_{ij}\right)$ is the pair energy between atoms i and j at the distance r_{ij}, $F(\rho_i)$ is the local embedded energy of the atom i with electron density ρ_i.

Each term in (33.1) has the analytical expression that was proposed by approximating the data obtained from the calculations from the first principles. Thus, the pair energy of the interatomic interaction has the form

$$\varphi(r) = \frac{A\cdot\exp\left[-\alpha\left(\frac{r}{r_e}-1\right)\right]}{1+\left(\frac{r}{r_e}-\kappa\right)^{20}} - \frac{B\cdot\exp\left[-\beta\left(\frac{r}{r_e}-1\right)\right]}{1+\left(\frac{r}{r_e}-\lambda\right)^{20}}, \tag{33.2}$$

where r_e—the equilibrium distance between the two atoms of the given type; A, B, α, β—approximation parameters; κ, λ—additional parameters for ensuring zero energy of interaction at significant interatomic distances.

The local embedding energy as a function of electron density $F(\rho_i)$ is calculated in several steps. First, the electronic density ρ_i is calculated as

$$\rho_i = \sum_{i,j\neq i} f\left(r_{ij}\right), \tag{33.3}$$

where $f(r_{ij})$—local electron density in the atomic region of atom i, calculated through the following expression

$$f(r) = \frac{f_e\cdot\exp\left[-\beta\left(\frac{r}{r_e}-1\right)\right]}{1+\left(\frac{r}{r_e}-\lambda\right)^{20}}, \tag{33.4}$$

that has the same form as the second term in formula (33.2) with the same parameter values β and λ. Then the electronic density function $F(\rho_i)$ should be calculated, from three following equations, depending on the value of ρ_i,

$$F(\rho) = \sum_{i=0}^{3} F_{ni}\left(\frac{\rho}{\rho_n} - 1\right)^i, \rho < \rho_n, \rho_n = 0,85\rho_e, \tag{33.5}$$

$$F(\rho) = \sum_{i=0}^{3} F_i\left(\frac{\rho}{\rho_e} - 1\right)^i, \rho_n \leq \rho < \rho_0, \rho_0 = 1,15\rho_e, \tag{33.6}$$

$$F(\rho) = F_e\left[1 - \ln\left(\frac{\rho}{\rho_e}\right)^\eta\right] \cdot \left(\frac{\rho}{\rho_e}\right)^\eta, \rho_0 \leq \rho. \tag{33.7}$$

Such a method for determining the electronic density function $F(\rho_i)$ is necessary for the realistic approximation of the embedding energy, and for reproducing the properties of the material in a wide range of values ρ.

To find the analytical expressions and numerical parameters for the functions used in expressions (33.1) and (33.2) refers to work [29].

To investigate the melting behavior of the nanoparticles the Lindemann index [28] that is a measure of the thermally driven disorder of atoms was computed. The local Lindemann index of an atom i can be defined as:

$$q_i = \frac{1}{N-1} \sum_{j \neq i} \frac{\sqrt{\langle r_{ij}^2 \rangle - \langle r_{ij} \rangle^2}}{\langle r_{ij} \rangle} \tag{33.8}$$

where angle brackets indicate time average at a constant temperature.

General Lindemann index for a system of N atoms can be estimated through averaging over all atoms in the sample. As it is known, with increasing of the sample's temperature, the Lindemann index growth linearly till the sharp threshold, which is close to the melting point [30]. This threshold can be considered as an indication of the crystal-amorphous phase transition. For bulk sample, the critical value of the Lindemann index which can be considered as melting criteria is expected to be in the range from 0.1 to 0.15, depending on the type of the materials [22]. However, recent research [22] had shown that melting of the nanoparticle may occur at the much smaller value of a Lindemann index q_c in the range from 0.03 to 0.05.

During the simulation, the system temperature was increased consequentially using Berendsen thermostat [31] from 300 to 2500 K.

The computational code is implemented using algorithms from LLAMPS software package [32].

33.3 Results

It is worth to mention that calculation of the Lindemann index for very large samples requires significant computational time. Thus in present work temperature dependences of the Lindemann indexes were calculated for silver nanoparticles with diameter from 4.8 to 13.0 nm. Obtained dependences are shown in Fig. 33.2. As can be seen from the figure, presented dependence exhibits typical behavior, slightly increasing in the pre-melting region, with the threshold where it exceeds the critical value around 1000 K for smallest NP a slightly shifting right up to approximately 1200 K for the sample with diameter 13.0 nm. With further temperature growth, Lindemann index continues to increase, and nanoparticle became completely melted above 1100–1400 K for smallest and largest NP radius respectively.

To investigate the melting behavior of the silver NP of larger sizes and to validate obtained results, temperature dependencies of the averaged potential energy of the studied samples, also were calculated (Fig. 33.3). Dependencies presented in Fig. 33.3 also show typical for the melting process behavior with regions of slow growth prior to melting and rapid increasing as the melting starts.

The snapshots of the typical atomistic configuration of three nanoparticles with a diameter of 4.8, 11.4, and 24.5 nm at different temperatures values are shown in Fig. 33.4. As it follows from visual analysis of the general views of nanoparticle, the heating leads to the rearrangement of the atoms from their initial FCC crystal structure to almost amorphous. At temperatures of approximately 1200 K, the longitudinal order in the crystalline structure of the samples begins to collapse and at temperatures above 2250 K, the melting processes intensively occur in the samples, causing the destruction of the crystalline structure.

To detect the changes in the structure of NP's depending on the size of the sample the radial distribution functions $g(r)$ [33] were calculated at temperature of 1500 K for nanoparticles with diameter 4.8, 11.4, and 24.5 nm (Fig. 33.5).

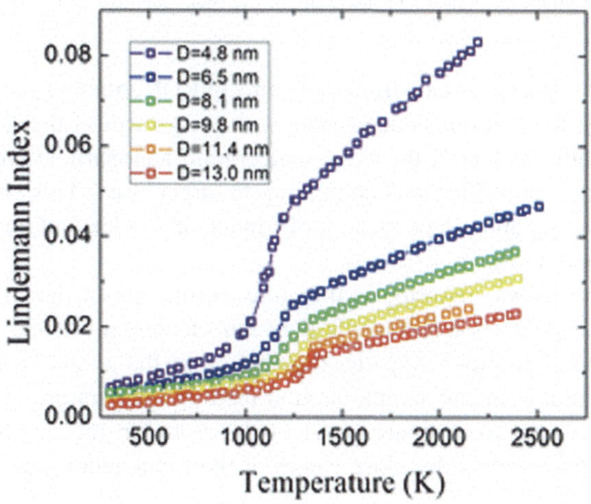

Fig. 33.2 Temperature dependence of the Lindemann index of the silver nanoparticles with different diameter

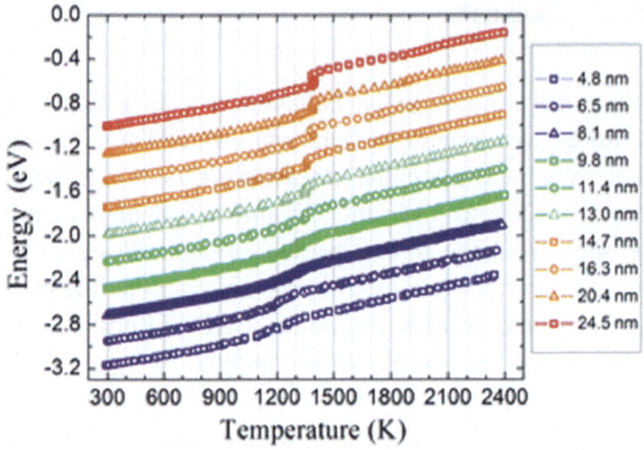

Fig. 33.3 Temperature dependence of the average potential energy of the silver nanoparticles with different diameter (curves are shifted vertically for clearance)

Fig. 33.4 Atomistic configurations of the Ag nanoparticles with diameter 4.8, 11.4, and 24.5 nm from top to bottom. Temperature of the snapshots from left to right: 900, 1200 and 2250 K

As can be seen from the figure, in RDF obtained at the same temperature intensity of fluctuations is decreasing with the growth of the size of nanoparticle. This may indicate that at the same temperature nanoparticle with smaller radius have more amorphous regions comparing to larger ones. This situation was also observed in temperature dependences of Lindemann index and potential energy (see Figs. 33.2 and 33.3).

As was reported in literature [6] melting of the silver nanoparticle of spherical shape starts at surface. To detect this feature in studied samples we plot the atomistic configuration of spatial distribution of the atoms on Lindemann index within the volume of the sample around melting temperature at 1280 K (Fig. 33.6). As can be seen from figure, most of the atoms in the middle section of the sample are characterized by close values of the Lindemann indexes, around value of $q \approx 0.01$,

Fig. 33.5 Radial distribution functions for silver nanoparticles with diameter 4.8, 11.4, and 24.5 nm at temperature 1500 K

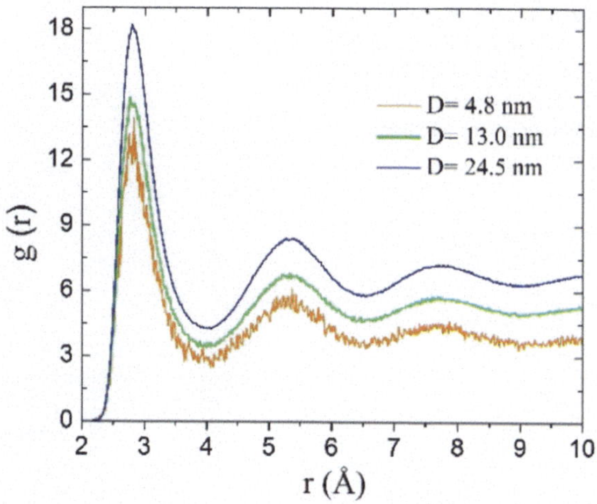

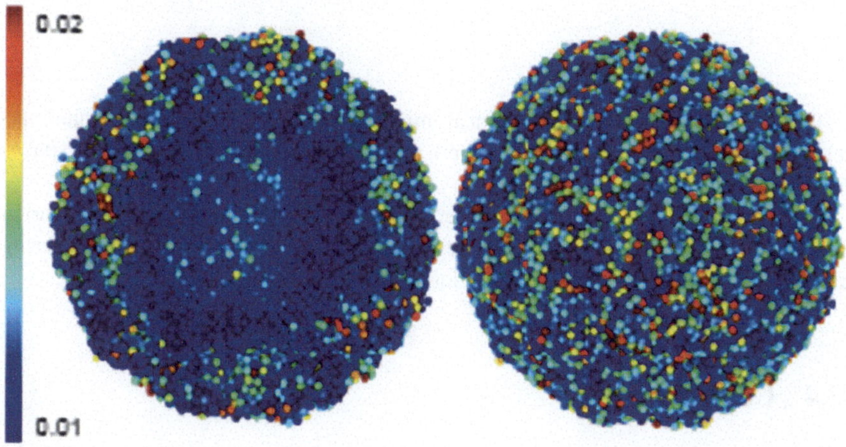

Fig. 33.6 Spatial distribution of the atoms on Lindemann index within the volume of the sample at T 1280 K for NP with diameter 13.0 nm. Atoms with different values of Lindemann index are shown in different colors according to scale

while some atoms in the center of the nanoparticle have slightly higher value of Lindemann index.

At the same time, atoms on the surface of nanoparticle are characterized by values of Lindemann index from the whole range, many of which are higher than atoms in middle section. Thus, some atoms on the surfaces move more intensively, comparing to atoms within the volume of the sample, and melting is expected to occur on surface first.

Fig. 33.7 Cross-section of the structure of the silver NP with diameter of 24.5 nm at 1400 (note the amorphous structure on the surface)

To validate this assumption, we plot atomistic configuration of the largest nanoparticle with a diameter nanoparticle 24.5 nm at 1400 K with visible interatomic spaces (Fig. 33.7).

As Fig. 33.7 shows, the process of thermal destruction of the crystal lattice start at the surfaces of the samples and at the beginning of the melting, long-range ordering is preserved at the center of nanoparticles.

33.4 Conclusions

Melting behavior of the silver nanoparticles with spherical shape and of the different radius was investigated by classical molecular dynamics simulations. To detect the structural changes in the atomic configuration of the nanoparticles the temperature dependencies of Lindemann indexes were calculated for six samples and average potential energies were calculated for all samples in the range from 300 K to 2500 K. As it follows from the computed parameters, the melting of the considered nanoparticles occurred at temperatures of about 1000 K for smallest NP, shifting to higher values with the growth of NP size. At that point, the Lindemann index exceeds the critical value and increasing rapidly with temperature growth, as well as the average energy. As it follows from the snapshots of atomistic configurations, and distribution of atoms on Lindemann index within the volume of the sample thermal destruction of the FCC crystal structure of spherical silver nanoparticles begins at the surface, as the atoms located in outmost from center layers lose their long-range ordering.

Then, with the temperature growth, thermal degradation propagates to the center of the samples. This situation was observed for all studied nanoparticles of different diameters.

Acknowledgements Presented work was financially supported by Ministry of Education and Science of Ukraine (Project No. 0117U003923).

References

1. R. Ferrando, J. Jellinek, R.L. Johnston, Nanoalloys: from theory to applications of alloy clusters and nanoparticles. Chem. Rev. **108**, 845–910 (2008). https://doi.org/10.1021/cr040090g
2. H. Beitollai, F. Garkani Nejad, S. Tajik, S. Jahani, P. Biparva, Int. J. Nano Dimens. **8**, 197 (2017)
3. M. Khorasani-Motlagh, M. Noroozifar, S. Jahani, Preparation and characterization of nano-sized magnetic particles $LaCoO_3$ by ultrasonic-assisted coprecipitation method. Synth. React. Inorg. Metal-Organ. Nano-Metal Chem. **45**, 1591–1595 (2015). https://doi.org/10.1080/15533174.2015.1031010
4. H.M. Moghaddam, H. Beitollahi, S. Tajik, S. Jahani, H. Khabazzadeh, R. Alizadeh, Voltammetric determination of droxidopa in the presence of carbidopa using a nanostructured base electrochemical sensor. Russ. J. Electrochem. **53**, 452–460 (2017). https://doi.org/10.1134/S1023193517050123
5. R.G. Chaudhuri, S. Paria, Core/shell nanoparticles: classes, properties, synthesis mechanisms, characterization, and applications. Chem. Rev. **112**, 2373–2433 (2012). https://doi.org/10.1021/cr100449n
6. H.A. Alarifi, M. Atis C. Özdoğan, A. Hu, M. Yavuz, Y. Zhou, Determination of complete melting and surface premelting points of silver nanoparticles by molecular dynamics simulation. J. Phys. Chem. C **117**, 12289–12298 (2013). https://doi.org/10.1021/jp311541c
7. M.H.S. Poor, M. Khatami, H. Azizi, H., Y. Abazari, Cytotoxic activity of biosynthesized Ag nanoparticles by Plantago major towards a human breast cancer cell line. Rend. Lincei **28**, 693–699 (2017). https://doi.org/10.1007/s12210-017-0641-z
8. M. Khatami, S.M. Mortazavi, Z.K. Farahani, A. Amini, E. Amini, H. Heli, Biosynthesis of silver nanoparticles using pine pollen and evaluation of the antifungal efficiency. Iran. J. Biotechnol. **15**, 95–101 (2017). https://doi.org/10.15171/ijb.1436
9. Z.U.H. Khan, A. Khan, Y. Chen, A.U. Khan, N.S. Shah, N. Muhammad, B. Murtaza, K. Tahir, F.U. Khan, P. Wan, Photo catalytic applications of gold nanoparticles synthesized by green route and electrochemical degradation of phenolic Azo dyes using AuNPs/GC as modified paste electrode. J. Alloys Compd. **725**, 869–876 (2017). https://doi.org/10.1016/j.jallcom.2017.07.222
10. A.I. López-Lorente, B.M. Simonet, M. Valcárcel, Analytical potential of hybrid nanoparticles. Anal. Bioanal. Chem. **399**, 43–54 (2011). https://doi.org/10.1007/s00216-010-4110-0
11. A. Zaleska-Medynska, M. Marchelek, M. Diak, E. Grabowska, Noble metal-based bimetallic nanoparticles: the effect of the structure on the optical, catalytic and photocatalytic properties. Adv. Colloid Interface Sci. **229**, 80–107 (2016). https://doi.org/10.1016/j.cis.2015.12.008
12. M. Tsuji, N. Miyamae, S. Lim, K. Kimura, X. Zhang, S. Hikino, M. Nishio, Crystal structures and growth mechanisms of Au@Ag core-shell nanoparticles prepared by the microwave-polyol method. Crys. Growth Des. **6**, 1801–1807 (2006). https://doi.org/10.1021/cg060103e
13. Z. Yang, X. Yang, Z. Xu, Molecular Dynamics simulation of the melting behavior of Pt–Au nanoparticles with core-shell structure. J. Phys. Chem. C **112**, 4937–4947 (2008). https://doi.org/10.1021/jp711702y

14. S.J. Mejía-Rosales, C. Fernández-Navarro, E. Pérez-Tijerina, Two-stage melting of Au–Pd nanoparticles. J. Phys. Chem. B **110**, 12884–12889 (2006). https://doi.org/10.1021/jp0614704
15. Q. Jiang, S. Zhang, M. Zhao, Size-dependent melting point of noble metals. Mater. Chem. Phys. **82**, 225–227 (2003). https://doi.org/10.1016/S0254-0584(03)00201-3
16. I.A. Lyashenko, V.N. Borysiuk, N.N. Manko, Statistical analysis of self-similar behaviour in the shear induced melting model. Condens. Matter Phys. **17**, 23003 (2014). https://doi.org/10.5488/CMP.17.23003
17. A.I. Olemskoi, O.V. Yushchenko, V.N. Borisyuk, T.I. Zhilenko, YuO Kosminska, V.I. Perekrestov, Hierarchical condensation near phase equilibrium. Phys. A **391**, 3277–3284 (2012). https://doi.org/10.1016/j.physa.2011.10.027
18. I.A. Lyashenko, A.V. Khomenko, A.M. Zaskoka, Hysteresis behavior in the stick-slip mode at the boundary friction. Tribol. Trans. **56**, 1019–1026 (2013). https://doi.org/10.1080/10402004.2013.819541
19. O.I. Olemskoi, S.M. Danyl'chenko, V.M. Borysyuk, I.O. Shuda, Metallofiz. Noveishie Tekhnol. **31**, 777 (2009)
20. I.A. Lyashenko, Tribological properties of dry, fluid, and boundary friction. Tech. Phys. **6**, 701–707 (2011). https://doi.org/10.1134/S1063784219080140
21. Y. Zhao, R.E. Smalley, B.I. Yakobson, Coalescence of fullerene cages: topology, energetics, and molecular dynamics simulation. Phys. Rev. B **66**, 195409 (2002). https://doi.org/10.1103/PhysRevB.66.195409
22. K. Zhang, G.M. Stocks, J. Zhong, Melting and premelting of carbon nanotubes. Nanotechnology **18**, 285703 (2007). https://doi.org/10.1088/0957-4484/18/28/285703
23. R. Huang, Y-H Wen, G.-F Shao, Z-Z Zhu, S.-G. Sun, Thermal stability and shape evolution of tetrahexahedral Au–Pd core-shell nanoparticles with high-index facets. J. Phys. Chem. C **117**, 6896–6903 (2013). https://doi.org/10.1021/jp401423z
24. L. Lu, G. Burkey, I. Halaciuga, D.V. Goia, Core-shell gold/silver nanoparticles: Synthesis and optical properties. J. Colloid Interface Sci. **392**, 90–95 (2013). https://doi.org/10.1016/j.jcis.2012.09.057
25. O.V. Maksakova, S.S. Grankin, O.V. Bondar, Ya.O. Kravchenko, D.K. Yeskermesov, A.V. Prokopenko, N.K. Erdybaeva, B. Zhollybekov, J. Nano-Electron. Phys. **7**, 04098 (2015)
26. A.D. Pogrebnjak, A.P. Shpak, N.A. Azarenkov, V.M. Beresnev, Structures and properties of hard and superhard nanocomposite coatings. Phys.-Usp. **52**, 29–54 (2009). https://doi.org/10.3367/UFNe.0179.200901b.0035
27. V.I. Lavrentiev, A.D. Pogrebnjak, High-dose ion implantation into metals. Surf. Coatings Technol. **99**, 24–32 (1998). https://doi.org/10.1016/S0257-8972(97)00122-9
28. W. Humphrey, A. Dalke, K. Schulten, VMD: Visual molecular dynamics. J. Mol. Graph. Model. **14**, 33–38 (1996). https://doi.org/10.1016/0263-7855(96)00018-5
29. X.W. Zhou, H.N.G. Wadley, R.A. Johnson, D.J. Larson, N. Tabat, A. Cerezo, A.K. Petford-Long, G.D.W. Smith, P.H. Clifton, R.L. Martens, T.F. Kelly, Atomic scale structure of sputtered metal multilayers. Acta Mater. **49**, 4005–4015 (2001). https://doi.org/10.1016/s1359-6454(01)00287-7
30. F.A. Lindemann, Physik Z **11**, 609 (1910)
31. H.J.C. Berendsen, J.P.M. Postma, W.F. Vangunsteren, A. Dinola, J.R. Haak, Molecular-dynamics with coupling to an external bath. J. Chem. Phys. **81**, 3684–3690 (1984). https://doi.org/10.1063/1.448118
32. S. Plimpton, Fast parallel algorithms for short-range molecular dynamics. J. Comput. Phys. **117**, 1–19 (1995). https://doi.org/10.1006/jcph.1995.1039
33. D.C. Rapaport, *The art of molecular dynamics simulation* (Cambridge University Press, New York, 2004)

Chapter 34
Fabrication and Pulsed Laser Modification of Antireflective Composite Materials for the Visible and Near-IR Ranges

F. F. Komarov, I. D. Parfimovich, O. V. Milchanin, A. G. Tkachev, A. V. Melezhik, N. R. Memetov, R. A. Stolyarov, O. R. Ludchik, M. N. Kolchevskaya and R. B. Miranovich

Abstract A pulsed laser treatment method has been developed and demonstrated for the formation of antireflection coatings based on composite materials with carbon nanotubes. The influence of structure surface of composite materials exposed to pulsed laser treatment on reflectivity has been carried out. The possibility of creating non-reflective surfaces of composite samples in the visible and near-IR ranges is demonstrated.

34.1 Introduction

Broadband antireflection materials have attracted much attention due to their relevance in many applications such as solar energy converters, thermal control devices, optical absorbers for photo and thermal detectors, as well as creating protective, shielding and masking coatings. It is well known that the reflectance of coatings could be reduced by adding absorbing species such as dyes, carbon particles, gold and platinum black particles [1]. However, the reflection at the air/surface interfaces would limit the fraction of incident radiation that can be absorbed. To address the problem, a special antireflective layer is created on the surface by forming subwave inhomogeneities [2], which by matching the refractive indices between air and the

---

F. F. Komarov · I. D. Parfimovich · O. V. Milchanin
A. N. Sevchenko Institute of Applied Physics Problems, Minsk, Belarus

F. F. Komarov (✉)
National University of Science and Technology MISIS, Moscow, Russia
e-mail: KomarovF@bsu.by; KomarovF@mail.ru

A. G. Tkachev · A. V. Melezhik · N. R. Memetov · R. A. Stolyarov
Tambov State Technical University, Tambov, Russia

O. R. Ludchik · M. N. Kolchevskaya · R. B. Miranovich
Belarussian State University, Minsk, Belarus

A. D. Pogrebnjak and O. Bondar (eds.), *Microstructure and Properties of Micro- and Nanoscale Materials, Films, and Coatings (NAP 2019)*, Springer Proceedings in Physics 240, https://doi.org/10.1007/978-981-15-1742-6_34

surface will minimize the reflection of incident radiation [3]. Currently, there are a number of methods to create antireflection coatings: ion and plasma etching [4–8], nanoimprint lithography [9], chemical methods for creating coatings based on aerogels and other porous materials [1, 10] and various methods of nanotube deposition on the substrate [11–13]. However, each of these methods has a number of disadvantages that prevent their use on an industrial scale. For materials, which formed by ion or plasma etching, this is the difficulty of controlling the uniform reproduction of a structure over large areas. For aerogels, issues related to the mechanical resistance and the toxicological effect of the reaction products can be problematic. The main disadvantage of the SWCNT arrays is the complexity of the technological process of nanotubes growing with specified parameters. The nanoimprint lithography is a technologically simple and well established process that does not require specific equipment and materials. In addition, a change of master mold structure allows controlling the structure of coatings. At the same time, the manufacture of the master mold itself is quite a long process, so this is also a disadvantage, since there is no possibility of quickly setting the parameters of the surface of the coatings. This paper presents a method for surface structuring of antireflection coatings by pulsed laser radiation. The essence of the method consists in forming a periodic pyramidal structure on the surface of the source material by "burning" a part of the material by laser radiation. This method allows controlling the height and width of the pyramids and generally automates the process of structuring the surface of the material. The source material was an epoxy polymer composite with 4 wt% additives of multi-walled carbon nanotubes. The use of multi-walled carbon nanotubes is determined by the intrinsic absorption properties of graphene-like structures, the developed technology of industrial production, and the ability to form inside the composite volume absorbing formations [14].

34.2 Experiment Details

The process of manufacturing an antireflection coating by pulsed laser radiation occurred in two stages: sample manufacturing of the composite material based on the epoxy polymer with MWCNT additives, and then pulsed laser surface treatment of the formed composite. "Taunit-M" multi-walled carbon nanotubes (NanoTechCenter, Tambov, Russia) characterized by the inner diameter of 4–8 nm, outer diameter of 8–12 nm, length from 1 to 10 microns and epoxy polymer resin were used in the coating composition.

The composite was manufactured using the ultrasonic homogenizer UP-400St. Multi-walled carbon nanotubes with a concentration of 4 wt% were previously dispersed in an alcohol solution of surfactants (100 mM) for 20 min at an ultrasound frequency of 24 kHz and a power of 40 W with a gradual and smooth change in amplitude from 20% to 70%. Then the mixture was added to the epoxy polymer and for 20 min at a frequency of 24 kHz and a power of 80 W was again subjected to ultrasonic treatment. After the completion of this procedure, a hardener was added to the polymer, and the finished mixture was set in a drying chamber for evaporation of the surfactant solution and polymerization.

The formation of a periodic surface structure was carried out by a Q-switched Nd:YAG laser (1064 nm, 50 Hz repetition rate). The focused laser beam repeatedly passed through the sample, forming parallel grooves on the surface. By varying parameters such as beam focusing, pump energy, pulse frequency and spatial step can easily control the shape and geometrical dimensions of the grooves. The distribution of MWCNTs in the polymer structure was studied using a Hitachi H-800 transmission electron microscope. The surface morphology of the structured composite material was examined on a Hitachi H-8010 scanning electron microscope. The reflectivity measurements of the coating were carried out in the range from 400 to 2500 nm on the scanning spectrophotometer Photon RT.

34.3 Results and Discussion

Figure 34.1 shows the TEM image of the composite materials with 4 wt% "Taunit-M" additives. The morphology of the formed composite material is a three-dimensional grid formed by the interlacing of carbon nanotubes, with numerous polymer cells with assize of about 10 nm. The formation of a three-dimensional grid in the composite structure leads to the appearance of inhomogeneous local regions with different conductive characteristics of the material, at the interfaces between which, multiple refraction and reflection of incident radiation occur. Thus, such a lattice structure will contribute to the effective absorption of electromagnetic radiation by the composite, which is confirmed by a number of works [1, 14].

As mentioned earlier, microscopic inhomogeneities in the internal environment of material contribute to an increase in radiation absorption, and in the presence of antireflection layer which will increase its effectiveness. The formation of an antireflection layer on the surface of the composite material was achieved by pulsed laser irradiation, which resulted in the creation of parallel extended grooves. With

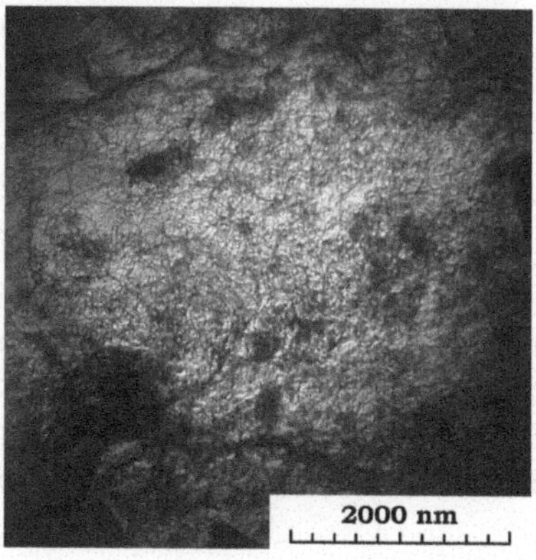

Fig. 34.1 Bright-field TEM image of the nanocomposites based on epoxy polymer with 4 wt% "Taunit-M" MWCNTs

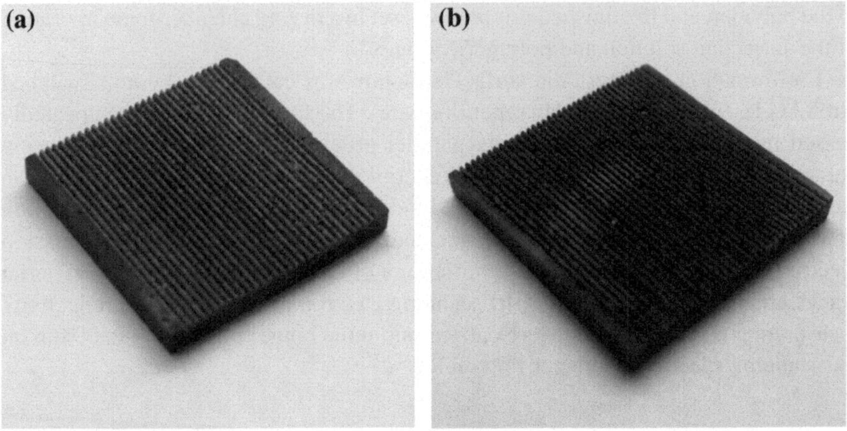

Fig. 34.2 Samples of antireflective composite materials with different pulsed laser treatment parameters: **a** $E_p = 12.5$ J, **b** $E_p = 13$ J

pulsed laser treatment in the perpendicular direction, the pyramidal structure of the material surface was set. The width and depth of the grooves were controlled by the magnitude of the laser pump energy, the focal length and the number of pass along one groove, the groove period was specified by the spatial repetition step of the laser pulse. As a result, a number of samples differing in topological and laser exposure parameters were formed. Figure 34.2 shows samples of composite materials exposed to laser treatment with various parameters.

The use of laser irradiation also leads to the formation on the lateral surface of grooves and pyramids of nanoscale inhomogeneous cells. These cells are numerous air cavities with sizes from 100 to 500 nm surrounded by molten composite material (Fig. 34.3).

Before proceeding to the demonstration and discussion of the results on the measurement of the formed coatings reflectivity, it is worth noting that the employment of the method of pulsed laser processing on pure epoxy polymer does not lead to the formation of grooves. And therefore it was not possible to observe the effect of the surface relief, without taking into account carbon additives, on reflection.

Figure 34.4 presents the results of measurements of the absolute reflection of composite materials filled with carbon nanotubes with a structured surface by the method of pulsed laser irradiation. The objects of comparison were also samples of pure epoxy polymer (sample N1) and composite material with carbon nanotubes (sample N2) without surface profiling (Fig. 34.4a).

The presence of structured surface, in turn, allows reducing the reflection from the material by 1–2 orders of magnitude, depending on the parameters of laser irradiation. So, for sample N3, the value of the absolute reflection coefficient does not exceed 0.1% in the visible (400–800 nm) range and varies from 0.04% to 0.07% in the near IR (1000–2500 nm). For sample N5, the reflection coefficient in the visible range

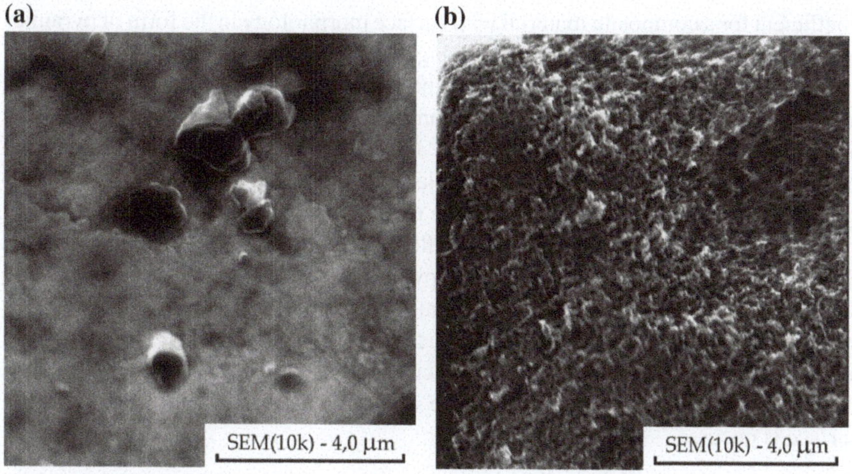

Fig. 34.3 SEM microphotographs of the nanocomposites based on epoxy polymer with 4 wt% "Taunit-M" MWCNTs. **a** Without pulsed laser treatment, **b** with pulsed laser treatment

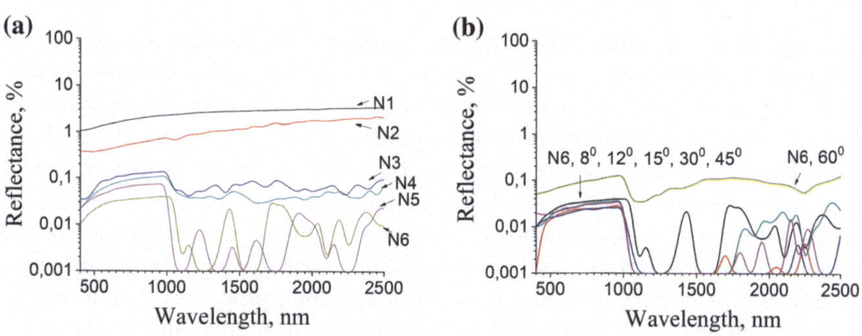

Fig. 34.4 **a** Specular reflection of MWCNT composite in the range of 400–2500 nm, **b** angular dependence of reflectance for composite material N6. N1—epoxy polymer, N2—MWCNT composite without pulsed laser treatment, N3—MWCNT composite with pulsed laser treatment (E_p = 12.5 J, grooves surface), N4—MWCNT composite with pulsed laser treatment (E_p = 12.5 J, pyramidal surface), N5—MWCNT composite with pulsed laser treatment (E_p = 13 J, grooves surface), N6—MWCNT composite with pulsed laser treatment (E_p = 13 J, pyramidal surface)

does not exceed 0.07%, and in the near IR it could not be registered at all correctly, due to the lack of sensitivity of the device.

It should be noted that the surface topology in the form of pyramids makes it possible to further weaken the reflection in comparison with the formed grooves (Fig. 34.4a). For example, when comparing samples N3 and N4, formed with the same parameters of laser exposure, but differing in surface morphology, there is a general decrease in the value of the reflection coefficient by 0.02%. A similar effect is also observed when comparing samples N5 and N6: the recorded reflection coefficient for a composite material with surface morphology in the form of pyramids has a 0.04% lower value in the visible range and is 0.03%.

An important feature observed for all profiled samples is a sharp increase in the absolute reflection coefficient recorded at an angle of 60°. A more detailed study of the angular characteristics for sample N6 (Fig. 34.4b) demonstrates that the manifestation of this effect occurs quite sharply—at angles of incidence up to 45°, the values of the coefficients are quite close, and at 60° the reflection coefficient sharply increased.

We associate a sharp increase in the reflection coefficient with the fact that when the radiation falls at an angle of 60°, the pyramidal structure stops working. Due to the geometry of the pyramids, the incident radiation should be repeatedly reflected and propagated deep into the formed structure, however, instead of this, at a given angle of incidence, the radiation is reflected outwards. Thus, for such structures, there exists a certain critical angle, above which the increase in reflection from the surface is occurs.

34.4 Conclusions

Thus, the ability to create "non-reflective" surfaces of composite samples in the visible and near-IR spectral ranges due to a number of factors has been demonstrated. First of all, this is the use of pulsed laser irradiation during the formation of an anti-reflective structure on the surface of a composite material, which provides multiple re-reflections and absorption of radiation. The main advantage of this method is flexibility in relation to the specified structural parameters of the sample surface. At the same time, due to the formation of microscopic inhomogeneities in the internal medium of the material due to the interweaving of bundles and clusters of multi-walled carbon nanotubes, as well as the intrinsic absorbing properties of multi-walled carbon Taunit-M nanotubes, the effectiveness of these materials can be improved.

Acknowledgements The present work is supported by the Russian Foundation for Basic Research and Belarusian Republican Foundation for Fundamental Research (grants No. T18P-249).

References

1. J. Zhu, X. Yang, Z. Fu, C. Wang, W. Wu, L. Zhanbg, Facile fabrication of ultra-low density, high-surface-area, broadband antireflective carbon aerogels as ultra-black materials. J. Porous Mater. **23**(5), 1217 (2016)
2. J.Y. Liu, M. Soltani, R.K. Dey, B. Cui, R. Lee, H. Podmore, Moth-eye antireflection nanostructure on glass for CubeSats. J. Vac. Sci. Technol. **36**(6), 06JG01 (2018)
3. Y. Lin, J. He, Recent progress in antireflection and self-cleaning technology – from surface engineering to functional surfaces. Prog. Mater. Sci. **61**, 94 (2014)
4. K. Amemiya, H. Koshikawa, T. Yamaki, Y. Maekawa, H. Shitomi, T. Numata, K. Kinoshita, M. Tanabe, D. Fukuda, Fabrication of hard-coated optical absorbers with microstructured surfaces

using etched ion tracks: toward broadband ultra-low reflectance. Nucl. Instr. Meth. Phys. Res. B **356**, 154 (2015)
5. S. Chuang, H. Chen, J. Shieh, C. Lin, C. Cheng, H. Liu, C. Yu, Nanoscale of biomimetic moth eye structures exhibiting inverse polarization phenomena at the Brewster angle. Nanoscale **2**, 799 (2010)
6. M. Steglich, D. Lehr, S. Ratzsch, T. Kasebier, F. Schrempel, E. Kley, A. Tunnermann, An ultra-black silicon absorber. Laser Photonics Rev. **8**(2), L13 (2014)
7. Y. Sun, J. Evans, F. Ding, N. Liu, Y. Zhang, S. He, Bendable, ultra-black absorber based on a graphite nanocone nanowire composite structure. Opt. Express **23**(15), 20115 (2015)
8. M. Otto, M. Algasinger, H. Branz, B. Gesemann, T. Gimpel, K. Fuchsel, T. Kasebier, S. Kontermann, S. Koynov, X. Li, V. Naumann, J. Oh, A. Sprafke, J. Ziegler, M. Zik, R. Wehrspohn, Black silicon photovoltaics. Adv. Optical. Mater. **3**(2), 147 (2015)
9. T. Uchida, M. Moro, S. Hiwasa, J. Taniguchi, in *Transfer properties of Moth-eye Structure Film by RTR UV-NIL*, 2015 International Conference on Electronics Packaging and iMAPS All Asia Conference (ICEP-IAAC), Kyoto, April 2015
10. W. Sun, A. Du, Y. Feng, J. Shen, S. Huang, J. Tang, B. Zhou, Super black material from low-density carbon aerogels with subwavelength structures. ACS Nano **10**, 9123 (2016)
11. C.J. Chunnilall, J.H. Lehman, E. Theocharous, A. Sanders, Infrared hemispherical reflectance of carbon nanotube mats and arrays in the 5–50 μm wavelength region. Carbon **50**, 5348 (2012)
12. K. Mizuno, J. Ishii, H. Kishida, Y. Hayamizu, S. Yasuda, N. Futaba, M. Yumura, K. Hata, A black body absorber from vertically aligned single-walled carbon nanotubes. PNAS **106**(15), 6044 (2009)
13. F. De Nicola, P. Hines, M. Crescenzi, N. Motta, Thin randomly aligned hierarchical carbon nanotube arrays as ultrablack metamaterials. Phys. Rev. B **96**, 045409 (2017)
14. F.F. Komarov, A.G. Tkachev, O.V. Milchanin, I.D. Parfimovich, M.V. Grinchenko, I.N. Parkhomenko, D.S. Bychenok, A Composite based on epoxy polymer and carbon nanotubes: structure, optical properties and interaction with microwave radiation. Adv. Mat. Tech. **2**, 19 (2017)

Chapter 35
Photo- and Electroluminescence of Layered Structures Based on Silicon Oxide and Nitride Films

I. Romanov, F. F. Komarov, I. Parkhomenko, L. Vlasukova, N. Kovalchuk, A. Mudryi and V. Zhivulko

Abstract In the present work, silicon nitride, and oxide films, as well as two-layered SiN_x/SiO_2 and three-layered $SiO_2/SiN_x/SiO_2$ structures, were fabricated on p-type Si-substrates. The structure and element composition was studied by scanning electron microscopy and Rutherford backscattering spectroscopy, respectively. It was shown, spectral positions of photoluminescence maxima, as well as edge absorption, were determined by stoichiometry of SiN_x layers. No electroluminescence signal was registered from single-layered silicon nitride films. The intense band at 1.9 eV dominated the electroluminescence spectra of single-layered silica film as well as two- and three-layered structures. It was attributed to silane groups in SiO_2 layers. The contribution of emission from silicon nitride layers into electroluminescence of the two- and three-layered structures is discussed.

35.1 Introduction

Since the early 1990s until today, research groups around the world have been working to fabricate efficient light-emitting diode (LED) structures based on current silicon technology. The synthesis of A_3B_5, A_2B_6 semiconductors with effective light-emitting characteristics is a laborious and high-priced technological process.

I. Romanov · I. Parkhomenko (✉) · L. Vlasukova
Belarussian State University, Minsk, Belarus
e-mail: parkhomenko@bsu.by

F. F. Komarov
A. N. Sevchenko Institute of Applied Physics Problems, Minsk, Belarus

National University of Science and Technology, Moscow, Russia

N. Kovalchuk
Joint Stock Company "Integral", Minsk, Belarus

A. Mudryi · V. Zhivulko
Scientific and Practical Materials Research Center, National Academy of Sciences of Belarus, Minsk, Belarus

A. D. Pogrebnjak and O. Bondar (eds.), *Microstructure and Properties of Micro- and Nanoscale Materials, Films, and Coatings (NAP 2019)*,
Springer Proceedings in Physics 240, https://doi.org/10.1007/978-981-15-1742-6_35

The creation of LED with Si or Si-based compounds as an active layer provides opportunities for development of silicon photonics. In this regard, our report is devoted to the possible application of silicon nitride and oxide films as active layers of light-emitting devices.

To date, various methods have been developed to improve the light-emitting properties of silicon nitride and oxide films such as variation of the stoichiometric composition, post-deposition heat treatment, formation of light-emitting nanocrystals inside them [1–6]. Mainly in these works, the photoluminescence (PL) technique is used to study the light-emitting characteristics and mechanisms of radiative recombination. However, the most important for practical application is the study of electroluminescence (EL) of these dielectric films. Though, there are very few articles devoted to EL of Si-based dielectrics [6–8]. In this regard, this report is focused on the electroluminescence study of structures based on silicon nitride and oxide films. EL was excited in the "electrolyte-dielectric-semiconductor" system. For the EL investigation of dielectric films, this system has some advantages such as the ability to create a strong electric field in layer, possibility to measure a spectral EL depth distribution via layer-by-layer etching, high spectral transparency of electrolyte.

35.2 Experiment Details

To study light-emitting properties of Si-based dielectrics five types of films were fabricated on Si-substrates (p-type): two single-layer silicon nitride films (samples N1 and N2), single-layer silicon oxide film (sample O1), two-layered SiN_x/SiO_2 (sample M2) and three-layered $SiO_2/SiN_x/SiO_2$ (sample M3) structures.

For all samples, the silicon nitride layers were deposited by low-pressure chemical vapor deposition at 830 °C. The ratio of reagent gases ($R = SiH_2Cl_2/NH_3$) in the reactor chamber was 1/8 for the samples N1 and M2 and 8/1 for the samples N2 and M3. The silicon oxide films for the samples O1 and M2 were deposited by thermal oxidation. The buffer and capping layer of silicon oxide for three-layered $SiO_2/SiN_x/SiO_2$ structure (sample M3) were deposited by plasma-enhanced chemical vapor deposition at 350 °C using a gaseous mixture of SiH_4 (250 sccm) and N_2O (1166 sccm) as precursors. The atomic ratio ($x = [N]/[Si]$) of as-deposited SiN_x films was determined by Rutherford backscattering spectrometry (RBS). The thicknesses of layers in the fabricated structures were estimated by scanning electron microscopy (SEM) using the Hitachi S-4800 microscope. Table 35.1 presents the stoichiometric composition (x) and thicknesses obtained from RBS and SEM data for fabricated layered structures.

Spectroscopic ellipsometry measurements of the refraction index (n) and the extinction coefficient (k) were performed using the spectroscopic ellipsometer HORIBA UVISEL 2. Photoluminescence spectra were recorded at room temperature in the spectral range of 1.5–3.5 eV (350–800 nm) with a He–Cd laser (3.8 eV) as an excitation source. EL was excited in the "electrolyte-dielectric-semiconductor" system and was registered in the range of 1.7–3.7 eV. 1 M aqueous solution of Na_2SO_4

Table 35.1 The structural characteristics of fabricated structures

The sample	The stoichiometric composition and thickness (nm)
N1	$SiN_{1.3}$(150 nm)/Si
N2	$SiN_{0.9}$(100 nm)/Si
O1	SiO_2(110 nm)/Si
M2	$SiN_{1.3}$(75 nm)/SiO_2(75 nm)/Si
M3	SiO_2(42 nm)/$SiN_{0.9}$(66 nm)/SiO_2(32 nm)/Si

was used as the electrolyte in galvanostatic mode with positive substrate voltage. The EL spectra were recorded at voltages above 80 V on the sample and current densities of 0.4–0.6 mA/cm^2. The registration of the PL and EL spectra was carried out at room temperature.

35.3 Results and Discussion

Figure 35.1 shows the normalized room-temperature PL spectra of the as-deposited single-layered silicon nitride films (samples N1 and N2). As can be seen, the position of the PL maxima depends on the film composition. The PL maxima are located in the red range at ~1.9 eV for the Si-rich film (sample N2) and in the blue range at ~2.65 eV for the stoichiometric one (sample N1). The PL signal from silicon oxide film (the sample O1) is substantially weaker compared to PL intensity of silicon nitride films (not shown here). It is due to insufficient energy of He–Cd laser (3.8 eV) to excite luminescence centers in SiO_2. It is therefore not surprising that shapes of PL spectra

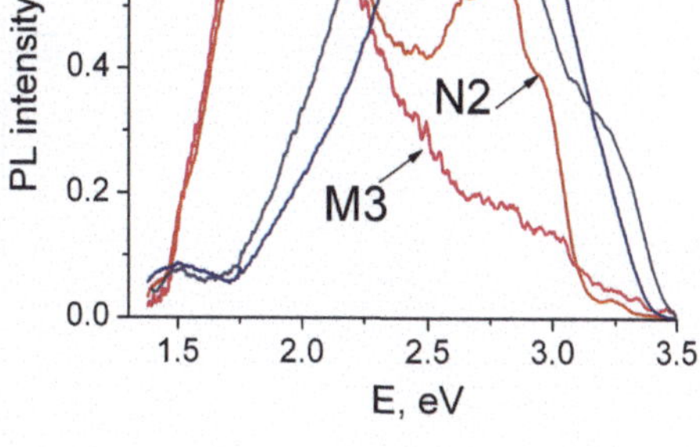

Fig. 35.1 Normalized PL spectra of single-layer silicon nitride films (samples N1 and N2), two-layered (sample M2) and three-layered (sample M3) structures

of the two- and three-layered structures (the samples M2 and M3) are determined largely by light-emitting properties of SiN_x layer.

According to spectroscopic ellipsometry data, the values of E_g calculated via the straight-line extrapolation in the linear region of Tauc's plots with the energy axis are equal to 4.65 eV and 3.5 eV for N1 and N2 samples, respectively (not shown here). The energy positions of the PL bands (1.9 and 2.65 eV) are quite far away from their band edges E_g. This suggests that PL arises from radiative recombination via band tail and/or defect states. The narrowing of band gap and, in turn, redshift of PL maximum with increasing Si content in film are typical features for SiN_x alloys [8–10]. It should be noted that red PL shift is also can be explained by the emission of Si nanocrystals formed in silicon nitride matrix due to quantum confinement effect. Taking into account inability to synthesis Si nanoclusters in stoichiometric silicon nitride film (as in the samples N1 and M2), the band-tail recombination is most suited as mechanism of PL emission.

Figure 35.2 shows normalized EL spectra of fabricated structures. We have failed to register perceptible EL signals from single-layered silicon nitride films (samples N1 and N2). On the other hand, the single-layered silicon oxide film exhibits an intense narrow band in the red spectral range (1.9 eV). However, this emission is not stable and degrades over time. On the contrary, the EL emission from the two- and three-layered structures is more stable. Both structures also exhibit intense red band at 1.9 eV. Below, we will discuss the intensity of other EL bands relatively this red

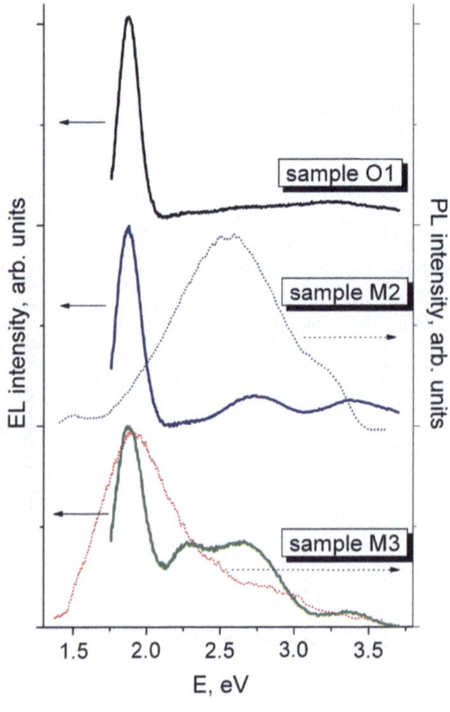

Fig. 35.2 Normalized EL spectra (solid lines) and PL spectra (dot lines) of single-layer silicon oxide films (sample O1), two-layered (sample M2) and three-layered (sample M3) structures

band taking into account that the total thickness of silica layer in the samples M2 and M3 is similar (~75 nm).

Except dominated red band at 1.9 eV, the bands in the broad spectral range from green to UV are observed for the samples M2 and M3. It should be noted that the EL spectral distribution in the range of (2.1–3.8) eV differs for the sample M2 and M3. Namely, the EL emission of the sample M3 is shifted to a shorter wavelength range in comparison with the emission of the sample M2. Also, EL intensity in the range of (2.1–3.8) eV is higher for the sample M3 than for the sample M2.

The red band at 1.9 eV is observed for all registered EL spectra and can be attributed to the presence of silanol (Si–OH) groups in SiO_2 layers [6]. According to [9], the band with a maximum at ~2.6 eV observed for the samples M2 and M3 can be attributed to recombination at silylene ($O_2 = Si$:) centers, which can be formed in intermediate oxynitride layer at the SiO_x–SiN_x interface. This does not contradict with higher intensity of this band relative to intensity of red band at 1.9 eV for the three-layered structures (sample M3), than for the sample M2, which has only one SiO_x–SiN_x interface.

Concerning the EL spectrum of the sample M3 it should be noted that the band-tail recombination in $SiN_{0.9}$ layer can contribute to the luminescence band at 2.6 eV. This band is shifted to blue region with respect to the corresponding PL band (1.9 eV). Such a shift to shorter wavelengths at replacing optical pumping by electrical excitation is observed in other [10, 11]. It is explained by higher energy of heated electrons (~5 eV for our structures) during EL registration than the energy of laser used in our experiment to excite PL (3.8 eV).

The green band at 2.3 eV is associated with the presence of three-coordinated silicon defects ($\equiv$ Si·) [12]. A high concentration of $\equiv$Si· defects is typically for Si-rich SiO_x films. In our case, three-coordinated silicon defects can be formed at the SiO_2–$SiN_{0.9}$ interfaces of M3 sample due to oxygen atoms diffusion from neighboring SiO_2 layers into a silicon-rich $SiN_{0.9}$ layer.

Such electric pumping is favorable for the radiative transition from higher energy states. It should be noted, the higher relative intensity of emission from silicon nitride layers and "oxide-nitride" interfaces (2.1–3.8 eV) (in respect with intensity of red band at 1.9 eV) for the sample M3 can be explained by increasing signal via interference effect in silicon nitride layer located between two SiO_2 layers.

35.4 Conclusions

The light-emitting properties of single-layered (SiN_x, SiO_2) structures as well as two- ($SiN_{1.3}/SiO_2$) and three-layered structures ($SiO_2/SiN_{0.9}/SiO_2$) were studied at optical and electric excitation of luminescence. In contrast to silicon oxide, at optical pumping (3.8 eV) silicon nitride films exhibit intense PL signal. The spectral position of signal depends on SiN_x stoichiometric composition. At electric pumping, unlike single-layered nitride films, the single-layered oxide film exhibits intense red band at 1.9 eV. This band is also manifested in the EL spectra of the two- and three-layered structures. The emission from nitride layers is also appeared in EL spectra of multilayered structures and is blue shifted in respect of PL maximum. Also, three-coordinated silicon defects located at the SiO_2–$SiN_{0.9}$ interfaces can contribute into radiative recombination under electrical excitation. In terms of practical application of the multilayered structures in fabrication of white LED, the $SiO_2/SiN_{0.9}/SiO_2$ is more perspective due to stability and higher emission intensity in wide visible range.

Acknowledgements The authors would like to acknowledge D. Zhigulin for the SEM diagnostics and N. Krekoten for the optical measurements.

References

1. D.Q. Shi, B.C. Hu, W. Xu, X.N. Li, C.Y. Ma, Q.Y. Zhang, J. Lumin. **175**, 67 (2016)
2. D.V. Shuleiko, S.V. Zabotnov, D.M. Zhigunov, A.A. Zelenina, I.A. Kamenskih, P.K. Kashkarov, Semiconductors **51**, 196 (2017)
3. T. Torchynska, L. Khomenkova, A. Slaoui, J. Electr. Mater. **47**, 3927 (2018)
4. B. Benyahia, F. Tiour, L. Guerbous, R. Chaoui, I. Menous, B. Mahmoudi, A. Mefoued, A. Guenda, J. Nano Res.-Sw. **49**, 163 (2017)
5. J. Alarcón-Salazar, R. López-Estopier, E. Quiroga-González, A. Morales-Sánchez, J. Pedraza-Chávez, I.E. Zaldívar-Huerta, M. Aceves-Mijares, in *Recent Advances and Applications in Optical, Solar Cells and Solid State Devices* (IntechOpen, London, 2016)
6. P. Baraban, D.V. Egorov, AYu. Askinazi, L.V. Miloglyadova, Tech. Phys. Lett. **28**, 14 (2002)
7. J. López-Vidrier, S. Gutsch, O. Blázquez, J. Valenta, D. Hiller, J. Laube, J. Blanco-Portals, L. López-Conesa, S. Estradé, F. Peiró, B. Garrido, S. Hernández, M. Zacharias Zhang, Adv. Electron. Mater. **4**, 1700666 (2018)
8. A. Ortiz-Santos, C. Ramos, J. Sastré-Hernández, G. Santana, A. Dutt, Mater. Tech. **in press** (2019)
9. J. Kistner, X. Chen, Y. Wenig, H.P. Strunk, M.B. Schubert, J.H. Werner, J. Appl. Phys. **110**, 023520 (2011)
10. W.A. Jackson, T.M. Searly, I.G. Austin, R.A. Gibson, J. Non-Cryst, Solids **77**, 909 (1985)
11. Z. Pei, Y.R. Chang, H.L. Hwang, Appl. Phys. Lett. **80**, 2839 (2002)
12. P. Baraban, P.P. Konorov, L.V. Malyavka, A.G. Troshikhin, Tech. Phys. **45**, 1042 (2000)

Chapter 36
Influence of Hydrogen Annealing on Ordering in FePd Films with Ag Underlayer

Mark N. Shamis, Pavlo V. Makushko, Tetiana I. Verbytska, Sergiy I. Sidorenko and Yurii N. Makogon

Abstract The comparison of the ordering process in FePd/Ag films during annealing in vacuum and hydrogen atmosphere was carried out. It is shown that annealing in hydrogen leads to acceleration of ordering process, as well as to formation of films with anisotropic magnetic properties and smooth surface, which are required for the data storage application.

36.1 Introduction

Currently, a lot of attention in terms of high energy of magnetocrystalline anisotropy (K_u) for magnetic data storage application is being given to $L1_0$-ordered FePt, PtCo and FePd based films, which have high values of K_u and saturation magnetization [1, 2]. Among these materials, the most well-studied is $L1_0$-FePt due to its highest K_u (6–10 MJ/m^3) among these compounds [2, 3]. At the same time, there are not so many studies devoted to tuning the structure and properties of $L1_0$-ordered FePd based films.

Evolution of crystal structure and magnetic properties of polycrystalline Fe_xPd_{1-x} (x = 37–64 at. %) films annealed at 550 °C for 3 h were investigated in the work [4]. It was found that the Pd-rich films (Fe concentration is less than 37 at. %) have disordered structure and soft magnetic properties, while films with higher Fe content become ordered at the same annealing conditions. In addition, in this work, it was established that the film with composition $Fe_{51}Pd_{49}$ possesses the optimal magnetic properties. The authors of the work [5] showed that the thickness of FePd alloy films has a significant effect on the ordering process. The increase in the film thickness promotes transformation from *fcc* to *fct* phase. It was shown that firstly coercivity rises with thickness increases from 22.5 to 67.5 nm, and then the value decreases with further increasing film thickness. However, in the work [6] the amount of the

M. N. Shamis (✉) · P. V. Makushko · T. I. Verbytska · S. I. Sidorenko · Y. N. Makogon
Department of Physics of Metals, National Technical University of Ukraine "Igor Sikorsky Kyiv Polytechnic Institute", Kiev, Ukraine
e-mail: m.n.shamis@gmail.com

A. D. Pogrebnjak and O. Bondar (eds.), *Microstructure and Properties of Micro- and Nanoscale Materials, Films, and Coatings (NAP 2019)*,
Springer Proceedings in Physics 240, https://doi.org/10.1007/978-981-15-1742-6_36

disordered phase in the FePd/MgO(001) films annealed at 500 °C during 2 h was reduced when the film thickness was decreased. Obtained 3 nm and 5 nm-thick films were mainly composed of the $L1_0$ phase with strong (001) texture. The films were continuous, and the surface roughness was around 0.5 nm. In the case of FePd films deposition on amorphous SiO_2 substrate and subsequent annealing, the films exhibited a predominant (111) grain orientation [7].

In the [8] was shown that the solid-state reaction in the Fe/Pd films (the crystallite size being 10–20 nm, Fe:Pd $\approx$ 46:54 at.%) started at 390 °C in the process of thermal heating at a rate of 4 °C/min. The temperatures of the order-disorder transition of the Fe-Pd system did not correspond to the phase equilibrium diagram for the given concentration range. A considerable order-disorder transition temperature broadening was observed (from 605 to 695 °C).

The presence of the Ag underlayer leads to the lowering of the ordering temperature by 100 °C in FePd thin films [9]. Also, the location of Ag atoms at the boundaries reduces exchange coupling of FePd grains. The top Ag layer could also promote the ordering process in FePd films [10]. It was found that the phase transformation from disordered to ordered state accelerates with the rise of the top Ag layer thickness. In this study the maximum coercivity value of FePd/Ag film is 2.8 kOe when the thickness of Ag is equal to 31 nm. Moreover, Ag addition to the FePd-Ag (5 nm) films grown on MgO (001) changed the surface morphology of the film from island to continuous [11]. The authors suggested that the formation of FePdAg alloy takes place.

The influence of the annealing atmosphere on phase transformations in thin films is also an attractive field for scientific studies. For instance, in [12] the influence of annealing atmosphere on ordering processes in FePt based films was investigated. It is shown that in the case of annealing in forming gas, the grain growth process during heat treatment is strongly suppressed compared with other annealing ambient.

Quite interesting results were reported in work [13]. The authors compared the influence of hydrogen annealing and vacuum annealing on ordering in AuCu nanoalloy. The transformation rate in the hydrogen atmosphere was found to be 100 times faster than in the vacuum.

And if empirical studies on the annealing ambient influence on ordering in FePt and some other $L1_0$-type alloys are widely represented in scientific literature, there are only a few theoretical studies on FePd. For instance, the effect of hydrogen absorption on the electronic structure was investigated [14] and on crystal structure in the works [15, 16].

In the present study, we will focus on FePd based thin films, particularly on the influence of hydrogen annealing on the ordering process in ultra-thin FePd films with Ag underlayer.

36.2 Experiment Details

Equiatomic $Fe_{50}Pd_{50}$ films with Ag underlayers were deposited by magnetron sputtering from individual Fe, Pd and Ag targets on SiO_2(100 nm)/Si(001) substrates at room temperature in an Ar atmosphere of about 3.5×10^{-1} Pa. The basic vacuum level was 3×10^{-5} Pa. The evaporation rates and the layer thicknesses were monitored in situ by a quartz microbalance. In addition, the composition of the as-grown films was checked using Rutherford backscattering spectrometry (RBS). Films have different thicknesses of FePd and Ag layers. The total thickness of films was 5 nm, and described by the formula: $Fe_{50}Pd_{50}$(5-x nm)/Ag(x nm)/Substrate, where x is 0.3, 0.6, 0.9 nm.

After sample preparation, thermal annealing in vacuum 10^{-3} Pa for 0.5–20 h and in hydrogen (under pressure of 1 atm) for 1 h in the temperature range of 600–700 °C was applied. The heating rates were about 5 and 1 °C/s in a vacuum and in the hydrogen atmosphere, respectively. The cooling rates were about 0.25 and 1 °C/s, correspondingly.

The phase composition and structural properties of as-deposited and post-annealed samples were studied by X-ray Diffraction (XRD) using the standard (θ–2θ) geometry (Cu Kα radiation). The surface morphology was observed by the atomic-force microscope (AFM). The magnetic measurements were carried out using superconducting quantum interference device—vibrating sample magnetometry (SQUID-VSM) in out-of-plane and in-plane geometry of the applied magnetic field up to 70 kOe at room temperature.

36.3 Results

On the XRD pattern of the as-deposited FePd/Ag film, there are no diffraction peaks from both the FePd alloy and silver, which indicates the x-ray amorphous state of the film. It should be noted that there are no peaks from Ag on to the XRD patterns that will be described in this paper, due to the small amount of silver in the films.

36.3.1 Annealing in the Vacuum

After the annealing of the FePd/Ag thin film in vacuum at a temperature of 600 °C for 0.5 h no peaks are present on the corresponding XRD patterns (Fig. 36.1). SQUID-measurements indicate that the film is in a soft magnetic (*A*1-FePd phase) state (Fig. 36.2).

The XRD analysis shows that after increasing temperature up to 650 °C, and a significant increase in the annealing duration up to 12 h, a low-intensity superlattice peak with an angular position of $2\theta = 24.07°$ (Fig. 36.1) appears, which may indicate

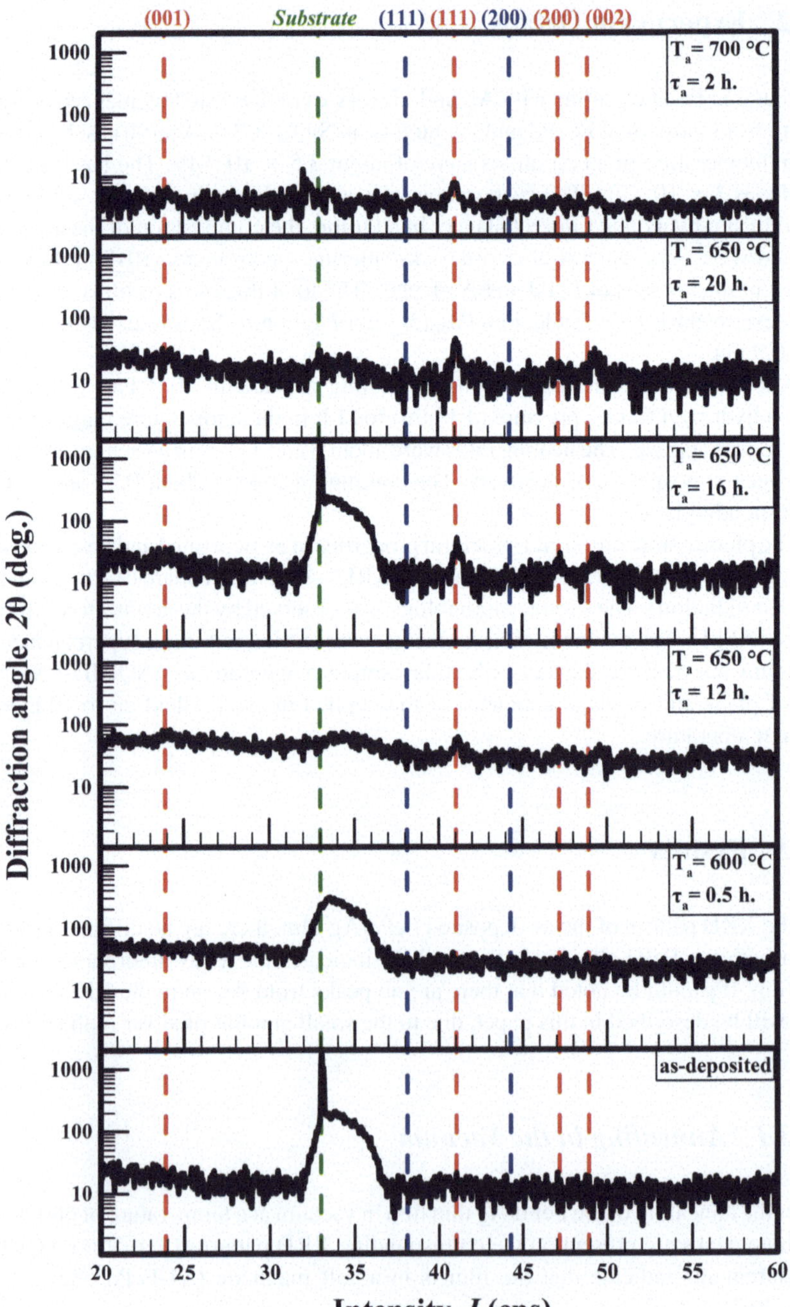

Fig. 36.1 The XRD patterns of FePd (4.7 nm)/Ag(0.3 nm) films in the as-deposited state and after annealing in vacuum

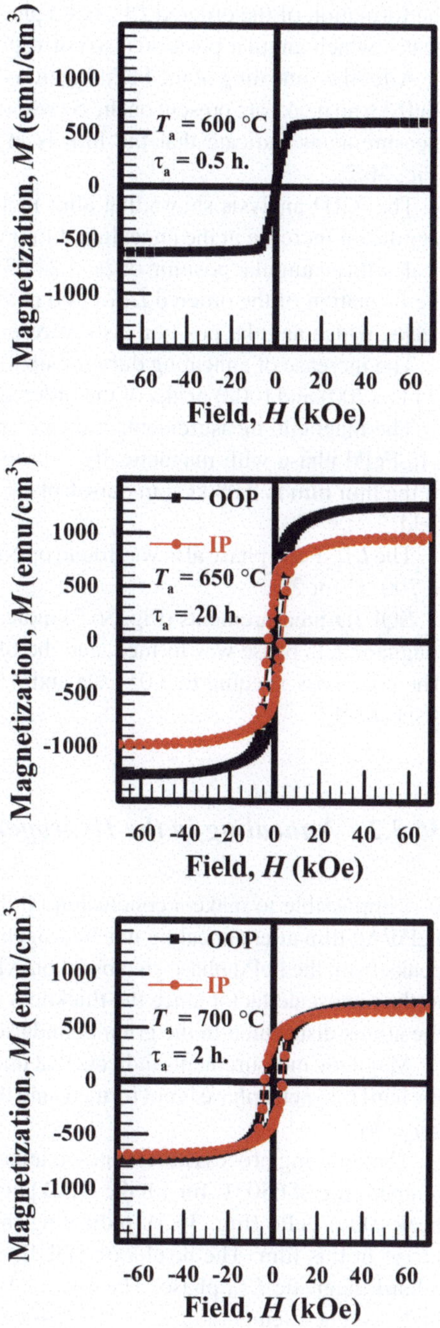

Fig. 36.2 *H-M* loops of FePd(4.7 nm)/Ag(0.3 nm) films after annealing in vacuum

the formation of the ordered $L1_0$-FePd phase. Also, there are small (111) and (200) peaks, which angular position also corresponds to the $L1_0$-FePd phase.

After the annealing of the FePd/Ag thin film in vacuum at a temperature of 600 °C for 0.5 h no peaks are present on the corresponding XRD patterns (Fig. 36.1). SQUID-measurements indicate that the film is in a soft magnetic ($A1$-FePd phase) state (Fig. 36.2).

The XRD analysis shows that after increasing temperature up to 650 °C, and a significant increase in the annealing duration up to 12 h, a low-intensity superlattice peak with an angular position of $2\theta = 24.07°$ (Fig. 36.1) appears, which may indicate the formation of the ordered $L1_0$-FePd phase. Also, there are small (111) and (200) peaks, which angular position also corresponds to the $L1_0$-FePd phase.

The increase of annealing duration up to 16 and 20 h results in the appearance of (111), (200) and (002) peaks of the ordered $L1_0$-FePd phase (Fig. 36.1).

The magnetic measurements indicate about the formation of the hard-magnetic $L1_0$-FePd phase with magnetically isotropic properties (Fig. 36.2). The coercivity of the thin film is 2.08 kOe in out-of-plane and 2.91 kOe in plane applied magnetic field.

The $L1_0$-FePd phase also was found on XRD patterns of the FePd/Ag film annealed at 700 °C for 2 h.

SQUID-measurements (Fig. 36.2) indicate that in the structure of the film a hard magnetic $L1_0$ phase was formed, and the film has magnetically isotropic properties. The coercivity is equal to 3.05 kOe and 3.93 kOe in the out-of-plane and in plane, respectively.

36.3.2 Annealing in the Hydrogen Atmosphere

It is impossible to make a conclusion on the structure or phase composition of the FePd/Ag film after annealing in a hydrogen atmosphere at 600 °C for one hour since peaks from the FePd phase are absent on XRD pattern (Fig. 36.3). This could be due to the nanoscale factor since the thickness of the film is only 5 nm, most amount of the atoms distributed in the grain boundaries.

Magnetic measurements indicate that under these conditions of thermal treatment, the hard magnetic phase is not formed, and the film composition is superparamagnetic (Fig. 36.4).

The ordering process accelerates at annealing in hydrogen. After annealing at the temperature of 650 °C for 1 h the XRD patterns reveal the presence of a superlattice peak (001) FePd (Fig. 36.3), indicating the formation of the ordered of $L1_0$-FePd phase in this film. The results of SQUID-magnetometry confirm the formation of a hard magnetic $L1_0$ phase. The coercivity of the FePd/Ag film after annealing in hydrogen at a temperature of 650 °C for 1 h in the perpendicular to plane applied magnetic field is 2.98 kOe (Fig. 36.4). In this case of parallel to plane applied magnetic field the coercivity of the film is equal to 0.86 kOe. Therefore, the film has a clearly expressed anisotropy of magnetic properties. Also, from the form of hysteresis loops,

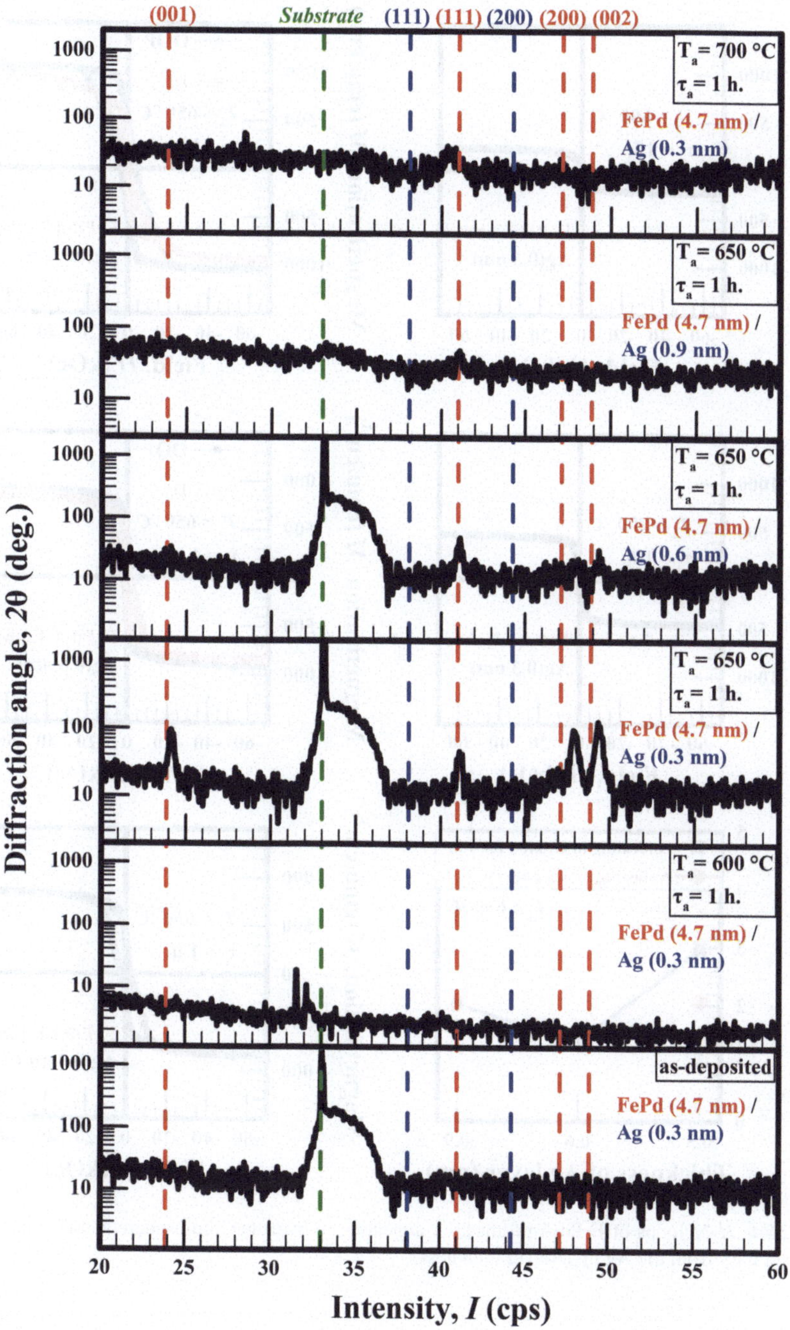

Fig. 36.3 The XRD patterns of FePd/Ag films after annealing in a hydrogen atmosphere

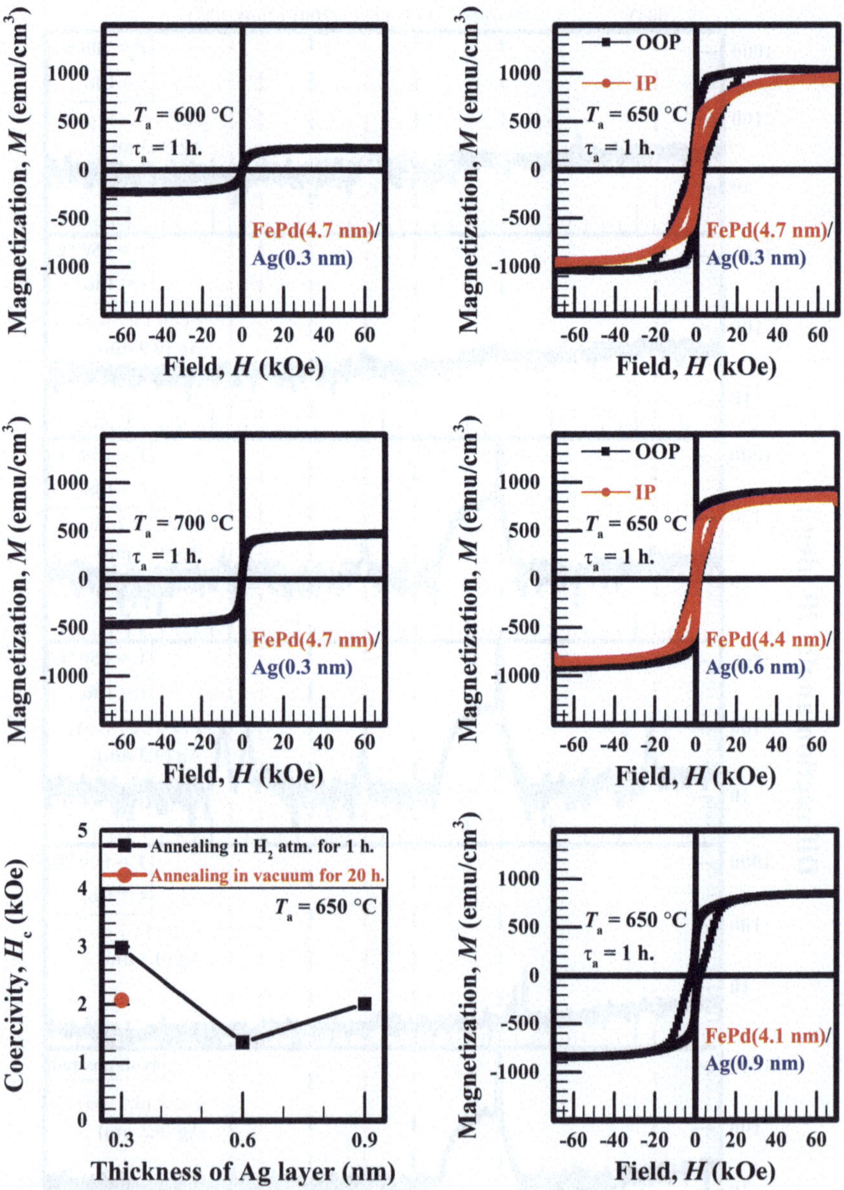

Fig. 36.4 *H-M* loops of FePd/Ag films after annealing in hydrogen atmosphere and dependence of coercivity from the Ag underlayer thickness

it could be concluded that in addition to the hard magnetic $L1_0$-FePd phase, there is still a soft magnetic disordered $A1$-FePd phase in the film.

The disordering process also accelerates at annealing in hydrogen and occurs with an increase in the temperature up to 700 °C for 1 h. On the XRD patterns of the film, the superlattice peak (001) L10-FePd is absent, and only the low-intense peak (111) $A1$-FePd is available (Fig. 36.3). The results of magnetic measurements also indicate the absence of the hard magnetic phase in the film (Fig. 36.4).

In addition, the properties of films with different Ag content after annealing in a hydrogen atmosphere at 650 °C for 1 h were investigated. With an increase in Ag underlayer thickness the diffraction patterns become less clear. In case of the FePd (4.4 nm)/Ag (0.6 nm) film, the superlattice peak (001) FePd is absent, and the (002) and (200) peaks are hardly noticeable (Fig. 36.3). On XRD patterns of the FePd(4.1 nm)/Ag(0.9 nm) there is only a low-intensive peak (111) FePd.

The study of magnetic properties showed that, irrespectively from the amount of silver in FePd/Ag films, the ordered $L1_0$-FePd phase forms after annealing in an atmosphere of hydrogen at a temperature of 650 °C for 1 h. However, the magnetic properties of the films deteriorate. The coercivity decreases and the films become more magnetically isotropic.

The temperature dependence of the surface roughness of the film is represented in Fig. 36.5. In the case of hydrogen annealing for 1 h, the roughness is greater after annealing at 650 °C than after annealing at 600 and 700 °C. Regardless of the silver amount, in the films annealed in hydrogen atmosphere at 650 °C for 1 h, the roughness is almost the same, in the margin of error, for all three films and is approximately equal to 9 nm. This is considerably lower value compared with annealing in vacuum at a temperature of 650 °C for 20 h.

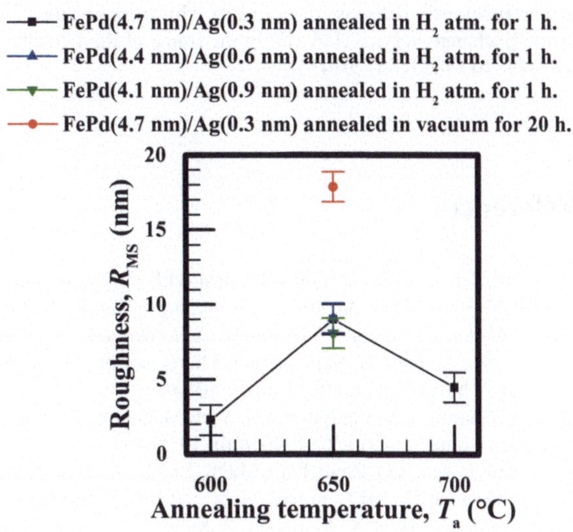

Fig. 36.5 FePd/Ag films roughness dependence via annealing temperature

We suggest that such a difference in processes of ordering and disordering at annealing in hydrogen in comparison with annealing in vacuum is caused by hydrogen influence. Atoms of hydrogen penetrate in FePd lattice and locate in octahedral and tetrahedral voids. This increases the volume of FePd unit cells and weakens the bonding between the Fe↔Pd atoms. This contributes to the ordering processes and formation of the ordered $L1_0$-FePd phase with anisotropy of magnetic properties. The superparamagnetic properties are caused by the influence of hydrogen atoms on electron structure of $A1$-FePd phase.

36.4 Conclusions

Annealing in hydrogen accelerates the ordering and disordering processes in FePd/Ag films in comparison with annealing in vacuum. The annealing at the temperature of 650 °C for 1 h leads to the formation of the ordered $L1_0$-FePd phase with better magnetic properties and expressed anisotropy of magnetic properties compared with films annealed in vacuum.

With the increase in a thickness of Ag underlayer from 0.3 to 0.9 nm the coercivity and anisotropy of magnetic properties slightly decrease.

In addition, after annealing in hydrogen the films have a much smaller surface roughness, which is an important factor for the practical application of FePd based films.

Acknowledgements The authors would like to thank Prof. Dr. M. Albrecht, Dr. G. Beddies, N. Schmidt and other team members of the Experimental Physics IV chair of the Institute of Physics of the University of Augsburg. (Germany) for sample preparation, assistance in conduction of investigations and discussion of results. This work was financially supported by the German Academic Exchange Service (DAAD) in the frame of the Leonard Euler scholarship program (Grant ID 57198300 and ID 57291435).

References

1. D. Weller, A. Moser, L. Folks, M. Best, M.F. Toney, M. Schwickert, J.-U. Thiele, M.F. Doerner, IEEE Trans. Magn. **36**, 10–15 (2000). https://doi.org/10.1109/20.824418
2. D. Weller, G. Parker, O. Mosendz, A. Lyberatos, D. Mitin, N.Y. Safonova, M. Albrecht, J. Vac. Sci. Technol. B, Nanotechnol. Microelectron. Mater. Process. Meas. Phenom. **34**, 060801 (2016). https://doi.org/10.1116/1.4965980
3. O.V. Shamis, I.A. Vladymyrskyi, Y.M. Makogon, S.I. Sidorenko, Usp. Fiz. Met. **19**, 337–363 (2018). https://doi.org/10.15407/ufm.19.03.337
4. T. Liu, L. Ma, S.Q. Zhao, D.D. Ma, L. Li, G. Cheng, G.H. Rao, J. Mater. Sci.: Mater. Electron. **28**, 3616–3620 (2017). https://doi.org/10.1007/s10854-016-5963-6
5. Z. Yanli, C. Gang, X. Xiaozong, P. Kuang, L. Lin, D. Yusong, Z. Xin, M. Lei, G. Zhengfei, Rare Met. Mater. Eng. **46**, 1788–1791 (2017). https://doi.org/10.1016/s1875-5372(17)30167-4
6. J. Ko, T. Bae, J. Hong, J. Appl. Phys. **112**, 113919 (2012). https://doi.org/10.1063/1.4769737

7. Y.J. Chiu, C.Y. Shen, H.W. Chang, S.R. Jian, Results Phys. **9**, 17–22 (2018). https://doi.org/10.1016/j.rinp.2018.02.024
8. S.M. Zharkov, E.T. Moiseenko, R.R. Altunin, J. Solid State Chem. **269**, 36–42 (2019). https://doi.org/10.1016/j.jssc.2018.09.009
9. B. Li, W. Liu, X.G. Zhao, S. Ma, W.J. Gong, J.N. Feng, F. Wang, Z.D. Zhang, Mater. Lett. **100**, 58–61 (2013). https://doi.org/10.1016/j.matlet.2013.02.102
10. Z. Yanli, C. Gang, X. Xiaozong, P. Kuang, L. Lin, D. Yusong, Z. Xin, M. Lei, G. Zhengfei, J. Wuhan Univ. Technol. Sci. Ed. **33**, 1082–1085 (2018). https://doi.org/10.1016/s1875-5372(17)30167-4
11. Y. Tokuoka, Y. Seto, T. Kato, S. Iwata, J. Appl. Phys. **115**, 1–4 (2014). https://doi.org/10.1063/1.4864251
12. A. Vladymyrskyi, M.V. Karpets, F. Ganss, G.L. Katona, D.L. Beke, S.I. Sidorenko, T. Nagata, T. Nabatame, T. Chikyow, G. Beddies, M. Albrecht, I.M. Makogon, J. Appl. Phys. **114** (2013). https://doi.org/10.1063/1.4827202
13. M. Yamauchi, K. Okubo, T. Tsukuda, K. Kato, M. Takata, S. Takeda, Nanoscale **6**, 4067–4071 (2014). https://doi.org/10.1039/c3nr06327e
14. E.A. Gonzalez, P.V. Jasen, N.J. Castellani, A. Juan, J. Phys. Chem. Solids **65**, 1799–1807 (2004). https://doi.org/10.1016/j.jpcs.2004.05.008
15. E.A. Gonzalez, P.V. Jasen, N.J. Castellani, A. Juan, Solid State Commun. **131**, 81–85 (2004). https://doi.org/10.1016/j.ssc.2004.04.046
16. P.V. Jasen, E.A. Gonzalez, N.J. Castellani, A. Juan, Phys. Rev. B—Condens. Matter Mater. Phys. **71**, 1–7 (2005). https://doi.org/10.1103/physrevb.71.235422

Chapter 37
Thermal Dispergation of Pb-In Alloys Films on the Molybdenum Substrate

S. V. Dukarov, S. I. Petrushenko, I. Churilov, A. Lyalka, Z. Bloshenko and V. Sukhov

Abstract The work is devoted to the study of island structures, which are formed during the melting of initially continuous vacuum condensates. Pb-In bilayer films deposited by the method of the sequent vacuum condensation onto the molybdenum substrate were chosen as the objects of the study. Films containing in the fusible alloy from 0 to 100 wt% of lead were studied. It was established that under films melting, the arrays of separate particles are formed, for which the unimodal size distribution is typical. It is shown that for Pb-In films, the value of the most probable particle radius and of the full width at half maximum of histograms of their size distribution increase linearly with the general thickness. For films of alloys with a fixed mass, these parameters do not practically dependent on the component concentration in the range from 20 to 80 wt% of Pb. At the same time, for one-component Pb/Mo and In/Mo films narrower histograms and a value decrease of the most probable particle radius are typical. The small value of the coverage, which is observed in (Pb-In)/Mo films is probably the result of the high excess energy of the studied samples.

37.1 Introduction

Functional nano-arrays are a perspective object of modern technologies. Due to collective effects, arrays of separated metal particles are used as biosensors, photocatalytic generators and light-generating devices [1–3]. More complex structures find their use in spintronics and sensor techniques, showing high technological parameters [4–6]. For the creation of functional arrays, a variety of different ways is proposed, each of which has its own particularities. Thus, the most common and cost-effective are chemical methods [7–10]. However, the nanoparticle precipitation from solutions requires the use of stabilizers that prevent a coalescence [11]. At the same time,

S. V. Dukarov · S. I. Petrushenko (✉) · I. Churilov · Z. Bloshenko · V. Sukhov
V. N. Karazin Kharkiv National University, Kharkiv, Ukraine
e-mail: petrushenko@univer.kharkov.ua

A. Lyalka
O. L. Pshenychky Vysochanska Secondary School, Kharkiv, Ukraine

A. D. Pogrebnjak and O. Bondar (eds.), *Microstructure and Properties of Micro- and Nanoscale Materials, Films, and Coatings (NAP 2019)*,
Springer Proceedings in Physics 240, https://doi.org/10.1007/978-981-15-1742-6_37

the presence of the third-party organic layer on the particle surface can change the sample behavior in a poorly predictable way.

The creation of ordered nanoparticle arrays with the use of vacuum condensates looks very perspective. Such methods are particularly topical for the study of the fundamental properties of functional structures. Island structures on the substrate can be obtained in the way of the substance condensation to both equilibrium and supercooled liquid phase or by the thermal dispergation of initially continuous films [12, 13]. In the last case, the solid-phase films dispergation can be carried out without melting of the main component [1, 14, 15]. The high purity of the obtained samples is an obvious advantage of using vacuum preparation methods. Also, an important advantage of this approach is providing the atomic contact of multilayer structures components. This occurs to be topical during the study of multicomponent systems.

On the other hand, the temperature evolution of the morphological structure of films is an independent general scientific interest. In particular, the study of thermal dispergation of initially continuous films allows obtaining unique information about their energy balance. This is due to the fact, that the de-wetting driving force is presented by the excess energy of the initially continuous film, which is connected with mechanical stresses and a branched network of grain boundaries.

The transition of an initially continuous film into an island one in the result of its melting is a known phenomenon, which clearly shows itself when the initial film is located on the poorly wettable substrate [12, 13, 16]. The value of the wetting angle is determined by the substrate [16–19] and for the films of fusible metals on the carbon sublayer an almost complete non-wetting is observed [12, 13, 16]. In this case, an array of islands of various sizes and of regular spherical shape is formed on the substrate. In the work [13] it is shown that the size distribution for particles, which appeared in the result of melting is normal and unimodal, and the most probable size and the distribution full width at half maximum (FWHM) are determined by the condensate thickness. For tin films on the amorphous carbon substrate, the most probable size and histograms FWHM decrease linearly with a mass thickness of the condensate [13]. At that, the coverage of the substrate with a film in a non-trivial manner depends on the thickness: at first, with a thickness decrease, it decreases very slowly, but at a certain thickness, a change in the dependence sign is observed and the coverage begins to grow rapidly. In the work [13] it is shown that this phenomenon is observed for films in which the formation of a continuous structure did not occur during the condensation process.

The evaluation, made in the framework of thermodynamics, shows that the excess energy, which actually provides the dispergation of initially continuous films, for samples with coverage, which is bigger than about 0.3 quickly tends to zero. This indicates that such samples in the condensation process remain of island-type, and the melting leads only to the change in the shape of the islands but is not accompanied by the further dispergation.

The authors of the most of works devoted to the films dispergation concentrate on the study of enough simple systems, in the first place single-component films on the inert substrate [12, 13]. The data of alloys dispergation, especially of those ones, which are located on a more active, for example, metal substrate exists in much less amount. At the same time, such systems find wide use in modern technologies. This work is devoted to the study of thermal dispergation of In-Pb alloy films, deposited on a molybdenum substrate. The object choice of study is due to both the applied significance, which fusible alloys contacting with metal layers have and methodological considerations. Thus, the components of the In-Pb contact pair form enough simple phase diagram with high solubility in the solid state [20, 21]. At that, indium and lead do not form new phases and chemical compounds with molybdenum. At the same time, molybdenum is wetted good enough by the components of the selected alloy [22]. This allows extending the results obtained in the work [13] to the case of alloys, located on a wettable substrate. At that, it is possible to disregard the effects, connected with possible chemical reactions and phase transformations that can be observed in the substrate during its interaction with the sample.

37.2 Experiment Details

Films (Pb-In)/Mo, which are obtained by the method of the sequent vacuum condensation in a vacuum of 10^{-6} mm Hg created by oil-free means of pumping were chosen as the objects of study. A set of samples, corresponding to the lead concentration in a fusible alloy in the range of 0–100% was obtained. Glass plates of approximately 10×15 mm size were used as the substrate. Fusible components evaporated from molybdenum boats, which were heated by the direct passing of electric current. A layer of molybdenum was deposited in the way of electron-beam sputtering. The layers thickness, and, consequently, the alloy component concentrations were determined in the process of condensation with the help of the quartz resonator. Additionally, the component concentration was determined during the SEM-studies by the EDS microanalysis method.

Two identical samples were obtained in the result of metals deposition in each experimental cycle. Then, the first of them in a short time was heated to the temperature, which was above the melting point of the condensed alloy. The second sample was not heated so that it was always under the room temperature. When the experiment was completed, the films were taken from the vacuum chamber and were examined with the use of scanning electron microscopy. The analysis of the obtained SEM images was performed with the use of specially developed software.

37.3 Results and Discussion

Under SEM-studies of films, it was found that, as it was to be expected, the samples consist of separate spherical islands after the melting (Fig. 37.1). The average value of the coverage of the substrate with a film is about 0.08 for samples with a thickness of 220 nm containing from 20 to 80 wt% lead in a fusible alloy. By the comparison of the obtained values with the values given in the work [13], we see that our samples

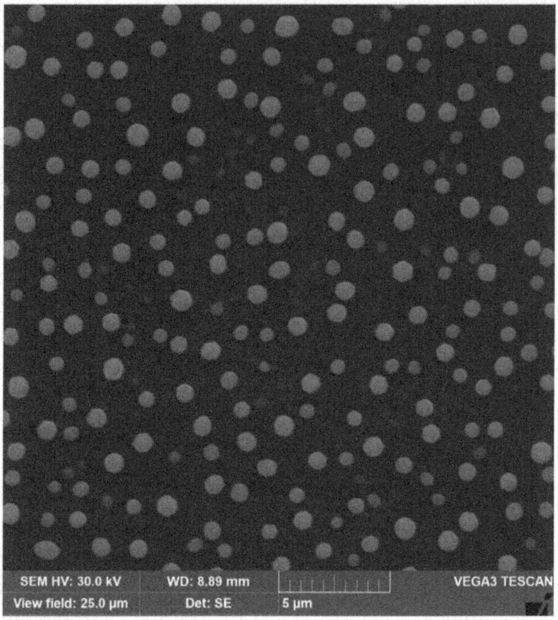

Fig. 37.1 SEM image of the island structure formed in the result of a Pb-In alloy film melting (65 wt% Pb) on the molybdenum substrate

are characterized by rather small values of the coverage. In accordance with [13], it indicates the essential excess energy of the original film. At that, it should be noted, that for single-component films of the same thickness, located on an amorphous carbon film, significantly bigger coverage [13] are typical, i.e. smaller values of excess energy. It appears that an increase of the excess energy that stimulates the decay is due to the metallic sublayer presence in our samples. According to [23] metal sublayer deposited onto the inert substrate till the metal film condensation prevents crystallites coalescing, which increases the internal dispersity of the structure and the excess energy, associated with it.

In structures, which were obtained by vacuum condensation methods, a violation of the homogeneity and the composition variations of the arising particles can be observed [1]. This can adversely impact the functional properties of the structures. The size of particles, which were observed in the samples is too small for building the maps of the elements' distribution by the method of EDS microanalysis. However, BSE images (Fig. 37.2) indicate the absence of fractionation and the approximate composition constancy for all particles. Thus, it can be stated, that the method based on the layer-by-layer condensation of components makes it possible to form an array of homogeneous nanoparticles of almost the same composition on the substrate.

As a rule, arrays of particles of as possible equal size are preferred for the practical applications [13]. That is why, for the series of samples corresponding to the lead concentration in In-Pb alloy from 0 to 100 wt%, histograms of size distributions of particles, which arise after the melting, were built.

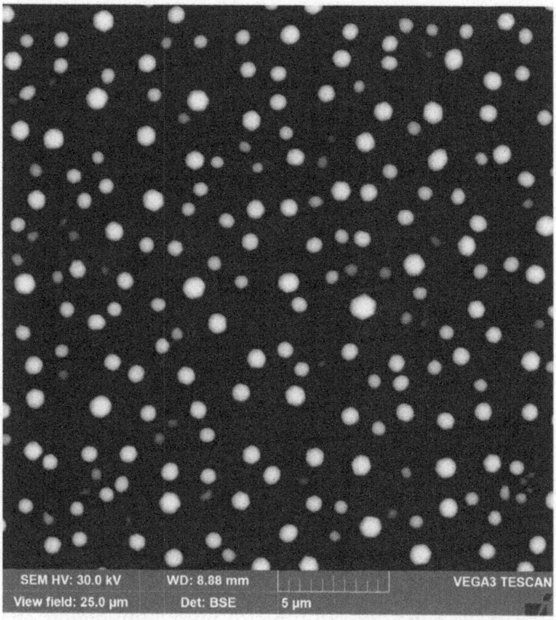

Fig. 37.2 SEM images of (Pb-In)/Mo films, obtained in the mode of the maximum contrast of BSE detector

At that, for different composition samples, the summary mass of lead and indium deposited per unit area of the substrate surface was being kept constant. This was necessary in order to emphasize the influence namely of components concentration on the distribution parameters. During histograms constructing on the vertical axis, such value was marked off

$$W = \frac{4}{3}\frac{\pi N(R)R^3}{A\Delta R},$$

where $N(R)$ is the number of particles whose radius falls into the selected step of particle size change ΔR. A is the area of the image space from which the histogram calculation was conducted. This choice of normalization allows obtaining the particle distribution, which appears from the film dispergation according to the amount of the substance contained in them. Namely this parameter is of prime importance for solving many technical problems [13]. The obtained distribution histograms corresponding to Pb-In alloys with lead concentrations from 20 to 80 wt%, are represented in Fig. 37.3

As we can see, there is enough narrow unimodal size distribution of particles resulted from the thermal dispergation. The average value of the most probable radius of particles and of the FWHM of distribution histograms in samples (Pb-In)/Mo with a summary thickness of fusible components layers of 220 nm is 450 nm and 200 nm, respectively. For In-Pb alloys containing from 20 to 80 wt% of lead, parameters of distributions do not practically dependent on the component concentrations. At the same time, for In/Mo and Pb/Mo films more sharper distributions are typical.

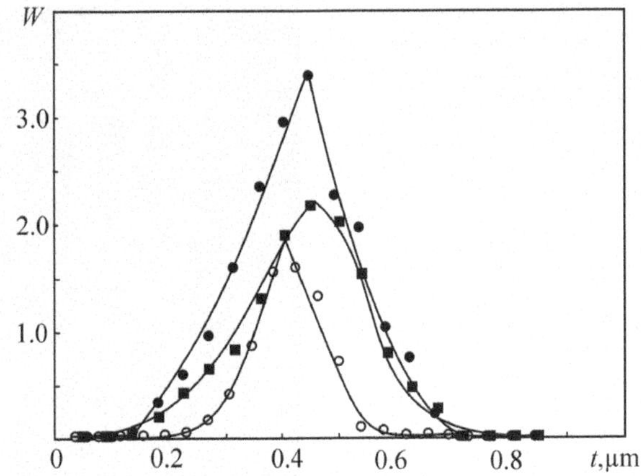

Fig. 37.3 A series of typical histograms of islands size distribution corresponding to Pb-In alloys with different components concentration: 27 wt% Pb (●), 65 wt% Pb (○) and 80 wt% Pb (■)

Smaller values of the most probable radius and of the FWHM, which are equal to 220 nm and 130 nm, respectively, correspond to them. Corresponding histograms are shown in Fig. 37.4. At that, since the mass of the fusible substance in all samples was being kept constant, the number density (i.e. the number of particles per unit area of the substrate) in films of pure components sharply increases. As a result of this, the coverage of the substrate with a film increases and becomes about 0.4.

It should be noted that, as for the films of pure metals on the non-wettable substrate, an increase of the sample thickness naturally leads to the structures coarsening, which appear during film melting (see Fig. 37.5).

The size dependence of the most probable radius of particles, which was observed in films (Pb-In)/Mo with a lead concentration in a fusible alloy of 65 wt% is shown in Fig. 37.6. It can be seen that, just as for one-component films on a carbon substrate [13], the value of the most probable radius is linearly connected with the thickness

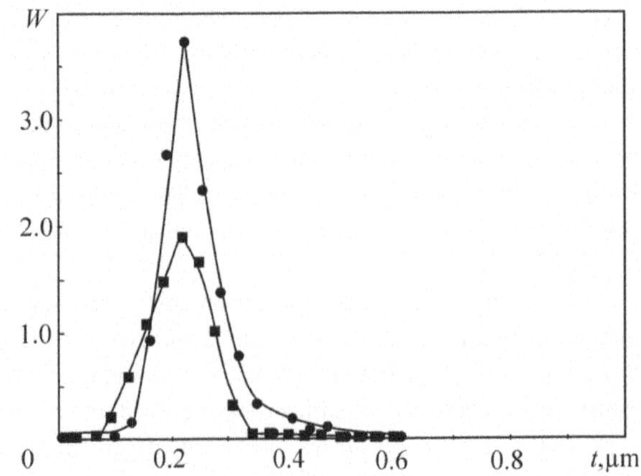

Fig. 37.4 Histograms of the particle size distribution, which are formed during the melting of one-component films of indium (●) and lead (■) on the molybdenum substrate

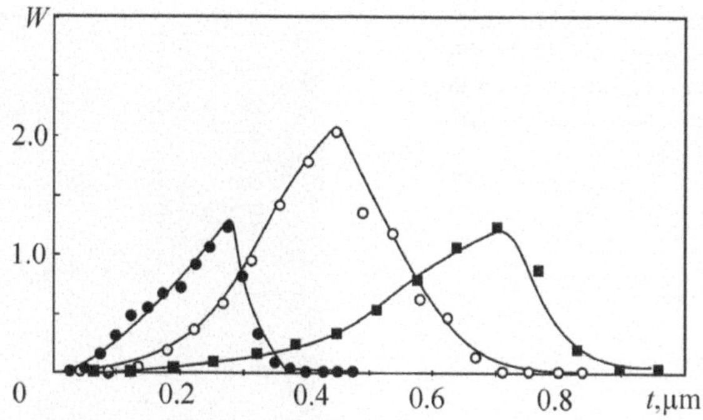

Fig. 37.5 Histograms of particle size distribution in (Pb-In)/Mo films with a summary thickness of fusible metals layers of 170 nm (●), 220 nm (○) and 300 nm (■). The lead concentration in the fusible alloy was 65 wt%

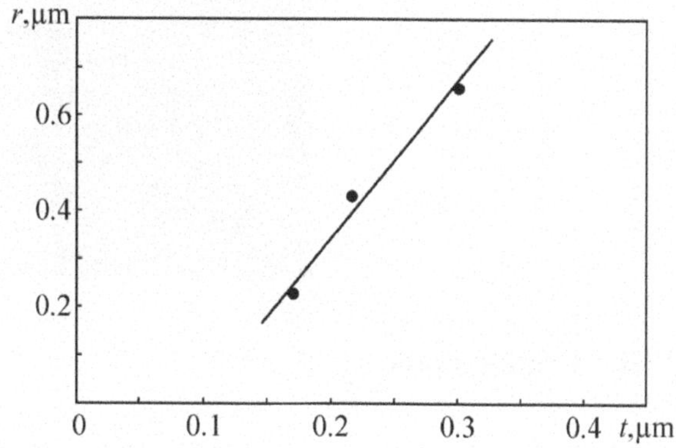

Fig. 37.6 Dependence of the most probable radius on the thickness of the fusible layer in the (Pb-In)/Mo films

of the condensate. In an analogical way, the FWHM of histograms of the particle size distribution changes.

Note, that the film thicknesses studied in work [13] in a guaranteed way provided the formation of continuous layers under the condensation. These layers transformed into the island film only during the melting. In our work, thinner samples were investigated. Moreover, indium, which is the component of In-Pb alloy, has a not high melting point and is supercooled well enough [13]. This leads to the fact, that island structures can be formed in the samples even before the melting. Indeed, as SEM-studies indicate, an extremely developed structure is observed in as-deposited samples that were not exposed to thermal impact (Fig. 37.7). This structure looks like the transitional one between continuous and island films. It is obvious that such structure formation is an effective way for excess energy relaxation. However, the low value of the coverage, which is typical for films after the melting, indicates that (Pb-In)/Mo samples have a significant value of the excess energy. This is probably due to the fine-crystalline structure of the objects observed in Fig. 37.7, which in turn can be explained by the presence of the surfactant molybdenum layer, which prevents the particles coalescence at the stage of early film growth.

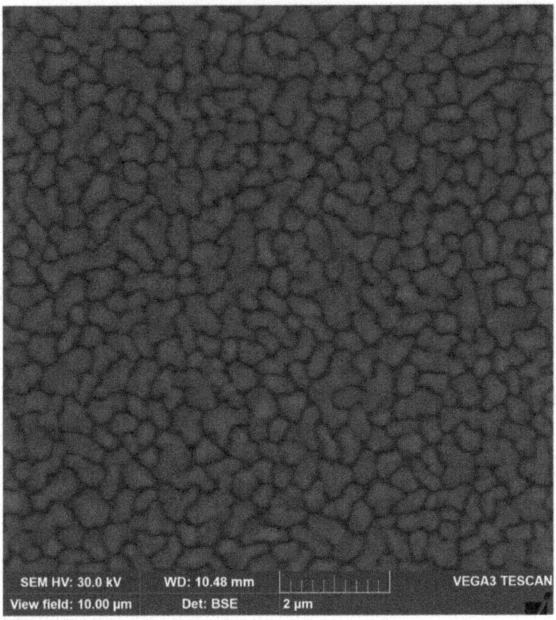

Fig. 37.7 SEM image of the as-deposited (Pb-In)/Mo film, showing the developed structure of the original condensate

37.4 Conclusions

It is shown, that the use of the method of sequential condensation with the subsequent melting of the films allows obtaining an array of homogeneous nanoparticles with small variations of the composition. The size distribution histograms of the particles arising under the thermal dispergation of In-Pb alloy films on a molybdenum substrate are unimodal ones. The full width at half maximum of the histogram and the value of the most probable radius are determined by the condensate thickness and do not depend on the alloy composition in a considerable interval of concentrations. The coverage of the substrate with a film occurs to be significantly lower than in the case of one-component films of the same thickness. This indicates a high value of the excess energy of the as-deposited films, which one stimulates their decay.

Acknowledgements This work was supported by the Ministry of Education and Science of Ukraine.

References

1. M. Proença, J. Borges, M.S. Rodrigues, R.P. Domingues, J.P. Dias, J. Trigueiro, N. Bundaleski, O.M.N.D. Teodoro, F. Vaz, Development of Au/CuO nanoplasmonic thin films for sensing applications. Surf. Coatings Technol. **343**, 178–185 (2018). https://doi.org/10.1016/j.surfcoat.2017.08.033

2. A.I. Barbosa, J. Borges, D.I. Meira, D. Costa, M.S. Rodrigues, R. Rebelo, V.M. Correlo, F. Vaz, R.L. Reis, Development of label-free plasmonic Au-TiO 2 thin film immunosensor devices. Mater. Sci. Eng., C **100**, 424–432 (2019). https://doi.org/10.1016/j.msec.2019.03.029
3. B. Hartiti, M. Siadat, E. Comini, H.M.M. Arachchige, S. Fadili, P. Thevenin, Acetone sensor based on Ni doped ZnO nanostructues: growth and sensing capability. J. Mater. Sci.: Mater. Electron. **30**(8), 7681–7690 (2019). https://doi.org/10.1007/s10854-019-01083-9
4. I.O. Shpetnyi, D.M. Kondrakhova, S.I. Vorobiov, B. Scheibe, V.I. Grebinaha, D.O. Derecha, Y.I. Gorobets, I.Y. Protsenko, The structural-phase state and magnetoresistive properties of thin film alloys obtained by co-evaporated Cu and Co. J. Magn. Magn. Mater. **474**, 624–631 (2019). https://doi.org/10.1016/j.jmmm.2018.12.013
5. Y. Xia, T. Yoo, Y. Xiang, Y. Zhang, J.J. Kim, T.S. Kuan, G.C. Wang, Uniaxial magnetic anisotropy in three-bilayer Co/Cu and Co/Al superlattices. Thin Solid Films **681**, 32–40 (2019). https://doi.org/10.1016/j.tsf.2019.04.048
6. D.I. Saltykov, Yu O. Shkurdoda, I.Yu. Protsenko, Influence of heat treatment conditions on the magnetoresistive properties of three-layer structures Fe0,2Co0,8/Cu/Fe0,2Co0,8. J. Nano-Electron. Phys. **10**(4), 04031 (2018). https://doi.org/10.21272/jnep.10(4).04031
7. M.S. Rodrigues, J. Borges, M. Proença, P. Pedrosa, N. Martin, K. Romanyuk, A.L. Kholkin, F. Vaz, Nanoplasmonic response of porous Au-TiO2 thin films prepared by oblique angle deposition. Nanotechnology **30**, 225701 (2019). https://doi.org/10.1088/1361-6528/ab068e
8. S.B. Dalavi, R.N. Panda, New synthetic methodology and magnetic properties of fcc Co–Ni nanostructured alloys embedded in KIT-6 matrix. J. Mater. Res. **31**(16), 2430–2437 (2016). https://doi.org/10.1557/jmr.2016.255
9. N.V. Bondar, M.S. Brodyn, N.A. Matveevskaya, Influence of surface and polarization effects on electronic excitation energy transfer in colloidal solutions and films of ZnSe quantum dots. Low Temp. Phys. **44**(11), 1198–1203 (2018). https://doi.org/10.1063/1.5062158
10. S.B. Van Der Meer, K. Loza, K. Wey, M. Heggen, C. Beuck, P. Bayer, M. Epple, Click chemistry on the surface of ultrasmall gold nanoparticles (2 nm) for covalent ligand attachment followed by NMR spectroscopy. Langmuir **35**(22), 7191–7204 (2019). https://doi.org/10.1021/acs.langmuir.9b00295
11. K. Loza, M. Epple, M. Maskos, in *Biological Responses to Nanoscale Particles. NanoScience and Technology*, ed. by P. Gehr, R. Zellner (Springer, 2019), pp. 85–100. https://doi.org/10.1007/978-3-030-12461-8_4
12. A.P. Kryshtal, Formation of island arrays by melting of Bi, Pb and Sn continuous films on Si substrate. Appl. Surf. Sci. **321**, 548–553 (2014). https://doi.org/10.1016/j.apsusc.2014.10.046
13. S.V. Dukarov, S.I. Petrushenko, V.N. Sukhov, I.G. Churilov, The formation of island structures during the melting process of tin films on an amorphous carbon substrate. Metallofiz. Noveishie Tekhnol. **41**(4), 445–459 (2019). https://doi.org/10.15407/mfint.41.04.0445
14. S.I. Petrushenko, S.V. Dukarov, V.N. Sukhov, Effect of lead on the thermal dispersion of continuous polycrystalline copper films. Vacuum **142**, 29–36 (2017). https://doi.org/10.1016/j.vacuum.2017.04.037
15. R.P. Domingues, M.S. Rodrigues, C. Lopes, P. Pedrosa, E. Alves, N.P. Barradas, J. Borges, F. Vaz, Thin films composed of metal nanoparticles (Au, Ag, Cu) dispersed in AlN: The influence of composition and thermal annealing on the structure and plasmonic response. Thin Solid Films **676**, 12–25 (2019). https://doi.org/10.1016/j.tsf.2019.02.047
16. S.I. Petrushenko, S.V. Dukarov, V.N. Sukhov, I.G. Churilov, Inner size effect in the polycrystalline metal films of fusible metals. J. Nano- Electron. Phys. **7**(2), 02033 (2015)
17. A.A. Minenkov, A.P. Kryshtal, S.I. Bogatyrenko, Effect of size on phase transformation temperatures in Ge/Bi/Ge films. J. Alloy. Compd. **756**, 50–56 (2018). https://doi.org/10.1016/j.jallcom.2018.04.335
18. S.V. Dukarov, S.I. Petrushenko, V.N. Sukhov, Supercooling during crystallization of a fusible component in Cu/(Bi-Sn) multilayer films. Mater. Res. Express **6**(1), 016403 (2019). https://doi.org/10.1088/2053-1591/aae0c7
19. S.I. Bogatyrenko, A.A. Minenkov, A.P. Kryshtal, Supercooling under crystallization of Bi-Sn eutectic alloy in contact with Bi. Sn and amorphous C. Vacuum **152**, 1–7 (2018). https://doi.org/10.1016/j.vacuum.2018.02.039

20. H. Okamoto, *Phase Diagrams for Binary Alloys* (ASM International Materials Park, 2010)
21. N.P. Lyakishev (ed.), *Diagrammy sostoyaniya dvoinykh metallicheskikh sistem (Phase diagrams of binary metallic systems)* (Mashinostroenie, Moscow, 2001)
22. S.I. Petrushenko, S.V. Dukarov, V.N. Sukhov, Stability limits of the liquid phase in the layered Mo/Pb/Mo, Mo/Bi/Mo and Mo/In/Mo film systems. J. Nano- Electron. Phys. **8**(2), 04073 (2016). https://doi.org/10.21272/jnep.8(4(2)).04073
23. R.I. Bihun, Z.V. Stasyuk, O.A. Baliskii, Crossover from quantum to classical electron transsport in ultrathin metal film. Physica B **487**, 73–77 (2016). https://doi.org/10.1016/j.physb.2016.02.003

Chapter 38
Structure and Low-Temperature Properties of U-15 at.% T Alloys (T = Mo, Nb, Pt, Ru, Ti)

Sylwia Sowa, N. -T. H. Kim-Ngan, Volodymir Buturlim, Ladislav Havela, Maciej Chrobak and Zbigniew Tarnawski

Abstract We show that the high-temperature cubic γ-U phase can be retained to room temperature by means of ultrafast-cooling technique (with a cooling rate in the range of 10^6 K/s) combined with alloying with selected 4*d* and 5*d* elements in the IV–VIII group of the periodic table Hofman et al. (Proceedings of the International meeting on Reduced Enrichment for Research and Test Reactors (RERTR), Sao Paulo, Brazil, pp. 18–23 1998 [1]). Most of U-15 at.% T alloys (atomic percentage, T = Mo, Nb, Pt, Ru, and Ti) revealed the single cubic phase (or dominant cubic phase). All our investigated samples become superconducting below 2.11 K.

38.1 Introduction

Pure uranium metal exists in three allotropic phases: α, β and γ [2]. The orthorhombic α-U phase stabilizes below 940 K down to room temperature, the tetragonal β-U phase exists between 942-1049 K [3] and the γ-phase with a body-centered cubic A2 structure is stable at high temperature of 1049–1409 K. The low-temperature properties of uranium have been mostly known for the orthorhombic α-U phase [4].

S. Sowa · N. .-T. H. Kim-Ngan
Institute of Physics, Pedagogical University of Cracow, 30-084 Cracow, Poland

S. Sowa (✉)
Łukasiewicz Research Network—Institute for Sustainable Technologies, 26-600 Radom, Poland
e-mail: sylwia.sowa@itee.radom.pl

V. Buturlim · L. Havela
Department of Condensed Matter Physics, Faculty of Mathematics and Physics, Charles University, 12116 Prague, Czech Republic

M. Chrobak · Z. Tarnawski
Faculty of Physics and Applied Computer Science, AGH University of Science and Technology, 30-059 Cracow, Poland

M. Chrobak
Academic Centre for Materials and Nanotechnology, AGH University of Science and Technology, 30-059 Cracow, Poland

A. D. Pogrebnjak and O. Bondar (eds.), *Microstructure and Properties of Micro- and Nanoscale Materials, Films, and Coatings (NAP 2019)*, Springer Proceedings in Physics 240, https://doi.org/10.1007/978-981-15-1742-6_38

Our earlier results demonstrated that using the ultrafast cooling technique (with a cooling rate of 10^6 K/s) we were able to retain the γ-U phase down to the room temperatures with a lower T alloying-content. In this work, we summarized our results obtained for selected U-T alloys with T = Mo, Nb, Pt, Ru and Ti [5–8]. We focus on stabilization of the γ-U phase in U-T alloys with 15 at.% T and on investigation of their low-temperature properties.

38.2 Experiment Details

The sample-ingots with a typical mass of 200–300 mg of U-T alloys with T concentration up to 15 at.% were prepared by arc melting of high-purified elements. Splat-cooled samples (splats) were produced from such ingots by means of a splat cooler (Vacuum Praha). This technique allows obtaining thin-foil splats with a disc shape of a diameter of 20 mm and a flat surface with a thickness of 100–200 μm (Fig. 38.1). Such samples were used directly for the XRD measurements or cut into proper-size pieces for other measurements. The crystal structure of splats was characterized by XRD using a Brucker D8 Advance diffractometer with Cu-K_α radiation. The resistivity measurements were performed in a standard four-probe configuration and the specific-heat measurements were performed using the relaxation method using a Quantum Design PPMS in the temperature range of 0.4–300 K and in applied magnetic fields up to 7 T. For obtaining more information about superconductivity at low temperatures, we performed the electrical resistivity measurements by means of Triton dilution refrigerator featuring a superconducting magnet in the temperature range from 70 mK to 30 K with magnetic fields up to 14 T. It gives an opportunity

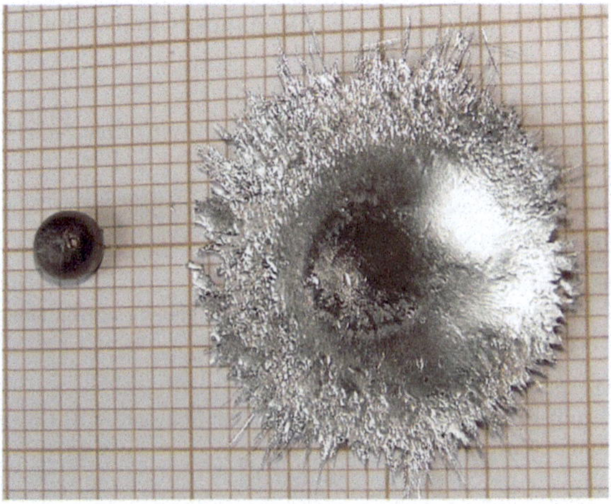

Fig. 38.1 Photograph of the bulk ingot (left) and the resulting splat (right) with a typical thickness of 100–200 μm produced by the splat cooler [8]

to follow the superconducting shift induced by magnetic fields below 0.4 K (which is the limit of PPMS measurements) [8–11].

38.3 Results

38.3.1 Crystal Structure

In most cases we found that the γ-U phase becomes very dominant by doping with 15 at.% T (T = Mo, Nb, Pt, Ru and Ti).

Figure 38.2 presents a comparison of the most intense reflection in the X-ray pattern around 37° for U-15 at.% T alloys. The very sharp single γ-U peak in U-15 at.% Mo indicated that this alloy crystallizes in an ideal cubic A2 structure [5]. The XRD pattern of U-15 at.% Ru splat is very similar to that with U-15 at.% Mo indicating the γ-U phase. The cubic γ-U phase could be also stabilized in U-15 at. % Pt, but a small trace of the orthorhombic α-U phase (e.g. α(021) peak) is still

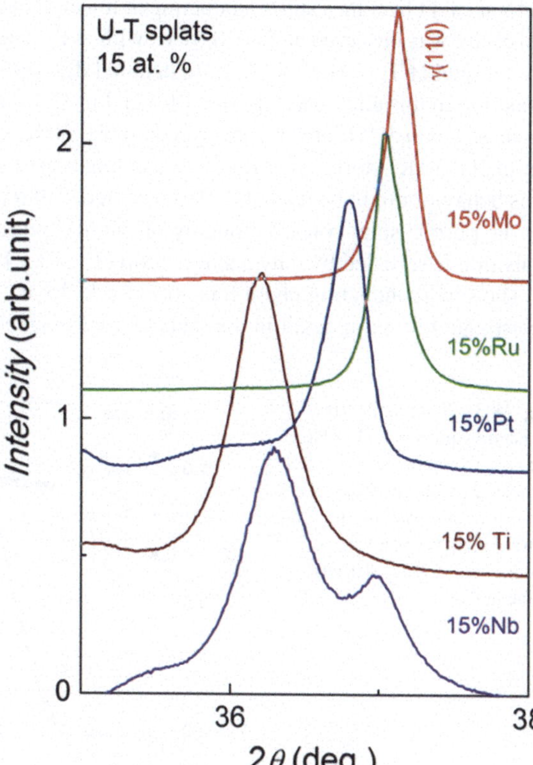

Fig. 38.2 Comparison of (normalized) XRD patterns of the γ(110) reflection of the U-T splats with the same T concentration of 15 at.%

persisted. U-15 at.% Ti alloy exhibits the mixed orthorhombic and cubic ($\alpha + \gamma$)-U phase. (The pure γ-U phase was obtained for U-30 at.% Ti). Alloying with 15 at.% Nb stabilizes the tetragonally distorted cubic phase (γ^o-U phase) revealed by a splitting of γ-reflections into doublets (which is similar to that of U-11/12 at.% Mo) [11].

38.3.2 Electrical Resistivity and Superconducting Transition

The temperature dependence of the electrical resistivity in the low temperatures in zero magnetic field is shown in Fig. 38.3. For comparison the data for pure U metal splat was also included. The superconducting phase transitions were observed in all investigated samples. U-15 at.% Mo reveals the highest critical temperature (T_c = 2.11 K) with the smallest transition width (ΔT_ρ = 0.02 K) [5, 6]. For U-15 at.% Ru, the superconducting phase transition was observed at T_c = 1.45 K with ΔT_ρ = 0.09 K. U-15 at.% Nb has revealed a single resistivity drop at T_c = 1.90 K with ΔT_ρ = 0.15 K. Except of a small difference in the T_c value, the resistivity drop in U-15 at.% Nb is very similar to that of U-11/12% Mo splats consisting of γ^o-U phase [11]. U-15 at. % Pt becomes superconducting at lower temperatures. A sharp drop to zero resistivity was observed at T_c = 0.61 K, with ΔT_ρ = 0.04 K and an extra weak drop can be seen at T_c = 0.95 K. A more detailed description of superconducting phase transition in this alloy was reported [9–11]. For U-15 at.% Ti, the resistivity starts to decrease below 0.9 K and reaches zero below 0.60 K, i.e. it reveals a broad transition width of 0.30 K. Moreover, it reveals a combination of two drops with different slops. This behavior might be related to the coexistence of α and γ-U phase. We notice that for the pure U splat consisted mainly of the α-U phase, the superconducting phase transition is revealed by only a single drop (T_c = 1.24 K) [5].

The superconducting phase transition for U-15 at. % Ru and U-15 at. % Ti were investigated in more detail in applied external magnetic fields using Triton, shown

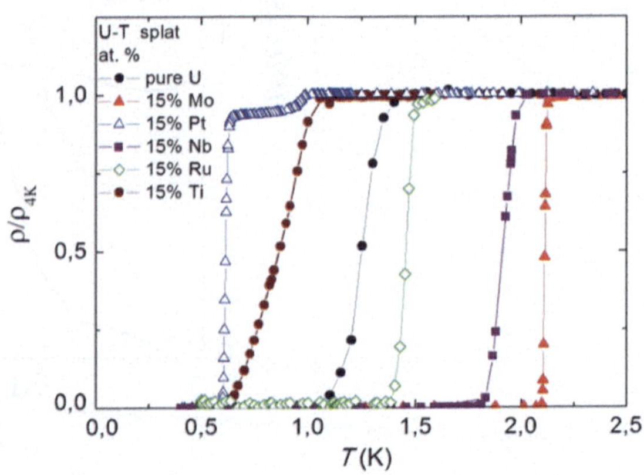

Fig. 38.3 Superconducting phase transition in U-T splats (in zero magnetic fields) indicated by abrupt drops of the electrical resistivity to zero. All curves were normalized to the resistivity value at 4 K

in Fig. 38.4. In both cases, the transitions move towards lower temperatures with an increasing magnetic field.

For U-15 at.% Ru, despite the shifting, the same shape of the sharp, single resistivity drop was maintained. As mentioned earlier, the resistivity drop for U-15 at.% Ti is a combination of two drops with different slops. In this case, we observe not only the temperature shift but also a small change of the resistivity drop-shape with increasing magnetic fields. Namely one part of the resistivity curve (the lower one) seems to move faster to lower temperatures than the second (the higher one). It certainly relates to the different response of the two different phases (α and γ-U phase) in U-15 at.% Ti in applied magnetic fields. The estimated value of the critical slope at T_c ($-\mu_0(dH_{c2}/dT)_{Tc}$) is 4.3 T/K for U-15 at.% Ru splat is only slightly smaller than that for U-15 at.% Mo one (4.7 T/K). In the case of U-15 at.% Ti, the critical-slope value is equal to 2.7 T/K, i.e. almost 1.5 times smaller than that found for the splat containing the pure γ-U phase superconductivity.

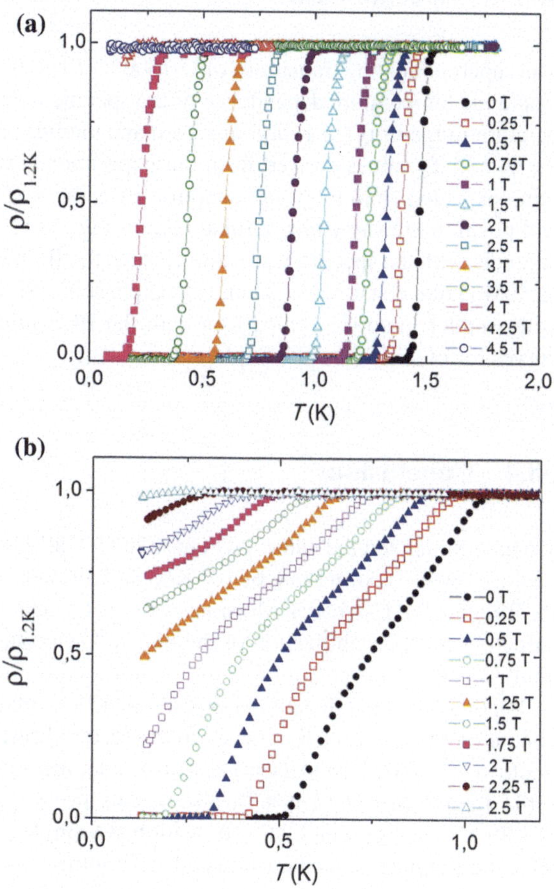

Fig. 38.4 Low-temperature resistivity of U-15 at.% Ru (**a**) and U-15at.% Ti (**b**) splat in applied magnetic fields. The curves were normalized to the resistivity values at 2.0 K and 1.2 K, respectively

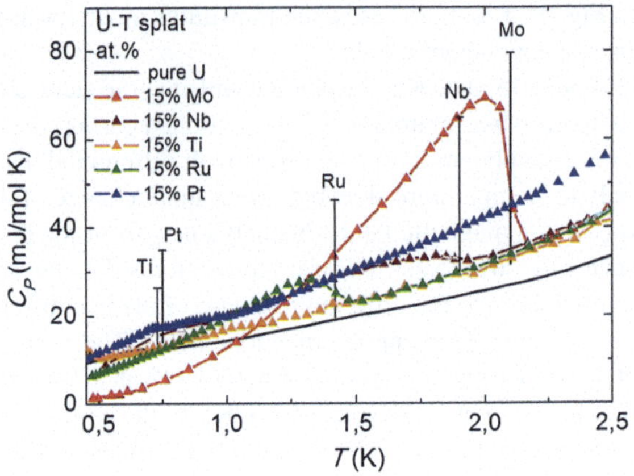

Fig. 38.5 The superconducting phase transition of selected U-15 at.% T alloys revealed in the specific-heat in zero magnetic field

38.3.3 Specific Heat

The superconducting transitions of U-15 at.% T splats were also revealed by anomalies in the temperature dependence of the specific heat. The C(T) curves in the temperature range of 0.4-4 K and in zero magnetic field are shown in Fig. 38.5. Using the γ_e and T_c values derived from our experiments, we estimated the specific-heat jump expected from the weak-coupling BCS theory ($\Delta C(T_c) = 1.43\gamma_e T_c$). Its size and position are shown by vertical bars in Fig. 38.5. The most pronounced λ-type anomaly with the specific-heat jump close to the BCS value was found only for U-15 at.% Mo splat consisted of a single γ-U phase [5–8]. Only a weak and broad bump with smaller jump (in comparison with the BCS jump) was observed in the C(T) curves for other U-15 at.% T splats.

38.4 Conclusions

We have stabilized the cubic γ-U phase in U-T alloys by a combination of ultrafast cooling (with a cooling rate of 10^6 K/s) and alloying with 15 at.% T (T = Mo, Nb, Pt, Ru, and Ti). U-15 at.% Mo splat consists of ideal cubic A2 structure. Ru also is sufficient for the stabilization of the pure γ-U structure at the composition level of 15 at.% Ru.

We demonstrated that all our investigated U-T splat-cooled alloys become superconducting below 2.11 K. The sharpest electrical resistivity drop (with a transition width $\Delta T_\rho = 0.02$ K) and the highest critical temperature ($T_c = 2.11$ K) was observed for U-15 at.% Mo. U-15 at.% Ru also reveals a very sharp drop ($\Delta T_\rho = 0.09$ K) at $T_c = 1.45$ K. The superconducting transition as a single λ-type peak in the specific-heat with the jump close to the estimated BCS jump was observed only for U-15 at.% Mo.

Acknowledgements We highly acknowledge the fruitful help with experiments of M. Krupska-Klimczak, I. Tkach and M. Paukov.

References

1. G.I. Hofman, M.K. Meyer, A.E. Ray, *in Proceedings of the International meeting on Reduced Enrichment for Research and Test Reactors (RERTR)*, Sao Paulo, Brazil, 18–23 October 1998
2. J. Grenthe, J. Drożdżynski, T. Fujjno, E.C. Buck, T.E. Albrecht-Schmitt, S.F. Wolf, in *The Chemistry of The Acinide and Transactinide Elements*, vol. 1, ed. by L.R. Morss, N.M. Edelstein, J. Fuger (Springer, Netherlands, 2006) pp. 253–698
3. H.I. Yakel, in *Proceedings of the Physical Metallurgy of Uranium Alloys Conference*, Vail, Colorado, 12–14 February 1974
4. P. Chotti, H.H. Klepfer, R.W. White, Transactions of American Society for Metals **51**, 772 (1959)
5. I. Tkach, N.-T.H. Kim-Ngan, S. Maskova, M. Dzvenko, L. Havela, A.D. Warren, C. Sitt, T.B. Scott, J. Alloys Compd. **534**, 101 (2012)
6. N.-T.H. Kim-Ngan, I. Tkach, S. Maskova, A.P. Goncales, L. Havela, J. Alloys Compd. **580**, 223 (2013)
7. N.-T.H. Kim-Ngan, M. Paukov, S. Sowa, M. Krupska, I. Tkach, L. Havela, J. Alloys Compd. **645**, 158 (2015)
8. S. Sowa, N.-T.H. Kim- Ngan, M. Krupska, M. Paukov, V. Buturlim, L. Havela, M. Chrobak, Z. Tarnawski, Physica B **536**, 546 (2018)
9. N.-T.H. Kim-Ngan, Z. Tarnawski, M. Chrobak, S. Sowa, A. Duda, M. Paukov, V. Buturlim, L. Havela, Phys. B **536**, 708 (2018)
10. M. Chrobak, Z. Tarnawski, S. Sowa, M. Krupska, N.-T.H. Kim- Ngan, M. Paukov, V. Buturlim, L. Havela, Advances in Natural Sciences **8**, 1 (2017)
11. N.-T.H. Kim-Ngan, S. Sowa, M. Krupska, V. Buturlim, M. Paukov, D. Drozdenko, P. Minarik, L. Havela, Phys. B **545**, 152 (2018)

Chapter 39
Multilayer PECVD Si–C–N Films

A. O. Kozak, V. I. Ivashchenko, O. K. Porada, L. A. Ivashchenko, O. O. Sytikov, V. S. Manzhara and T. V. Tomila

Abstract Abrasive wear resistance, optoelectronic and mechanical properties of multilayer (ML) and monolayer (SL) silicon carbide nitride (Si–C–N) films were investigated. The Si–C–N films were prepared by plasma to enhance chemical vapor deposition using a mixture of hexamethyldisilazane vapor, hydrogen, and nitrogen. Two types of the SL films were deposited at high and low substrate biases of −250 V and −5 V. The films demonstrate hardness ~33 GPa and ~13 GPa, respectively. The ML films represent the sequence of hard and soft layers. The correlation between the deposition time of a single layer (t_D) and wear resistance was revealed: the smaller the deposition time, the higher the wear resistance is. The optoelectronic properties of the films were found to be insensitive to t_D. All the ML films have a lower energy gap (2.18–2.35 eV) than the SL films (4.1–2.42 eV). A strong quenching of photoluminescence was observed for the ML films.

39.1 Introduction

Silicon carbide nitride (Si–C–N) films are known for their remarkable properties: high hardness, low friction coefficient, chemical stability, and oxidation resistance. In addition, hydrogenated Si–C–N films are of wide interest as Si-based semiconductors that combine the properties of silicon carbide and silicon nitride [1–8]. The combination of these properties makes hydrogenated Si–C–N films widely used in optoelectronic devices, for example, as UV detectors [9] or high-temperature pressure sensors [10].

A. O. Kozak (✉) · V. I. Ivashchenko · O. K. Porada · L. A. Ivashchenko · O. O. Sytikov · T. V. Tomila
Institute for Problems of Materials Sciences, NAS of Ukraine, Kiev 03142, Ukraine
e-mail: andri.kozak@ipms.kiev.ua

V. S. Manzhara
Institute of Physics, NAS of Ukraine, Kiev 03028, Ukraine
e-mail: fizyka@iop.kiev.ua

A. D. Pogrebnjak and O. Bondar (eds.), *Microstructure and Properties of Micro- and Nanoscale Materials, Films, and Coatings (NAP 2019)*,
Springer Proceedings in Physics 240, https://doi.org/10.1007/978-981-15-1742-6_39

It follows from the literature, the Si–C–N films are successfully produced by using chemical vapor deposition (CVD) [11], plasma-enhanced CVD (PECVD) [5, 12], RF- and DC- magnetron sputtering (MS) methods [1, 13]. Significant attention is paid for the preparation of the homogeneous films by atomic layer deposition [14]. On the other hand, the PECVD and MS apparatus are most promising ones for the deposition of multilayer (ML) films that represent the sequence of the layers with strong different physical and chemical properties [1, 2, 15–19]. In the case of Si–C–N films, the ML films can be synthesized by periodically changing some deposition parameters during deposition [20–22]. In such so-called gradient or nanogradient films, the structure and composition change continuously [23–28]. Usually, the first layer is used to improve the adhesion properties of sequent layers.

Regardless of the fact that ML coatings with periodic structure exhibit good tribological properties [24], high hardness [25], high oxidation resistance [26] and bright photoemission due to quantum confinement effect [27], we found only one work in which ML coatings consisting of the Si–C–N layers were investigated [28]. But, in that study only the mechanical properties of the ML coatings configured by a combination of two and three layers with different hardness were studied.

Previous to our study [16] showed that substrate bias (U_d) had a significant effect on film properties. In particular, by changing U_d from −5 to −250 V, the energy gap (E_g) decreased from 4.15 to 2.38 eV; nanohardness and elastic modulus increased from 14 GPa to 24 GPa and from 147 GPa to 190 GPa, respectively. The aim of this work is to investigate the ML hydrogenated Si–C–N films with periodic structure deposited by PECVD. Fourier transform infrared spectroscopy (FTIR), X-ray photoelectron spectroscopy (XPS), X-ray diffraction analysis (XRD), photoluminescence (PL), ultraviolet-visible (UV-VIS) transmittance spectroscopy, Knop hardness (HK) test, and ball-on-plane (calowear) tests were used to characterize the deposited ML films.

39.2 Experimental Details

The Si–C–N films were deposited using the mixture of hexamethyldisilazane (HMDSN, $C_6H_{19}NSi_2$) vapor, nitrogen, and hydrogen by PECVD on the polished Si(001), quartz and glasses substrates. Details of the deposited system have been described elsewhere [17, 29]. The liquid precursor HMDSN was vaporized from a thermostated bubbler heated to 40 °C and delivered into the reaction chamber by hydrogen. The flow rate of hydrogen through the thermostated bubbler with HMDS (F_{H+HMDS}) was 12 sccm. The nitrogen flow rate (F_{N2}) was 1 sccm. The substrate temperature (T_d), gas mixture pressure in the reactor chamber (P_C), discharge power (P_W) were 400 °C, 0.2 Torr, and 0.2 W/cm^3, respectively. Substrate bias was applied to a substrate by 5.27 MHz generator. The Si–C–N ML films were deposited by means of the periodic switches of the value of U_d from −5 to −250 V. A time slot between switching of U_d (deposition time of one layer, ts) was held unchanged during the same deposition. The ML Si–C–N films with 720, 120, 80 and 20 layers were deposited using ts equal to 15 s, 30 s, 45 s and 180 s for 60 min, respectively. The SL Si–C–N films were deposited at substrates biases of −5 and −250 V for 60 min.

In order to characterize the structure of the deposited films, XRD analysis was performed by "DRON-3 M" apparatus. The composition of the deposited films was studied by X-ray Photoelectron Spectroscopy (XPS) (UHV-Analysis-System assembled by SPECS). The chemical bonds were studied by FTIR with the help of a spectrometer "FSM 1202" LLC "Infraspek". Measurements of the transmission spectra in the range from 200 to 900 nm wavelengths were carried out using a two-beam UV-VIS spectrometer SPECORD-M40. The optical band gap (E_g) was deduced from a UV-VIS transmittance spectrum of the film deposited on quartz and glass substrate. The film's thickness (d) was estimated by an optical profilometer "Micron-alpha". The photoluminescence (PL) spectra were recorded by the automated monochromator SPM-2 (Carl Zeiss). A low powerful LED laser ($\lambda_{em} = 405$ nm) was used to excite photoemission. Knoop hardness (HK) was determined by a device "MICROMET 2103 Microhardness Tester" (BUEHLER) at a load of 100 mN. The wear resistance of the thin films was estimated as the ratio of the wear coefficient *k* determined from calowear tests of coated and uncoated substrates (k_s/k_c). The calowear test was carried out using a device in which a ball of hard steel (diameter 31 mm) rotated against a sample in the presence of diamond paste at the normal load of 111 mN.

39.3 Results and Discussion

The denotation of the deposited films as well as their thickness, energy gap, Knoop hardness and wear resistance are listed in Table 39.1. We see that the SL250 films have a higher Knoop hardness than the SL5 films. Given these results, below, the layers deposited at low and high U_d will be denoted as the soft and hard layers, respectively.

One can see from Table 39.1 that, in spite of the same total deposition time of the hard and soft layers for all the ML films, their thicknesses are different. Considering the thickness of the SL250 and SL5 samples, we supposed that the total thickness of

Table 39.1 Sample denotation and properties of the films

Sample	U_D (ts)	d (nm)	Eg (eV)	HK (GPa)	k_s/k_c
SL250	$U_d = -250$ V	1220	2.42	32.9	3.1
SL5	$U_d = -5$ V	330	4.15	13.2	1.5
ML5	ts = 5 s	450	2.18	27.4	4.0
ML15	ts = 15 s	700	2.35	26.0	2.8
ML30	ts = 30 s	765	2.29	26.1	2.0
ML45	ts = 45 s	620	2.28	23.7	1.8
ML180	ts = 180 s	815	2.19	15.3	1.6

the ML films should be about 775 nm (expected thickness). However, the thicknesses of the films ML5–ML45 are lower than 775 nm, and the thickness of the ML180 film is higher than the expected thickness. This indicates that the structure of the ML films cannot be considered as the average network of the SL5 and SL250 monolayer films.

39.3.1 Film Structure and Composition

There are no reflexes in the XRD spectra of the deposited films that could be assigned to crystallite phases, for which reason we assumed that the films are X-ray amorphous.

The concentrations of Si, C, N and O in the SL5 and SL250 films are 36, 36, 14 and 14 and 37, 41, 13 and 9 at.%, respectively. The concentrations of these elements in the ML films are about 34, 47, 11 and 8 at.%, respectively. Unfortunately, the XPS analysis does not give a possibility to estimate the hydrogen content. The oxygen in the films may come from the ambient air after deposition or due to the residual oxygen absorbed on the reactor chamber walls [1, 2, 16].

39.3.2 Chemical Bonding Configurations

The chemical bonds in both the SL and ML films deposited on the silicon substrate were studied by FTIR spectroscopy. Figure 39.1 shows the FTIR spectra of these films. The broad absorption band between 550 cm^{-1} and 1250 cm^{-1} inherent to Si–C–N films [1, 2, 30–33] could be interpreted as a superposition of several absorption bands. In particular: the stretching mode of Si-C at 815 cm^{-1}, Si–N at 930 cm^{-1} and

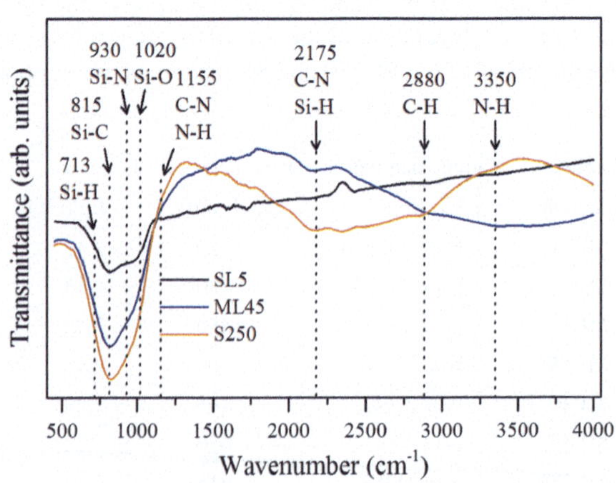

Fig. 39.1 FTIR spectra of the deposited Si–C–N films

Si–O at 1020 cm^{-1}, the wagging mode of SiH_n at 713 cm^{-1} and the C–N stretching and/or N–H bending modes in Si–NH-C and/or Si–NH–Si at 1155 cm^{-1} [34]. Figure 39.1 shows that the main bonds in the deposited monolayer and multilayer films are Si–C, Si–N and Si–O ones. Also, Si–Si, and C–C bonding could be present in these films [30], but it is difficult to detect them in the FTIR measurements due to the weak IR coupling of these homopolar bonds [31]. The relatively weak stretching modes of Si–H at 2175 cm^{-1}, C–H at 2880 cm^{-1} [32] and -NH- in Si–NH–Si and/or Si–NH–C at 3350 cm^{-1} [33] are also seen in the FTIR spectra (cf. Figure 39.1). The availability of hydrogen bonds clearly indicates the hydrogenated nature of the deposited films.

39.3.3 Optical Properties

The main parameter that characterizes amorphous semiconductors is the energy gap, E_g. It can be estimated from the Tauc plot: $(\alpha h\nu)^{1/2} = A(h\nu - E_g)$, where A is a constant, hν is the incident photon energy and α is the absorption coefficient determined from UV-VIS transmittance spectra. These spectra for the deposited films are shown in Fig. 39.2. The values of E_g are presented in Table 39.1. For the ML films, E_g varies in the range of 2.05–2.35 eV, whereas for the SL5 and SL250 films, the band gaps are 4.15 eV and 2.42 eV, respectively. Given these results and considering the spectra shown in Fig. 39.2, one can suppose that the values of E_g for the ML films will not exceed the band gap related to the layer deposited at the highest substrate bias.

The PL spectra of the ML and SL Si–C–N films are depicted in Fig. 39.3. The PL spectra of the SL films appear in one- and two- broad bands structure, with maximums near 575 nm (SL5) and 575, 675 nm (SL250). A possible nature of this PL bands was explained in [16, 35]. It was shown that the one-peak and two-peak structures of the PL spectra of the Si–C–N films are caused by electron-hole recombination between

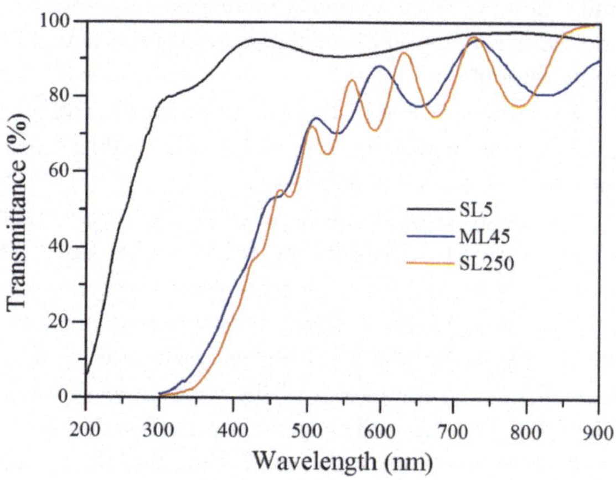

Fig. 39.2 Optical transmission spectra of the Si–C–N films

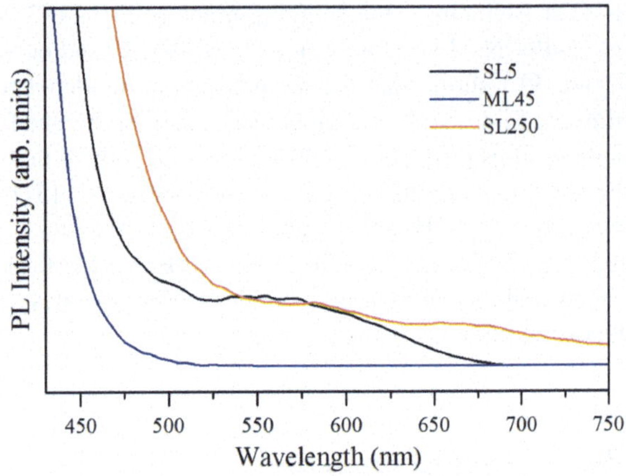

Fig. 39.3 PL spectra of the Si–C–N film

tail states within the energy gaps of the N-rich and C-rich Si–C–N–O–H networks, respectively. Figure 39.3 shows that the PL bands are absent in the typical PL spectra of the ML films. This can be explained by a larger number of dangling-bond atoms and over-coordinated atoms in the ML films compared to the SL ones owing to the availability of the transitional zones between layers in the ML films.

39.3.4 Hardness and Wear Resistance

We measured the Knoop hardness and carried out wear resistance tests to estimate the mechanical characteristics of the deposited films. These characteristics are listed in Table 39.1. The hardness of the SL films increases from 13.2 to 32.9 GPa, as the negative subtract bias increases from −5 to −250 V. It was shown [16, 17] that the hardness of Si–C–N films can increases if a number of Si–O bonds decreases and a number of Si–C bonds increases. The lowering of a number of Si–H bonds and the densification of the high-biased films can also lead to the observed strength enhancement.

The values of HK of the ML films are noticeably less than the hardness of the SL250 film. A possible reason for this could be the presence of soft transitional interlayer zones in the ML films.

In order to estimate the wear resistance against abrasive wear, the calowear test was performed. Its results are presented in Table 39.1. It is clearly seen that an increase in U_d promotes an increase in wear resistance of the SL films. The ML films show a gradual decrease in wear resistance with increase of ts. It is seen that wear resistance of the ML5 film is considerably higher than that of the SL films. It is confirmed by the photos of the wear craters of the films and substrate shown in Fig. 39.4. The size and shape of the craters clearly indicate that the ML films have smaller abrasive wear than the SL films and the silicon substrate.

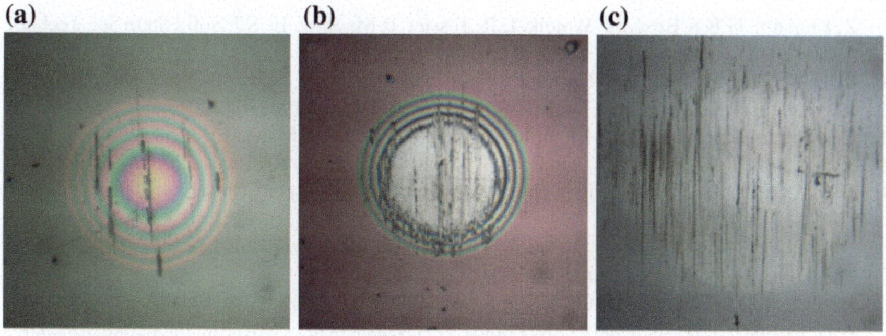

Fig. 39.4 Tupical wear craters of the SL250 (**a**), ML5 (**b**) films and the Si substrate (**c**)

It should be noted that multi-layered films deposited at low ts of 5–15 s as well the SL films deposited at Ud = −250 V demonstrate close the ks/kc ratio to that of a-SiC films deposited by PECVD [36].

39.4 Conclusions

Amorphous monolayer and multilayer Si–C–N films were deposited by PECVD using hexamethyldisilazane as the main precursor at substrate biases of −5 and −250 V. The multilayer films were prepared by periodic changing the substrate biases. It is found that an increase in the deposition time of layers leads to a decrease in the hardness and abrasive wear of the films. The main chemical bonds in the films are Si–N, Si–C and Si–O ones which form Si–C–N–O matrix. The presence of C–H, Si–H and N–H bonds in the infrared spectra indicates that the films are hydrogenated. The energy gap of the multilayer films was found to be comparable with that of the highly biased film. The monolayer films demonstrate photoemission with maxima at about 575 and 675 nm. The formation of the multilayer structure leads to quenching photoemission, which can be explained by the presence of the strongly disordered transitional zones in the multilayer films.

References

1. A.O. Kozak, O.K. Porada, V.I. Ivashchenko, L.A. Ivashchenko, P.L. Scrynskyy, T.V. Tomila, V.S. Manzhara, Appl. Surf. Sci. **425**, 646 (2017)
2. V.I. Ivashchenko, A.O. Kozak, O.K. Porada, L.A. Ivashchenko, O.K. Sinelnichenko, O.S. Lytvyn, T.V. Tomila, V.J. Malakhov, Thin Solid Films **569**, 57 (2014)
3. Z. Chen, J. Zhou, X. Song, X. Xu, Y. Liu, J. Song, Z. Ma, E. Xie, Opt. Mater. (Amst) **32**, 1077 (2010)
4. M.A. Abdul Rahman, W.S. Chiu, C.Y. Haw, R. Badaruddin, F.S. Tehrani, M. Rusop, P. Khiew, S.A. Rahman, J. Alloys Compd. **721**, 70 (2017)

5. Z. Khatami, G.B.F. Bosco, J. Wojcik, L.R. Tessler, P. Mascher, ECS J. Solid State Sci. Technol. **7**, N7 (2018)
6. C. Huber, B. Stein, H. Kalt, Thin Solid Films **634**, 66 (2017)
7. Z. Khatami, P.R.J. Wilson, J. Wojcik, P. Mascher, J. Lumin. **196**, 504 (2018)
8. C.W. Chen, M.-H. Lee, L.C. Chen, K.H. Chen, Diam. Relat. Mater. **13**, 1158 (2004)
9. T.-H. Chou, T.-W. Kuo, C.-Y. Lin, F.-S. Lai, Sensors Actuators A Phys. **279**, 462 (2018)
10. H.K.E. Latha, A. Udayakumar, V. Siddeswara Prasad, Acta Metall. Sin. (English Lett.) **27**, 168 (2014)
11. D. Kumar, U. Rizal, S. Das, B.S. Swain, B.P. Swain, in *Advances in Electronics, Communication and Computing* (Springer, Singapore, 2018), pp. 77–83
12. B. Swatowska, T. Stapinski, Phys. Status Solidi Curr. Top. Solid State Phys. **7**, 758 (2010)
13. A.O. Kozak, V.I. Ivashchenko, O.K. Porada, L.A. Ivashchenko, O.K. Sinelnichenko, S.N. Dub, O.S. Lytvyn, I.I. Tymofeeva, G.N. Tolmacheva, J. Superhard Mater. **37**, 300 (2015)
14. R.A. Ovanesyan, N. Leick, K.M. Kelchner, D.M. Hausmann, S. Agarwal, Chem. Mater. **29**, 6269 (2017)
15. V.I. Ivashchenko, P.L. Scrynskyy, O.S. Lytvyn, O.O. Butenko, O.K. Sinelnichenko, L. Gorb, F. Hill, J. Leszczynski, A.O. Kozak, J. Superhard Mater. **36**, 381 (2014)
16. A.O. Kozak, V.I. Ivashchenko, O.K. Porada, L.A. Ivashchenko, T.V. Tomila, V.S. Manjara, G.V. Klishevych, Mater. Sci. Semicond. Process. **88**, 65 (2018)
17. O.K. Porada, Mater. Sci. Nanostructures **3–4**, 3 (2014)
18. A.D. Pogrebnjak, V.I. Ivashchenko, P.L. Skrynskyy, O.V. Bondar, P. Konarski, K. Załęski, S. Jurga, E. Coy, Compos. Part B Eng. **142**, 85 (2018)
19. A.A. Onoprienko, V.I. Ivashchenko, A.O. Kozak, A.K. Sinelnichenko, T.V. Tomila, J. Superhard Mater. **41**, 90 (2019)
20. A.O. Kozak, V.I. Ivashchenko, O.K. Porada, L.A. Ivashchenko, V.Y. Malakhov, T.V. Tomila, J. Nano- Electron. Phys. **7**, 03040 (2015)
21. A.O. Kozak, V.I. Ivashchenko, O.K. Porada, L.A. Ivashchenko, T.V. Tomila, J. Nano- Electron. Phys. **6**, 04047 (2014)
22. A.O. Kozak, V.I. Ivaschenko, O.K. Porada, L.A. Ivashchenko, in *Proceeding 2017 IEEE 7th International Conference on Nanomaterials Application, Properties* (IEEE, 2017), Odessa, Ukraine, 10–15 September 2017
23. I. Mrkvica, M. Neslušan, R. Čep, V. Sléha, Teh. Vjesn. - Tech. Gaz. **23**, 569 (2016)
24. M. Lattemann, S. Ulrich, Surf. Coatings Technol. **201**, 5564 (2007)
25. M. Lattemann, S. Ulrich, H. Holleck, M. Stüber, H. Leiste, Diam. Relat. Mater. **11**, 1248 (2002)
26. K.-E. Bae, K.-W. Chae, J.-K. Park, W.-S. Lee, Y.-J. Baik, Surf. Coatings Technol. **276**, 55 (2015)
27. S. Nihonyanagi, K. Nishimoto, Y. Kanemitsu, J. Non. Cryst. Solids **299–302**, 1095 (2002)
28. S.K. Mishra, D. Verma, S. Bysakh, L.C. Pathak, J. Nanomater. **2013**, 1 (2013)
29. O.K. Porada, V.I. Ivashchenko, L.A. Ivashchenko, A.O. Kozak, O.O. Sytikov, J. Superhard Mater. **41**, 32 (2019)
30. F. Neri, P. Tripodi, S. Trusso, G. Faggio, Radiat. Eff. Defects Solids **165**, 754 (2010)
31. S.W. King, J. Bielefeld, M. French, W.A. Lanford, J. Non. Cryst. Solids **357**, 3602 (2011)
32. V. Dřínek, T. Strašák, F. Novotný, R. Fajgar, Z. Bastl, Appl. Surf. Sci. **292**, 413 (2014)
33. S. Saloum, B. Alkhaled, A.C.T.A. Phys, Pol. A **119**, 369 (2011)
34. I. Blaszczyk-Lezak, A.M. Wrobel, M.P.M. Kivitorma, I.J. Vayrynen, A. Tracz, Appl. Surf. Sci. **253**, 7211 (2007)
35. O.K. Porada, V.S. Manzhara, A.O. Kozak, V.I. Ivashchenko, L.A. Ivashchenko, J. Nano-Electron. Phys. **9**, 02022 (2017)
36. V.I. Ivashchenko, S.N. Dub, O.K. Porada, L.A. Ivashchenko, P.L. Skrynskyy, A.I. Stegniy, Surf. Coat. Technol. **200**, 6533 (2006)